Shashank Khatri 0908908

Nor

Topography and Bathymetry

45N

35N

25N

15N

0W 70W

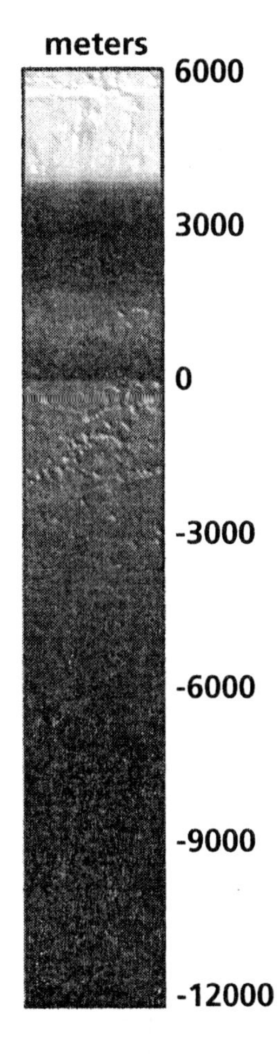

Shaded raised relief map of topography and water depth (bathymetry) for North America and surrounding regions. The map illustrates the concept of isostasy. The North American continent and its shelves are elevated because those regions have relatively thick (continental) crust. Areas with relatively thin (oceanic) crust lie well below sea level. Map compiled by A. Jon Kimerling from National Oceanic and Atmospheric Administration, ETOPO5 digital elevation model data.

Whole Earth Geophysics

An Introductory Textbook for Geologists and Geophysicists

Robert J. Lillie
Oregon State University

Prentice Hall
Upper Saddle River, New Jersey 07458

Library of Congress Cataloging-in-Publication Data

Lillie, Robert J.
Whole earth geophysics : an introductory textbook for geologists and geophysicists / Robert J. Lillie
p. cm.
Includes bibliographical references and index.
ISBN 0-13-490517-2
1. Geophysics I. Title.
QC806.L48 1999 97-42092
550—dc21 CIP

To my parents and grandmother,
for their inspiration and example.
To my son, Ben,
for his inspiration and example.
The Earth is a circle.

Executive Editor: Robert A. McConnin
Art Director: Jayne Conte
Cover Designer: Bruce Kenselaar
Manufacturing Manager: Trudy Pisciotti
Production Supervision/Composition: WestWords, Inc.
Cover Illustration: Robert J. Lillie
Inside Cover Maps: A. Jon Kimerling
Text Illustrations: Robert J. Lillie

A Pearson Education Company
Upper Saddle River, NJ 07458

Printed in the United States of America

ISBN 0-13-490517-2

Prentice-Hall International (UK) Limited, *London*
Prentice-Hall of Australia Pty. Limited, *Sydney*
Prentice-Hall Canada Inc., *Toronto*
Prentice-Hall Hispanoamericana, S.A., *Mexico*
Prentice-Hall of India Private Limited, *New Delhi*
Prentice-Hall of Japan, Inc., *Tokyo*
Prentice-Hall Asia Pte. Ltd., *Singapore*
Editora Prentice-Hall do Brasil, Ltda., *Rio de Janerio*

Contents

Foreword

The tendency for specialists in one discipline of science to become isolated in that discipline and hence unaware of potentially useful techniques, developments, and information from another discipline is a disturbing problem that pervades and grows in modern science. The problem is exacerbated by a paucity of written material designed to bridge such gaps.

Bob Lillie's book is at once a welcome contribution on this score and a fine textbook in the conventional sense. It is designed to introduce the principal geophysical phenomena and techniques, namely seismology in its various forms, gravity, magnetism, and heat flow, to students whose primary training is in geology and who have a basic, but not necessarily an advanced, knowledge of physics. The book's abundant figures (scores and scores of them) and exercises, coupled with the straightforward and tightly organized style of the text, put the subject of geophysics well within reach of such students.

But that very style will also make the book highly useful to others, such as: (a) beginning students of geophysics who seek an orientation and a ready reference volume, and (b) advanced students of physics who already understand the nature of phenomena such as elastic wave propagation and gravity but who lack familiarity with the manner in which such phenomena are utilized by geophysicists as they explore the earth's interior from core to crust.

Some other noteworthy aspects of the book include: (a) a brief orientation in plate tectonics as that subject relates to geophysics, and (b) an appendix that outlines an exercise in good writing based on the subject matter of the book. Science courses, unfortunately, often leave the teaching of writing to others, even though good communication is an essential component of good science. This book makes no such omission.

All in all, the book's organization, style, abundant illustrative figures, exercises, and lists of critical definitions at the head (not the end) of each chapter reflect the experience, dedication, and attention to detail of an author well versed in teaching such a course. Lillie has produced a most useful volume for those who seek to understand the great adventure that is exploration of the earth's interior through application of the principles and techniques of geophysics.

Jack Oliver
Emeritus Professor
of Geological Sciences
Cornell University

Preface

Whole Earth Geophysics illustrates how different types of geophysical observations provide constraints on Earth's gross structure and composition. It introduces geology students with a basic grounding in math and physics to the fundamentals of various geophysical techniques, but does not stop there. Specific observations illustrate how each technique constrains certain aspects of the plate tectonic framework that is fundamental to the study of Earth science today.

This book is designed for a Solid Earth Geophysics course at the advanced undergraduate or beginning graduate level. While the primary objective is to illustrate the utility of geophysical observations to geology students, geophysics students may benefit from the book by seeing the basics of their discipline in the context of geology.

The material is presented in a format that takes advantage of the visual learning skills students commonly develop in geology courses. The text and mathematics are purposefully kept at a minimum; students follow the development of geophysical concepts and their applications to the Earth through numerous illustrations and captions.

Mathematics is employed at a level mastered by most geology students. Equations are commonly accompanied by graphic illustrations, so that students can visualize how the equations relate to aspects of the Earth. Applying understandable mathematics to geology encourages students to employ quantitative analysis in their study of the Earth.

Exercises at the end of chapters are designed to enhance students' skills at illustration, quantitative problem solving, and the relationship between geophysical observations and geology. Students are asked to plot various types of geophysical observations along with geological cross sections at plate tectonic scales; constraints offered by geophysics can thus be analyzed and appreciated. Other exercises employ quantitative aspects of geophysics that are best learned by working through problems.

Appendix A is a sequenced writing assignment designed to accompany the text. The assignment enhances the understanding of geophysical techniques by creating a context; each student does literature research on a region of the Earth and writes about its crustal structure and tectonic evolution. A course using this book may thus be part of the "Writing Intensive" or "Writing Across the Curriculum" program of a college or university. The text and accompanying writing assignment thus serve as a vehicle to bring students to the professional world of geology.

Whole Earth Geophysics begins with an overview of geophysical techniques, discussing how measurements made at Earth's surface relate to properties investigated within the Earth. Physical aspects of Earth's crust, mantle, and core are then presented, with focus on the lithosphere/asthenosphere system. Subsequent chapters emphasize constraints on Earth's plate tectonic framework offered by different geophysical techniques, including seismic refraction and reflection, earthquake seismology, gravity, magnetics, and heat flow. Particular attention is paid to the interrelationship between topography, the crust/mantle boundary, and the lithosphere/asthenosphere transition; students learn how configurations of the latter two features are constrained largely by geophysical observations.

Ideas for *Whole Earth Geophysics* evolved over many years as a student at various universities, and through 14 years of teaching geology and geophysics. I owe a great deal to many individuals over those years. My mentors, Bill Payne of the University of Southwestern Louisiana, Dick Couch of Oregon State University, and Jack Oliver of Cornell University, inspired me to study the Earth, each in his own way. John Green and Dick Standard of CONOCO, Inc., made exploration of the interior of the Earth fun. I am especially grateful to the other two Bobs at Oregon State University (Bob Yeats and Bob Lawrence) for involving me in research that applied geophysical observations to the study of structural geology and tectonics.

Gravity modeling for this book was done with the interactive GM-SYS software, developed by Northwest Geophysical Associates, Inc. I am grateful to Gerry Connard for permission to use these programs and for his help and support over the years.

Vicki Collins of the Writing Intensive Course (WIC) program at Oregon State University was instrumental in emphasizing how the learning of concepts in any course, including geophysics, can be enhanced through writing.

Many of my colleagues have offered comments and encouragement during the development of this book. The following people gave helpful reviews and suggestions during the writing of various chapters: Bob Butler, Randy Richardson, and Clem Chase, University of Arizona; Kevin Pogue, Whitman College; Bob Karlin, University of Nevada, Reno; Rob McCaffrey, Rensselaer Polytechnic Institute; Greg Moore, University of Hawaii at Manoa; Dave Blackwell, Southern Methodist University; Katherine Favret and Walter Mooney, U.S. Geological Survey, Menlo Park, CA; Rick Saltus, U.S. Geological Survey, Denver; Joe Kruger, Idaho State University; Carol Simpson, Boston University; and Gary Huftile, Lisa McNeill, and Jeff Templeton, Oregon State University. Miroslav Bielik, Cestmir Tomek, Dan Davis, Jarka Plomerová, Vladi Babuška, Russell Nazarullah, Ashok Srivastava, Abul Farah, David Gee, Dan Dyrelius, and Paul Ryan provided stimulating research interaction that helped to clarify basic concepts of geophysics and how those concepts relate to the Earth. I am grateful to students who, over the years, have kept me focused on the objective of the book (to teach *them*); comments, suggestions, and literature research by Chris Boyette, Chris Davey, Ben Jacob, Gregg Lambert, Joe Licciardi, Colin Poellot, Don Reeder, Nicole Mare-Shue, Rachel Sours-Page, Jennifer Tatten, and Zach Washburn were especially helpful.

Many applications incorporated into the text were developed during research and teaching interaction with different organizations. I am grateful to individuals at Bucharest University, Crater Lake National Park, the Czech Academy of Sciences, the Geological Survey of Pakistan, Geofyzica Brno, the Hydrocarbon Development Institute of Pakistan, the National Institute of Oceanography (Pakistan), the Oil and Gas Development Corporation (Pakistan), the Oil and Natural Gas Commission (India), the Slovak Academy of Sciences, University College (Galway), the University of Uppsala, and the Wadia Institute of Himalayan Geology.

This book was penned mostly over cups of coffee. I am grateful to the workers and patrons of The Beanery, Java Rama, New Morning Bakery, M's Tea and Coffee House, Price Creek Bakery, Sam's Station, Boccherini's Coffee and Tea House, Java Stop, and Sunriver Coffee Company for providing atmosphere and inspiration that resulted in a book.

Robert J. Lillie

C H A P T E R 1

Introduction

geology *(jē äl' ə jē)* ***n.***, [< *Gr.* geō, *the Earth;* < *Gr.* logos, *science*], *the study of the Earth.*
physics *(fiz' iks)* ***n.***, [< *L.* physica, *physics*], *the study of matter and energy and their interactions.*
geophysics *(jē' ō fiz' iks)* ***n.,*** *the application of the principles of physics to study the Earth.*

Geology is visual. We are attracted to features on Earth's surface because we see them; our imagination helps us visualize processes within the Earth that form mountains, continents, and oceans. Students of geology commonly develop skills based on visualization; the first impulse of a geologist is often to "make a sketch," much like plays diagramed in the dirt during sandlot ball games.

The movement of objects or the passage of energy as waves occurs in predictable ways; physics lends itself to the formulation of mathematical expressions that describe these phenomena. The first impulse of a physicist might be to write a formula that portrays, concisely, a pattern or process.

Geophysics, as the hybrid of geology and physics, requires the ability to view the same problem from both visualization and mathematical formulation (Fig. 1.1). Most geophysics textbooks rely heavily on the latter approach, explaining concepts mainly through mathematics. That style can lead to two problems in introductory geophysics courses: 1) students with geology backgrounds may be lost in the abstract world of mathematical equations, without visualization of how the equations explain things about the Earth; 2) physics students may understand the equations, but without a feel for aspects of the Earth modeled by the equations.

Whole Earth Geophysics is the outgrowth of a two-term course consisting of undergraduate and graduate geology students, along with a few physics and geophysics majors. While having a qualitative feel for the Earth, geology students often lack advanced-level courses in math and physics. This book explains concepts through numerous graphic illustrations; equations, where necessary, are developed with mathematics that most geology students have mastered. The book presents geophysical techniques, but the focus is on how each technique provides information on the internal structure and tectonic development of the Earth.

Geophysics students may not have been exposed to their subject in a graphic and systematic way; concepts in many geophysics courses are revealed through equations and illustrations from the literature intended for advanced-level researchers. The visual approach employed in this book may help geophysics students see how ideas developed mathematically in other courses relate to the "real world."

The book presents plate tectonic theory in an early chapter, explaining the development of continental rifts, ocean basins, continental margins and various types of mountain ranges. Illustrations in later chapters portray the expressions of geophysical data in different tectonic settings. Simple models predict the appearance of geologic structures on seismic reflection profiles, and show the form of gravity anomalies developed during stages of opening and closing of ocean basins.

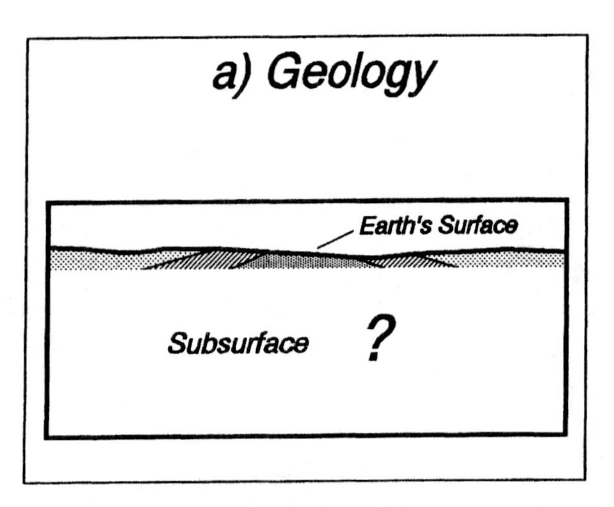

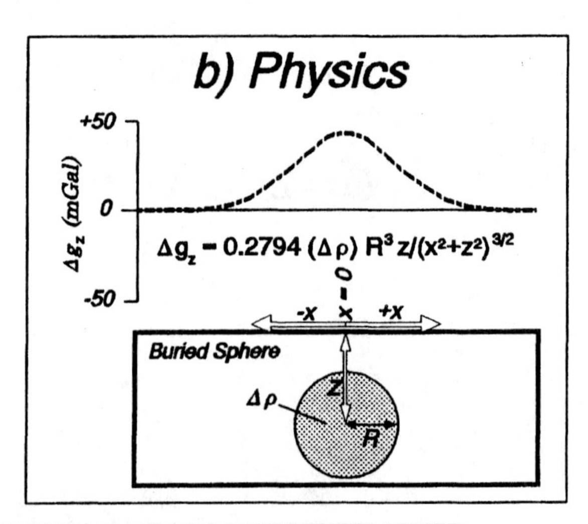

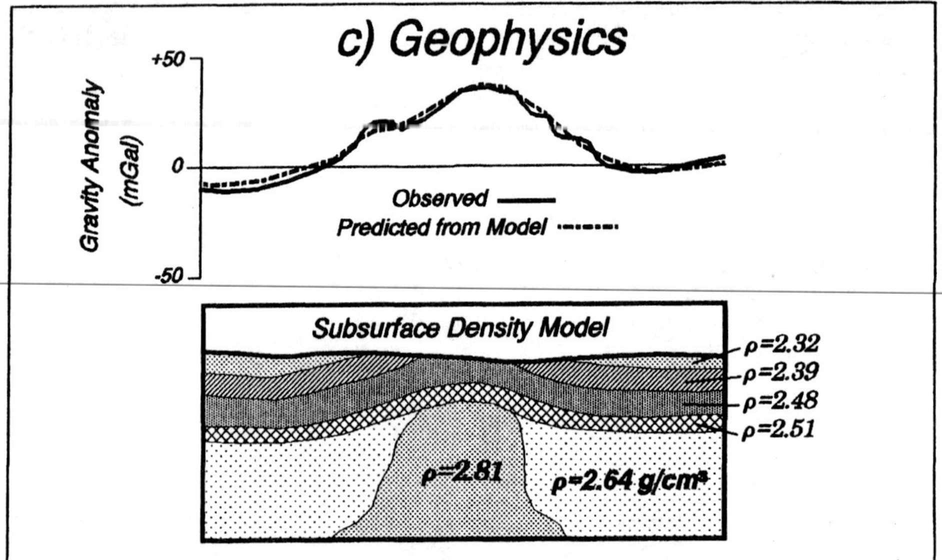

FIGURE 1.1 Geophysics aims to interpret the subsurface by combining observations of Earth materials (geology) with observations of physical phenomena (physics). a) Cross-section illustrating surface geology, with no subsurface information. b) Model of a physical parameter, the change in Earth's gravity field (Δg_z) that would result from a sphere of radius (R) and density contrast ($\Delta\rho$). The mathematical equation predicts the change in gravity caused by such a mass buried depth (z) below the surface. c) Observed change in gravity field, along with a model of subsurface density distribution (ρ) that might cause such a change. The model agrees with the observed surface geology and shows density distributions that result in a predicted gravity anomaly close to that observed.

A primary focus is the relationship between topography, the crust/mantle boundary, and the lithosphere/asthenosphere transition; these three features balance through *isostasy,* whereby pressure equalizes at a certain depth within the Earth. Students thus appreciate the utility of geophysical measurements to constrain interpretations of the crust and lithosphere/asthenosphere system in different portions of the Earth.

The geophysical methods covered in this book (refraction, reflection, earthquake, gravity, magnetics, and heat flow) are basic to the education of geology students, particularly in helping them appreciate Earth's gross structure and plate tectonics. Other methods (for example, geochronology, radioactivity, well logging, electrical methods) are important, but they may be addressed better in geochemistry or more advanced geophysics courses.

OVERVIEW OF GEOPHYSICAL TECHNIQUES

Measurements of natural or induced properties are commonly made at the surface of the Earth (for example, gravitational acceleration). Applied geophysics interprets those observations in terms of properties within the Earth (for example, density distributions that locally change the gravitational acceleration). Geophysical techniques employed at or near Earth's surface include *seismic, potential field* and *heat flow* measurements (Fig. 1.2).

Seismic

Relatively small and rapid, up-and-down or sideways movements of Earth's surface, measured by a seismometer, relate to the passage of seismic waves through the Earth. The "ground motion" (*displacement, velocity,* or *acceleration* of the

Geophysical Technique			Property Measured at Earth's Surface	Property Investigated within Earth
SEISMIC	Natural Source: Earthquake		*Ground Motion (Displacement, Velocity or Acceleration)*	Seismic Velocity (V) and Attenuation (Q)
SEISMIC	Controlled Source	Refraction	*Ground Motion (Displacement, Velocity or Acceleration)*	Seismic Velocity (V)
SEISMIC	Controlled Source	Reflection	*Ground Motion (Displacement, Velocity or Acceleration)*	Acoustic Impedance (Seismic Velocity, V, and Density, ρ)
POTENTIAL FIELD	Gravity		*Gravitational Acceleration ($\vec{g}$)*	Density (ρ)
POTENTIAL FIELD	Magnetics		*Strength and Direction of Magnetic Field ($\vec{F}$)*	Magnetic Susceptibility (χ) and Remanent Magnetization ($\vec{J}_{rem}$)
HEAT FLOW			*Geothermal Gradient (∂T/∂z)*	Thermal Conductivity (k) and Heat Flow (q)

FIGURE 1.2 Geophysical techniques measure properties at Earth's surface. Interpretation of the measurements suggests properties within the Earth.

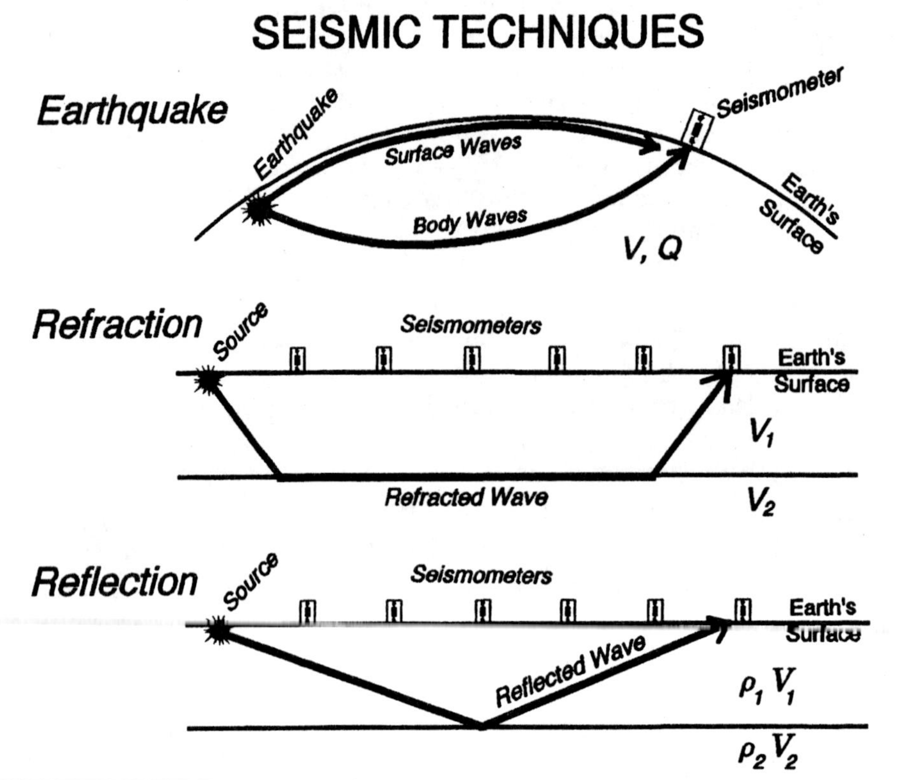

FIGURE 1.3 Seismic techniques employ seismometers to measure movement of the ground resulting from waves generated by an earthquake or artificial source. The time of travel from the source to a receiver is a function of the seismic velocity (V) of the material along the wave's path. The amount and type of ground motion may reveal other properties within the Earth, such as seismic attenuation (Q) and acoustic impedance (the product of density, ρ, and seismic velocity).

seismometer) reveals properties of the materials that the waves encountered (Fig. 1.3). The time it takes for the waves to get from their source to a seismometer (*travel time*) is a function of the speed the waves passed through a region of the Earth (*seismic velocity, V*). The amount and type of ground motion reveals how readily the region absorbed or scattered wave energy (*attenuation,* or "quality factor," Q).

Seismic waves are generated naturally by *earthquakes.* They travel through the Earth as *body waves,* or follow Earth's outermost regions as *surface waves.* Seismic waves can also be generated from explosions or other controlled sources, facilitating techniques to measure the *refraction* of waves as they encounter regions of changing *velocity* (V), or *reflections* due to changes in *acoustic impedance* (density, ρ, times velocity, V).

Potential Field

Potential fields, like those due to Earth's *gravitational attraction* and *magnetism,* change strength and direction depending on the position of observation within the field. The strength of a potential field generally decreases with distance from the source of the field. When the broad effects of Earth's rotation, equatorial bulge, and topography are subtracted, observations of *gravitational acceleration* ($\vec{\mathbf{g}}$) relate to nearby mass distributions (that is, subsurface *density* changes, $\Delta\rho$; Fig. 1.4). Earth's *magnetic field* ($\vec{\mathbf{F}}$) is changed locally by the ability of nearby rocks to be magnetized

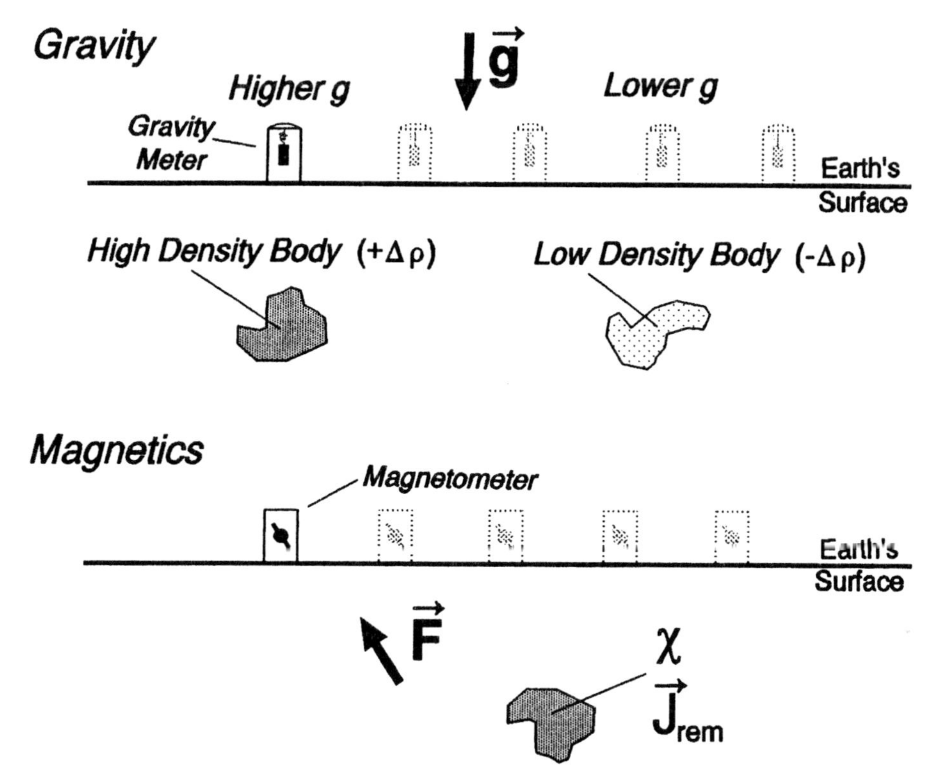

FIGURE 1.4 *Potential field techniques.* A gravity meter measures gravitational acceleration ($\vec{\mathbf{g}}$), sensitive to local density changes ($\Delta\rho$). Magnetometers reveal the Earth's total magnetic field ($\vec{\mathbf{F}}$), influenced by the magnetic susceptibility (χ) and remanent magnetization ($\vec{\mathbf{J}}_{\mathbf{rem}}$) of subsurface materials.

(*magnetic susceptibility,* χ) or by the rocks having been magnetized as they formed (*remanent magnetization*, $\vec{J}_{rem}$).

Heat Flow

Heat constantly flows outward, from hotter regions to Earth's surface. The change in temperature (T) can be measured from the surface downward in drillholes (Fig. 1.5). Knowing the *thermal conductivity* (k) of the rocks in the area, the *geothermal gradient* ($\partial T/\partial z$) can be used to calculate the rate at which heat escapes from that region of the Earth (*heat flow*, q). Without heat, Earth's interior would be completely solid and motionless. Heat softens up a portion of the upper mantle (asthenosphere); movement of rigid plates (lithosphere) over this softer zone is a product of the flow of heat from Earth's interior.

INTERPRETATION: METHODS AND CONSTRAINTS

Various methods are used to interpret aspects of the Earth from geophysical data. The quality of interpretations depends on how well the problem is constrained by other criteria, such as additional geological and geophysical observations, or assumptions based on models.

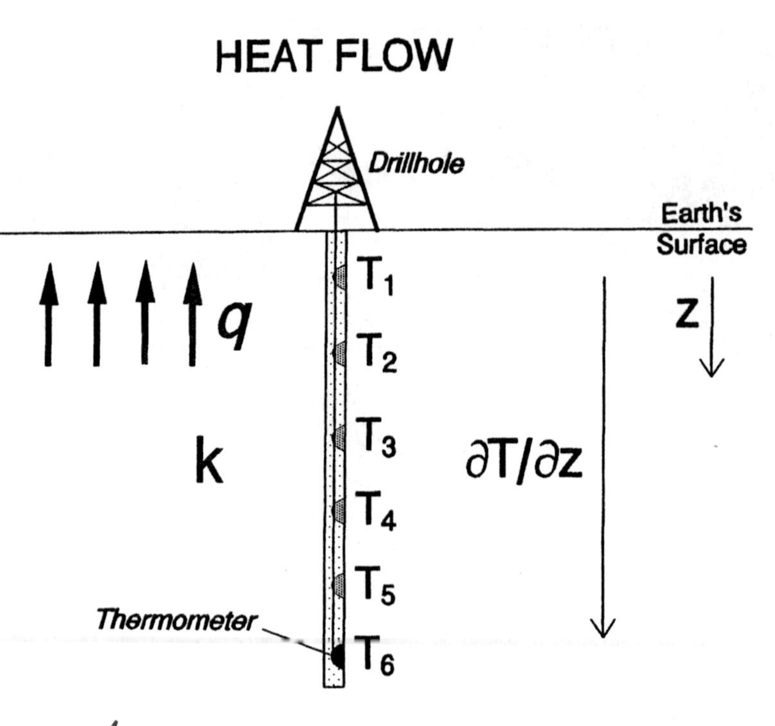

FIGURE 1.5 *Heat flow technique.* Temperatures (T_1, T_2, etc.) are measured at various depths in a drillhole. The change in temperature (T) with depth (z), or geothermal gradient ($\partial T/\partial z$), is a function of the thermal conductivity (k) and the flux of heat through the surface (heat flow, q).

Methods

Both inverse and forward methods are used to interpret geophysical observations. In each case we ask, "What caused what we observed?" *Inversion* uses mathematical equations to calculate a subsurface model from observed data; *forward modeling* assumes a subsurface model and calculates observations that would result.

Fig. 1.6 shows an interpretation (model) that results from the inversion of seismic refraction observations. Observed parameters are inserted into mathematical equations that yield a model of the seismic velocities and thicknesses of layers. We are accustomed to using inversion in math and physics courses, when we "plug into" formulas to yield results.

The forward modeling of an observed gravity profile is illustrated in Fig. 1.7. Layers with different densities are assumed. Calculations from the subsurface model predict the gravity profile that would result. Thicknesses and densities of layers are then adjusted until the predicted ("calculated," or "computed") profile matches the observations; we might consider the adjusted model as one potential interpretation of the observed data.

Constraints

No matter what methods are used to interpret geophysical data, we should not lose sight of the objective: *We make observations of certain properties of the Earth in order to interpret the nature of the Earth.* Geologists often use rock hammers to crack open the Earth, to see what's there and analyze how it got to be that way. In many respects, *geophysical techniques are just another type of rock hammer;* we bang on the Earth and listen to what the Earth tells us about itself.

Alone, a geophysical technique may not tell us much about the Earth. When that technique is combined with other observations, however, we may learn a great deal about a region's subsurface geology and evolution. Those other observations, in the form of geological and other geophysical data, are *constraints.* The more con-

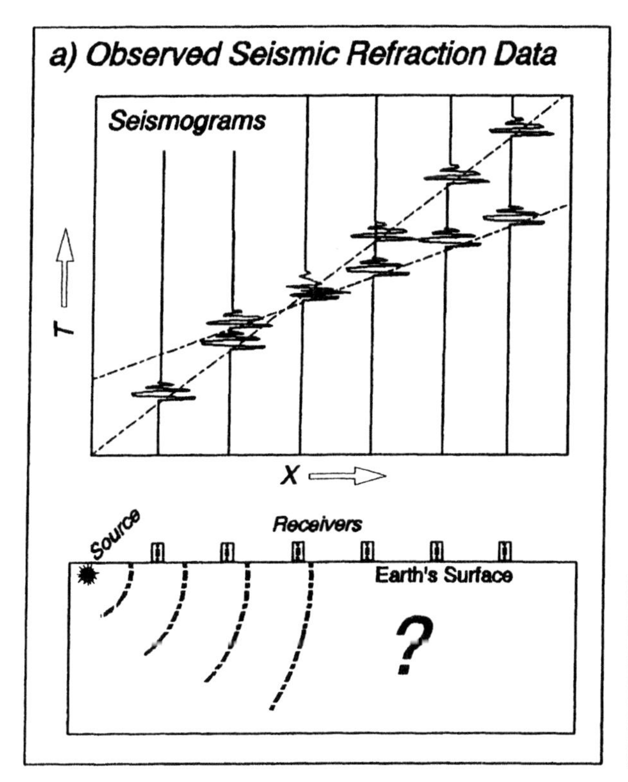

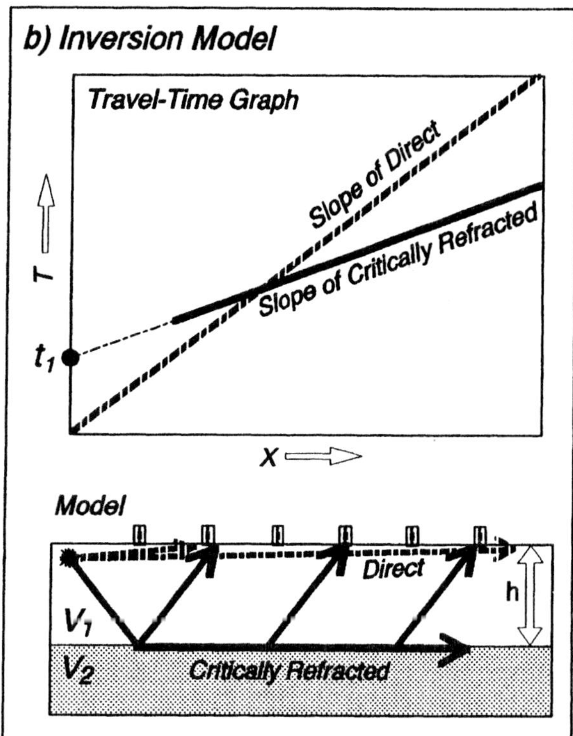

c) Inversion Equations

$V_1 = 1/(\text{Slope of Direct Wave})$

$V_2 = 1/(\text{Slope of Refracted Wave})$

$\theta_c = \text{Critical Angle} = \sin^{-1}(V_1/V_2)$

$t_1 = \text{T-Intercept} = (2h\cos\theta_c)/V_1$

$h = V_1 t_1/(2\cos\theta_c)$

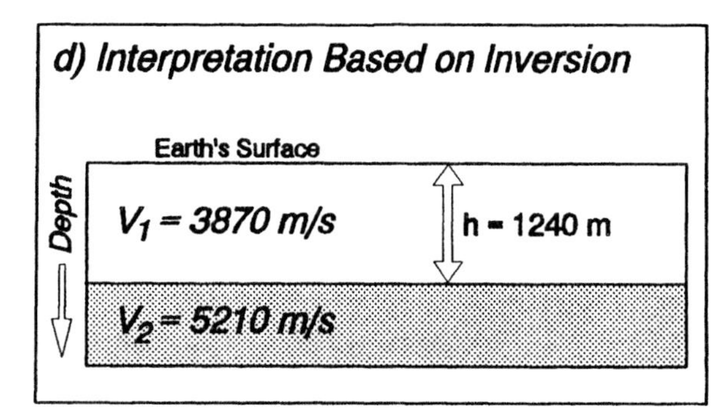

FIGURE 1.6 *Inversion example.* a) Observed seismograms showing times of arrival (T) of direct and critically refracted waves at seismometers placed increasing distance (X) from the source. b) Two-layer model showing parameters that can be read from the observed data: slope of the direct arrival; T-axis intercept time (t_1); and slope of the critically refracted arrival. c) Equations that can be used to invert observations of slopes and T-axis intercept time (see Chapter 4). d) Model of seismic velocities for two layers (V_1, V_2), and the thickness of the upper layer (h), that result from inserting the slopes and intercept time into the inversion equations.

straints we have on a problem, the more likely we are to come to a unique solution. Rarely is a region so well constrained that we can perceive of only one possibility for its subsurface interpretation; we therefore say that most interpretations of geophysical data are "nonunique."

Nonuniqueness means that it is possible to offer more than one interpretation that agrees with all available information. The problem is like that of the "blind men and the elephant;" the interpretation depends largely on our sample locations and the overall density of sampling. In approaching a problem, it is wise to keep in mind a *hierarchy of constraints* (Fig. 1.8).

Level 1: (*Firm Constraints*):
Direct observations of the Earth that you can put in your hand:
a) Outcrop samples.
b) Drill cores.

Level 2: (*Softer Constraints*):
Indirect or inferred observations about the Earth.
a) Map interpretations based on scattered direct observations.
b) Geophysical observations.

Level 3: (*Reasonable Assumptions*):
Theoretical considerations, based on logic and common sense (for example, "modeling").

Thinking should flow from higher (1) to lower (2, 3) levels. For example, surface geological mapping of a region (Level 1) can constrain interpretations of geophysical data (Level 2); the geophysical interpretations further suggest models (Level 3) for overall structure or processes in the region. Thinking in the other

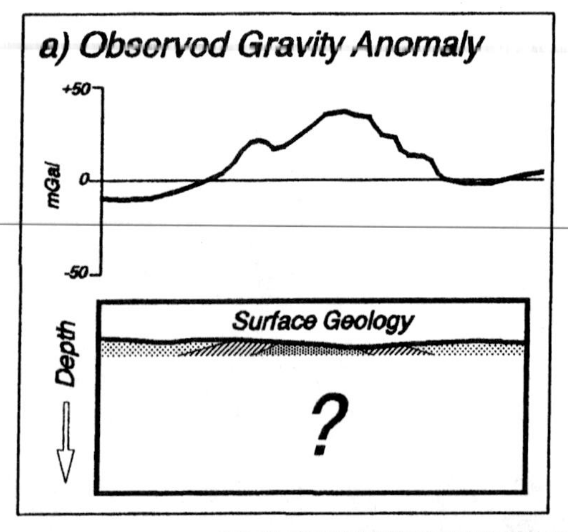

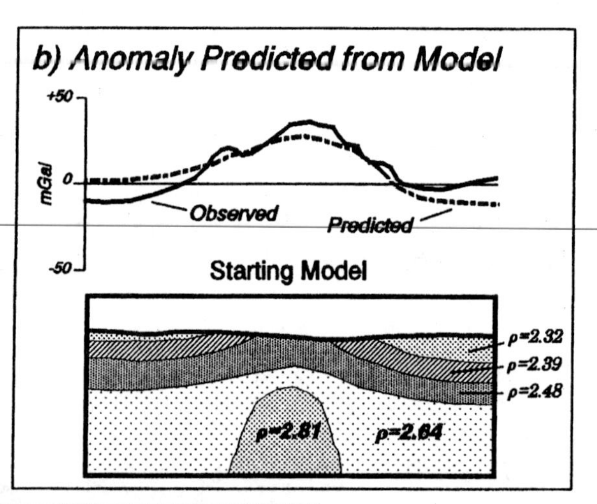

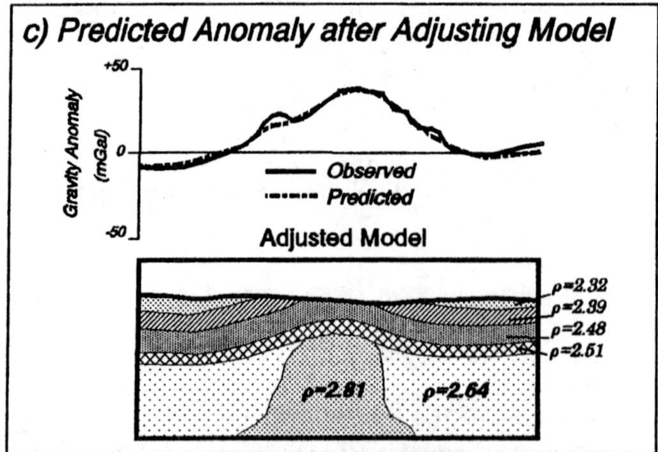

FIGURE 1.7 *Forward modeling example.* a) Gravity anomaly measured at several stations on the surface. Surface outcrop and dips serve as constraints on interpretation. b) Preliminary model testing subsurface densities and geometries, resulting in a predicted gravity anomaly (see Chapter 8). ρ = density (g/cm^3). c) Densities and geometries adjusted to achieve closer agreement between observed and predicted anomalies. Note that the constraints offered by surface geology are not changed.

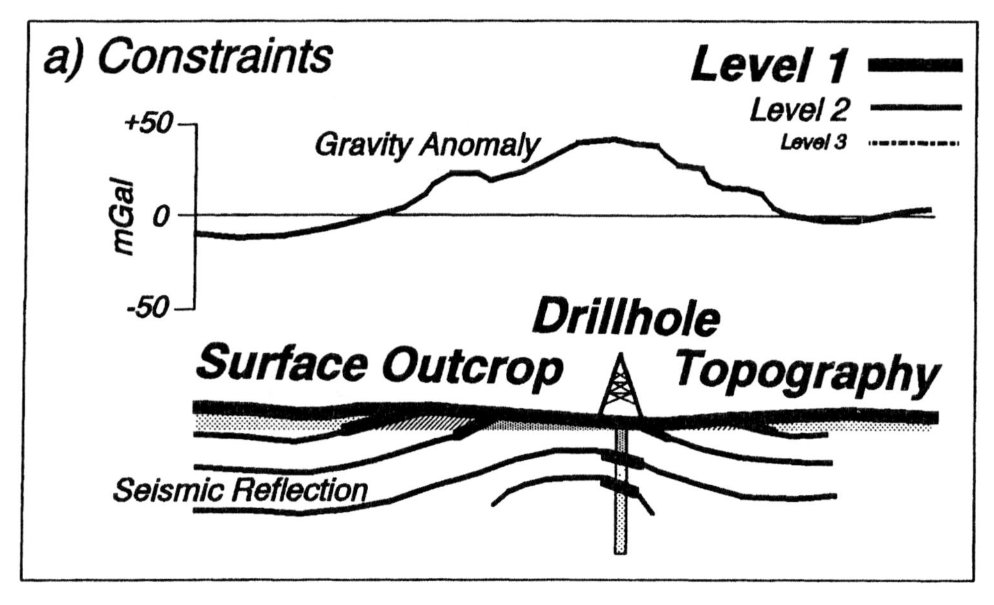

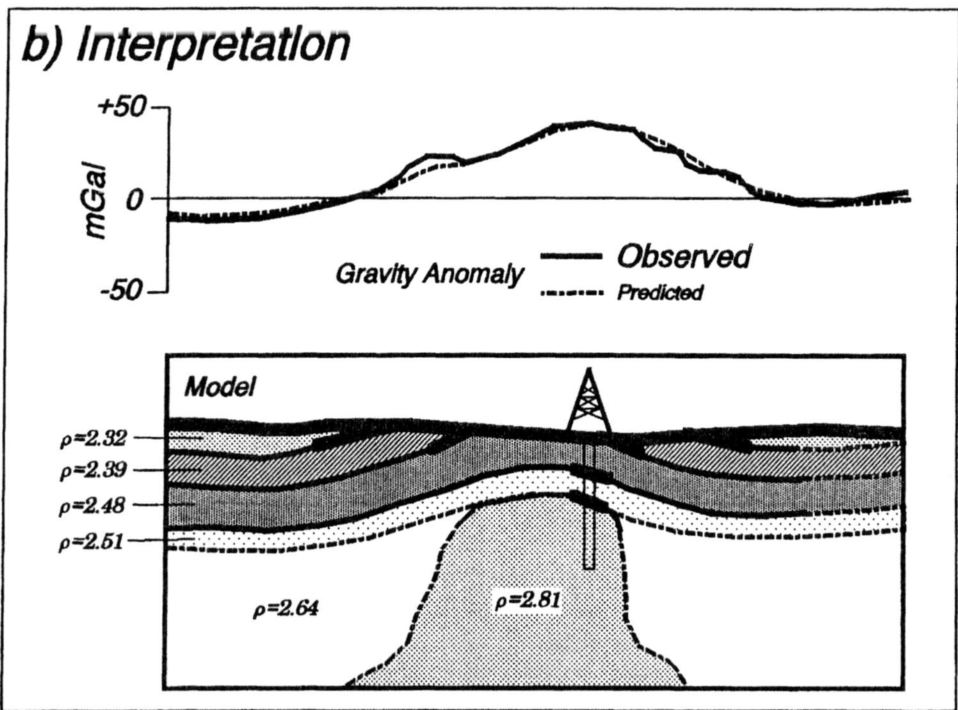

FIGURE 1.8 *Constraints and interpreted cross section.* a) The firm (Level 1) constraints include topography, types of rocks, and dips observed at the surface and in the drillhole. Less firm (Level 2) constraints come from the orientations of reflectors interpreted from seismic reflection profiles, and from the observed gravity anomaly. b) An interpreted cross section places strict value on the Level 1 constraints, less on those from Level 2. Level 3 constraints (shown by the dashed lines) result from a model of density configurations that brings the predicted gravity anomaly close to the observed.

GEOLOGIC TIME	THE EARTH	RESOLUTION (What We Know)
Our Lifetime	*Earth's Surface*	*Great Detail*
Recorded History (0 - 5,000 years)	*Crust (0 - 30 km)*	*Quite a Bit*
Phanerozoic (0 - 600,000,000 years)	*Lithosphere - Asthenosphere System (0 - 700 km)*	*Some*
Precambrian (600,000,000 - 4,600,000,000 years)	*Lower Mantle and Core (700 - 6300 km)*	*Not Much*

FIGURE 1.9 Analogy showing diminishing resolution going back in time and going deeper into the Earth.

direction is not wise (for example, models or geophysical observations do not give us better information about surface geology than we could get from a detailed geologic mapping project).

WHOLE EARTH KNOWLEDGE

Our knowledge of Earth's interior is analogous to looking back in time (Fig. 1.9). We know a lot about events that happened in our lifetimes, far less about things from our parents' and grandparents' time. As we look farther back in time or deeper within the Earth, the quality of observation deteriorates; not only do we have less information, but the detail of the information diminishes. The concept of *resolution* is therefore important in appreciating what is known about the Earth, both temporally and spatially. In our own lifetimes, we know about events that shaped history (the fall of the Berlin Wall), but also great details of some less important observations (the route from our homes to school or work). We know of some important events that shaped our parents and grandparents lives (landing on the moon; World War II), but far less about what daily life was like for them. As we continue back through recorded history, some events stand out (Europeans coming to America; development of Roman, Greek, and Egyptian civilizations), but the time between "important" events becomes longer and vague. Likewise, the scale of features we can interpret becomes larger as we probe deeper and deeper into the Earth. We know a great deal about Earth's surface and upper crust, a fair amount about its lower crust and uppermost mantle. Going deeper, we have only general appreciation of the composition and boundaries of the lower mantle and core.

Knowledge of the Earth comes almost entirely from observations made at or near Earth's surface (Fig. 1.10). Direct observations, in the form of actual rock or magma, sample only the upper 200 or so km, about 1/30th of Earth's 6300 km radius. Surface exposures are almost entirely rocks formed within Earth's crust, with occasional pieces of uppermost mantle brought to the surface during deformation; those

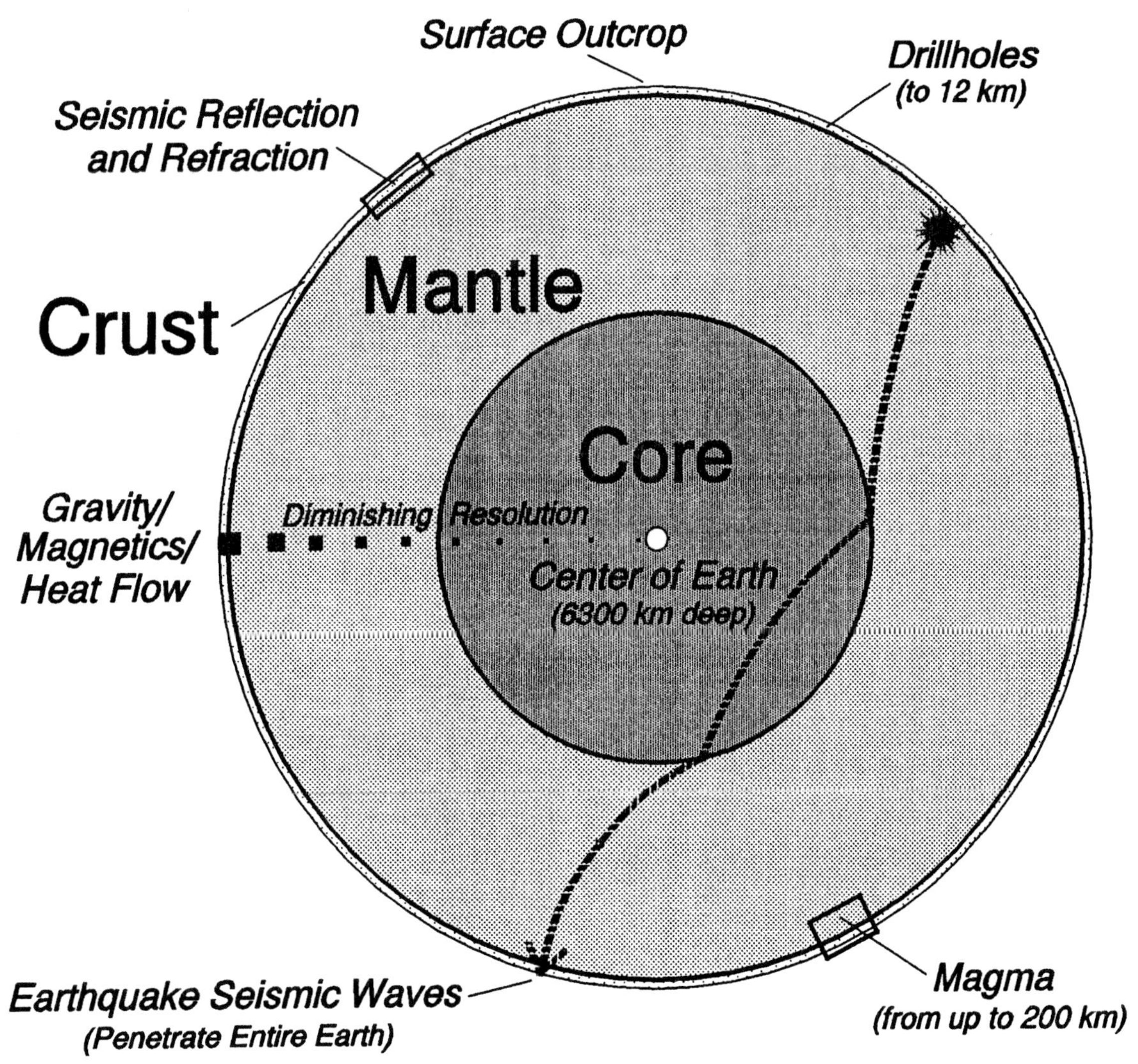

FIGURE 1.10 *Constraints on nature of Earth's interior.* Direct observations (surface outcrops, drillholes, magma reaching surface) generally sample only the crust and uppermost mantle. Geophysical observations (controlled-source seismic, potential field, heat flow) provide further constraints on the outer shells; earthquake seismic waves give most of the information on the lower mantle and core.

materials generally formed in the uppermost 50 km. The deepest drillhole penetrates to about 12 km depth, less than half the thickness of typical continental crust. Volcanic eruptions and igneous intrusions (with the exception of deep-seated kimberlites) come from magma that originated at lower crustal or upper mantle depths, generally within the upper 200 km.

Geophysical data allow us to look deeper into the Earth and sample more widely, but with varying degrees of resolution (Fig. 1.10). Our knowledge is limited by the maximum depth that particular techniques can probe, effectively, within the Earth. Seismic reflection data show details within sedimentary basins (upper 10 km) and, in recent years, provide information about the lower crust and the crust/mantle transition ("Moho"). Seismic refraction data provide constraints on crustal thickness changes and, in some cases, seismic velocities within the crust and uppermost mantle.

Most of our constraints on the deep interior of the Earth are due to the fact that seismic waves from large earthquakes travel through the entire Earth, where they are recorded on the other side. Changes in seismic wave velocity with depth are derived from analysis of the travel times and paths of various earthquake waves. The seismic velocities in turn give constraints on the composition and physical state of portions of the Earth.

Gravity and magnetic measurements constrain the size and positions of anomalous bodies within the Earth, but their resolution decreases with the depth of the bodies (potential field strength lessens with increasing distance from the source). Similarly, heat flow pinpoints shallow intrusions and suggests changes in the depth to the lithosphere/asthenosphere transition zone. For the Earth as a whole, gravity, magnetic, and heat flow data provide constraints on gross properties (density, magnetism, and thermal state, respectively), though not at the detail given by earthquake seismic studies.

This book first presents a general framework for study of gross features of the Earth (plate tectonics), then examines how each geophysical technique contributes to whole Earth knowledge.

SELECTED BIBLIOGRAPHY

General Geophysics

Bott, M. H. P., 1982, *The Interior of the Earth: Its Structure, Constitution and Evolution* (2nd ed.), New York: Elsevier Science Pub. Co., 403 pp.

Chapman, R. E., 1995, *Physics for Geologists,* London: UCL Press Limited, 143 pp.

De Bremaecher, J. C., 1985, *Geophysics: The Earth's Interior,* New York: John Wiley and Sons, Inc., 342 pp.

Fowler, C. M. R., 1990, *The Solid Earth: An Introduction to Global Geophysics,* Cambridge: Cambridge University Press, 472 pp.

Garland, G. D., 1979, *Introduction to Geophysics* (2nd ed.), Toronto: W. B. Saunders Comp., 494 pp.

Parasnis, D. S., 1997, *Principles of Applied Geophysics* (5th ed.), New York: Chapman and Hall, 429 pp.

Steinhart, J. S., and T. J. Smith (editors), 1966, *The Earth Beneath the Continents,* American Geophysical Union, Geophysical Monograph 10, 663 pp.

Stacey, F. D., *Physics of the Earth*, 1992, Brisbane, Australia: Brookfield Press, 513 pp.

Telford, W. M., L. P. Geldart, R. E. Sheriff, and D. A. Keys, 1976, *Applied Geophysics,* Cambridge: Cambridge University Press, 860 pp.

Turcotte, D. L., and G. Schubert, 1982, *Geodynamics: Applications of Continuum Physics to Geological Problems,* New York: John Wiley and Sons, 450 pp.

Earthquake Seismology

Bolt, B. A., 1988, *Earthquakes,* New York: W. H. Freeman and Comp., 282 pp.

Bullen, K. E., 1965, *An Introduction to the Theory of Seismology* (3rd ed.), Cambridge: Cambridge University Press, 381 pp.

Lay, T., and T. C. Wallace, 1995, *Modern Global Seismology,* International Geophysics Series Number 58, New York: Academic Press, 521 pp.

Yeats, R. S., K. Sieh, and C. R. Allen, 1997, *The Geology of Earthquakes,* New York: Oxford University Press, 568 pp.

Exploration Seismology

Anstey, N. A., 1977, *Seismic Interpretation: The Physical Aspects,* Boston: International Human Resources Development Corporation, 625 pp.

Badley, M. E., 1985, *Practical Seismic Interpretation,* Boston: International Human Resources Development Corporation, 266 pp.

Bally, A. W. (editor), 1983, *Seismic Expression of Structural Styles,* American Association of Petroleum Geologists, Studies in Geology Series 15, (3 volume set).

Burger, H. R., *Exploration Geophysics of the Shallow Subsurface,* 1992, Englewood Cliffs, N.J.: Prentice Hall, Inc., 489 pp.

Dobrin, M. B., 1976, *Introduction to Geophysical Prospecting,* (3rd ed.), New York: McGraw–Hill, 446 pp.

Dohr, G., 1981, *Applied Geophysics: Introduction to Geophysical Prospecting,* Geology of Petroleum, v. 1, New York: Halsted Press, 231 pp.

Halbouty, M. T. (editor), 1982, *The Deliberate Search for the Subtle Trap,* Tulsa: American Association of Petroleum Geologists, Memoir 32, 351 pp.

Kearey, P. and M. Brooks, 1984, *An Introduction to Geophysical Exploration,* Boston: Blackwell Scientific Publications, 296 pp.

Payton, C. E. (editor), 1977, *Seismic Stratigraphy: Applications to Hydrocarbon Exploration,* Tulsa: American Association of Petroleum Geologists, Memoir 26, 516 pp.

Robinson, E. S., and C. Çoruh, 1988, *Basic Exploration Geophysics,* New York: John Wiley and Sons, 562 pp.

Robinson, E. A., and S. Treitel, 1980, *Geophysical Signal Analysis,* Englewood Cliffs, N.J.: Prentice Hall, Inc., 466 pp.

Sengbush, R. L., 1983, *Seismic Exploration Methods,* Boston: International Human Resources Development Corp., 296 pp.

Sheriff, R. E., 1978, *A First Course in Geophysical Exploration and Interpretation,* Boston: International Human Resources Development Corp., 313 pp.

Sheriff, R. E., 1984, *Encyclopedic Dictionary of Exploration Geophysics* (2nd ed.), Tulsa: Society of Exploration Geophysicists, 323 pp.

Sheriff, R. E., and L. P. Geldart, 1995, *Exploration Seismology* (2nd ed.), Cambridge: Cambridge University Press, 592 pp.

Deep Seismic Reflection Studies

Barazangi, M. and L. Brown (editors), 1986, *Reflection Seismology: A Global Perspective,* Washington, D.C.: American Geophysical Union, Geodynamics Series, v. 13, 311 pp.

Barazangi, M. and L. Brown (editors), 1986, *Reflection Seismology: The Continental Crust,* Washington, D.C.: American Geophysical Union, Geodynamics Series, v. 14, 339 pp.

Matthews, D., and C. Smith (editors), 1987, *Deep Seismic Reflection Profiling of the Continental Lithosphere,* London: Geophysical Journal of the Royal Astronomical Society, v. 89, No. 1, 495 pp.

Watkins, J. S., and C. L. Drake (editors), 1982, *Studies in Continental Marine Geology,* Tulsa: American Association of Petroleum Geologists, Memoir 34, 801 pp.

Potential Field and Electrical Methods

Blakely, R., *Potential Theory in Gravity and Magnetic Applications,* 1995, Cambridge: Cambridge University Press, 441 pp.

Butler, R. F., 1992, *Paleomagnetism: Magnetic Domains to Geologic Terranes,* Boston: Blackwell Scientific Publications, 319 pp.

Griffiths, D. H., and R. F. King, 1981, *Applied Geophysics for Geologists and Engineers: The Elements of Geophysical Prospecting* (2nd ed.), New York: Pergamon Press, 230 pp.

Merrill, R. T., and M. W. McElhinny, 1983, *The Earth's Magnetic Field,* London: Academic Press, 401 pp.

National Research Council, 1986, *The Earth's Electrical Environment,* Washington, D.C.: National Academy Press, Studies in Geophysics, 263 pp.

Nettleton, L. L., 1971, *Elementary Gravity and Magnetics for Geologists and Geophysicists,* Tulsa: Society of Exploration Geophysicists, Monograph Series 1, 121 pp.

Heat Flow

Elder, J., 1981, *Geothermal Systems,* New York: Academic Press, 508 pp.

Jessop, A. M., *Thermal Geophysics,* 1990, Amsterdam: Elsevier Science Publishers, Developments in Solid Earth Geophysics, No. 17, 306 pp.

Lachenbruch, A. H., and J. H. Sass, 1977, Heat flow in the United States and the thermal regime of the crust, in: *The Earth's Crust, its Nature and Physical Properties,* edited by J. G. Heacock, Washington, D.C.: American Geophysical Union, Monograph 20, pp. 626–675.

Lee, W. H. K., (editor), 1965, *Terrestrial Heat Flow,* Washington, D.C.: American Geophysical Union, Monograph 8, 1965.

Sclater, J. G., and J. Francheteau, 1970, The implications of terrestrial heat flow observations on current tectonic and geochemical models of the crust and upper mantle of the earth, *Geophysical Journal Royal Astronomical Soc.,* v. 20, pp. 509–542.

CHAPTER 2

Plate Tectonics

plate *(plāt)* ***n.***, [< *Gr.* platys, *flat*], *a smooth, flat, thin piece of metal or other material.*
tectonic *(tek tän′ ik),* ***adj.***, [< *Gr.* tektōn, *a builder*], *pertaining to large-scale features on Earth's surface and the internal processes that led to their development.*
plate tectonics *(plāt tek tän′ iks),* ***n.***, *a modern idea that Earth's outer shell is made of rigid plates; large-scale features on Earth's surface result from movement of the plates relative to one another.*
litho- *(lith′ ō),* [< *Gr.* lithos, *a stone*], *stone, rock.*
sphere *(sfir),* ***n.***, [< *Gr.* sphaira, *sphere*], *a round body with a surface equally distant from the center at all points.*
lithosphere *(lith′ ə sfir′),* ***n.***, *the strong (rigid), outer part of the Earth.*
asthenia *(as thē ′nē ə),* ***n.***, [< *Gr. a, without;* < *Gr.* sthenos, *strength*], *bodily weakness.*
asthenosphere *(as thē′ nə sfir),* ***n.***, *a relatively soft (less rigid) region of the Earth underlying the lithosphere.*

Large features on Earth's surface, such as continents, ocean basins, and mountain ranges, result from Earth's internal and external processes. Early geologists (most notably James Hall in the 1850's) developed geosynclinal theory, whereby long, narrow troughs subsided and were filled with sedimentary and volcanic strata (Fig. 2.1a). The deeply-buried layers, subjected to high temperatures, were metamorphosed and expanded, much like a cake rising in an oven (Fig. 2.1b). The expansion caused deformation and uplift of Earth's surface as mountain ranges (Kay, 1951). *Geosynclinal theory* involves vertical movements; mountains result from materials moving up and down without large horizontal displacements, which were thought unlikely.

Observations that blocks of Earth's crust had, indeed, moved laterally over long distances led to drastic revision of mountain building ideas. *Plate tectonic the-*

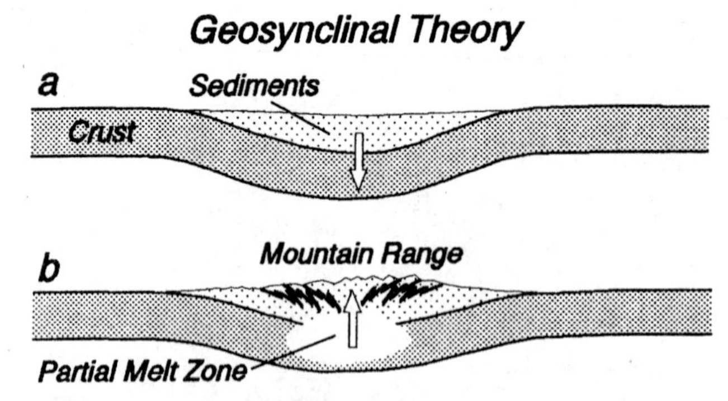

FIGURE 2.1 Geosynclinal theory suggests that mountain ranges result from vertical movement, without significant horizontal displacement. a) Crust warps downward, forming depression that fills with sediments. b) Heated sediments and crust expand and uplift, forming mountain range.

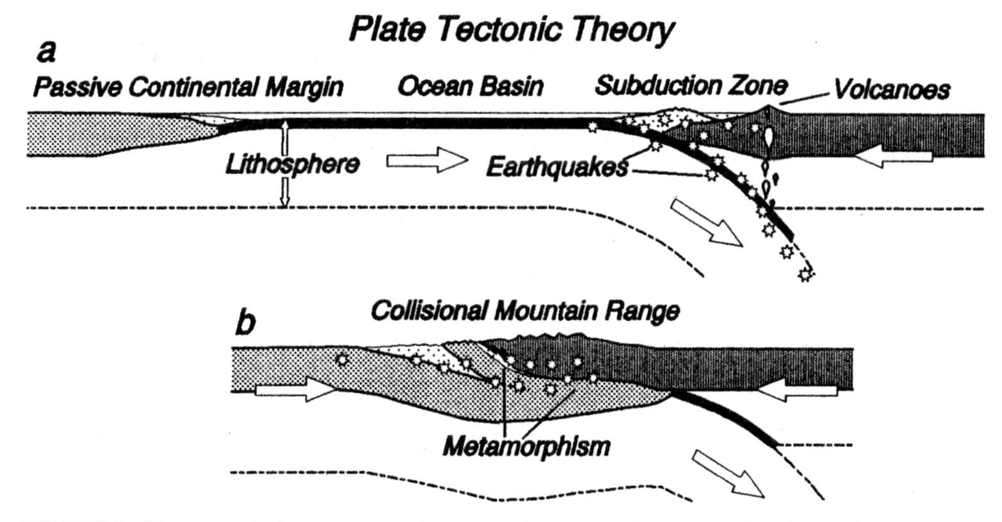

FIGURE 2.2 Plate tectonic theory suggests that mountain ranges, volcanoes, earthquakes, and metamorphism result from large horizontal displacements. a) Ocean basin closes as lithospheric plates converge. b) Horizontal displacements lead to vertical uplift as continents collide.

ory, with horizontal motion as a major premise, more adequately explains the origins of many mountain ranges, as well as the distribution of earthquakes, volcanoes, and the metamorphism of rocks (Fig. 2.2).

Plate tectonic theory had its origins in the early part of the 20th century, with the idea of "continental drift," proposed by the German meteorologist Alfred Wegener. The inspiration came from the "jig-saw puzzle" fit of continents across the Atlantic Ocean, recognized on maps as far back as the 1500's. The most severe objections to Wegener's theory involved his mechanism, that blocks of continental crust drifted over the mantle. Harold Jeffreys pointed out (correctly) that it was inconceivable that blocks of crustal material could plow their way over mantle, known from seismic wave studies to be much stronger. Geophysical observations thus disputed the notion of large-scale horizontal movements. Later geophysical observations made during the 1950's and 1960's caused geologists to realize that the continents are passengers on large plates comprised mostly of mantle, information critical to the acceptance of plate tectonic theory. The pressure and temperature conditions in the mantle create a zone of relatively soft mantle, sandwiched between harder mantle above and below. The softer mantle (*"asthenosphere"*) allows the rigid plates of mantle and crust (*"lithosphere"*) to move horizontally for thousands of kilometers.

MAJOR DIVISIONS OF THE EARTH

Appreciation of the origin of large features observed at Earth's surface (*tectonics*) requires information about the overall structure and composition of the Earth. Geophysical data gathered at the surface provide the bulk of observations on Earth's plate tectonic system and deeper regions.

Classical Divisions

Earth's deep interior is known primarily from observations of the types and velocities of seismic waves traversing various regions (Fig. 1.10); Earth's gravity field further constrains density distribution. The classical view of Earth's interior, developed

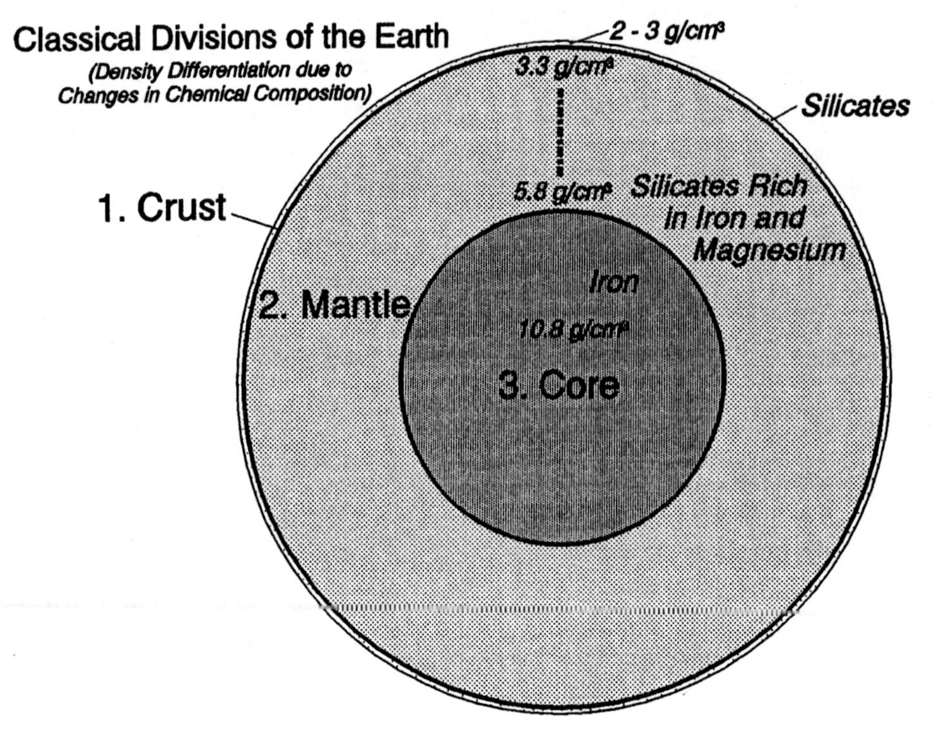

FIGURE 2.3 *Classical divisions of the Earth.* Three zones of differing density correspond to changes in chemical composition. 1) The *crust* is mainly minerals rich in silica (silicon and oxygen), with densities between 2 and 3 g/cm^3 and thickness between 2 and 70 km. It is less than 2% of Earth's volume, 1% of its mass. 2) The *mantle* is silicate minerals rich in iron and magnesium. Extending from about 30 to 2900 km depth, mantle density increases from 3.3 to 5.8 g/cm^3. It comprises 82% of Earth's volume, 68% of its mass. 3) The *core* is predominately iron, with other heavy elements like nickel. It has an average density of about 10.8 g/cm^3 and extends from 2900 km to the center of the Earth at 6300 km. The core has 16% of Earth's volume and 32% of its mass.

in the early part of the 20th century, differentiated the Earth into three spheres according to density, the denser material concentrated toward the center (Fig. 2.3). Drastic differences in density correspond to changes in chemical composition, defining the *crust, mantle* and *core.*

Modern Divisions

The installation of more and better seismographs in the 1960's resulted in finer resolution of seismic wave velocities within the three spheres. A more modern division describes portions of the three spheres according to their physical state (*hard solid*, relatively *soft solid*, or *liquid*). Five zones thus recognized are the *lithosphere, asthenosphere, lower mantle* (or *mesosphere), outer core,* and *inner core* (Fig. 2.4).

Fig. 2.5 illustrates that the classical scheme is not out of date. Major divisions are still along chemical boundaries, between the crust and the mantle, and the mantle and the core (Fig. 2.3). The modern scheme simply describes the physical state of those chemicals under conditions of increasing temperature and pressure within the Earth (Fig. 2.4). Silicates comprising the crust are generally so cold that they are rigid, forming the top part of the lithosphere. The iron/magnesium-rich silicates of the uppermost mantle are also relatively cold and rigid, forming the remainder of the lithosphere. At depths below about 150 km, those same mantle materials undergo slight partial melting, forming the softer asthenosphere. The

Physical state of the modern subdivisions of Earth

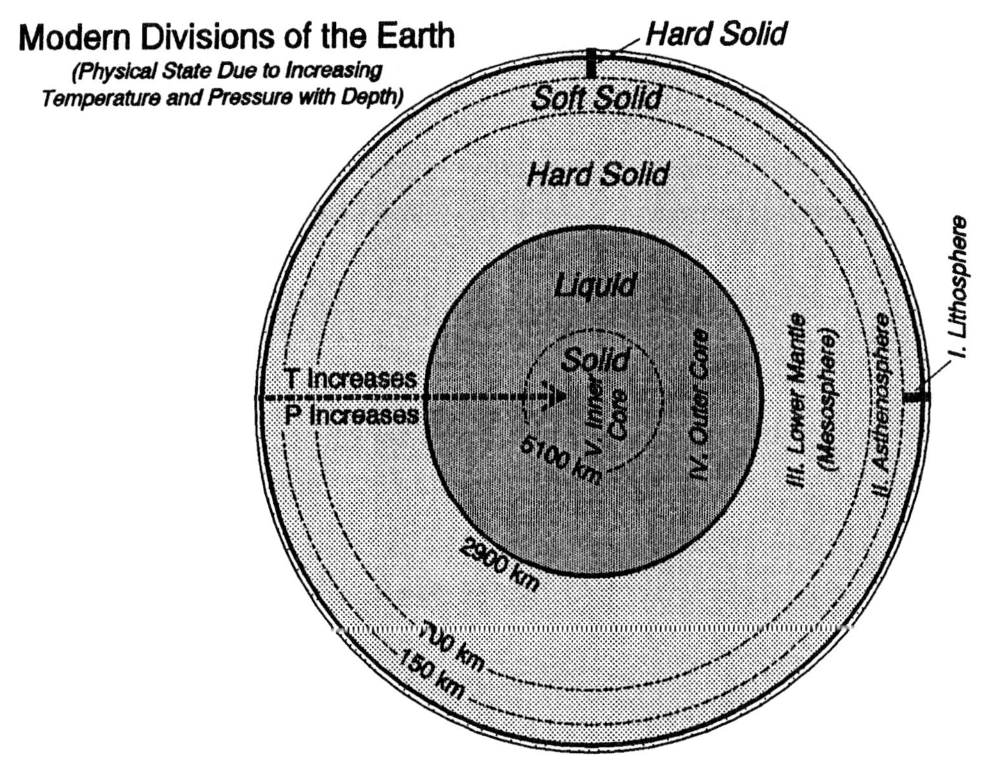

FIGURE 2.4 *Modern divisions of the Earth.* Five zones of differing physical state relate to increasing temperature (T) and pressure (P) with depth. I) The *lithosphere* is a hard solid, extending from the surface to about 100 to 200 km depth. It includes the crust and uppermost mantle. II) The *asthenosphere* is solid with a slight amount of partial melt. Relative to the material above and below, the upper part (from about 150 to 400 km depth) is a soft (plastic) solid; the lower part gradually becomes harder to 700 km depth. III) The *lower mantle* (or mesosphere) is a hard solid, extending from about 700 to 2900 km depth. IV) The *outer core*, from 2900 to 5100 km depth, is liquid. V) The solid *inner core* extends from about 5100 to 6300 km depth.

pressure becomes so great below depths of 350 to 700 km that the lower mantle is a hard solid. The heavy (iron-rich) material of the outer core is liquid at the temperatures and pressures encountered between 2900 and 5100 km depth. The pressure is so great near Earth's center, however, that the same material exists as a solid inner core.

LITHOSPHERE/ASTHENOSPHERE SYSTEM

Ideas about large lateral displacements of Earth's crust originated in the early 1900's. Development and widespread acceptance of plate tectonic theory did not occur until the 1960's and 1970's, however, when a vast number of geological and geophysical observations were made in the ocean basins. Drilling and magnetic anomaly patterns suggest that the basaltic rocks of the ocean crust form at mid-ocean ridges and are progressively older away from the ridge axes (Vine, 1966). The resulting concept of "sea-floor spreading" provides an acceptable mechanism for continents to drift apart, leaving new oceanic crust in their wakes. A worldwide seismograph network reveals that most earthquakes occur along narrow bands outlining the boundaries of lithospheric plates (Isaacs, Oliver, and Sykes, 1968). Bands of earthquakes extend to great depths in regions landward of deep-sea trenches, where one plate appears to descend beneath another.

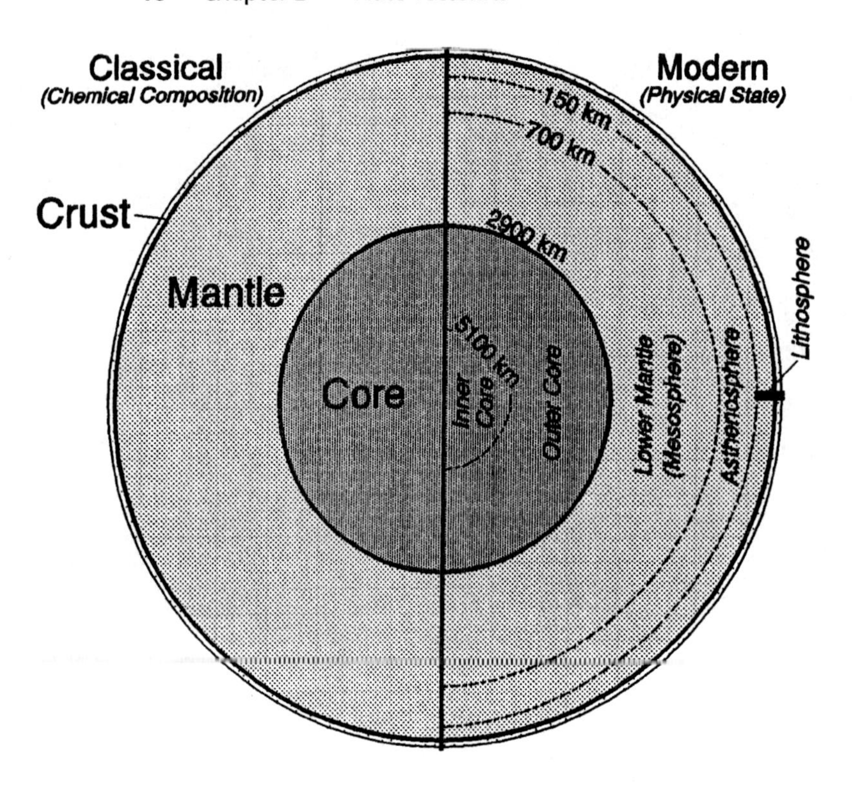

FIGURE 2.5 Comparison of classical and modern divisions of the Earth.

Plate tectonic theory concludes that the outer, rigid layer of the Earth, the lithosphere, is broken into a mosaic of large plates. The plates ride on a softer substratum, the asthenosphere, drifting laterally a few cm/year; at those very slow rates, the asthenosphere is ductile. Most large-scale geologic phenomena result from the relative motion of plates, as they interact at their boundaries.

Physical State of Upper Mantle

The graphs in Fig. 2.6 show how pressure and temperature increase from the surface to 700 km depth within the Earth. Notice that, due to the constantly increasing overburden, pressure increases at a roughly constant (linear) rate with depth. Temperature increase is nonlinear; it rises abruptly at shallow depths, more gradually with increasing depth.

The physical state of materials depends on pressure (P) and temperature (T) conditions. A phase diagram illustrates the P-T conditions under which Earth materials go from solids, to partially melted solids, to liquids. The phase diagram in Fig. 2.7 illustrates the physical state of the igneous rock peridotite (see Appendix C) under conditions of pressure and temperature encountered in the upper mantle. With increasing temperature peridotite will melt, but the temperature to melt the rock is greater with increasing pressure (that is, increasing depth within the Earth). Near the surface (at very low pressure), melting of high-silica minerals within the rock begins at about 1200 °C. The rock undergoes increasing degrees of partial melting until it becomes totally liquid at about 1900 °C. At 300 km depth (greater pressure) considerably higher temperatures are required to initiate partial and complete melting (2000 °C and 2700 °C, respectively).

The dashed line in the P-T diagram (Fig. 2.7) shows a normal geothermal gradient in the upper mantle. The nonlinear gradient means that phase boundaries are

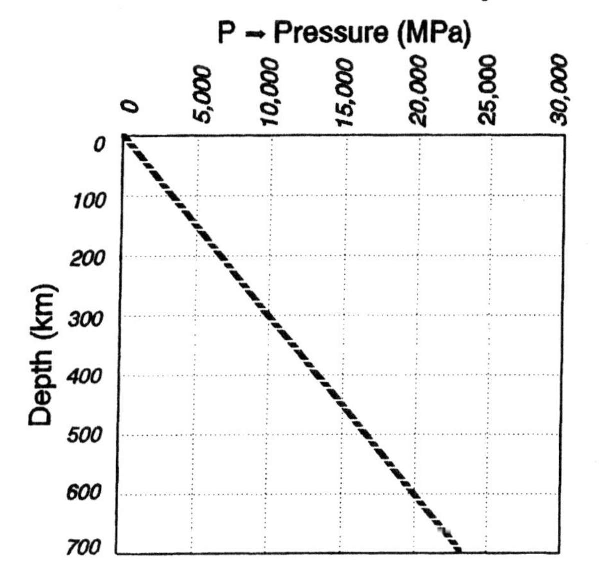

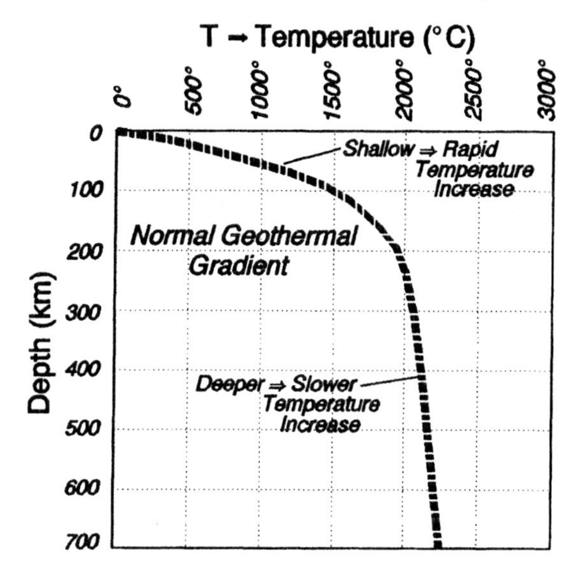

FIGURE 2.6 Pressure and temperature in Earth's upper 700 km. Modified from *Physical Geology* by Skinner/Porter, © 1987, with permission of John Wiley and Sons, Inc., New York.

crossed, giving three distinct regions: (a) in the upper 100 km, the peridotite is cold and rigid, resulting in a solid *lithosphere*; (b) between about 100 and 350 km the temperature rise causes a small amount of partial melt, giving the softer *asthenosphere*; (c) below about 350 km the pressure is so great that, even though the temperature is hotter, there is a transition (increasing strength) to the solid *mesosphere*.

The lithosphere consists of both the crust and uppermost mantle. The boundary between the crust and mantle, called the Mohorovičić discontinuity, or "Moho," thus lies within the lithosphere (Fig. 2.8). The Moho was recognized in the early part of the 20th century because compressional seismic waves travelling at about 6.5 km/s in the lower crust were refracted along the higher velocity mantle (≈ 8.2 km/s). The boundary is pronounced because it separates regions of different chemical composition; the lower crust is generally gabbroic (≈ 50% silica; see Appendix C), while the upper mantle is composed of peridotite (≈ 30% silica).

Unlike the abrupt crust/mantle boundary, boundaries within the mantle are more subtle and gradational. Peridotite comprises the three zones of the mantle: lower lithosphere, asthenosphere, and mesosphere. Slight changes in the velocity of seismic waves allow recognition of the three zones. Compressional waves travelling 8.2 km/s in the uppermost mantle slow down to about 7.8 km/s at depths of 75 to 200 km, indicating a transition to the softer asthenosphere. The velocity rises gradually with increasing strength in the lower asthenosphere, culminating in a jump to about 12 km/s in the mesosphere, at about 700 km depth. Depths for boundaries within the mantle (Figs. 2.5, 2.8) are therefore rough approximations to the depths of subtle phase changes from hard, to softer, to hard material of the lower lithosphere, asthenosphere, and mesosphere, respectively.

The Earth's outer shell consists of seven major and several minor lithospheric plates (Fig. 2.9). The plates move at a few centimeters per year (cm/yr) relative to

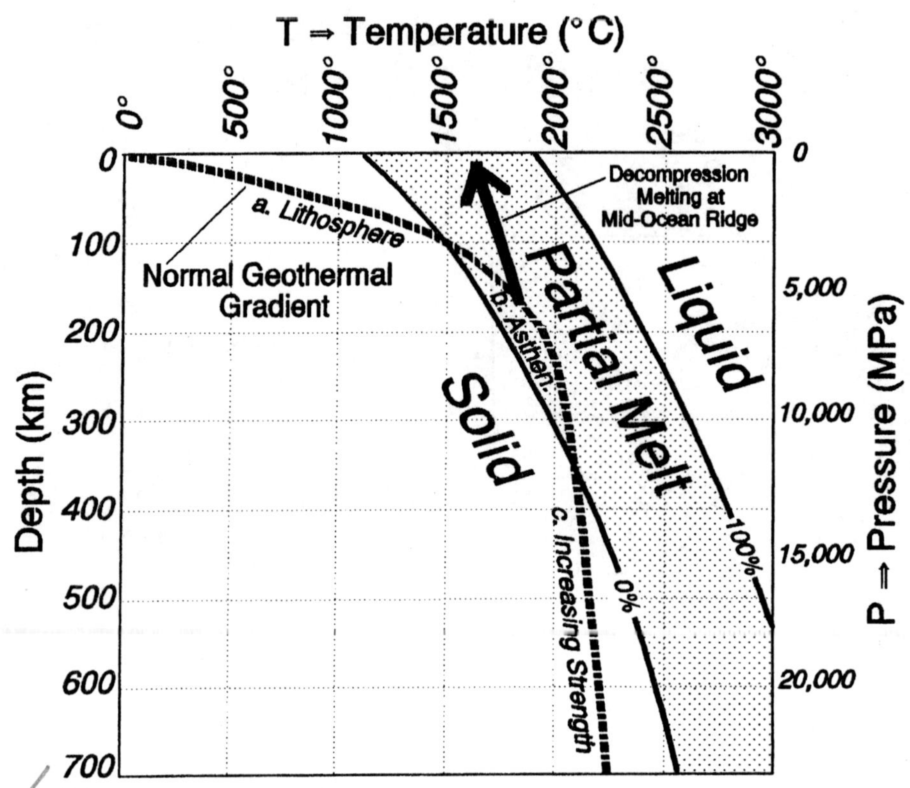

FIGURE 2.7 Phase diagram for peridotite (Earth's mantle). Vertical scale can be viewed as either pressure or depth, because of the linear relationship between the two variables (Fig. 2.6a). Dashed line shows normal increase in temperature with depth for upper mantle (Fig. 2.6b). Percentages refer to the volume of liquid versus solid in the partial melt. Modified from *Physical Geology* by Skinner/Porter, © 1987, with permission of John Wiley and Sons, Inc., New York.

one another. This rate is important not only because it is small, but also because it results in plates moving large distances over geologic time. It is small in that our fingernails grow at about the same rate. Yet taken over a million years, 1 cm/yr results in a displacement of 10 km. It is thus easy to see how strain rates slow enough for the asthenosphere to behave ductilly can, over hundreds of millions of years, produce features as extensive as ocean basins and mountain ranges.

Lithospheric plates are thought to be driven by convection currents within the upper mantle (Fig. 2.10). Where convection currents rise, they split plates apart, generating new magma from the mantle; the magma cools to form new lithosphere. Plates converge where convection currents descend; one plate commonly extends downward into the mantle, destroying lithosphere. The motion of lithospheric plates thus constantly recycles rock materials, with new lithosphere created at mid-ocean ridges and lithosphere deformed and consumed at deep-sea trenches (subduction zones).

Plate motion may also be driven by horizontal density changes within the plates, giving rise to the *ridge push* and *slab pull* hypotheses. According to the first idea, gravity acting on relatively high elevations at mid-ocean ridges may push plates away from divergent boundaries. At convergent boundaries, basaltic oceanic crust on the downgoing plate is metamorphosed to higher-density eclogite, which pulls the lithospheric slab deeper into the mantle.

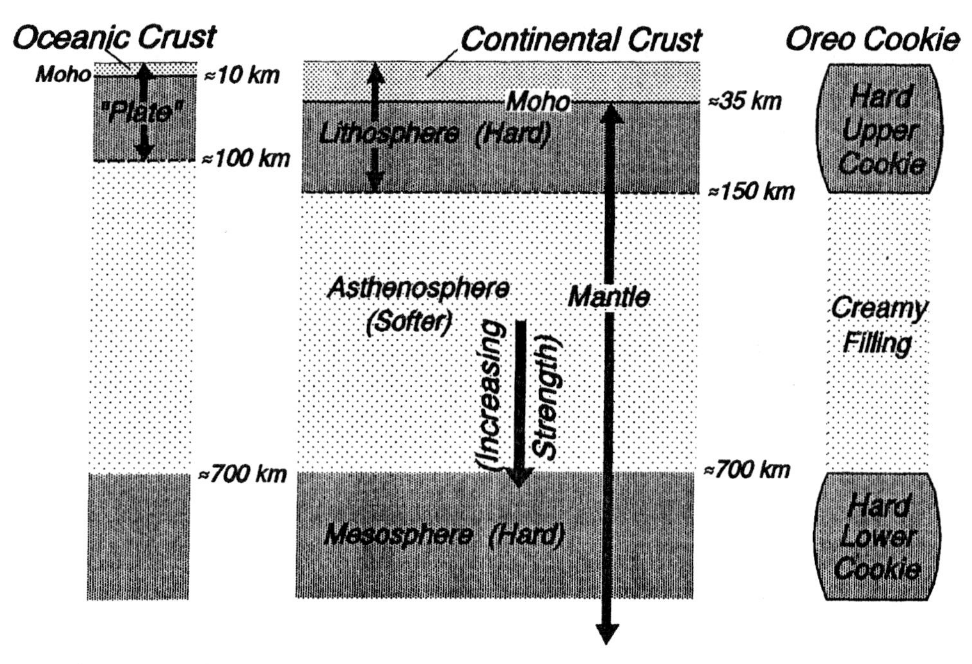

FIGURE 2.8 The relatively high temperatures and low pressures of the mantle between about 150 and 400 km create a special situation (Fig. 2.7), analogous to the soft creamy filling of an Oreo® cookie. Lithospheric plates ride over the softer asthenosphere. Crust forms the top part of the lithosphere; oceanic crust is substantially thinner than continental crust. While the chemical boundary between the crust and mantle ("Moho") is sharp, the change in physical state between the lithosphere and asthenosphere is more subtle. The transition from asthenosphere to mesosphere is a broad zone of increasing strength.

TYPES OF PLATE BOUNDARIES

There are three types of lithospheric plate boundaries: *divergent*, *convergent* and *transform* (Fig. 2.11). Most large-scale geologic structures (mountains, continents, ocean basins) and processes (earthquakes, volcanism, metamorphism) are associated with interactions along the boundaries.

Some volcanoes occur far from plate boundaries, in the interior of a plate (Fig. 2.12). Many of these volcanoes are thought to be caused by the movement of lithospheric plates over *hotspots* that remain fixed within the deeper mantle.

Divergent Plate Boundaries

Where plates move away from one another the lithosphere thins, so that underlying, buoyant asthenosphere elevates a broad region (Fig. 2.13). The elevated regions are *continental rift zones* or *mid-ocean ridges*, depending on whether the lithosphere is capped by continental or oceanic crust.

Divergent plate boundaries are characterized by tensional forces that produce fissures, normal faults, and rift valleys (Fig. 2.14a). If a continent completely rips apart, the two fragments can drift away as parts of different lithospheric plates (Fig. 2.14b). New oceanic lithosphere is created between the continents, at a mid-ocean ridge. If the process continues long enough, a large ocean basin forms (Fig. 2.14c). The plate boundary is then at the mid-ocean ridge, far from the margins separating continental from oceanic crust; such margins are termed "passive continental margins."

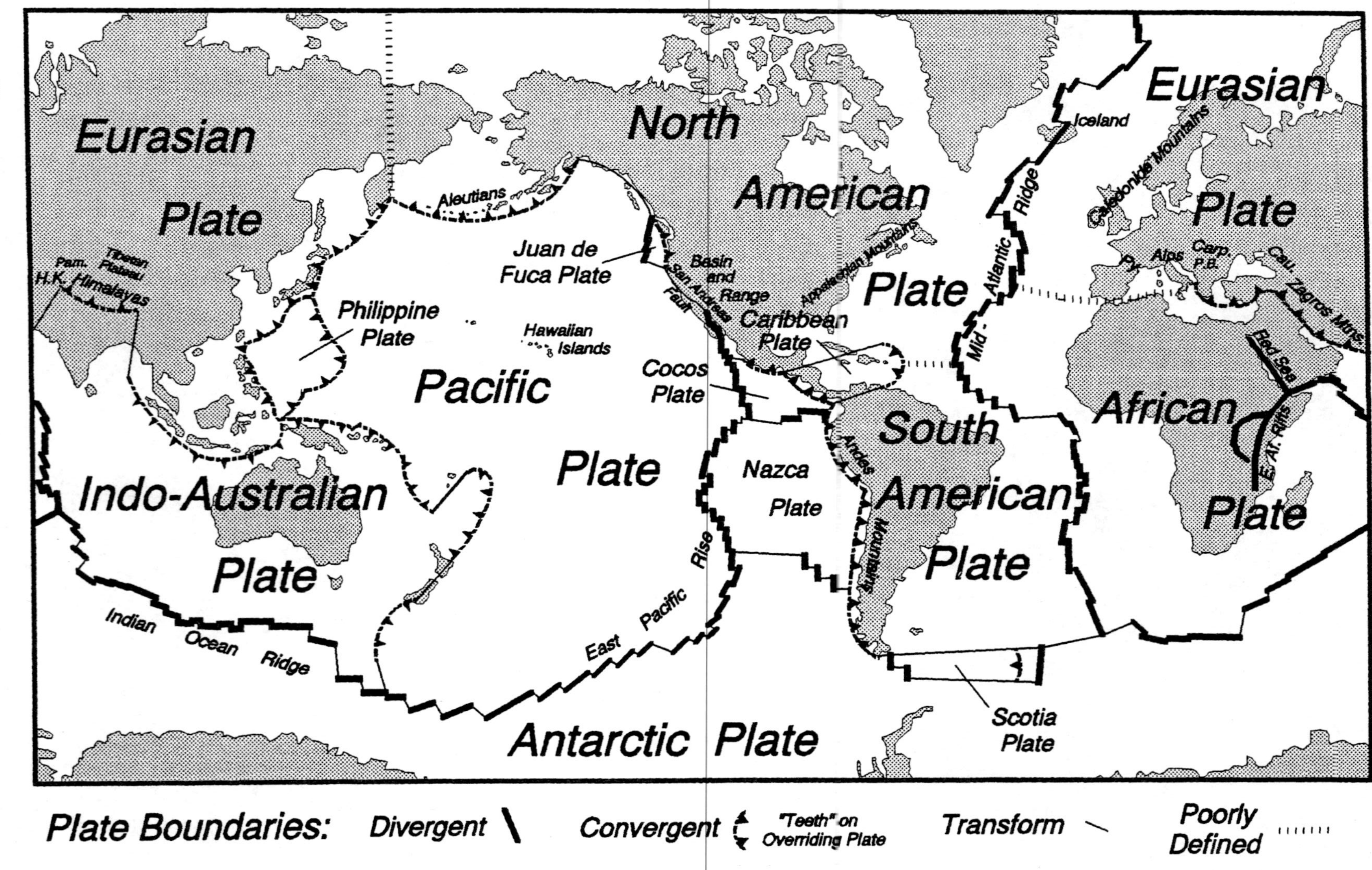

FIGURE. 2.9 *Lithospheric plates and their boundaries.* Most plates have portions capped by thick continental crust and portions with thinner oceanic crust. Geographic features discussed in text are labeled. Cau. = Caucasus Mountains; Carp. = Carpathian Mountains; H.K. = Hindu Kush; Pam. = Pamirs; P.B. = Pannonian Basin; Py. = Pyrenees. Modified from *Laboratory Manual Physical Geology,* 3/e, by AGI/NAGT, © 1993, with permission of Prentice-Hall, Inc, Upper Saddle River, NJ.

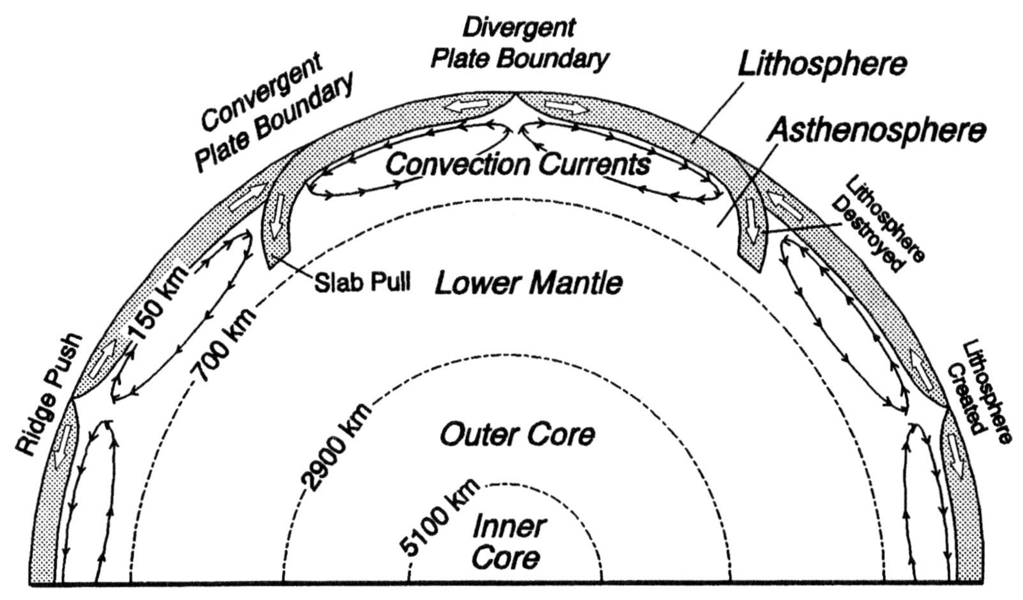

FIGURE 2.10 Lithospheric plates ride over the asthenosphere, perhaps driven by a combination of mass transfer of heat (convection currents) and horizontal density changes (ridge push; slab pull). Lithosphere is created where plates diverge, destroyed where they converge.

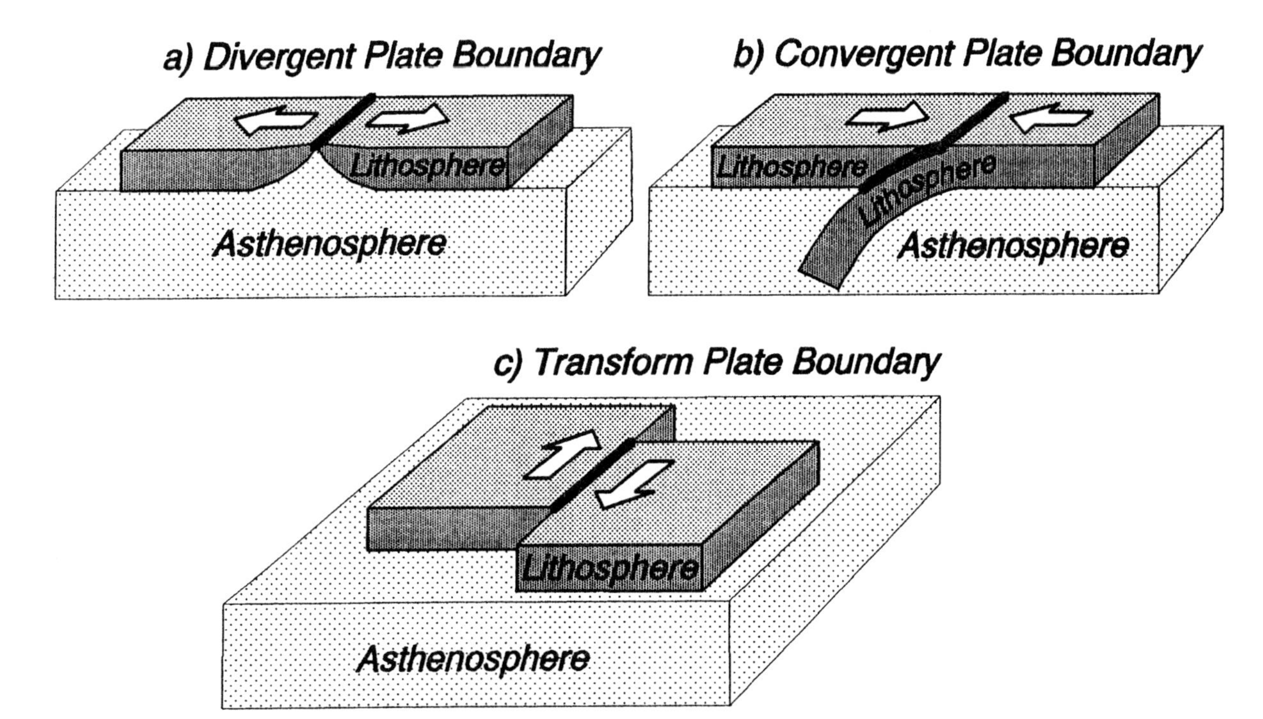

FIGURE 2.11 The three general types of plate boundaries, highlighted by bold lines. Plates rip apart and grow at divergent boundaries (a), compress and are destroyed at convergent boundaries (b), and slide past one another at transform boundaries, neither creating nor destroying plate material (c).

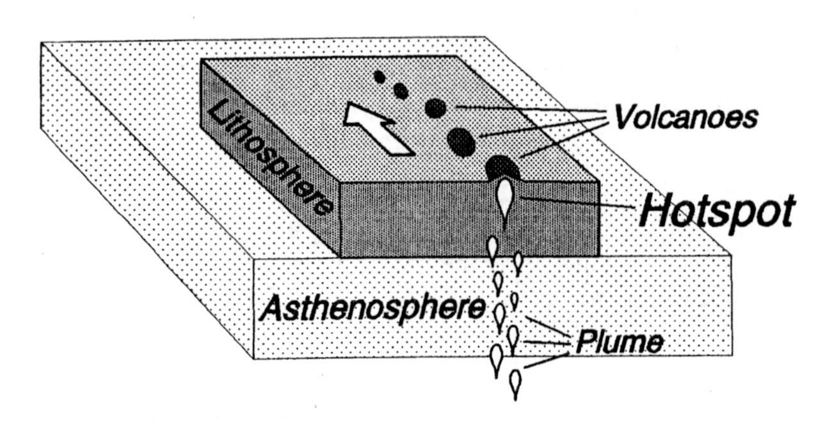

FIGURE 2.12 Lines of volcanoes may be due to the movement of a lithospheric plate over a hotspot, originating from a plume of hot material deep within the mantle.

Continental Rift Zones As a continent pulls apart it stretches, thinning the crust and entire lithosphere (Fig. 2.13b). The region is raised to high elevation because the underlying asthenosphere is hot and buoyant. The upper part of the crust deforms in a cold, brittle fashion, causing earthquakes and elevated ridges (German word "horst"), separated by down-dropped valleys (German "graben"). The grabens fill with up to 8 km of sedimentary and volcanic strata as they subside, forming basins; the adjacent horst blocks remain high as mountain ranges (Fig. 2.14a). A region of fault block mountains in North America, comprising all of Nevada and portions of Utah, Idaho, Oregon, California, Arizona, New Mexico, and Mexico, is thus called the Basin and Range Province (Fig. 2.15). Other areas of active continental rifting are the Pannonian Basin of central Europe and the East African Rifts (Fig. 2.9).

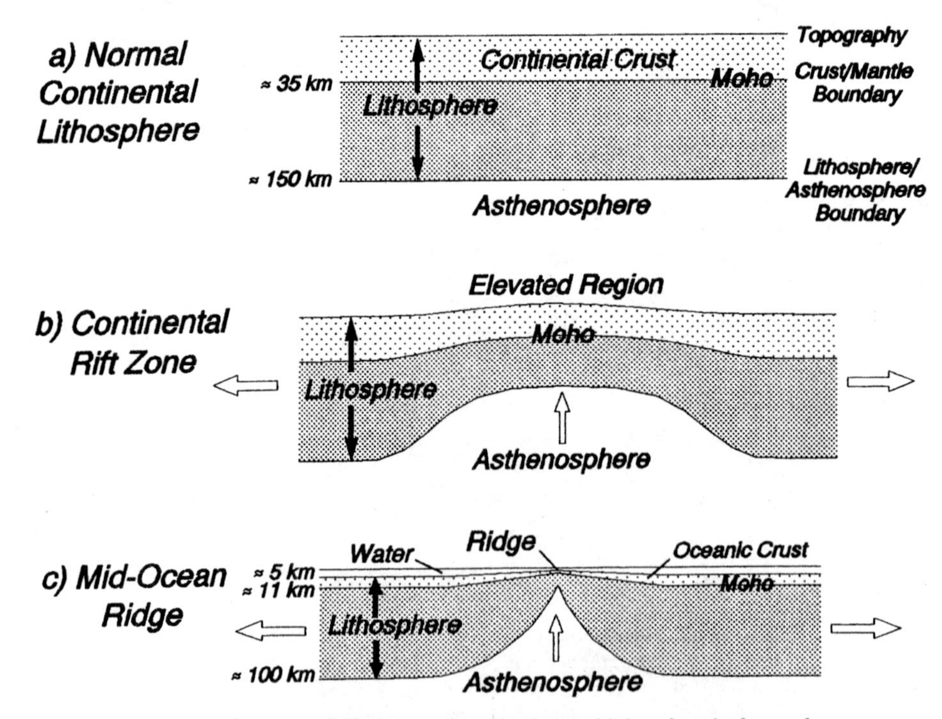

FIGURE 2.13 *Topography, crust/mantle boundary (Moho) and lithosphere/asthenosphere boundary at divergent plate boundaries.* (a) Typical crustal and lithospheric thicknesses on continental craton, with topography near sea level. (b) Crust and entire lithosphere thin as continent rifts apart; all three boundaries elevate. (c) Newly-formed lithosphere at mid-ocean ridge contains crust about 1/6 the thickness of typical continental crust.

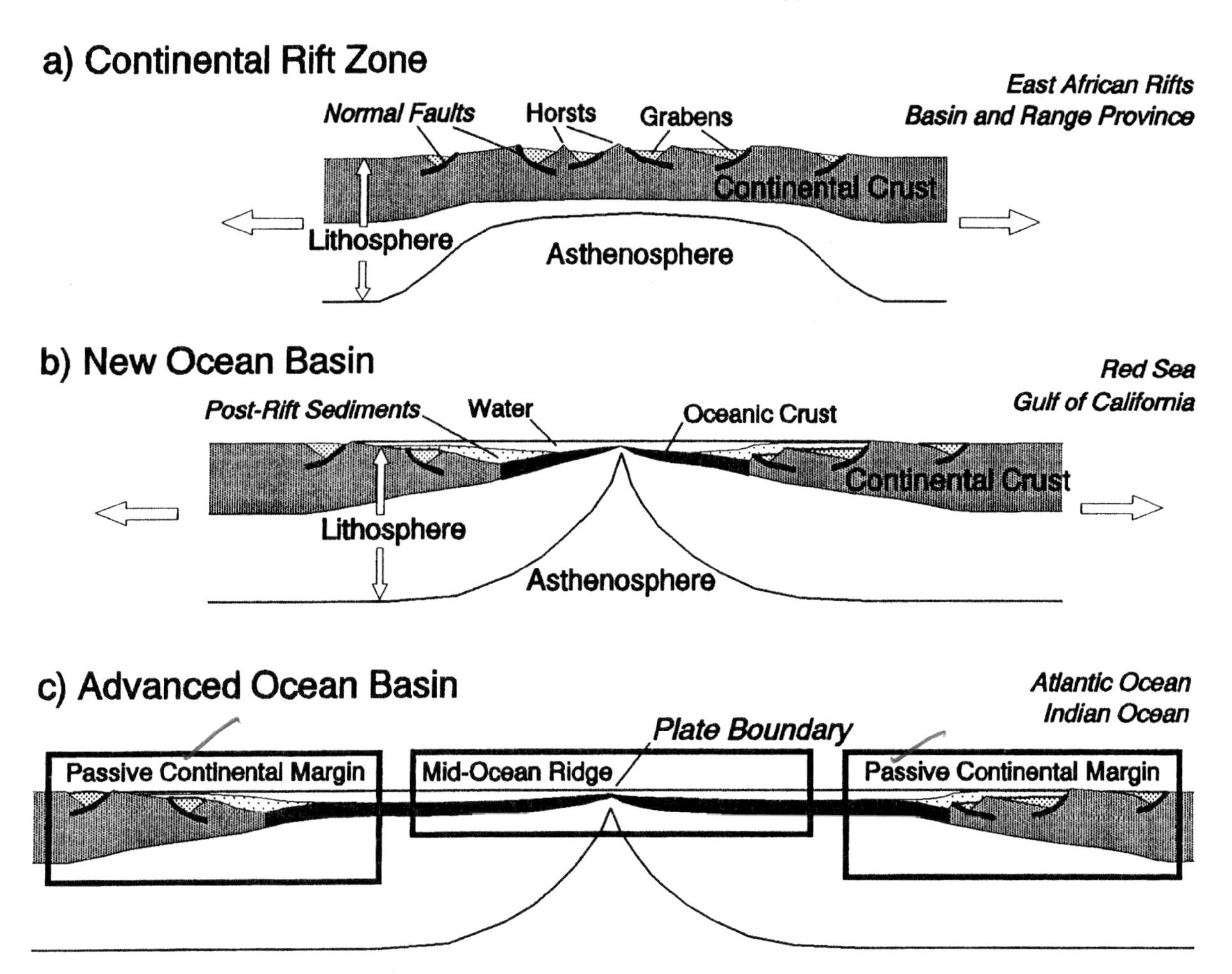

FIGURE 2.14 *Evolution of structures at divergent plate boundary.* (a) Brittle deformation of upper crust at continental rift zone causes normal faulting. Elongate mountain ranges (horsts) separate valleys (grabens) that fill with sedimentary and volcanic strata. (b) As lithosphere moves apart a new ocean basin forms. Horsts and grabens are covered by post-rift sedimentary strata along continental to oceanic crustal transition zone. (c) At advanced ocean basin the divergent plate boundary is along a mid-ocean ridge system; passive continental margins are some distance away, within the interiors of the diverging plates.

Mid-Ocean Ridges When continents completely rift apart, new oceanic lithosphere forms, as in the Red Sea separating Saudi Arabia from Africa (Fig. 2.14b). With continued divergence the buoyant asthenosphere elevates a ridge on the seafloor that may be a few hundred to as much as 4000 km wide, depending on how fast the plates move apart (Fig. 2.14c). Although the region of the ridge is hot, the upper part of the oceanic crust can be cold and brittle, causing earthquakes and normal faults. A chain of mid-ocean ridges, including the Mid-Atlantic Ridge, the East Pacific Rise, and the Indian Ocean Ridge, extends in a more or less continuous fashion for over 50,000 km (Fig. 2.9). Though mostly covered by water, the mid-ocean ridge system is thus the longest mountain range on Earth.

Convergent Plate Boundaries

Where lithospheric plates converge, the plate with thinner, less buoyant crust commonly descends beneath the other plate. The region where a lithospheric plate descends deeply within the mantle is called a *subduction zone*. Two types of subduction zones are common, depending on whether the overriding plate is capped by

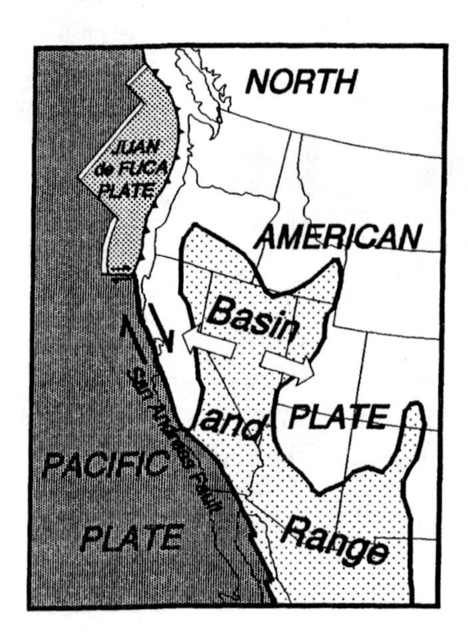

FIGURE 2.15 Shaded area shows active rifting of the North American continent in the Basin and Range Province.

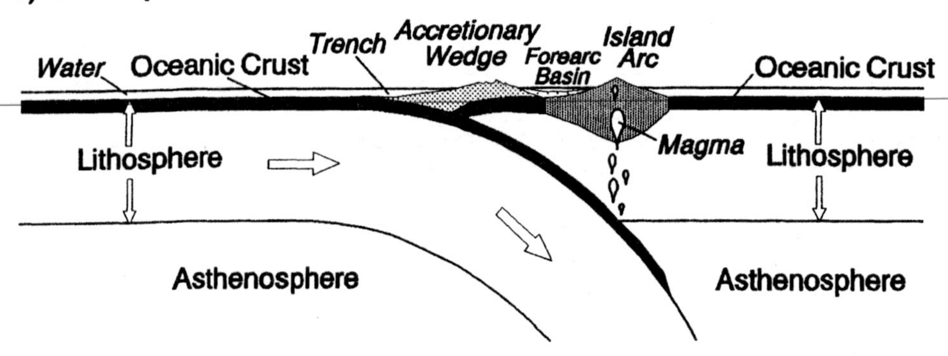

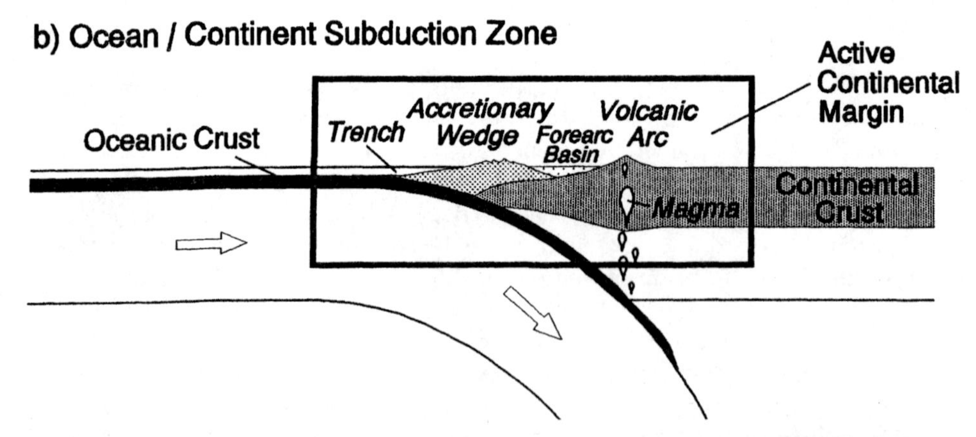

FIGURE 2.16 *Formation of topography and structures at subduction zones.* Two parallel mountain ranges form, one structural ("accretionary wedge") and one volcanic ("island arc" or "volcanic arc"). a) Both plates capped by oceanic crust. b) Oceanic lithosphere subducts beneath more buoyant continental lithosphere. Box shows that the transition from continental to oceanic crust is at the plate boundary, hence an "active continental margin."

thin (oceanic) or thick (continental) crust. Sometimes entire oceans close through subduction, causing blocks of continental crust to collide; *collisional mountain ranges* form as the crust is compressed and thickened.

Ocean/Ocean Subduction Zones If both of the converging plates contain oceanic crust, one plate subducts beneath the other (Fig. 2.16a). Two chains of mountains, one structural and one volcanic, form parallel to the *deep-sea trench* at the surface juncture of the plates. Just landward of the trench, where the top of the plate is shallow and cold, some of the sediments and underlying rock are scraped off and deformed into a wedge shape. These materials attach (or "accrete") to the overriding plate; portions of this *accretionary wedge* may rise above sea level as the sediments and rock are compressed, folded and faulted, forming long ridges and valleys. Prominent islands that are part of accretionary wedges at subduction zones include Barbados, in the Caribbean Sea, and Timor, north of Australia.

Farther from the trench the top of the descending plate may reach depths of 100 to 150 km, where it is so hot that fluids are driven from its crust. The fluids rise, melting silicate minerals from the mantle and crust of the overriding plate. The resulting magma thickens the crust of the overriding plate to two to three times that of normal oceanic crust. Magma that makes it to the surface erupts as a (straight or curved) chain of volcanic islands, called an *island arc*. A depression between an accretionary wedge and an island arc, which often accumulates sediments, is called a *forearc basin*. The Aleutian, Lesser Antilles, and Philippine islands (Fig. 2.9) are examples of trench/accretionary wedge/forearc basin/island arc systems built on oceanic crust.

Ocean/Continent Subduction Zones Continental crust is thicker, and therefore more buoyant, than oceanic crust; a plate with oceanic crust will subduct beneath one capped by continental crust (Fig. 2.16b). The resulting morphology is similar to that developed in an ocean/ocean subduction zone, including a deep sea trench, accretionary wedge, and forearc basin. The volcanic arc is on the continental crust, because that crust is part of the overriding plate.

Ocean/continent subduction zones are often called *"convergent"* or *"active" continental margins*. Examples include Japan, western South America, and the Pacific Northwest of the United States. For the latter, the accretionary wedge includes the coastal ranges of Washington, Oregon and northern California (Fig. 2.17). The volcanic arc is the Cascade Mountains, and the intervening forearc basin is the Great Valley, Willamette Valley and Puget Sound. The Sierra Nevada Mountains in northern California are the roots of a volcanic arc; the volcanoes have eroded away, exposing now solidified magma chambers.

Continental Collision Zones At collisional mountain ranges, two plates that both have thick (continental or island arc) crust converge (Fig. 2.18a). Collision occurs after the thinner, oceanic crustal part of the downgoing plate is consumed through subduction (Fig. 2.18b). When the thicker, more buoyant crusts meet, both plates are too light to subduct into the deeper mantle. The thick crusts are deformed by compression, their rocks metamorphosed and uplifted. If convergence continues, one thick crustal block may thrust underneath the other (as the Indian subcontinent extends beneath Asia today); the result is a broad region of high elevation.

The highest mountains on Earth, the Himalayas, are part of a chain of mountains extending from south-central Asia to western Europe (Fig. 2.9). The chain is

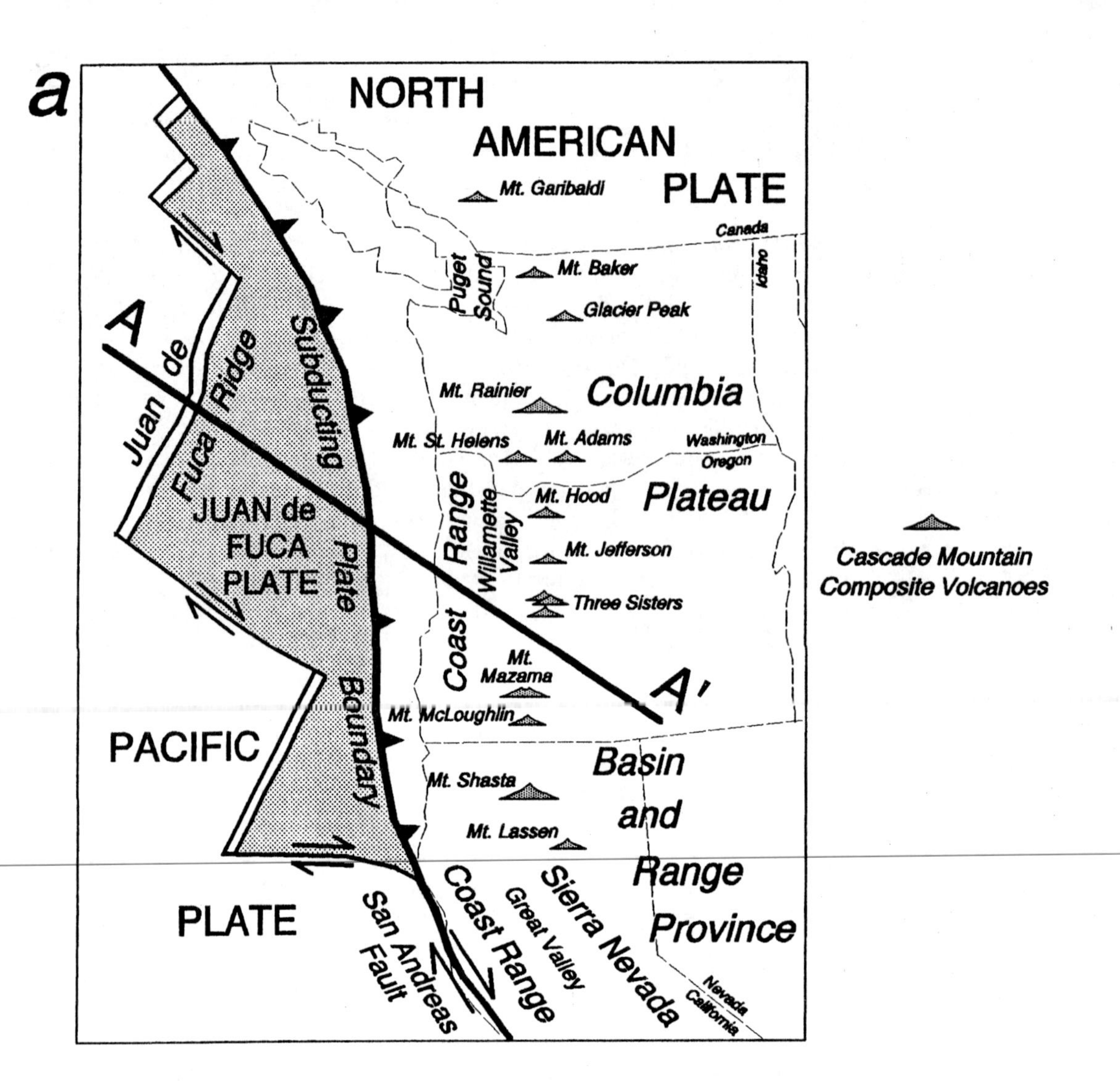

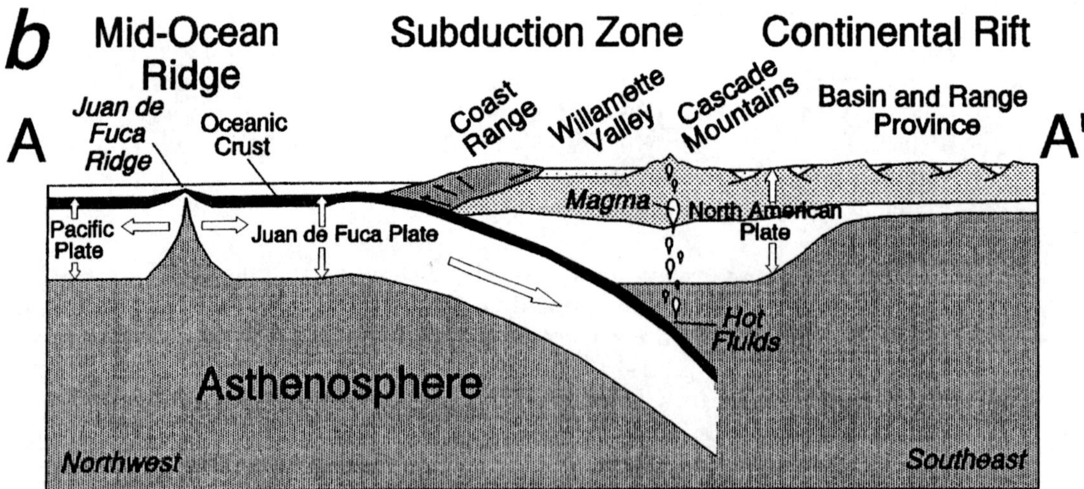

FIGURE 2.17 a) Map of Pacific Northwest of the United States and southwestern Canada. All three types of plate boundaries occur. *Divergent*: Juan de Fuca Ridge (mid-ocean ridge); Basin and Range Province (continental rift). *Convergent*: Cascadia Subduction Zone ("subducting plate boundary" to Cascade volcanoes). *Transform*: San Andreas Fault. b) Schematic cross-section A-A′.

a) Ocean Basin Closes During Subduction

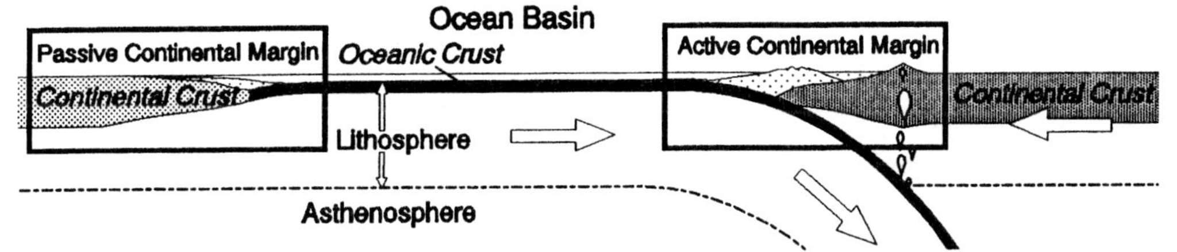

b) Thick Continental Crusts Collide

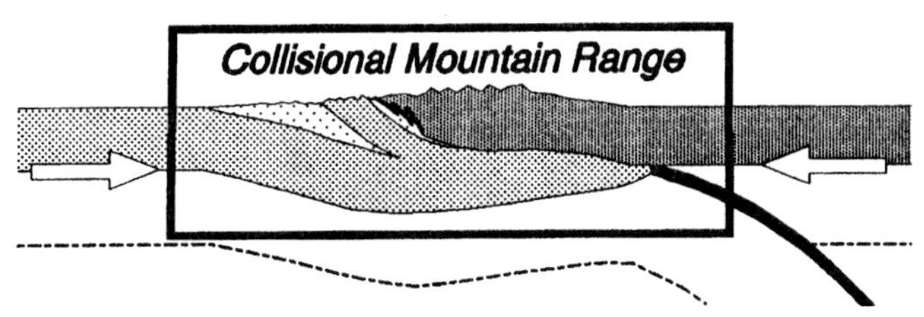

FIGURE 2.18 *Development of continental collision zone.* a) Continental margins approach as intervening oceanic lithosphere is consumed by subduction. b) Collisional mountain range forms as thick continental crusts collide. High elevations are due to a combination of thrust faulting and isostatic uplift (see Chapter 8).

formed by convergence of Eurasia with other continental fragments, including India, Saudi Arabia and smaller pieces of thick crust swept northward as Africa moves toward Europe. Other mountains in the collision zone are the Pamirs, Hindu Kush, Zagros, Caucasus, Carpathians, Alps and Pyrenees.

The Appalachian Mountains in the United States and Canada, and the Caledonide Mountains in southern Greenland, the British Isles, and Scandinavia, are parts of a continental collision zone that formed during the Paleozoic era, about 300 to 400 million years ago. Subsequent continental rifting opened the Atlantic Ocean, isolating the mountains as separate ranges on different continents.

Transform Plate Boundaries

Where plates slide horizontally past one another, lithosphere is neither created nor destroyed. Such boundaries are called "transform" because they connect other plate boundaries in various combinations, transforming the site of plate motion. A common example is an offset connecting segments of a mid-ocean ridge (Fig. 2.19). Relative movement between the plates occurs only at and between the ridge segments, where earthquakes occur. Relative movement outside the ridge segments has ceased, so that no earthquakes occur there; inactive extensions of transform boundaries are called *fracture zones*. Prominent examples of transform plate boundaries that extend on land include the Anatolian Fault in Turkey, the Alpine Fault of New Zealand, and the San Andreas Fault in California (Fig. 2.20).

Hotspots

Some volcanoes lie within the interior of lithospheric plates, rather than along the edges. Commonly, the volcanoes get progressively older away from the largest and most active volcanoes (Fig. 2.21). The volcanoes are thought to form over narrow "plumes" of heat that rise from deep within the mantle. A "hotspot" is a region in the mantle where magma forms due to a plume. As a lithospheric plate moves over a hotspot, the line of volcanoes forms. Examples of such hotspot tracks include the

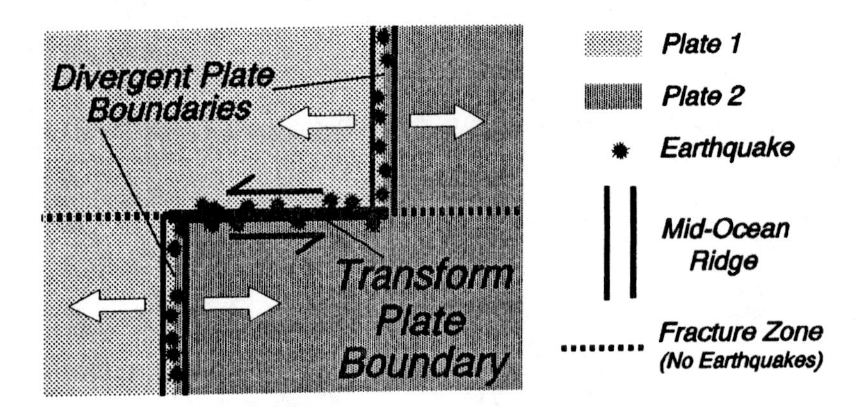

FIGURE 2.19 Map view of transform plate boundary connecting two mid-ocean ridge segments.

FIGURE 2.20 Transform plate boundary on the West Coast of the United States. Across the San Andreas Fault, the Pacific Plate moves north-northwest, relative to the North American Plate.

Hawaiian Islands within the Pacific Plate and the Columbia Plateau/Snake River Plain/Yellowstone area of Washington, Oregon, Idaho, and Wyoming (discussed below).

Mantle plumes and associated hotspots are thought to be fixed relative to the deep mantle, as well as to one another. Hotspots are thus important because they provide a framework through which absolute motions of plates can be determined.

MANIFESTATIONS ALONG PLATE BOUNDARIES

Lithospheric plates are generally strong enough to ride slowly over the ductile asthenosphere without internal deformation. Materials within the interior of plates do not undergo much vertical movement, so that the temperature and pressure of those materials does not change appreciably. Horizontal interactions build stresses near the edges of plates, where materials rise or sink to regions of different pressure and temperature. Most tectonic activity, including earthquakes, volcanism, meta-

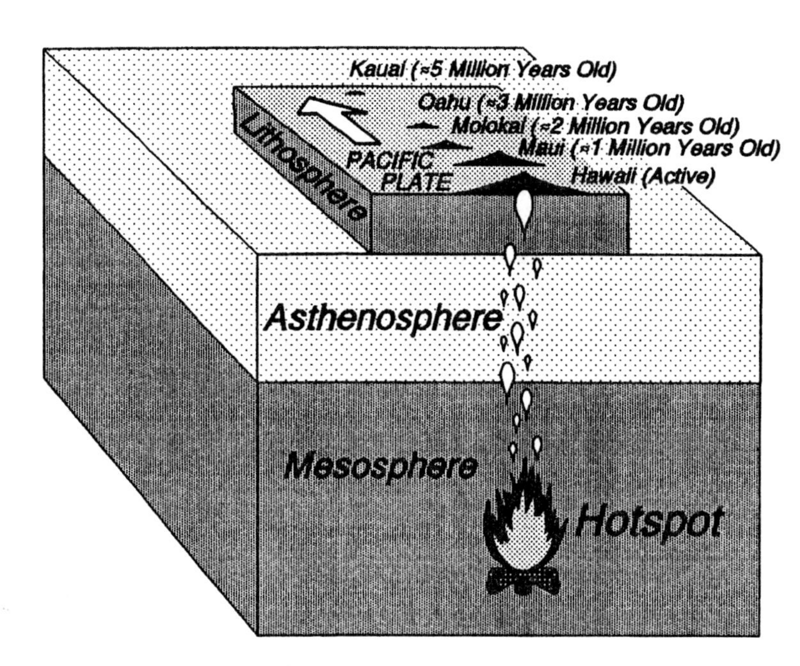

FIGURE 2.21 Cartoon example of hotspot beneath the interior of the Pacific Plate. Volcanoes of the Hawaiian Islands record the passage of the plate over the hotspot.

morphism, mountain formation, and the creation and destruction of lithosphere, is thus concentrated at plate boundaries or over hotspots.

Earthquakes

Earthquakes occur because materials are stressed to their breaking point. Two factors are important: 1) the presence of brittle material; and 2) motion that builds stress in the brittle material. Practically the only part of the Earth that meets both of these conditions is where the rigid lithospheric plates are in motion, building stresses where they are in contact with each other. Most earthquakes therefore occur along or near plate boundaries, within the brittle regime near the top of the rigid plates (Fig. 2.22).

At divergent and transform boundaries, the rigid lithospheric plates normally do not extend deeper than about 100 km (Fig. 2.22a,b). The cold, brittle part of the plates is generally in the upper 20 km, so that only shallow earthquakes occur.

Virtually all deep, and most very large, earthquakes occur at convergent plate boundaries, where a rigid plate can extend downward to as deep as 700 km (Fig. 2.22c). Shallow earthquakes (upper 70 km) are associated with compression and other contortions on the tops of both plates. Very large earthquakes occur due to sudden stress release where the two plates are locked together, at their boundary. Moderate to large earthquakes can occur deeper, if the lower plate descends so quickly that it is still cold enough to undergo brittle failure.

Volcanism

Earth materials commonly melt in two situations (Fig. 2.23): a) the pressure on hot material drops; or b) cold material is subjected to higher temperature. Decompression melting occurs when asthenosphere rises as plates rip apart at continental rift zones or mid-ocean ridges (Fig. 2.23a). Cold crustal materials heat up when a lithospheric plate descends into a subduction zone (Fig. 2.23b). Most volcanic eruptions are therefore

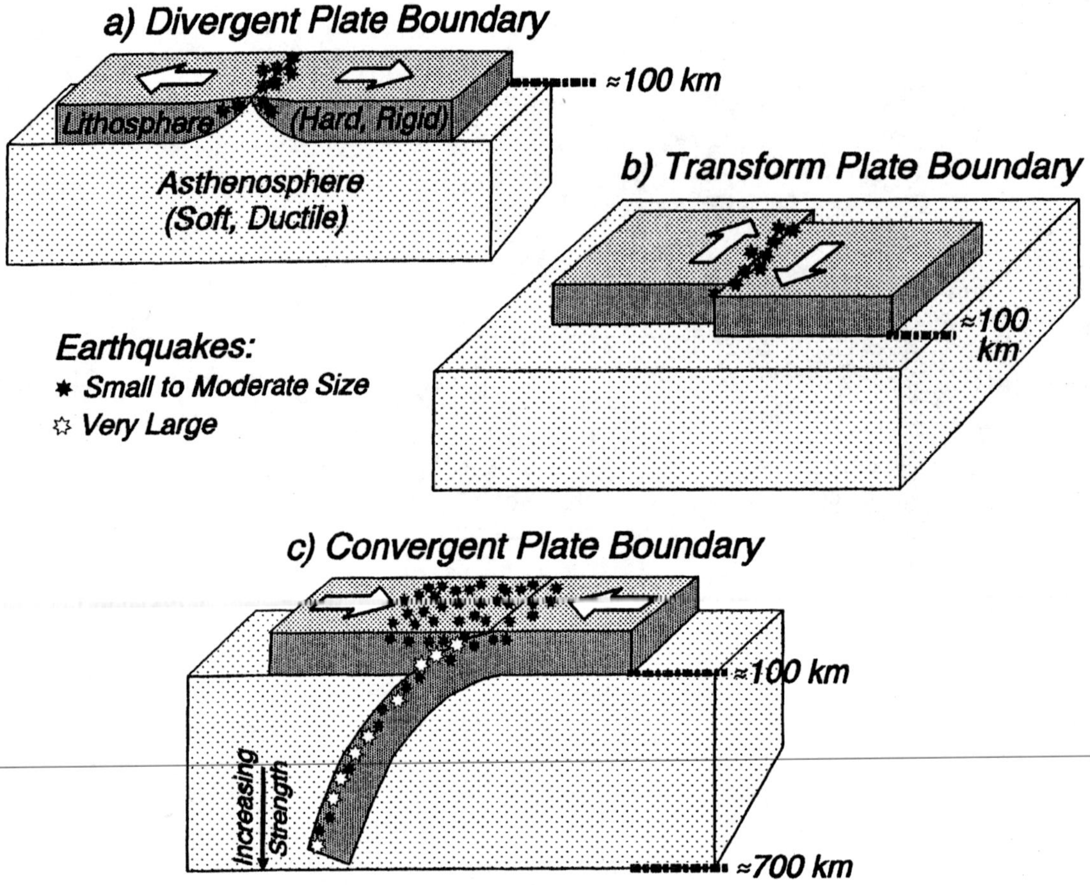

FIGURE 2.22 *Occurrence of earthquakes at plate boundaries.* a,b) Shallow earthquakes of small to moderate size occur along divergent and transform boundaries (see also map view, Fig. 2.19). c) Small to moderate size earthquakes at convergent plate boundaries occur at shallow depths in the descending and overriding plates. Very large earthquakes occur where the plates lock together, and at greater depths in the upper portion of the descending plate.

associated with divergent or convergent plate boundaries. Volcanism is normally absent from transform plate boundaries because materials remain at their normal depths; there is no significant temperature rise or pressure drop (Fig. 2.23c). Some volcanism occurs in plate interiors, where the plates ride over hotspots that may originate from the deep mantle or mantle/core boundary (Fig. 2.23d).

Divergent Plate Boundary Volcanism The drop in pressure on hot asthenosphere rising at divergent boundaries causes an increase in partial melting, as shown by the bold arrow on the phase diagram (Fig. 2.7). High-silica minerals commonly melt at lower temperatures than low-silica minerals (Appendix C); the partial melt that comes off the peridotite of the asthenosphere (30% silica) is therefore a basalt/gabbro composition of somewhat higher silica (50%).

At mid-ocean ridges, partially melted asthenosphere produces relatively low-silica magma, resulting in a characteristic sequence of igneous rocks (Fig. 2.24a). Lavas that pour out onto the ocean floor cool rapidly, forming pillows of *basalt* (50% silica). Material of the same (50% silica) composition that does not make it to the surface forms intrusive dikes of *gabbro*, comprising the lower crust. The high density material that remains below is still essentially *peridotite* (30% silica), form-

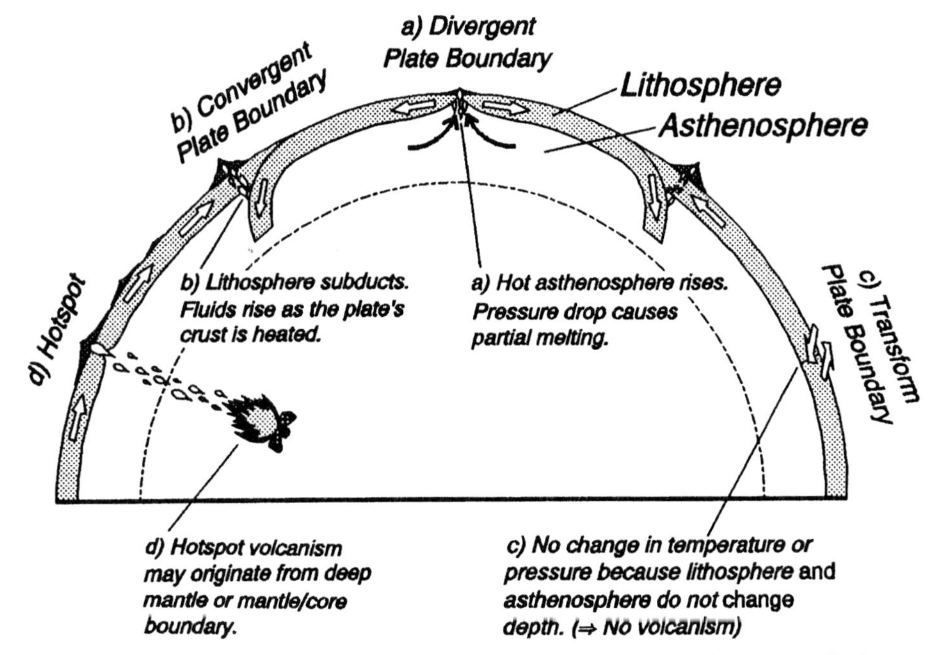

FIGURE 2.23 Schematic cross-section of Earth illustrating the generation of magma that leads to volcanism.

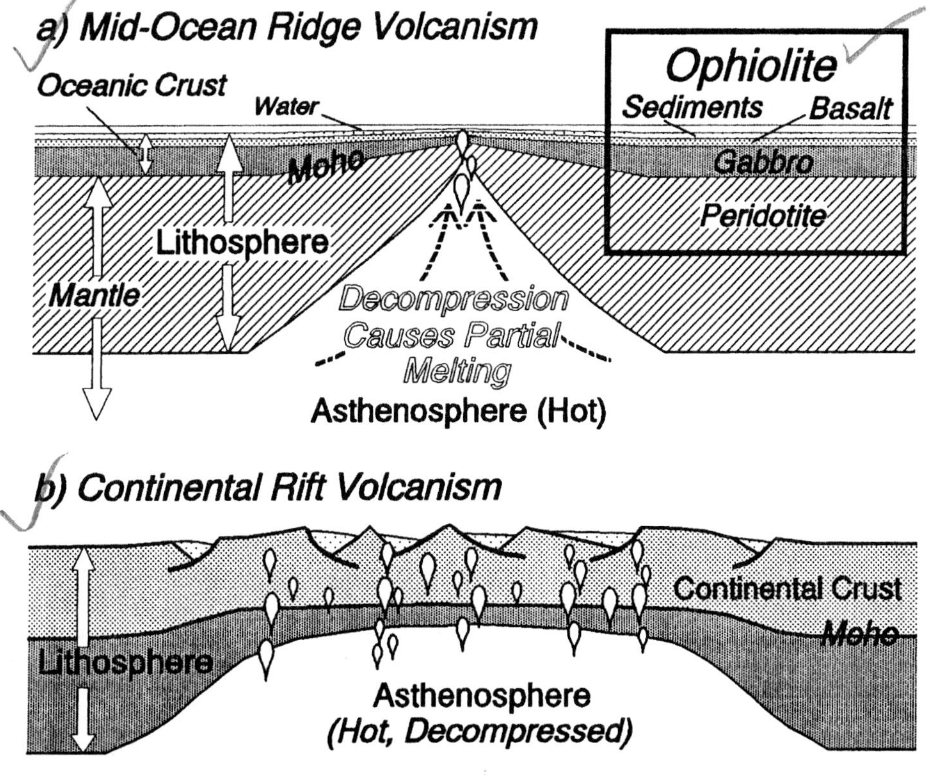

FIGURE 2.24 *Volcanism at divergent plate boundaries.* a) Midocean ridge illustrating decompression melting and the development of an ophiolite sequence. b) Continental rift zone showing that magma must initially melt through continental crust, enriching the silica content of the magma.

ing the mantle portion of the new lithospheric plate. Together with overlying sediments, this sequence of oceanic crust and uppermost mantle is called an *ophiolite*. Volcanoes at mid-ocean ridges form mostly below sea level. Iceland is an exception, where the Mid-Atlantic Ridge is above the water; the low-silica (basaltic) magmas are very fluid, giving broad, gently sloping shield volcanoes (Appendix C).

Continental rift zones commonly exhibit two-stage (bimodal) volcanism (Fig. 2.24b); 1) an early stage, where ascending magma melts a lot of continental crust, producing high-silica (rhyolitic) volcanism; and 2) an advanced stage, where magma comes more directly from the asthenosphere, producing lower-silica (*basaltic*) volcanism. The later stage may evolve to a mid-ocean ridge (Fig. 2.14). Examples of continental rift volcanism include Mt. Kenya and Mt. Kilimanjaro in the east African rift system, and Newberry Volcano in the Basin and Range Province in central Oregon.

Convergent Plate Boundary Volcanism As a lithospheric plate descends into a subduction zone, it gets hot enough for fluids to be driven from the crust and sediments on top of the plate. Those fluids migrate toward the surface, inducing melting of the mantle and crustal materials in their path. Minerals high in silica tend to melt first, so that magmas generated in this process can range in silica content from low (basaltic; 50%) to intermediate (andesitic; 60%) to very high (rhyolitic; 70%). A volcanic arc grows on the overriding plate, in the position where the crust of the lower plate is at 100 to 150 km depth, thereby hot enough to release fluids. The high-viscosity of the higher-silica lavas results in steep sided, composite volcanoes; trapped gasses can result in violent eruptions, as occurred at Mt. St. Helens in Washington state in 1980 (Fig. 2.17a).

At an ocean/ocean subduction zone (Fig. 2.25a) the volcanic (island) arc develops on one of the oceanic plates about 100 to 500 km from the plate boundary (trench). The rising fluids melt through mantle and thin (island arc) crust on the upper plate; magmas are therefore low-to-intermediate silica (basaltic to andesitic) in composition. Examples of active volcanoes are Krakatau (near Java, in the Indian Ocean) and Mt. Pelee (West Indies).

At ocean/continent subduction zones (active continental margins) the plate with thin (oceanic) crust descends below the (more buoyant) plate with thick (continental) crust (Fig. 2.25b). Volcanic mountain chains thus form along the edges of some continents. Rising fluids melt mantle and continental crust in their path, producing lavas generally of intermediate-to-high silica content (andesite to rhyolite). In the deeper crust, magma chambers cool to batholiths of high-silica (granitic) composition. Examples of composite volcanoes at active continental margins include Mt. Mazama (which formed Crater Lake) in Oregon (Fig. 2.17a), Mt. Vesuvius in Italy, and Mt. Fuji in Japan.

At continental collision zones the crust is too thick and buoyant to subduct. Generation of fluids is commonly low, so that little or no volcanism occurs (Fig. 2.25c). Continental crust on the lower plate, however, can extend to 50 km or deeper, causing some partial melting of high-silica material; granitic magma can thus form in the lower crust.

Hotspot Volcanism Hotspots provide a framework to track the motion of lithospheric plates over deeper portions of the mantle (Figs. 2.21, 2.23d). One volcano after another forms as a plate moves over a hotspot. The resulting chain of volcanoes is; 1) parallel to the direction of plate motion; and 2) older in a direction away from the hotspot. By mapping the changing age of volcanism, the rate and

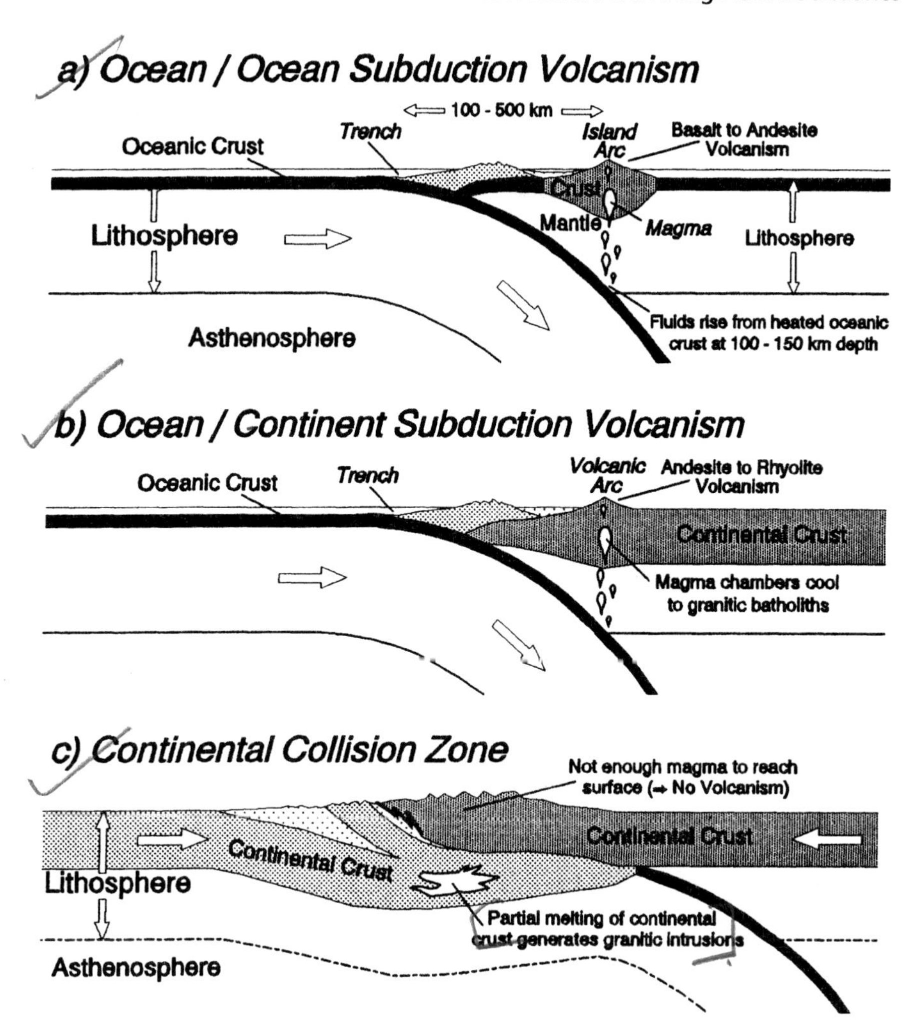

FIGURE 2.25 *Volcanism at convergent plate boundaries.* a) Subduction zone where oceanic crust caps both plates. b) Boundary where plate with oceanic crust subducts beneath plate with continental crust. c) Volcanism ceases during continental collision, although high-silica intrusions may form due to partial melting as crustal material extends below normal continental depths.

direction of movement of the lithospheric plate over the deeper mantle can be calculated.

Hotspot volcanism can occur where either oceanic or continental crust is on the overriding plate. Hotspot magma intruding only thin oceanic crust, as in Hawaii, results in fluid lavas of basaltic composition, forming shield volcanoes (Fig. 2.21). In India, a broad region of basaltic rock, known as the Deccan Trap, is thought to have formed during the initial surfacing of hotspot magma beneath the Indian subcontinent. As the Indian plate moved northward, volcanoes of the Chago-Laccadive Ridge, in the Indian Ocean, have progressively formed over the hotspot.

The Pacific Northwest of the United States reveals a region where continental lithosphere apparently moved over a hotspot. Southeast Washington and northeast Oregon are dominated by layers of basalt, comprising the Columbia Plateau (Fig. 2.17a). This massive volume of rock resulted from mantle-derived lavas that extruded 15 to 18 million years ago. Along the Snake River Plain in Idaho, lavas are progressively younger from west to east, where there is active volcanism in Yellowstone National Park (Fig. 2.26). It appears that hotspot magma migrated

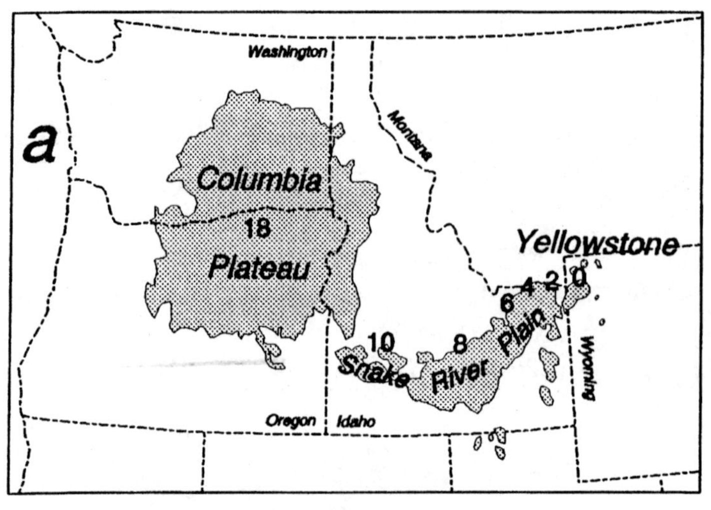

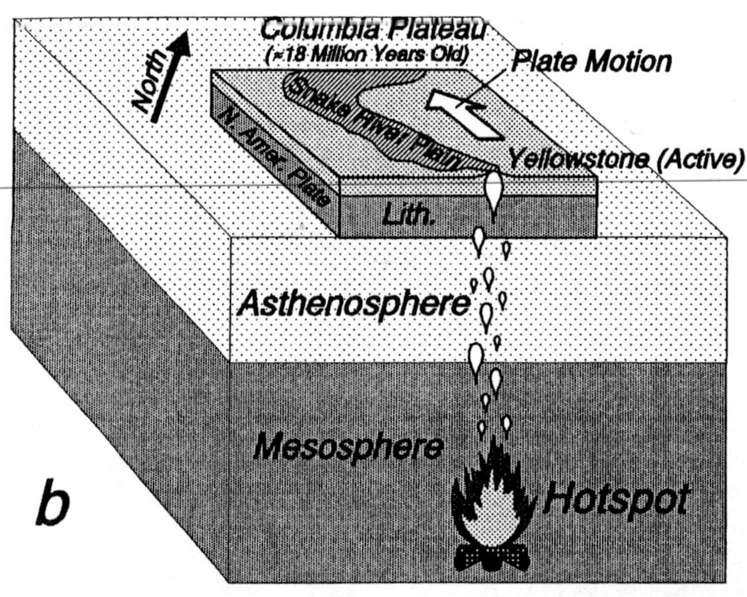

FIGURE 2.26 *Interpretation of Columbia Plateau/Snake River Plain/Yellowstone volcanism.* a) Numbers show age of volcanism from the Columbia Plateau to Yellowstone National Park (from A. Grunder, personal communication, 1995). b) Observed ages of volcanism are consistent with the westward movement of the North American lithospheric plate over a hotspot within the deeper mantle.

upward from the deep mantle in the shape of a "mushroom cloud." The broad head of the cloud struck in Oregon and Washington 18 million years ago, resulting in massive outpourings of basalt forming the Columbia Plateau (Fig. 2.26a). As the North American Plate moved over the hotspot, the thin "stem" remained. This stem resulted in far less volcanism, with a larger portion of the magma enriched in silica due to the melting of continental crust as the magma migrated upward. The resulting lavas erupted in a narrow, steady stream between Oregon and Yellowstone, as the plate moved westward (Fig. 2.26b).

PLATE TECTONIC CONSTRAINTS OFFERED BY GEOPHYSICAL OBSERVATIONS

The plate tectonic framework owes much of its overall form and detail to geophysical observations. Each technique offers potential constraints on the composition and physical state of materials in a given setting, as well as on the three-dimensional structure of the region. The remaining chapters present the application of different geophysical techniques to tectonics, including the potential constraints, observations, and interpretations outlined below. Research on the crustal structure and evolution of a tectonic feature can be done in conjunction with this textbook by following the "Sequenced Writing Assignment" outlined in Appendix A.

Geophysical observations are particularly useful in constraining the depths to two fundamental boundaries: the crust/mantle boundary (Moho) and the lithosphere/asthenosphere boundary. Together with a third fundamental boundary (topography on land and bathymetry at sea), the two subsurface boundaries facilitate understanding of plate tectonic structure and associated processes. The concept of *isostasy* (discussed in Chapter 8) illustrates the inter-relationship of the elevations and depths of the three fundamental boundaries.

Seismic Refraction (Chapter 4)

The seismic refraction method is useful in resolving relatively flat interfaces separating abrupt increases in seismic velocity, like the Moho. The method can therefore map the pattern of crustal thickness changes in a region as well as gross seismic velocities for thc crust and uppermost mantle.

Refraction observations often reveal that the crust is thin at continental rifts and mid-ocean ridges (Fig. 2.14), and thick at mountain ranges (Fig. 2.18). They show the transition from thicker continental to thinner oceanic crust at continental margins (Figs. 2.14c, 2.16b). Refraction surveys suggest that the upper mantle has relatively low seismic velocities where hot asthenosphere is shallow at continental rifts and mid-ocean ridges (Fig. 2.13). The characteristic layering within oceanic crust, thought to correspond to an "ophiolite" sequence (Fig. 2.24a), was first revealed by refraction studies.

Seismic Reflection (Chapters 5 and 6)

Seismic reflection profiles image stratigraphic and structural features that are characteristic of certain tectonic settings; some features developed in one setting are identified in the subsurface at another. For example, continental rift grabens are revealed by wedge-shaped zones of reflections from sedimentary and volcanic strata (Fig. 2.14a). Similar reflections suggest rift grabens beneath younger ("post-rift") sedimentary strata on passive continental margins (Fig. 2.14b).

At continental rift zones, seismic reflection profiles often reveal a brittle upper crust that is nonreflective, overlying a layered, ductile lower crust. A deeper zone of horizontal reflections can be used to interpret the depth and nature of the Moho. This reflection signature of the continental crust and Moho is also observed beneath some passive continental margins.

Oceanic crust developed at mid-ocean ridges has block-faulted topography that results in a characteristic reflection pattern. This pattern is observed to extend to the edge of passive continental margins and beneath deformed strata of accretionary wedges.

Other observations with tectonic implications offered by reflection profiles include seaward-dipping, wedge-shaped sequences of volcanic strata that occur along the boundary between continental and oceanic crust on some passive margins. Strong reflections beneath the axes of mid-ocean ridges suggest the presence of magma chambers. The extent of horizontal detachment surfaces beneath accretionary wedges and collisional mountain ranges is often imaged, as well as the zone of suturing of continents in collision zones.

Earthquake (Chapter 7)

Earthquake observations reveal the positions of lithospheric plate boundaries and the types of stresses in a region. They can be used to determine the strike and dip of a fault responsible for an earthquake, as well as the type of faulting (normal, reverse, or strike-slip). The early or late arrival of seismic waves from distant earthquakes can be used to map changes in lithosphere thickness in a region.

Earthquake observations can be used to map the brittle, top portion of lithospheric plates. Studies show that virtually all earthquakes occurring at divergent and transform plate boundaries are shallow, while earthquakes may extend to considerable depths at convergent boundaries (Fig. 2.22). Earthquakes reveal predominately normal faults (extensional stresses) in areas of plate divergence, and a preponderance of strike-slip faults (shearing stresses) in transform settings. Where plates converge, reverse faults (compressional stresses) are prevalent, but normal and strike-slip faults may be found in various parts of the region. A delay in the arrival of seismic waves at continental rifts and mid-ocean ridges suggests that the lithosphere is thin, while thick lithosphere at subduction zones and collisional mountain ranges speeds up seismic waves.

Gravity (Chapter 8)

Changes in the density of rocks cause local changes in Earth's gravity field. Crustal material is less dense than mantle, so that gravity observations can be used to map crustal thickness changes. Asthenosphere is slightly less dense than the mantle part of the lithosphere, making it possible for gravity data to constrain changes in lithosphere thickness. The state of isostatic equilibrium in a region can also be tested through gravity observations.

Gravity observations are consistent with the presence of thick crust beneath the high topography of some mountain ranges (Fig. 2.18). At continental margins, a characteristic gravity signature reveals that the water deepens where the crust thins (Figs. 2.14c, 2.16b). Mass distributions from interpretations of gravity reveal that shallow asthenosphere supports the weight of elevated mantle and topography at continental rifts and mid-ocean ridges (Fig. 2.13). Details of gravity observations highlight deviations from simple models of isostasy, suggesting that in places lithospheric plates have enough strength to support the weight of topographic features.

Magnetics (Chapter 9)

Changes in Earth's magnetic field relate to the size and orientation of magnetic bodies, and to how susceptible rocks are to being magnetized. Rock magnetism is also dependent on temperature; rocks attain their strongest magnetization after cooling below a certain temperature. Earth's magnetic field periodically reverses in direction; the reversals preserved as magnetization in some rocks can be used as a record of geologic time.

Crystalline rocks are generally more magnetic than sedimentary rocks, so that magnetic observations can be used to map igneous intrusions and the changing depth to crystalline basement beneath sedimentary basins. The increase in temperature with depth in a region may be inferred from magnetic observations, as cooler regions will have a greater thickness of rocks that are strongly magnetized. The record of magnetic reversals preserved within rocks can be used to determine the age of seafloor created at mid-ocean ridges, the age of some sedimentary deposits, and the latitudes where rocks formed.

Heat Flow (Chapter 10)

The rate of flow of heat out of the Earth is a function of the change in temperature with depth and the ability of rocks in the region to conduct heat. Heat flow observations relate to the transfer of heat from one region to another.

Heat flow can be used to determine the proximity of hot materials to Earth's surface. Areas of high heat flow reveal shallow asthenosphere beneath continental rift zones and mid-ocean ridges (Fig. 2.13); low heat flow is associated with thick slabs of (relatively cold) lithosphere extending into subduction zones and beneath collisional mountain ranges (Figs. 2.16, 2.18). Shallow magma sources beneath hotspots, continental rifts, mid-ocean ridges, and volcanic arcs are revealed by very high heat-flow readings (Figs. 2.21, 2.24, 2.25).

EXERCISES

2-1 The Earth's crust, lower lithosphere, and asthenosphere are thin in comparison to deeper zones of the Earth (mesosphere, outer core, and inner core). Textbooks commonly exaggerate the outer zone thicknesses, for clarity on page-size illustrations. Using a millimeter ruler and compass with a very fine pencil, redraw the whole Earth cross-section (Fig. 2.5) at a scale of 1:100,000,000 (1 cm = 1,000 km). Illustrate the following boundaries exactly to scale, at the depths indicated: a) oceanic Moho at 12 km; b) continental Moho at 35 km; c) transition from lithosphere to asthenosphere at 150 km; d) bottom of very soft asthenosphere at 350 km; e) top of hard mesosphere at 700 km; f) core/mantle boundary at 2900 km; g) outer core/inner core boundary at 5100 km; h) center of the Earth at 6300 km.

2-2 Draw a cross section along the line of section (B-B′) shown on the map on the following page. Notice that the line extends from the Pacific Plate, across the East Pacific Rise, Peru-Chile Trench, South America, the Mid-Atlantic Ridge and all the way across the East African Rift.

a) Portray how the depths to the following boundaries change along the length of the cross section: i) topography onshore and bathymetry offshore; ii) crust/mantle boundary (Moho); iii) lithosphere/asthenosphere boundary.

b) Put a series of X's within the cross-section to represent zones where significant earthquake activity might be expected to occur.

c) Put Δ's on the surface of the cross-section to illustrate where you would expect volcanic activity to occur.

d) Identify and label the positions of the following features on the cross-section: i) plate boundaries and their types; ii) lithospheric plates and their names; iii) continental margins and their types (passive or active).

2-3 Based on examination of maps and globes, give at least three examples of features on Earth's surface that represent each of these tectonic settings: a) "normal" continental lithosphere; b) continental rift zone; c) mid-ocean ridge; d) small ocean basin; e) advanced ocean basin; f) ocean/ocean subduction zone; g) ocean/continent subduction

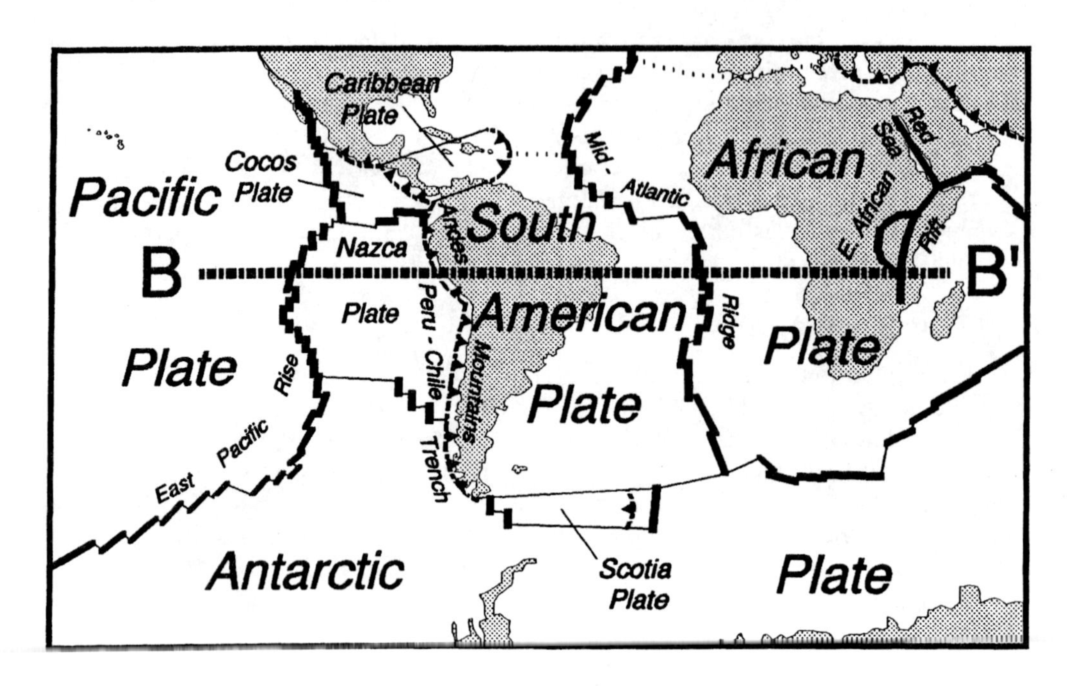

zone; h) active continental collision zone; i) ancient continental collision zone; j) transform boundary connecting ridge segments; k) transform boundary connecting ridges to trenches; l) hotspot trace.

2-4 A popular novel from the 1970's contains the following passage:

Earth scientists had just discovered something fascinating about the continent Patty Keene was standing on, incidently. It was riding on a slab about forty miles thick, and the slab was drifting around on molten glurp. And all the other continents had slabs of their own. When one slab crashed into another one, mountains were made. (Breakfast of Champions by Kurt Vonnegut, Jr., © 1973, Delacorte Press).

a) Explain what appears right and what appears wrong with the passage.

b) Rewrite the passage so that it is technically correct (though perhaps not as poetic).

2-5 Explain why pressure increases with depth within the Earth (Fig. 2.6a). Why is the increase roughly linear?

2-6 The phase diagram for peridotite (Fig. 2.7) explains why rising asthenosphere partially melts at a mid-ocean ridge. Using Fig. 10.15a, redraw the phase diagram for granite. Illustrate on the diagram and explain in words why granitic intrusions form in the lower crust during the late stages of continental collision (Fig. 2.25c).

2-7 **a)** Based on the ages of volcanism depicted in Fig. 2.21 and the distances between the Hawaiian Islands, determine the *absolute velocity* and direction of the Pacific Plate. **b)** The south Atlantic ocean began to open about 100 million years ago. Estimate the *relative velocity* between the African and South American plates.

SELECTED BIBLIOGRAPHY

General Tectonics

AGI/NAGT, 1993, *Laboratory Manual in Physical Geology* (3rd ed.), American Geological Institute and National Association of Geology Teachers, New York: Macmillan Pub. Comp., 260 pp.

Cox, A. (editor), 1973, *Plate Tectonics and Geomagnetic Reversals: Readings with Introductions by Allan Cox*, San Francisco: W. H. Freeman and Co., 702 pp.

Isaacs, B., J. Oliver, and L. R. Sykes, 1968, Seismology and the new global tectonics, *Journal of Geophysical Research*, v. 73, pp. 5855–5899.

Jeffreys, H., 1976, *The Earth* (6th ed.), Cambridge: Cambridge University Press.

Kay, M., 1951, *North American Geosynclines*, Boulder, Colorado: Geological Society of America, Memoir 48, 143 pp.

Kearey, P., and F. J. Vine, 1996, *Global Tectonics*, 2nd ed., Oxford: Blackwell Science Ltd., 333 pp.

Moores, E. M., and R. J. Twiss, 1995, *Tectonics*, New York: W. H. Freeman and Comp., 415 pp.

Skinner, B. J., and S. C. Porter, 1987, *Physical Geology*, New York: John Wiley and Sons, Inc., 750 pp.

Vine, F. J., 1966, Spreading of the ocean floor: New evidence, *Science*, v. 154, pp. 1405–1415.

Wegener, A. L., 1924, *The Origin of Continents and Oceans* (translated by J. G. A. Skerl), London: Nethuen and Co.

Wilson, J. T. (editor), 1976, *Continents Adrift and Continents Aground*, New York: W. H. Freeman and Co.

Extensional Tectonic Settings

Continental Rifts

Anderson, R. E., M. L. Zoback, and G. A. Thompson, 1983, Implications of selected subsurface data on the structural form and evolution of some basins in the northern Basin and Range province, Nevada and Utah, *Geological Society of America Bulletin*, v. 94, pp. 1055–1072.

Baker, B. H., P. A. Mohr, and L. A. J. Williams, 1972, *Geology of the eastern rift system of Africa*, Boulder, Colorado: Geological Society of America, Special Paper 136, pp. 1–66.

Chapola, L. S., and C. E. Kaphwiyo, 1992, The Malawi rift: Geology, tectonics, and seismicity, *Tectonophysics*, v. 209, pp. 159–164.

Cochran, J. R., A model for development of the Red Sea, 1983, *American Association of Petroleum Geologists Bulletin*, v. 67, pp. 41–69.

Horváth, F., 1994, Towards a mechanical model for the formation of the Pannonian Basin, *Tectonophysics*, v. 226, pp. 333–357.

Keller, G. R., E. G. Lidiak, W. J. Hinze, and L. W. Braile, 1983, The role of rifting in the tectonic development of the midcontinent, U.S.A., *Tectonophysics*, v. 94, pp. 391–412.

Keller, G., C. Prodehl, J. Mechie, K. Fuchs, M. Khan, P. Maguire, W. Mooney, U. Achauer, P. Davis, R. Meyer, L. Braile, I. Nyambok, and G. Thompson, 1994, The East African rift system in the light of KRISP 90, *Tectonophysics*, v. 236, pp. 465–483.

McClay, K. R., 1989, Physical models of structural styles during extension, in: *Extensional Tectonics and Stratigraphy of the North Atlantic Margins*, edited by A. J. Tankard and H. R. Balkwill, Tulsa: American Association of Petroleum Geologists, Memoir 46, pp. 95–110.

Plomerová, J., V. Babuška, M. Dobrath and R. Lillie, 1993, Structure of the lithosphere beneath the Cameroon Rift from seismic delay time analysis, *Geophysical Journal*, v. 115, pp. 381–390.

Royden, L. H., 1988, Late Cenozoic tectonics of the Pannonian Basin system, in: *The Pannonian Basin: A study in Basin Evolution*, edited by L. H. Royden and F. Horváth, Tulsa: American Association of Petroleum Geologists, Memoir 45, pp. 27–48.

Woelk, T. S., and J. Hinze, 1991, Model of the midcontinent rift system in northeastern Kansas, *Geology*, v. 19, pp. 277–280.

Zandt, G., S. C. Myers, and T. C. Wallace, 1995, Crust and mantle structure across the Basin and Range-Colorado Plateau boundary at 37° N latitude and implications for Cenozoic extensional mechanism, *Journal of Geophysical Research*, v. 100, pp. 10529–10548.

Passive Continental Margins

Boillot, G., D. Mougenot, J. Girardeau, and E. L. Winterer, 1989, Rifting processes on the west Galicia margin, Spain, in: *Extensional Tectonics and Stratigraphy of the North Atlantic Margins*, edited by A. J. Tankard and H. R. Balkwill, Tulsa: American Association of Petroleum Geologists, Memoir 46, pp. 336–377.

deCharpal, P., P. Guennoc, L. Montadert, and D. G. Roberts, 1982, Rifting, crustal attenuation and subsidence in the Bay of Biscay, *Nature*, v. 275, pp. 706–710.

Gerrard, I., and G. C. Smith, 1983, Post-Paleozoic succession and structure of the southwestern African continental margin, in: *Studies in Continental Margin Geology*, edited by J. S. Watkins and C. L. Drake, Tulsa: American Association of Petroleum Geologists, Memoir 34, pp. 49–74.

Grantz, A., and S. D. May, 1983, Rifting history and structural development of the continental margin north of Alaska, in: *Studies in Continental Margin Geology*, edited by J. S. Watkins and C. L. Drake, Tulsa: American Association of Petroleum Geologists, Memoir 34, pp. 77–100.

Grow, J., R. Mattlick, and J. Schlee, 1979, Multichannel seismic depth sections and interval velocities over outer continental slope between Cape Hatteras and Cape Cod, in: *Geological and Geophysical Investigations of Continental Margins*, edited by J. S. Watkins, L. Montadert, and P. W. Dickinson, Tulsa: American Association of Petroleum Geologists, Memoir 29, pp. 65–83.

Hall, D. J., 1983, The rotational origin of the Gulf of Mexico based on regional gravity data, in: *Studies in Continental Margin Geology*, edited by J. S. Watkins and C. L. Drake, Tulsa: American Association of Petroleum Geologists, Memoir 34, pp. 115–126.

Klemperer, S. L., and N. White, 1989, Coaxial stretching or lithospheric simple shear in the North Sea? Evidence from deep seismic profiling and subsidence, in: *Extensional Tectonics and Stratigraphy of the North Atlantic Margins*, edited by A. J. Tankard and H. R. Balkwill, Tulsa: American Association of Petroleum Geologists, Memoir 46, pp. 511–522.

Lu, R. S., and K. J. McMillen, 1983, Multichannel seismic survey of the Columbia Basin and adjacent margins, in: *Studies in Continental Margin*

Geology, edited by J. S. Watkins and C. L. Drake, Tulsa: American Association of Petroleum Geologists, Memoir 34, pp. 395–410.

Naini, B. R., and M. Talwani, 1983, Structural framework and the evolutionary history of the continental margin of western India, in: *Studies in Continental Margin Geology*, edited by J. S. Watkins and C. L. Drake, Tulsa: American Association of Petroleum Geologists, Memoir 34, pp. 167–191.

Tankard, A. J., and H. J. Welsink, 1989, Mesozoic extension and styles of basin formation in Atlantic Canada, in: *Extensional Tectonics and Stratigraphy of the North Atlantic Margins*, edited by A. J. Tankard and H. R. Balkwill, Tulsa: American Association of Petroleum Geologists, Memoir 46, pp. 175–195.

Trehu, A. M., K. D. Klitgord, D. S. Sawyer and R. T. Buffler, 1989, Atlantic and Gulf of Mexico continental margins, 1989, in *Geophysical Framework of the Continental United States*, edited by L. C. Pakiser and W. D. Mooney, Boulder, Colorado: Geological Society of America, Memoir 172, pp. 349–382.

Mid Ocean Ridges

Ewing, J. I., and M. Ewing, 1959, Seismic refraction measurements in the Atlantic Ocean basins, in the Mediterranean Sea, on the mid-Atlantic ridge, and in the Norwegian Sea, *Geological Society of America Bulletin*, v. 70, pp. 291–318.

Francis, T. J. G., 1968, Seismicity of mid-oceanic ridges and its relation to properties of the upper mantle and crust, *Nature*, v. 220, pp. 899–901.

Riddihough, R. P., M. E. Beck, R. L. Chase, E. E. Davis, R. D. Hyndman, S. H. Johnson, and G. C. Rogers, 1983, Geodynamics of the Juan de Fuca Plate, in: *Geodynamics of the Eastern Pacific Region, Caribbean and Scotia Arcs*, edited by S. J. Ramón Cabré, Washington: American Geophysical Union, Geodynamics Series 9, pp. 5–21.

Rohr, K. M. M., 1988, Asymmetric deep crustal structure across the Juan de Fuca Ridge, *Geology*, v. 16, pp. 533–537.

Schlater, J. G., R. N. Anderson, and M. L. Bell, 1971, Elevation of ridges and evolution of the central eastern Pacific, *Journal of Geophysical Research*, v. 76, pp. 7888–7915.

Sykes, L. R., 1967, Mechanism of earthquakes and nature of faulting on the mid-ocean ridges, *Journal of Geophysical Research*, v. 72, pp. 2131–2153.

Vera, E. E., J. C. Mutter, P. Buhl, J. A. Orcutt, A. J. Harding, M. E. Kappus, R. S. Detrick, and T. M. Brocher, 1990, The structure of 0 to 2 m.y. old oceanic crust at 9° N on the East Pacific Rise from expanded spread profiles, *Journal of Geophysical Research*, v. 96, pp. 15529–15556.

Convergent Tectonic Settings

Ocean/Ocean Subduction Zones

Brown, K., and G. K. Westbrook, 1988, Mud diapirism and subcretion in the Barbados ridge accretionary complex: The role of fluids in accretionary processes, *Tectonics*, v. 7, pp. 613–640.

Byrne, D. E., D. M. Davis, and L. R. Sykes, 1988, Loci and maximum size of thrust earthquakes and the mechanics of the shallow region of subduction zones, *Tectonics*, v. 7, pp. 833–857.

Grow, J. A., 1973, Crustal and upper mantle structure of the central Aleutian Arc, *Geological Society of America Bulletin*, v. 84, pp. 2169–2192.

McCarthy, J., and D. W. Scholl, 1985, Mechanisms of subduction accretion along the central Aleutian Trench, *Geological Society of America Bulletin*, v. 96, pp. 691–701.

Moore, J. C., B. Biju-Duval, and 16 others, 1982, Offscraping and underthrusting of sediment at the deformation front of the Barbados Ridge: Deep Sea Drilling Project Leg 78A, *Geological Society of America Bulletin*, v. 93, pp. 1065–1077.

Westbrook, G. K., J. W. Ladd, P. Buhl, N. Bangs, and G. J. Tiley, 1988, Cross section of an accretionary wedge: Barbados Ridge complex, *Geology*, v. 16, pp. 631–635.

Zhao, W. L., D. M. Davis, F. A. Dahlen, and J. Suppe, 1986, Origin of convex accretionary wedges: Evidence from Barbados, *Journal of Geophysical Research*, v. 91, pp. 10246–10258.

Ocean/Continent Subduction Zones

Aoki, Y., T. Tamano, and S. Kato, 1983, Detailed structure of the Nankai Trough from migrated seismic sections, in: *Studies in Continental Margin Geology*, edited by J. S. Watkins and C. L. Drake, Tulsa: American Association of Petroleum Geologists, Memoir 34, pp. 309–322.

Couch, R. W., and R. P. Riddihough, 1989, The crustal structure of the western continental margin of North America, 1989, in *Geophysical Framework of the Continental United States*, edited by L. C. Pakiser and W. D. Mooney, Boulder, Colorado: Geological Society of America, Memoir 172, pp. 103–128.

Davis, D. M., Dahlen, F. A., and Suppe, J., 1983, Mechanics of fold-and-thrust belts and accretionary wedges, *Journal of Geophysical Research*, v. 88, pp. 1153–1172.

Harris, S. L., 1988, *Fire Mountains of the West: The Cascade and Mona Lake Volcanoes*, Missoula, Montana: Mountain Press Publishing Company, 379 pp.

Hayes, D. E., 1966, A geophysical investigation of the Peru-Chile trench, *Marine Geology*, v. 4, pp. 309–351.

Ibrahim, A. K., G. Latham, and J. Ladd, 1986, Seismic refraction and reflection measurements in the Middle America Trench offshore Guatemala, *Journal of Geophysical Research*, v. 84, pp. 5643–5649.

Schweller, W. J., and L. D. Kulm, 1978, Extensional rupture of oceanic crust in the Chile Trench, *Marine Geology*, v. 28, pp. 271–291.

Snavely, P. D., Jr., H. C. Wagner, and D. L. Lander, 1980, *Geologic cross section of the central Oregon continental margin*, Boulder, Colorado: Geological Society of America, Map and Chart Series MC-28J.

Shreve, R. L., and M. Cloos, 1986, Dynamics of sediment subduction, melange formation, and prism accretion, *Journal of Geophysical Research*, v. 91, pp. 10229–10245.

Stauder, W., 1975, Subduction of the Nazca Plate under Peru as evidenced by focal mechanisms and by seismicity, *Journal of Geophysical Research*, v. 80, pp. 1053–1064.

Yorath, C. J., and 7 others, 1985, Lithoprobe, southern Vancouver Island: Seismic reflection sees through Wrangellia to the Juan de Fuca plate, *Geology*, v. 13, pp. 759–762.

Collisional Mountain Ranges

Ando, C. J., F. A. Cook, J. E. Oliver, L. D. Brown and S. Kaufman, 1983, Crustal geometry of the Appalachian orogen from seismic reflection studies, in: *Contributions to the Tectonics and Geophysics of Mountain Chains*, edited by R. D. Hatcher, H. Williams and I. Zietz, Boulder, Colorado: Geological Society of America, Memoir 158, pp. 113–124.

Baker, D. M., R. J. Lillie, R. S. Yeats, G. D. Johnson and M. Yousuf, 1988, Development of the Himalayan frontal thrust zone: Salt Range, Pakistan, *Geology*, v. 16, pp. 57–70.

Bally, A. W., P. L. Gordy and G. A. Stewart, 1966, Structure, seismic data and orogenic evolution of southern Canadian Rockies, *Bulletin Canadian Society of Petroleum Geologists*, v. 14, pp. 337–381.

Birkenmajer, K., 1986, Stages of structural evolution of the Pieniny Klippen Belt, Carpathians, *Studia Geologica Polonica*, v. 88, pp. 7–32.

Bowin, C., G. M. Purdy, C. Johnston, G. Shor, L. Lawver, H. M. S. Hartono and P. Jezek, 1980, Arc-continent collision in Banda Sea region, *American Association of Petroleum Geologists Bulletin*, v. 64, pp. 868–915.

Burchfiel, B. C., 1980, Eastern European Alpine system and the Carpathian orocline as an example of collisional tectonics, *Tectonophysics*, v. 63, pp. 31–61.

Cook, F. A., D. S. Albaugh, L. D. Brown, S. Kaufman, J. E. Oliver and R. D. Hatcher, Jr., 1979, Thin-skinned tectonics in the crystalline southern Appalachians: COCORP seismic-reflection profiling of the Blue Ridge and Piedmont, *Geology*, v. 7, pp. 563–567.

Davis, D. A., and R. J. Lillie, 1994, Changes in mechanical response during continental collision: Active example from the foreland thrust belts of Pakistan, *Journal of Structural Geology*, v. 16, pp. 21–34.

Hamilton, W., 1979, *Tectonics of the Indonesian region*, Reston, Virginia: U. S. Geological Survey, Professional Paper 1078, 345 pp.

Horváth, F., 1984, Neotectonics of the Pannonian Basin and surrounding mountain belts: Alps, Carpathians and Dinarides, *Annales Geophysics*, v. 2, pp. 147–154.

Jadoon, I. A. K., R. D. Lawrence, and R. J. Lillie, 1994, Seismic data, geometry, evolution, and shortening in the active Sulaiman fold-and-thrust belt of Pakistan, southwest of the Himalayas, *American Association of Petroleum Geologists Bulletin*, v. 78, pp. 758–774.

Lefeld, J., and Jankowski, J., 1985, Model of deep structure of the Polish Inner Carpathians, *Publications of the Institute of Geophysics, Polish Academy of Sciences*, v. A-16, pp. 71–99.

Lillie, R. J., K.D. Nelson, B. deVoogd, J. A. Brewer, J. E. Oliver, L. D. Brown, S. Kaufman and G. W. Viele, 1983, Crustal structure of Ouachita Mountains, Arkansas: A model based on integration of COCORP reflection profiles and regional geophysical data, *American Association of Petroleum Geologists Bulletin*, v. 67, pp. 907–931.

Lillie, R. J., G. D. Johnson, M. Yousuf, A. S. H. Zamin and R. S. Yeats, 1987, Structural development within the Himalayan foreland fold-and-thrust belt of Pakistan, in *Sedimentary Basins and Basin-Forming Mechanisms*, edited by C. Beaumont and A. J. Tankard, Calgary: Canadian Society of Petroleum Geologists, Memoir 12, pp. 379–392.

Molnar, P., 1984, Structure and tectonics of the Himalaya: Constraints and implications of geophysical data, *Annual Reviews of Earth and Planetary Science*, v. 12, pp. 489–512.

Piffner, O. A., W. Frei, P. Valased, M. Stauble, L. Levato, L. DuBois, S. M. Schmid, and S. B. Smithson, 1990, Crustal shortening in the Alpine orogen, Results from deep seismic reflection profiling in the eastern Swiss Alps, line NFP 20-east, *Tectonics*, v. 9, pp. 1327–1355.

Ratschbacher, L., W. Frisch, L. Hans-Gert, and O. Merle, 1991, Lateral extrusion in Eastern Alps, 2: Structural analysis, *Tectonics*, v. 10, pp. 257–271.

Royden, L. and M. Săndulescu, 1988, The Carpathian-Pannonian region with Outer Carpathian units, in: *The Pannonian Basin: A study in Basin Evolution*, edited by L. H. Royden and F. Horváth, Tulsa: American Association of Petroleum Geologists, Memoir 45, Map 2 (scale 1:2,000,000).

Royden, L., 1993, The tectonic expression of slab pull at continental convergent boundaries, *Tectonics*, v. 12, pp. 303–325.

Săndulescu, M., 1988, Cenozoic tectonic history of the Carpathians, in: *The Pannonian Basin: A study in Basin Evolution*, edited by L. H. Royden and F. Horváth, Tulsa: American Association of Petroleum Geologists, Memoir 45, pp. 17–25.

Stockmal, G. S., C. Beaumont, and R. Boutilier, 1986, Geodynamic models of convergent margin tectonics: Transition from rifted margin to overthrust belt and consequences for foreland-basin development, *American Association of Petroleum Geologists Bulletin*, v. 70, pp. 181–190.

Von der Borch, C. C., 1979, Continent-Island Arc Collision in the Banda Arc, *Tectonophysics*, v. 54, pp. 169–193.

Transform Settings

Allis, R. G., 1981, Continental underthrusting beneath the Southern Alps in New Zealand, *Geology*, v. 9, pp. 303–307.

Brune, J. N., T. L. Henyey, and R. F. Roy, 1969, Heat flow, stress, and rate of slip along the San Andreas Fault, California, *Journal of Geophysical Research*, v. 74, pp. 3821–3827.

Healy, J. H., and L. G. Peake, 1975, Seismic velocity structure along a section of the San Andreas Fault near Bear Valley, California, *Bulletin Seismological Society of America*, v. 65, pp. 1177–1197.

Huftile, G. J., 1991, Thin-skinned tectonics of the upper Ojai Valley and Sulphur Mountain area, Ventura Basin, California, *American Association of Petroleum Geologists Bulletin*, v. 75, pp. 1353–1373.

Lemiszki, P. J., and L. D. Brown, 1988, Variable crustal structure of strike-slip fault zones as observed on deep seismic reflection profiles, *Geological Society of America Bulletin*, v. 100, pp. 665–676.

Mooney, W. D., and R. H. Colburn, 1985, A seismic-refraction profile across the San Andreas, Sargent, and Calaveras faults, west-central California, *Bulletin Seismological Society of America*, v. 75, pp. 175–191.

Sylvester, A. G., 1988, Strike-slip faults, *Geological Society of America Bulletin*, v. 100, pp. 1666–1703.

Wilson, J. T., 1965, A new class of faults and their bearing on continental drift, *Nature*, v. 207, pp. 343–347.

Yeats, R. S., 1986, Faults related to folding with examples from New Zealand, *Royal Society of New Zealand Bulletin*, v. 24, pp. 273–291.

Yeats, R. S., and K. R. Berryman, 1987, South Island, New Zealand, and the Transverse Ranges, California: A seismotectonic comparison, *Tectonics*, v. 6, pp. 363–376.

Hotspot Settings

Blackwell, D. D., 1989, Regional implications of heat flow of the Snake River Plain, northwestern United States, *Tectonophysics*, v. 164, pp. 323–343, 1989.

Bryan, C. J., and C. E. Johnson, 1991, Block tectonics of the island of Hawaii from a focal mechanism analysis of basal slip, *Bulletin Seismological Society of America*, v. 81, pp. 491–507.

Duncan, R. A., 1991, Hotspots in the southern oceans—An absolute frame of reference for motion of the Gondwana continents, *Tectonophysics*, v. 74, pp. 29–42.

Duncan, R. A., J. Backman and L. Peterson, 1989, Reunion hotspot activity through Tertiary time: Initial results from Ocean Drilling Program, Leg 115, *Journal of Volcanology and Geothermal Research*, v. 36, pp. 193–198.

Duncan, R. A., and D. G. Pyle, 1988, Rapid eruption of the Deccan flood basalts at the Cretaceous/Tertiary boundary, *Nature*, v. 333, pp. 841–843.

Evans, J. R., 1982, Compressional wave velocity structure of the upper 350 km under the eastern Snake River Plain near Rexburg, Idaho, *Journal of Geophysical Research*, v. 87, pp. 2654–2670.

Filmer, P. E., M. K. McNutt, H. F. Webb, and D. J. Dixon, 1993, Volcanism and archipelagic aprons in the Marquesas and Hawaiian islands, *Marine Geophysical Research*, v. 16, pp. 385–406.

Jackson, E. D., and others, 1980, Introduction and summary of results from DSDP leg 55, the Hawaiian-Emperor hot-spot experiment, *Initial Reports Deep Sea Drilling Project*, v. 55, pp. 3–31.

Molnar, P., and J. Stock, 1987, Relative motions of hotspots in the Pacific, Atlantic and Indian Oceans since late Cretaceous time, *Nature*, v. 327, pp. 587–591.

Morgan, W. J., 1971, Convection plumes in the lower mantle, *Nature*, v. 230, pp. 42–43.

Morgan, W. J., 1983, Hotspot tracks and the early rifting of the Atlantic, *Tectonophysics*, v. 94, pp. 123–139.

Smith, R. B., and L. W. Braile, 1994, The Yellowstone Hotspot, *Journal of Volcanology and Geothermal Research*, v. 61, pp. 121–187.

Walcott, R. I., 1970, Flexure of the lithosphere at Hawaii, *Tectonophysics*, v. 9, pp. 435–446.

Watts, A. B., U. S. ten Brink, P. Buhl and T. M. Brocher, 1985, A multichannel seismic study of lithospheric flexure across the Hawaiian-Emperor seamount chain, *Nature*, v. 315, pp. 105–111.

CHAPTER 3

Seismic Waves

seismic *(sīz mik') **adj.,*** [< *Gr.* seismos, *an earthquake* < seiein, *to shake*], *relating to an earthquake or artificial shaking of the Earth.*
wave *(wāv) **n.,*** [< *OE.* wafian], *motion that periodically advances and retreats as it is transmitted progressively from one particle in a medium to the next.*
seismic wave *(sīz mik' wāv) **n.,** propagation of energy through the Earth caused by earthquakes or artificial vibrations.*
elastic *(i las'tik) **adj.,*** [< *Lgr.* elastikos < *Gr.* elauneim, *to drive*], *ability of a material to return immediately to its original size, shape, or position after being squeezed, stretched, or otherwise deformed.*

A great deal of what we know about the interior of the Earth comes from the recording of seismic waves that have traveled through various portions of the Earth. Controlled source seismic techniques provide seismic velocity information, as well as some detail of layering, for the crust. Seismic refraction data (Chapter 4) are particularly useful for mapping depth to bedrock, crustal thickness, and uppermost mantle velocity. Seismic reflection profiles (Chapters 5 and 6) show details of layering within sedimentary basins and gross structure of the deeper crust. The velocity structure for deeper parts of the Earth are determined from the study of earthquake seismic waves (Chapter 7).

ELASTIC WAVES

Earth materials must behave elastically in order to transmit seismic waves. The degree of elasticity thus determines how well a material transmits a seismic wave. When a material is subjected to *stress* (compression, tension, or shearing), it undergoes *strain* (distortion in volume and/or shape; Fig. 3.1a). Elastic behavior means

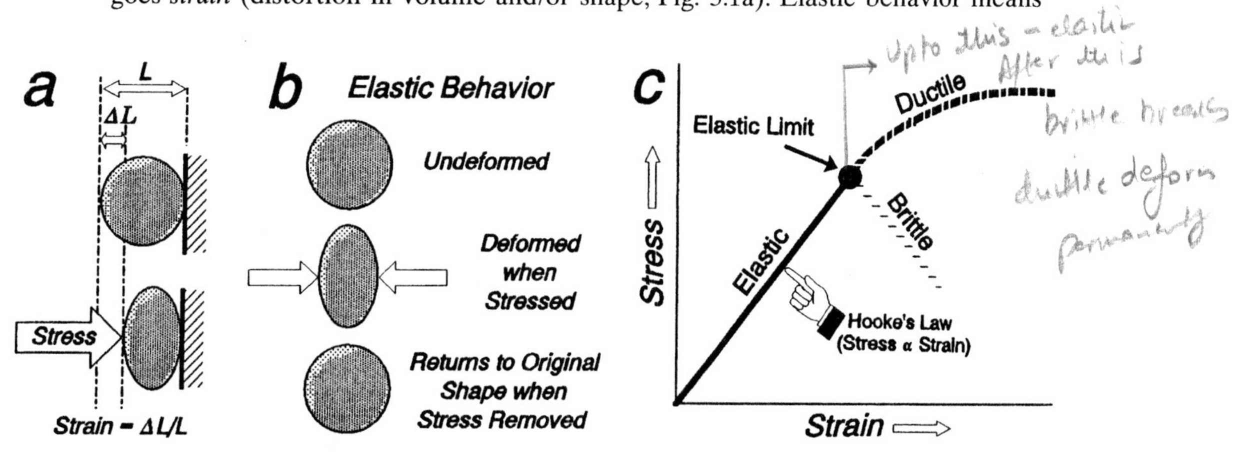

FIGURE 3.1 a) The deformation of a material (strain) results from a force per unit area (stress) acting on the material. b) An elastic material returns to its original shape and volume when the deforming stress is removed. c) Materials are elastic only up to a certain level of stress, the elastic limit; beyond the elastic limit, they undergo some degree of permanent deformation, through brittle or ductile failure.

that the material returns to its original volume and shape when the stress is removed (Fig. 3.1b). Hooke's Law describes elastic behavior where the amount of strain is linearly proportional to the amount of stress (Fig. 3.1c).

Earth materials may be elastic under some conditions, inelastic under others. Two important factors are: 1) the magnitude and orientation of the deforming stress (amount of *compression, tension,* or *shearing*); and 2) the length of time the material takes to achieve a certain amount of distortion (*strain rate*). Fig. 3.1c shows that a material may be elastic (obeying Hooke's Law) only up to a certain level of stress, known as the *elastic limit*. Beyond that limit the material fails, so that particles do not return to their original positions when the stress is removed. Deformation beyond the elastic limit may be *ductile,* whereby the material flows like silly putty, or *brittle,* like a pencil breaking or the lithosphere rupturing as an earthquake (Fig. 3.2).

The importance of time (that is, strain rate) is illustrated by the behavior of the asthenosphere (Fig. 3.3). Small but very rapid movement of particles results from an earthquake; at these very high strain rates the asthenosphere behaves elastically and is thus able to transmit seismic waves. Lithospheric plates move at only a few cm/year; at the very slow strain rates associated with this motion, the underlying asthenosphere deforms in an inelastic (ductile) fashion.

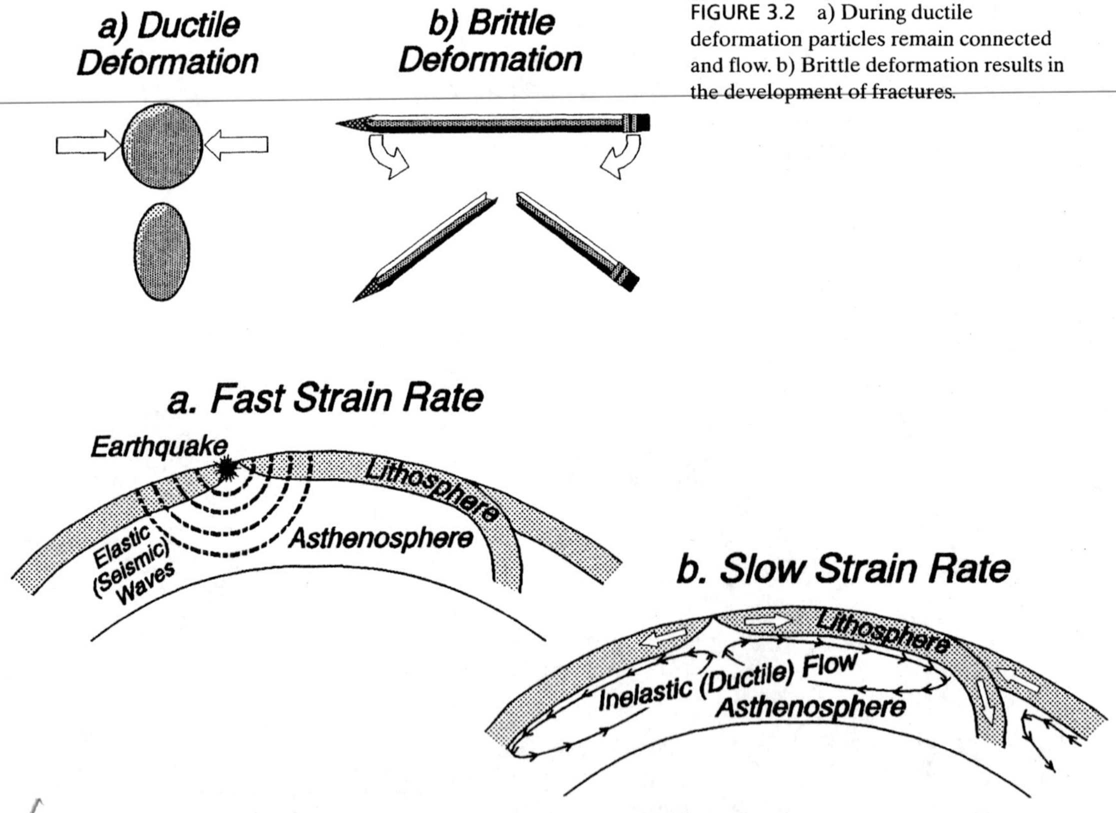

FIGURE 3.2 a) During ductile deformation particles remain connected and flow. b) Brittle deformation results in the development of fractures.

FIGURE 3.3 *Elastic and inelastic behavior of the asthenosphere.* a) Elastic behavior occurs at very high strain rates, as particles are vibrated by an earthquake; the vibrations result in the passage of seismic waves. b) Inelastic (ductile) behavior results from very slow straining of the asthenosphere; viscous flow facilitates the movement of overlying lithospheric plates.

Body Wave Propagation

Seismic waves (Fig. 3.4) can be categorized by whether they travel through the Earth (*body waves*) or along Earth's surface (*surface waves*). There are many factors that influence the propagation of body waves (see Telford et al., 1976, pp. 222–230); their analysis is simplified, however, if the elastic material can be considered isotropic and unbounded. *Isotropic* means that the material has the same physical properties (for example, density and rigidity) in all directions. *Unbounded* refers to that portion of the material with no free surfaces or interfaces. The analysis below thus pertains to vibrations moving through the "body" of such a homogeneous material.

Elastic Constants An *elastic constant* describes the strain of a material under a certain type of stress. The *bulk modulus* (or "*incompressibility*") describes the ability to *resist* being compressed (Fig. 3.5). Under pressure that is equal in all directions ("hydrostatic" for material under water; "lithostatic" for material within the Earth), the stress is the change in pressure (ΔP). The strain is the change in volume (ΔV) divided by the original volume (V). The bulk modulus (k) is the stress divided by the strain:

$$k = \frac{\text{stress}}{\text{strain}} = \frac{\Delta P}{\Delta V / V}$$

where:

$\Delta P = P' - P$ = pressure change (applied stress)
P = original confining pressure
P' = confining pressure under the applied stress
$\Delta V = V - V'$ = change in volume caused by ΔP
V = original volume
V' = volume under the applied stress.

The above equation illustrates that if a material undergoes no volume change ($\Delta V = 0$) when subjected to compressive stress (ΔP), the material is said to be

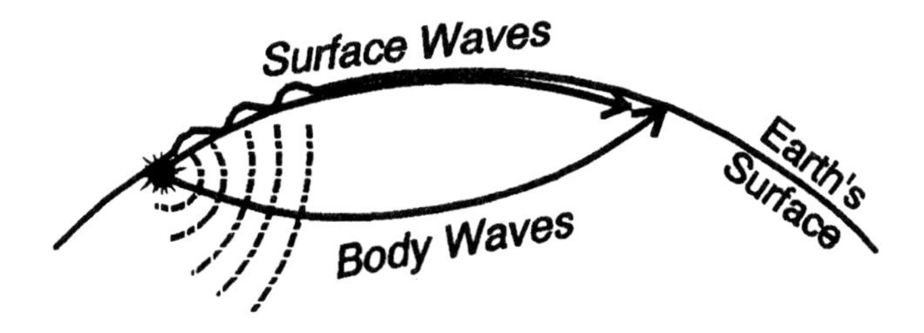

FIGURE 3.4 Surface wave energy is confined to a thin, outer shell of the Earth. Body wave energy radiates in three dimensions, like the surface of a balloon expanding through the Earth.

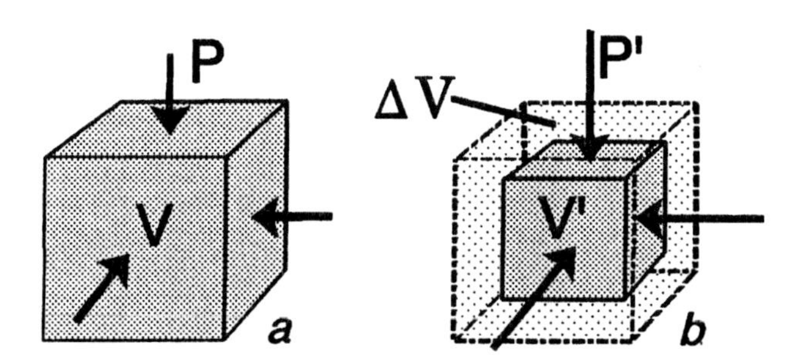

FIGURE 3.5 *Bulk modulus.* a) Material of volume V under confining pressure P, shown by equal arrows. b) Material compressed to volume V′ as pressure increases to P′ (longer arrows). The bulk modulus (k) determines the change in volume (ΔV) of the material.

incompressible (k = ∞). Conversely, materials that are easy to compress (k very small) undergo large changes in volume (large ΔV) when subjected to relatively small compressive stresses (small ΔP).

The *shear modulus* (or "*rigidity*") refers to the ability of a material to *resist* shearing (Fig. 3.6). When a cube of material is subjected to shearing, the stress is the tangential force (ΔF) divided by the area over which the force is applied (A). The strain is the shear displacement (Δl) divided by the length (l) of the area acted upon by ΔF. For such a stress, the shear modulus (μ) is:

$$\mu = \frac{\text{stress}}{\text{strain}} = \frac{\Delta F/A}{\Delta l/l}$$

A material that shows strong resistance to shearing (Δl ≈ 0) is very rigid (μ ≈ ∞). A fluid, on the other hand, has no resistance to shearing (Δl = ∞) and therefore lacks rigidity (μ = 0).

For an unbounded, isotropic material, the elastic constants k and μ, along with the density (ρ), determine how fast body waves travel through the material. It may not be practical, however, to measure those two elastic constants directly. Other constants may be more readily measured and used to calculate k and μ (Fig. 3.7). *Young's modulus* (or the "*stretch modulus*") describes the behavior of a rod that is pulled or compressed, according to the equation:

$$E = \frac{\text{stress}}{\text{strain}} = \frac{F/A}{\Delta L/L}$$

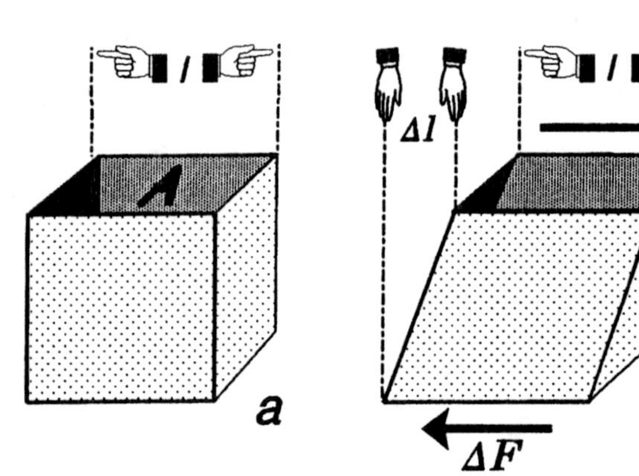

FIGURE 3.6 *Shear modulus.* a) Configuration of material before change in shear force. Note cube with sides of area A and length l. b) The change in shear force ΔF acts across the area A. One side of the cube (A) is displaced a distance Δl relative to the opposite side, according to the shear modulus of the material.

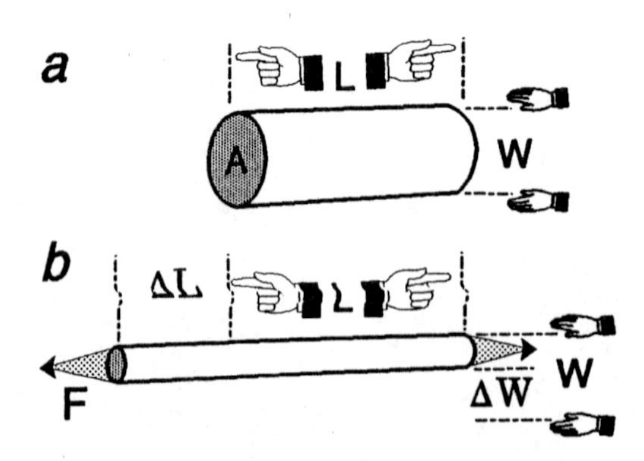

FIGURE 3.7 a) A rod of length (L), width (W) and cross-sectional area (A) can be used to measure Young's modulus and Poisson's ratio. The rod is subjected to a longitudinal stress (force, F, acting over the cross-sectional area, A). *Young's modulus* determines the resulting longitudinal strain (change in length, ΔL, divided by the original length, L). *Poisson's ratio* is the transverse strain (ΔW/W) divided by the longitudinal strain (ΔL/L).

where:

E = Young's modulus
F/A = force per unit area applied to end of rod
L = original length of rod
ΔL = change in length of rod.

Poisson's ratio states that, for a stretched rod, the ratio of transverse strain ($\Delta W/W$) to longitudinal strain ($\Delta L/L$) is:

$$\upsilon = \frac{\Delta W/W}{\Delta L/L}$$

where:

υ = Poisson's ratio
W = original width of rod
ΔW = amount by which width contracts.

Lame's constant (λ) illustrates the relationship between the four constants discussed above, according to:

$$\lambda = k - \frac{2\mu}{3} = \frac{\upsilon E}{(1 + \upsilon)(1 - 2\upsilon)}$$

Relatively easy measurements of E and υ can be used to determine λ for a material; λ can then be used as one of the parameters describing the velocity of seismic waves through the material.

Types of Body Waves Body waves propagate by a series of compressions and dilatations of the material (compressional wave) or by shearing the material back and forth (shear wave). A *compressional wave* is a "primary" or "P" wave because compressional waves arrive first from earthquakes; they are also called "longitudinal" and "push-pull" waves because particles of the material move back and forth, parallel to the direction the wave is moving (Fig. 3.8a). An example is a sound wave

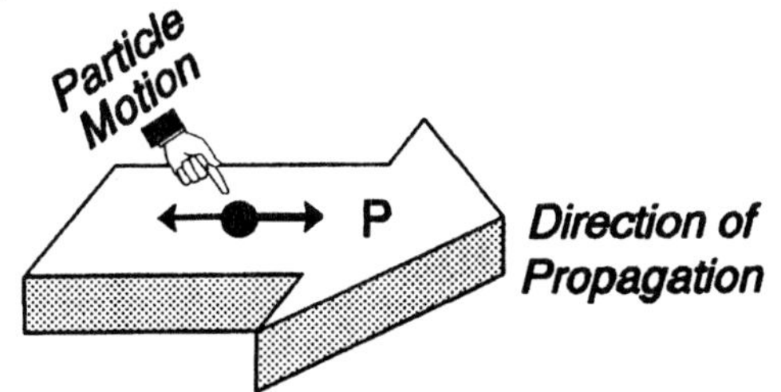

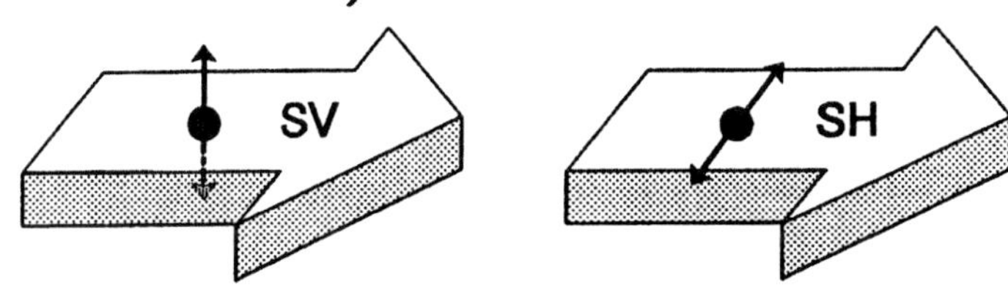

FIGURE 3.8 *Particle motions for body waves.* a) In a compressional wave, particles of the material move back and forth, parallel to the direction the wave energy moves. b) Particles of the material move perpendicular to the direction of propagation of shear wave energy. For wave energy moving horizontally: SV = shear wave with vertical particle motion; SH = shear wave with horizontal particle motion.

traveling through air. The sound propagates as the air is compressed then dilated; a person's eardrum senses the density changes in the air.

Particle motions for a *shear wave* are perpendicular to the direction of propagation (Fig. 3.8b). Shear waves are also referred to as "secondary" or "S" waves because they arrive from an earthquake after the initial compressional waves, and as "transverse" waves because of their particle motions.

Velocity of Body Waves

The term *velocity* refers to a vector, with both magnitude and direction. In seismology "velocity" is commonly used to refer to the magnitude component, or "speed," with direction not necessarily implied.

For an elastic, isotropic, unbounded material, seismic velocities depend on the elastic constants (k, μ, E and υ) and density (ρ) of the material. The more resistant the material is to deformation (that is, the greater the material's incompressibility or rigidity), the faster the waves travel. Body wave velocities are given by:

$$\boxed{V_P = \sqrt{\frac{k + 4/3\,\mu}{\rho}} = \sqrt{\frac{\lambda + 2\mu}{\rho}}}$$

and:

$$\boxed{V_S = \sqrt{\frac{\mu}{\rho}}}$$

where:

V_P = velocity of the compressional wave

V_S = velocity of the shear wave.

The velocity equations lead to the following generalizations. 1) *For the same material, shear waves will always travel slower than compressional waves.* 2) *The more rigid the material, the higher the P and S wave velocities.* For example, seismic waves speed up as they travel from the asthenosphere into the mantle part of the lithosphere, suggesting that the mantle lithosphere is more rigid (higher μ) than the asthenosphere. 3) *Fluids (liquids or gasses) have no shear strength* (μ = 0). This lack of rigidity means two things: a) shear waves cannot travel through fluids, like air, water and Earth's outer core; and b) compressional waves travel slower through the liquid state than through the solid state of the same material (the fluid outer core, for example, has lower V_P than the solid inner core).

Observations suggest that the following factors generally lead to an *increase in seismic velocity*. 1) *Increasing depth within the Earth.* As materials are better compacted and cemented, they become more rigid and incompressible, increasing both k and μ. 2) *Increase in density.* This generalization is commonly true for rocks, even though the equations show an inverse relationship between velocity and density. As rocks become more dense, they generally become more incompressible and more rigid; the corresponding elastic constants (k and μ) commonly increase more than the density (ρ). 3) *Decreasing porosity.* With decreasing pore space, a rock becomes more dense. At the same time, it is generally more lithified and has less fluid content, thereby increasing both k and μ. 4) *Change from liquid (or partial melt) to solid.* The rigidity (μ) increases, raising both V_P and V_S. Seismic waves therefore speed up while traveling from the asthenosphere to the mantle lithosphere, and from the outer core to the inner core.

Fig. 3.9 gives typical physical properties for various substances, listed downward according to increasing P-wave velocity. The fluids (air and water) have some

	ELASTIC CONSTANTS				SEISMIC VELOCITIES	
	10^9 N / m²		kg / m³	g / cm³	km / s	
	Bulk Modulus (k)	Shear Modulus (μ)	Density (ρ)	Density (ρ)	Compres. Wave (V_p)	Shear Wave (V_s)
Air	0.0001	0	1.0	0.001	0.32	0
Water	2.2	0	1000	1.0	1.5	0
Ice	3.0	4.9	920	0.92	3.2	2.3
Shale	8.8	17	2400	2.4	3.6	2.6
Sandstone	24	17	2500	2.5	4.3	2.6
Salt	24	18	2200	2.2	4.7	2.9
Limestone	38	22	2700	2.7	5.0	2.9
Quartz	33	39	2700	2.7	5.7	3.8
Granite	88	22	2600	2.6	6.7	2.9
Peridotite	139	58	3300	3.3	8.1	4.2

FIGURE 3.9 Typical values for elastic constants, density, and seismic velocities for selected materials, listed according to increasing compressional wave velocity (V_p). Compiled from Kinsler et al. (1982) and other sources. SI units for density are kg/m³; the literature, however, commonly gives densities in g/cm³.

resistance to compression (k ≠ 0), supporting P-wave travel. Lack of rigidity ($\mu = 0$) in the fluids contributes to low P-wave velocities and results in no propagation of S-waves ($V_S = 0$). Going down the list the materials generally become stronger (higher k and μ) and more dense with increasing velocity.

The graphs of P- and S-wave velocity vs. density (Fig. 3.10) were compiled from the information in Fig. 3.9. The solid portions of the superimposed curves highlight the crudely linear relationship between density and seismic velocity for crustal rocks and minerals (shale, sandstone, salt, limestone, quartz, granite).

Fig. 3.11 presents common ranges of compressional wave velocities for materials encountered at Earth's surface. The values on the chart illustrate that it is difficult to identify rocks based only on velocity information, because the wide ranges result in sizeable overlaps (see also Fig. 4.22 of Telford et al., 1976). It is clear, however, that seismic velocities are generally lowest for unconsolidated sediments, higher for sedimentary rocks, still higher for crystalline rocks of the crust, and highest for ultramafic rocks that formed in the mantle.

Surface Wave Motions

Surface waves can be generated when a medium has a free boundary, such as the surface of the Earth. As energy travels along the outer shell of the Earth, the disturbance is largest at the surface, decreasing exponentially with depth. Fig. 3.12 shows the particle motions for two types of surface waves, with body wave motions for comparison (see also Fig. 6 in Bolt, 1988). *Rayleigh waves* have retrograde elliptical

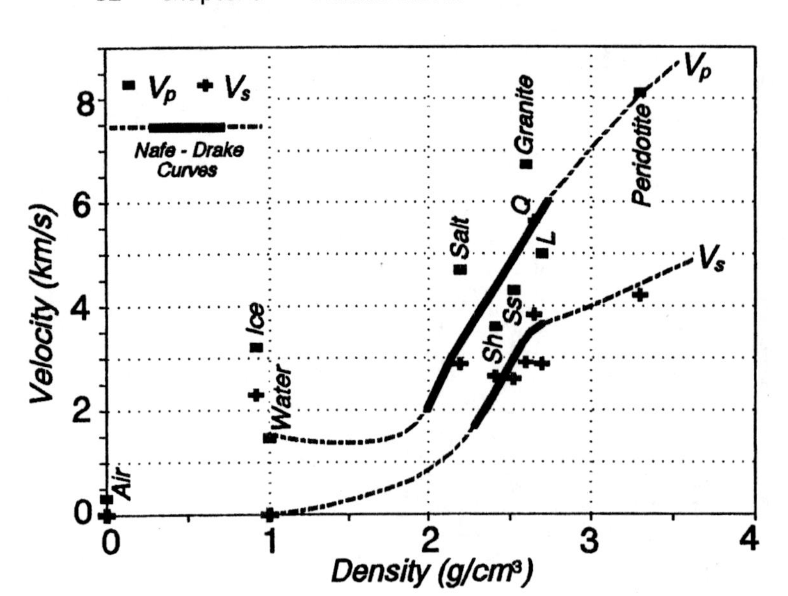

FIGURE 3.10 Graph of seismic velocity vs. density for materials presented in Fig. 3.9. L = limestone; Q = quartz; Sh = shale; Ss = sandstone. Solid rectangles with labels are compressional wave velocities (V_P); corresponding shear wave velocities (V_S) shown directly beneath by plus's. Empirical ("Nafe-Drake") curves, developed through analysis of numerous rock and sediment samples, are superimposed on the graph (Ludwig, Nafe, and Drake, 1971). Portions of these curves, highlighted by the solid lines, show roughly linear relationship between seismic velocities and densities for crustal rocks. 1 g/cm^3 = 10^3 kg/m^3.

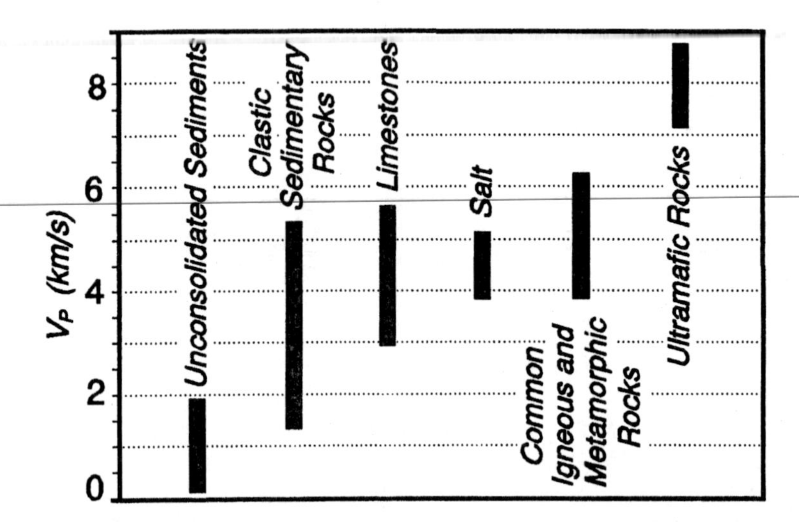

FIGURE 3.11 Approximate ranges of compressional wave velocity (V_P) for some materials encountered at Earth's surface (from Griffiths and King, 1981).

motion; at the top of the ellipse, particles move opposite to the direction of wave propagation. *Love waves* are surface waves that behave like shear waves; the particles move horizontally in directions perpendicular to the direction of propagation (SH motion, Fig. 3.8).

CONTROLLED SOURCE SEISMIC TECHNIQUES

Seismic waves can be used to determine depths to interfaces within the Earth and velocities of layers between the interfaces. The velocities in turn may be used as one parameter to interpret the nature of Earth materials. The discussion below develops techniques that can be used for studies of the crust and the top of the mantle, where refraction and reflection methods are applicable (Chapters 4, 5, and 6).

The sources of the seismic waves can be natural (earthquake) or produced artificially (controlled source). For the latter, the source is at (or just below) Earth's surface and an array of instruments are laid out on the surface to receive the direct, reflected, or refracted signals (Fig. 3.13).

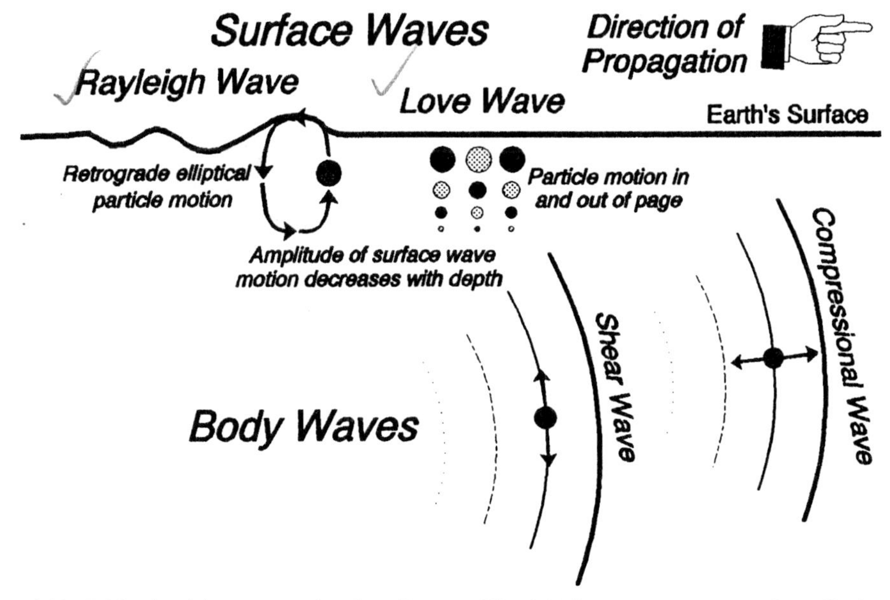

FIGURE 3.12 Particle motions of surface (Love and Rayleigh) waves, compared to that of body (compressional and shear) waves. Solid dots represent Love wave motions out of the page; gray dots are motions into the page.

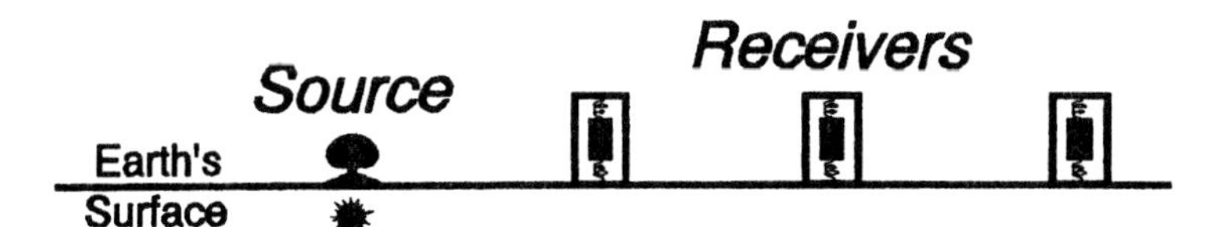

FIGURE 3.13 Receivers that record seismic waves are generally laid out in a line away from the artificial source. The source could be an explosion at the surface or in a shallow drillhole. The receivers depicted are geophones.

The artificial sources for surveys conducted on land may be from explosives, such as dynamite placed in drillholes, quarry blasts, or even nuclear tests. Alternatively, plates coupled to the ground beneath large trucks may massage the ground in a continuous, predetermined fashion, using a technique known as Vibroseis (registered trademark of Conoco, Inc). Seismic studies at sea (or on rivers or lakes) commonly employ air guns towed behind a survey boat.

The receivers used on land, called *geophones,* measure ground movement (either the displacement, the velocity, or the acceleration of the ground surface). A common geophone is a magnet on springs, suspended within a coil of wire (Fig. 3.13); movement of the magnet through the coil generates a measurable electrical current. At sea, *hydrophones* measure changes in water pressure caused by passing seismic waves.

Travel-Time Curves

The *travel time* of a seismic wave from a source to a receiver depends on the seismic velocities of the Earth materials traversed, the distance from the source to the receiver, and the geometry of boundaries separating Earth materials. A receiver commonly records more than one arrival of seismic energy because the energy may radiate from the source as various body and surface waves, and because the body waves are refracted and reflected along different paths when they encounter boundaries.

Fig. 3.9 shows that the bulk and shear moduli (k and μ) are generally of the same order of magnitude. The equations:

$$V_P = \sqrt{\frac{k + 4/3\,\mu}{\rho}} \quad \text{and} \quad V_S = \sqrt{\frac{\mu}{\rho}}$$

therefore illustrate that, for many Earth materials, $\mathbf{V_S \approx 0.6\ V_P}$. Rayleigh waves are slightly slower than shear waves, so that the Rayleigh wave velocity $\mathbf{V_R \approx 0.9\ V_S}$. Rayleigh waves thus travel about half the speed of compressional waves ($\mathbf{V_R \approx 0.5\ V_P}$).

A *wave front* is a surface along which portions of a propagating wave are *in phase.* For example, arrival of the wave in Fig. 3.14 occurs where particles first move as the wave approaches (wave front at 0° phase). The maximum amplitude of particle motion occurs along the 90° phase wave front. Other wave fronts correspond to positions where the wave goes from positive to negative amplitude (180°) and at minimum amplitude (270°).

Consider the wave fronts that represent the leading edges of oncoming P, S, and Rayleigh waves (Fig. 3.15a). In a homogeneous medium (constant seismic velocities), the body waves (P and S) radiate outward along spherical wavefronts, while Rayleigh waves (R) roll along the surface.

Seismic energy travels along trajectories perpendicular to wavefronts, known as *raypaths* (Fig. 3.15b). Variations in body wave velocity cause wave fronts to deviate from perfect spheres, thus bending or "refracting" the raypaths. Raypaths can be used to analyze portions of seismic waves that make it back to the surface, as discussed below for direct, critically refracted, and reflected waves.

A *seismic trace* is the recording of ground motion by a receiver, plotted as a function of time (Fig. 3.16). A seismic wave takes a certain amount of time to travel from the source to a receiver, depending on the distance to the receiver, the path taken by the wave, and the wave's velocity. Arrival of each of the P, S, and Rayleigh waves starts as initial movement of the ground, followed by reverberations that die out with time (Fig. 3.16b). A body wave arrives as a relatively short burst of energy. Surface waves, however, are commonly *dispersed;* broad (low-frequency) waves arrive first, followed by progressively narrower (higher-frequency) arrivals.

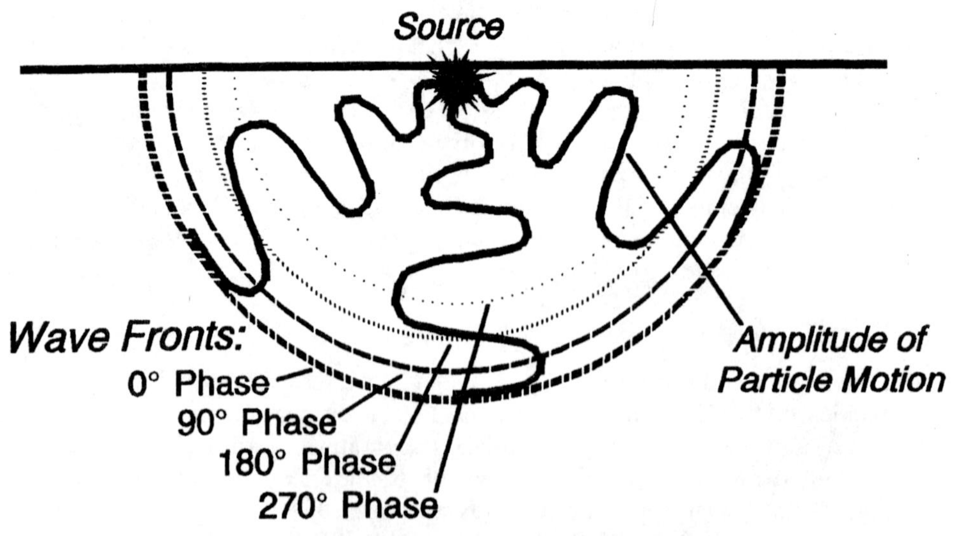

FIGURE 3.14 Wave fronts are surfaces along which particle motions of the propagating wave are in phase (one complete oscillation is 360° of phase). For example, a surface where particle motions reach their maximum positive amplitude is 90° phase; where they are maximum negative amplitude is 270° phase.

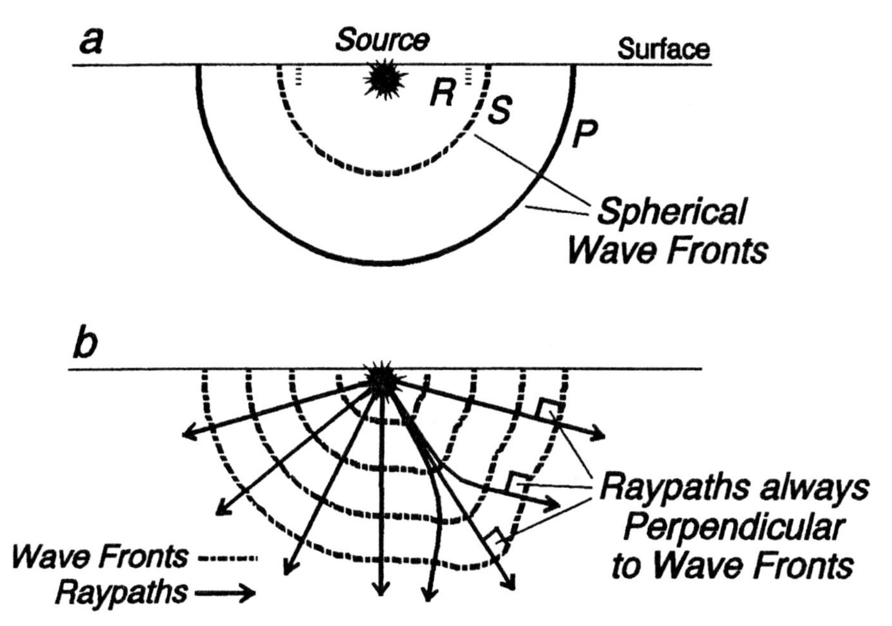

FIGURE 3.15 a) Initial wave fronts for compressional (P), shear (S), and Rayleigh (R) waves. b) Wave fronts for propagating P-wave. Changes in velocity cause segments of wave fronts to speed up or slow down, distorting the wave fronts from perfect spheres. Raypaths thus bend (refract) as velocity changes.

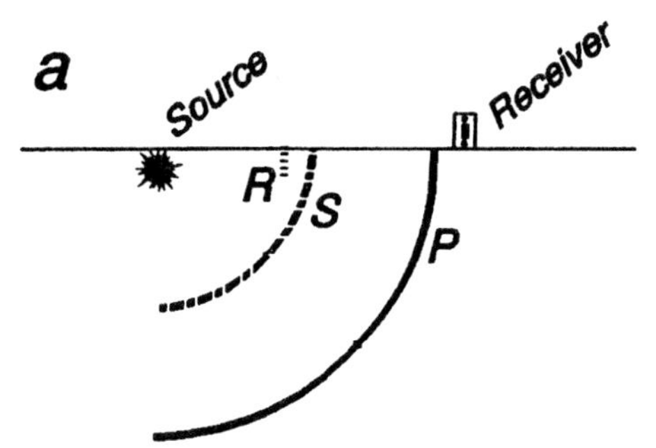

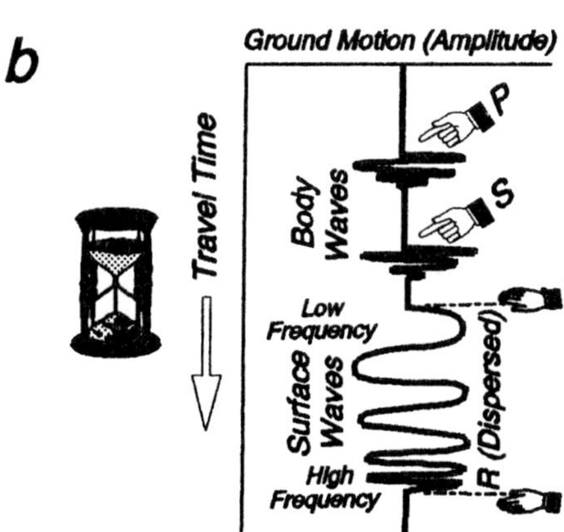

FIGURE 3.16 a) Seismic waves radiating from a source to one receiver. b) Seismic trace recording ground motion by the receiver, as a function of the travel time from the source to the receiver. For controlled source studies (seismic refraction and reflection), the travel time is commonly plotted positive downward.

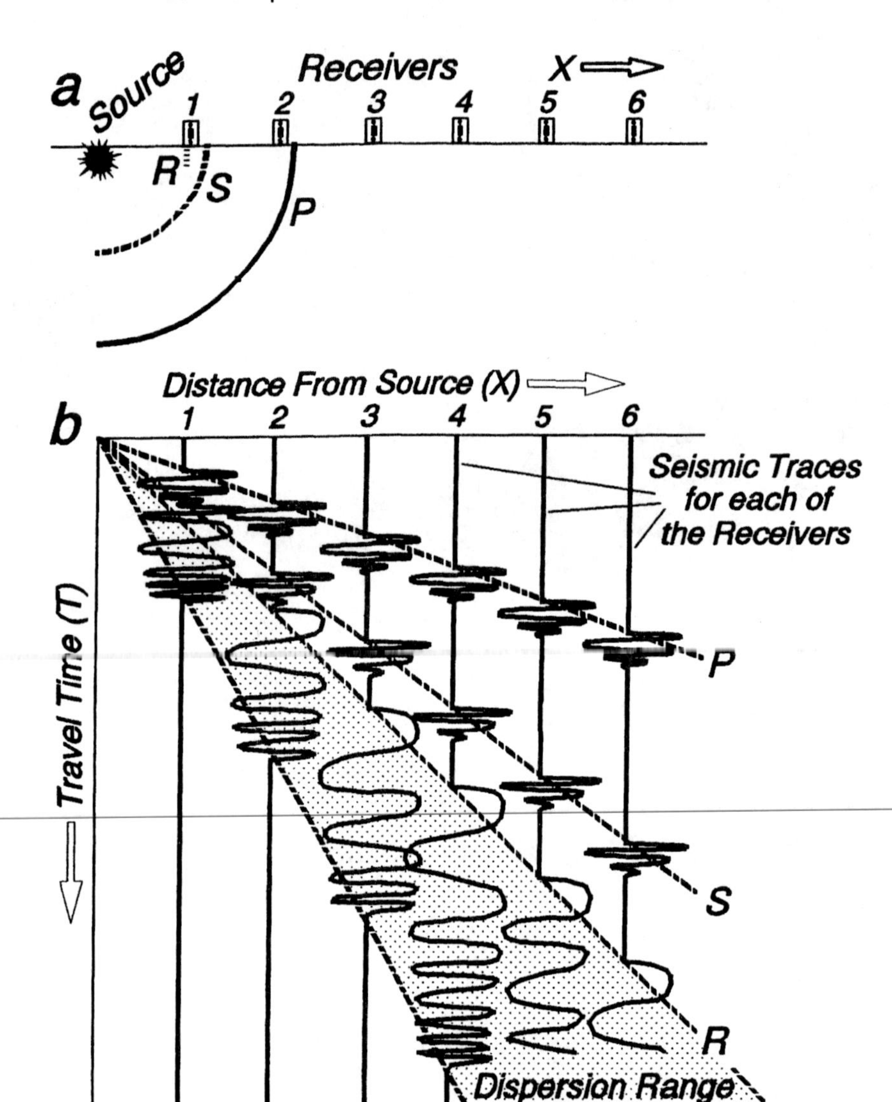

FIGURE 3.17 a) Initial wave fronts for P, S, and R waves, propagating across several receivers at increasing distance from the source. b) Travel-time graph. The seismic traces are plotted according to the distance (X) from the source to each receiver. The elapsed time after the source is fired is the travel time (T).

On a *travel-time graph* seismic traces from several receivers are plotted side by side, according to the horizontal distance (X) from the source to each receiver (Fig. 3.17). Travel time (T) is commonly plotted as *increasing downward* in refraction and reflection studies, because T often relates to depth within the Earth. For each of the initial P-wave, S-wave, or R-wave arrivals, the travel time from the source to a receiver is linear, expressed by the *travel-time curve:*

$$T = \frac{X}{V}$$

where:

T = total time for the wave to travel from the source to the receiver
X = distance from the source to the receiver, measured along the surface
V = seismic velocity of the P, S, or R arrival.

The slope of the line is the elapsed time (ΔT) divided by the distance traveled during that time (ΔX):

$$\text{Slope} = \frac{\Delta T}{\Delta X}$$

The slope at a given distance (X) can also be determined by taking the first derivative at that point on the travel time curve (Fig. 3.18a):

$$\frac{dT}{dX} = \lim_{\Delta t \to 0} \left(\frac{\Delta T}{\Delta X} \right)$$

The travel time can be written:

$$T = \frac{1}{V}(X)$$

so that:

$$\frac{dT}{dX} = \frac{1}{V}$$

The velocities for each type of wave can thus be calculated by taking the inverse of the slope (Fig. 3.18b):

$$\boxed{\mathbf{V} = \frac{\mathbf{1}}{\mathbf{dT/dX}}}$$

The first derivative, or slope, is thus useful in determining the velocity represented at any point on travel-time curves for different arrivals.

Direct, Critically Refracted, and Reflected Waves

Seismic energy is partitioned when waves encounter materials of different *acoustic impedance* (the product of seismic velocity and density). For example, when a compressional wave traveling in one material strikes the boundary of another material at an oblique angle, the energy separates into four *phases*: reflected P-wave, reflected S-wave, refracted P-wave, and refracted S-wave (Fig. 3.19).

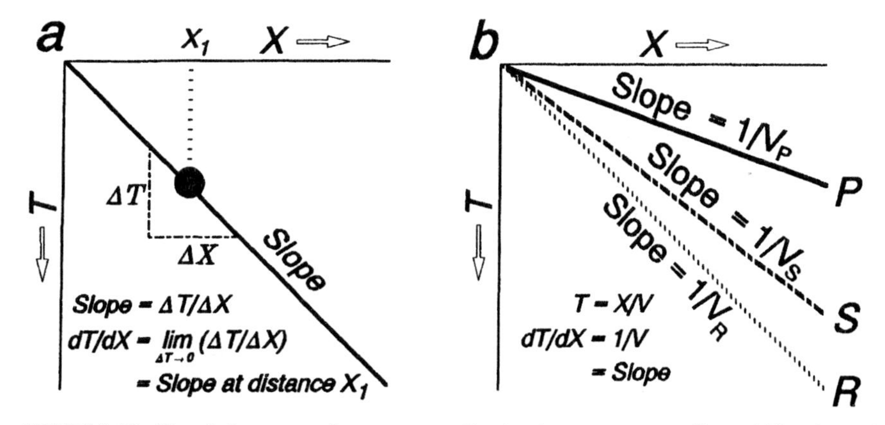

FIGURE 3.18 Travel-time curves for wave traveling in a homogenous medium. a) The slope of the line for each arrival is the first derivative (dT/dX). b) The slope of the travel time for each of the P, S, and R arrivals (Figure 3.17) is the inverse of the velocity.

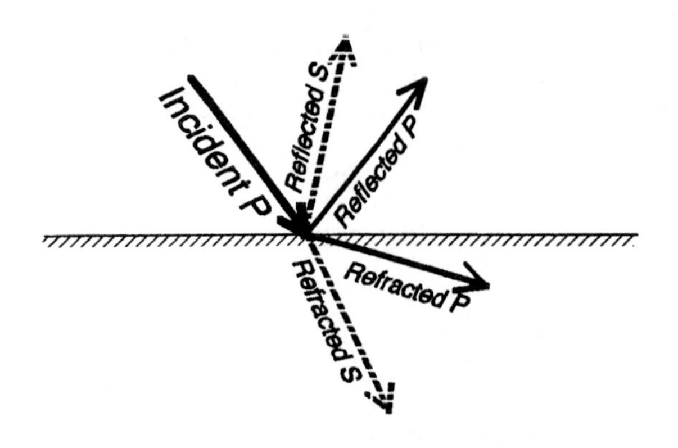

FIGURE 3.19 A compressional wave, incident upon an interface at an oblique angle, is split into four phases: P and S waves reflected back into the original medium; P and S waves refracted into the other medium.

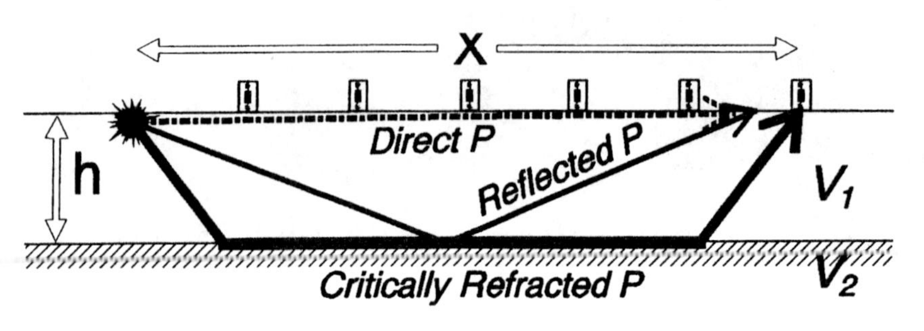

FIGURE 3.20 Raypaths for direct, reflected, and critically refracted waves, arriving at receiver a distance (X) from the source. The interface separating velocity (V_1) from velocity (V_2) material is a distance (h) below the surface.

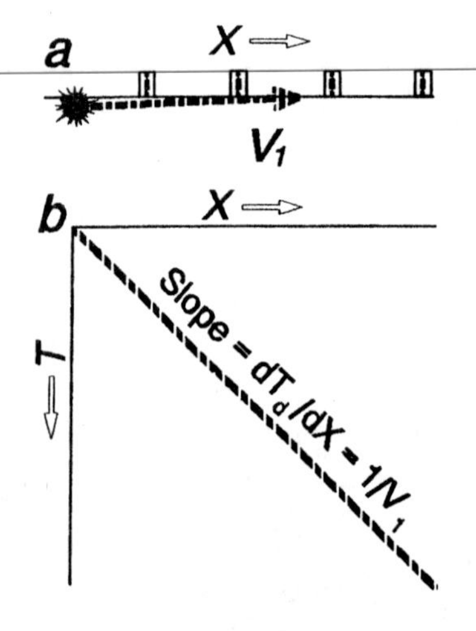

FIGURE 3.21 Selected raypath (a) and travel-time curve (b) for direct wave. The slope, or first derivative, is the reciprocal of the velocity (V_1).

The discussion below simplifies the problem by considering only P-wave phases for an upper layer with velocity V_1, separated by a horizontal interface from a layer of velocity V_2 (Fig. 3.20). Raypaths and travel-time curves developed for the direct, critically refracted, and reflected arrivals are helpful as starting points to interpret both controlled source and earthquake seismic data.

Direct Arrival The compressional wave that goes directly from the source to a receiver is a body wave traveling very close to the surface (Figs. 3.20, 3.21a). The velocity of the wave (V_1) is the distance from the source to the receiver (X) divided by the time it takes the wave to travel directly to the receiver (T_d):

$$V_1 = \frac{X}{T_d}$$

The equation for the straight line representing the direct arrival on a travel-time graph (Fig. 3.21b) is therefore:

$$\boxed{T_d = \frac{X}{V_1}}$$

Note that the velocity of the near-surface material can be determined by taking the inverse of the slope of the direct arrival (dT_d/dX) on the travel-time graph:

$$\frac{dT_d}{dX} = \frac{1}{V_1}$$

so that:

$$V_1 = \frac{1}{dT_d/dX}$$

Critically Refracted Arrival When seismic energy encounters material of different velocity, some of the energy may be transmitted into the second material (Fig. 3.19). An increase in velocity speeds up wave fronts in the second material; to remain perpendicular to wave fronts, raypaths bend across the interface (Fig. 3.22a). *Refraction* describes the bending of raypaths as seismic waves travel from one material to another. For a wave traveling from material of velocity V_1 into velocity V_2 material (Fig. 3.22b), raypaths are refracted according to *Snell's Law*:

$$\boxed{\frac{\sin\theta_1}{V_1} = \frac{\sin\theta_2}{V_2}}$$

where:

θ_1 = angle of incidence
θ_2 = angle of refraction
V_1 = seismic velocity of incident medium
V_2 = seismic velocity of refracting medium.

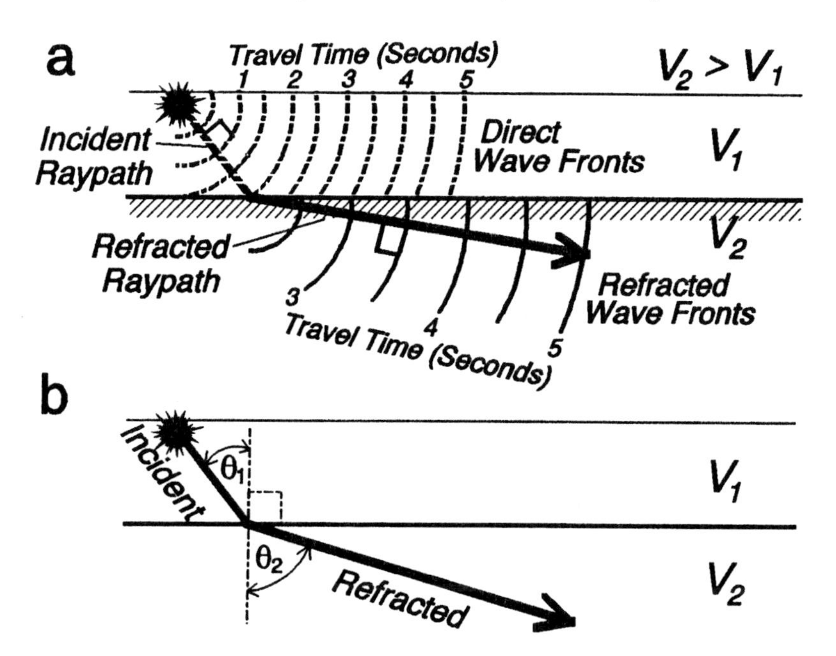

FIGURE 3.22 Refraction from layer of velocity (V_1) to one of velocity (V_2). a) Wave fronts are distorted from perfect spheres as energy is transmitted into material of different velocity (Fig. 3.15b). Raypaths thus bend ("refract") across an interface where velocity changes. b) The incident (θ_1) and refracted (θ_2) angles are measured from a line drawn perpendicular to the interface between the two layers.

Fig. 3.23 illustrates how Snell's Law describes three situations: 1) if the velocity *decreases* across the interface, the ray is refracted *away* from the interface; 2) if the velocity remains the same, the ray is not bent; 3) if the velocity *increases* across the interface, the ray is bent *toward* the interface.

In all cases, as the angle of incidence increases, so does the angle of refraction (Fig. 3.24a). A special situation, known as *critical refraction,* occurs when the angle of refraction (θ_2) reaches 90° (Fig. 3.24b). Note that critical refraction can occur only when $V_2 > V_1$, as in Fig. 3.23c.

The angle of incidence (θ_1) necessary for critical refraction is called the *critical angle* (θ_c). Setting the angle of refraction (θ_2) in Snell's Law to 90° shows how the critical angle relates to the velocities of the two materials:

$$\frac{\sin\theta_c}{V_1} = \frac{\sin(90°)}{V_2}$$

$$\frac{\sin\theta_c}{V_1} = \frac{1}{V_2}$$

$$\sin\theta_c = \frac{V_1}{V_2}$$

$$\boxed{\theta_c = \sin^{-1}\left(\frac{V_1}{V_2}\right)}$$

A critically refracted wave, traveling with velocity V_2 at the top of the lower layer, excites particle motions in the overlying layer. Energy returns to the surface along raypaths inclined at the critical angle (Fig. 3.25). Critically refracted waves are sometimes referred to as *headwaves* because, at distance, they arrive ahead of the direct waves that travel only in the upper, lower-velocity layer.

Fig. 3.26 shows that the travel time for a critically refracted wave (T_r) can be calculated by adding the time spent in each of the three segments of the travel path: down through layer 1 (T_1); horizontally along the top of layer 2 (T_2); and back up through layer 1 (T_3).

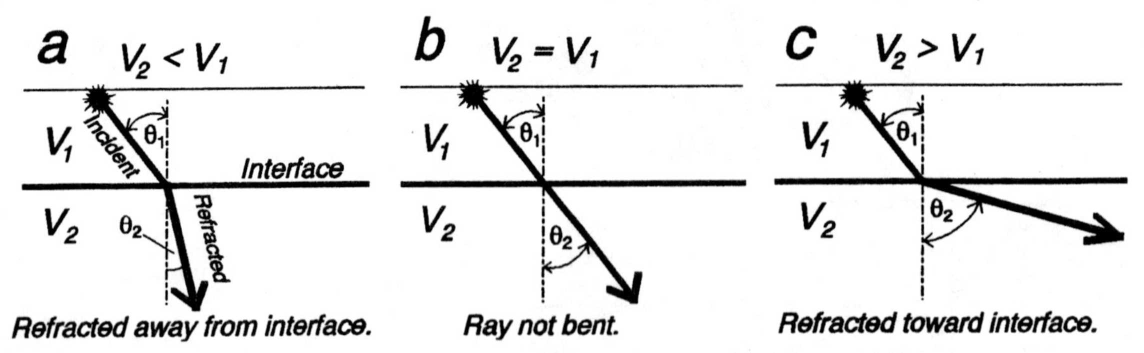

FIGURE 3.23 Behavior of refracted ray when velocity (a) decreases, (b) remains the same, and (c) increases across an interface.

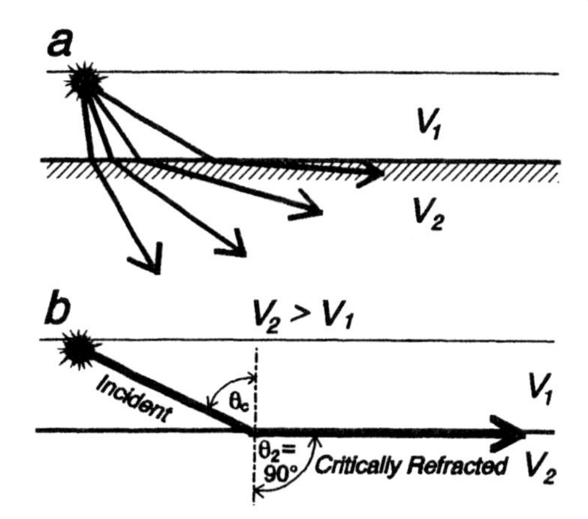

FIGURE 3.24 a) The angle of refraction increases as the angle of incidence increases. b) If $V_2 > V_1$, the angle of refraction (θ_2) can reach 90°. Critical refraction then occurs, with energy following the top part of the higher velocity layer. For such a case the angle of incidence (θ_1) is called the critical angle (θ_c).

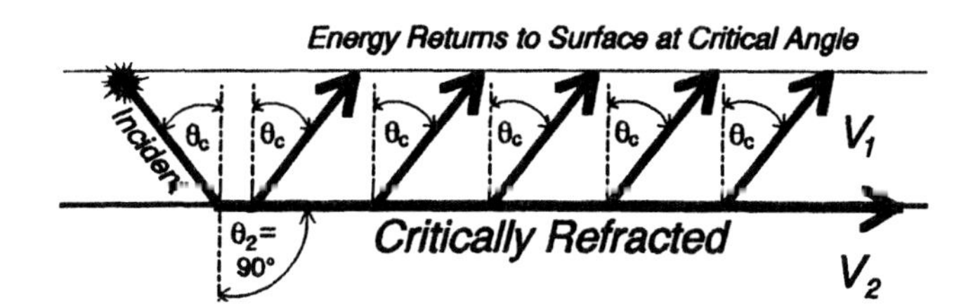

FIGURE 3.25 A critically refracted wave, traveling at the top of the lower layer with velocity V_2, leaks energy back into the upper layer at the critical angle (θ_c).

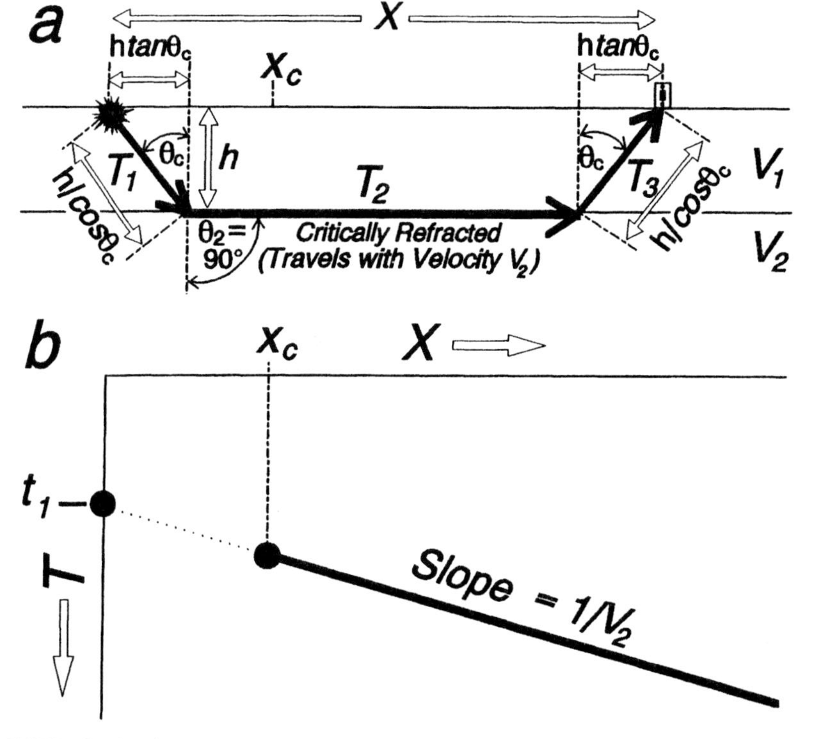

FIGURE 3.26 a) Geometry showing the three segments (T_1, T_2, T_3) comprising the total time path for a critically refracted ray that returns to the surface. b) Travel-time curve for critically refracted wave. The wave arrives at the surface only at and beyond the critical distance (X_c). The intercept time (t_1) is the projection of the curve to the T-axis.

The time for each segment is the distance travelled divided by the velocity for that segment. For an upper layer of thickness (h) the travel times in the segments are:

$$T_1 = \frac{h/\cos\theta_c}{V_1}$$

$$T_2 = \frac{X - 2h\tan\theta_c}{V_2}$$

$$T_3 = \frac{h/\cos\theta_c}{V_1}$$

where:

X = horizontal distance from source to receiver.

The total travel time from source to receiver is:

$$\begin{aligned} T_r &= T_1 + T_2 + T_3 \\ &= \frac{h/\cos\theta_c}{V_1} + \frac{X - 2h\tan\theta_c}{V_2} + \frac{h/\cos\theta_c}{V_1} \\ &= \frac{2h/\cos\theta_c}{V_1} + \frac{X}{V_2} - \frac{2h\tan\theta_c}{V_2} \\ &= 2h\left(\frac{1}{V_1\cos\theta_c} - \frac{\tan\theta_c}{V_2}\right) + \frac{X}{V_2} \end{aligned}$$

But: $\tan\theta_c = \dfrac{\sin\theta_c}{\cos\theta_c}$

so that:

$$\begin{aligned} T_r &= 2h\left(\frac{1}{V_1\cos\theta_c} - \frac{1}{V_2}\frac{\sin\theta_c}{\cos\theta_c}\right) + \frac{X}{V_2} \\ &= \frac{2h}{V_1\cos\theta_c}\left(1 - \frac{V_1}{V_2}\sin\theta_c\right) + \frac{X}{V_2} \end{aligned}$$

By Snell's Law: $\dfrac{V_1}{V_2} = \sin\theta_c$

so that:

$$\begin{aligned} T_r &= \frac{2h}{V_1\cos\theta_c}(1 - \sin^2\theta_c) + \frac{X}{V_2} \\ &= \frac{2h}{V_1\cos\theta_c}(\cos^2\theta_c) + \frac{X}{V_2} \end{aligned}$$

$$\boxed{\mathbf{T_r = \frac{2h\cos\theta_c}{V_1} + \frac{X}{V_2}}}$$

Let: $t_1 = \dfrac{2h\cos\theta_c}{V_1}$ (t_1 is a constant)

$$\boxed{\mathbf{T_r = t_1 + \frac{X}{V_2}}}$$

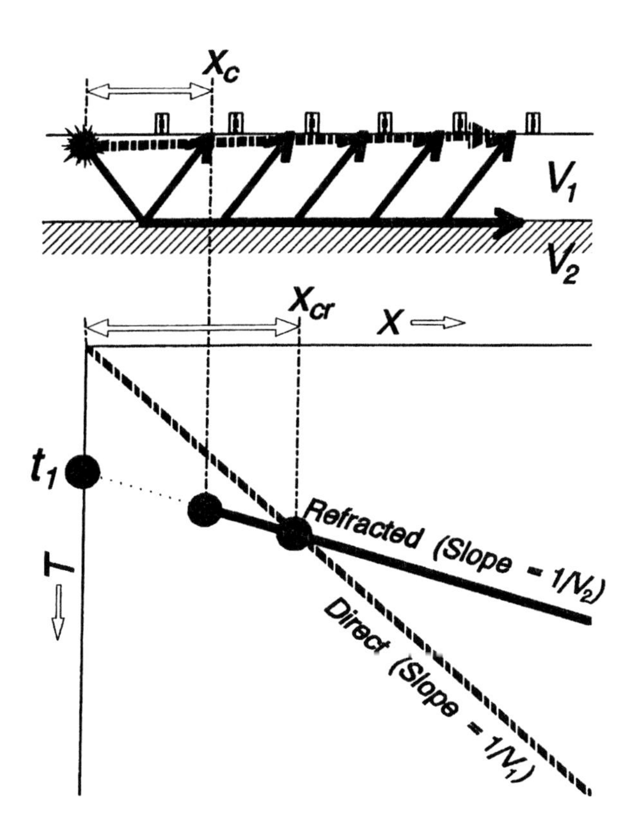

FIGURE 3.27 Selected raypaths and travel-time graph for the direct (dashed) and critically refracted (solid) arrivals. X_c = critical distance; X_{cr} = crossover distance.

The above equation is a straight line (Fig. 3.26b) with:

$$\mathbf{t_1 = T\text{-intercept time}}$$

and

$$\mathbf{\frac{1}{V_2} = slope}$$

The raypaths and travel-time curves in Fig. 3.27 show relationships between the direct and critically refracted arrivals. Note that critically refracted waves are observed only at and beyond a certain distance from the source, known as the *critical distance* (X_c). The dotted line segment extending from X_c to the T-intercept at t_1 is therefore fictitious, because no critically refracted waves arrive at less than the critical distance. Fig. 3.26 shows that the critical distance is a function of the critical angle (θ_c) and the thickness of the upper layer (h):

$$\mathbf{X_c = 2h\tan\theta_c}$$

Near the source the refracted wave, where present, arrives after the direct wave. The refracted wave appears before the direct wave at distant receivers, however, because proportionally more of the refracted travel path is through the layer with higher velocity (V_2). At the *crossover distance* (X_{cr}) the direct and critically refracted waves arrive at the same time; beyond the crossover distance the critically refracted wave arrives before the direct wave. The crossover distance is:

$$\mathbf{X_{cr} = 2h\sqrt{\frac{V_2 + V_1}{V_2 - V_1}}}$$

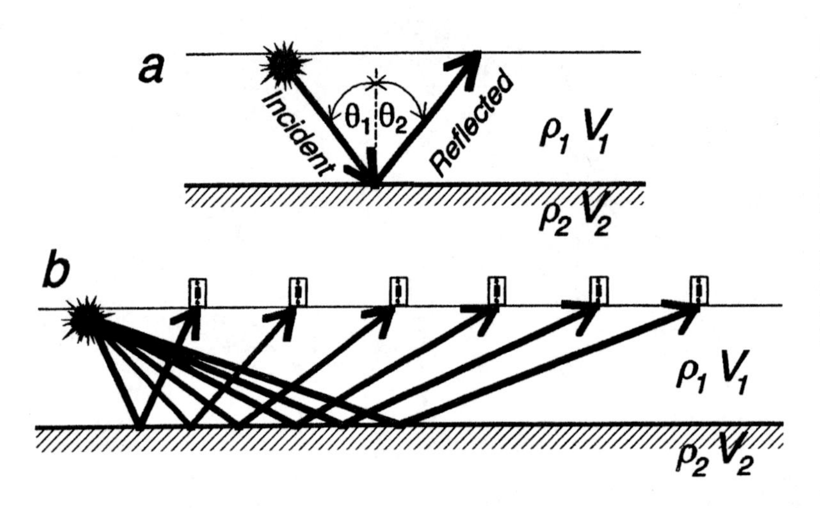

FIGURE 3.28 a) A compressional wave is reflected back at an angle (θ_2) equal to the incident angle (θ_1). ρ_1, V_1 and ρ_2, V_2 represent the densities and compressional wave velocities of the two layers. Reflection occurs when the acoustic impedance of the lower layer ($\rho_2 \times V_2$) differs from that of the upper layer ($\rho_1 \times V_1$). b) V-shaped raypaths for a compressional wave from a source to six receivers, reflected from a horizontal interface.

Reflected Arrival When seismic energy traveling in one layer encounters a layer with different acoustic impedance, some of the energy is reflected back into the first layer (Fig. 3.19). Snell's law illustrates that the angle of reflection (θ_2) is exactly the same as the angle of incidence (θ_1), because both rays travel at the same velocity (V_1; Fig. 3.28a):

$$\frac{\sin\theta_1}{V_1} = \frac{\sin\theta_2}{V_1}$$

so that:

$$\boxed{\theta_1 = \theta_2}$$

Reflected waves therefore follow *V-shaped raypaths* (Fig. 3.28b).

Fig. 3.29a shows that the travel time for a reflected wave, from the source to a receiver at horizontal distance (X), can be calculated by determining the total time spent in the constant velocity (V_1) layer. The total length of the reflected path (L), is the sum of the raypath segments for the incident (L_1) and reflected (L_2) segments:

$$\begin{aligned} L &= L_1 + L_2 \\ &= \sqrt{h^2 + (X/2)^2} + \sqrt{h^2 + (X/2)^2} \\ &= 2\sqrt{h^2 + (X/2)^2} \\ &= \sqrt{4h^2 + X^2} \end{aligned}$$

The total travel time (T_f) from source to receiver is:

$$\begin{aligned} T_f &= L/V_1 \\ &= \frac{\sqrt{4h^2 + X^2}}{V_1} \\ T_f^2 &= \frac{4h^2 + X^2}{V_1^2} \\ &= \frac{4h^2}{V_1^2} + \frac{X^2}{V_1^2} \\ T_f^2 &= \left(\frac{2h}{V_1}\right)^2 + \left(\frac{1}{V_1}\right)^2 X^2 \end{aligned}$$

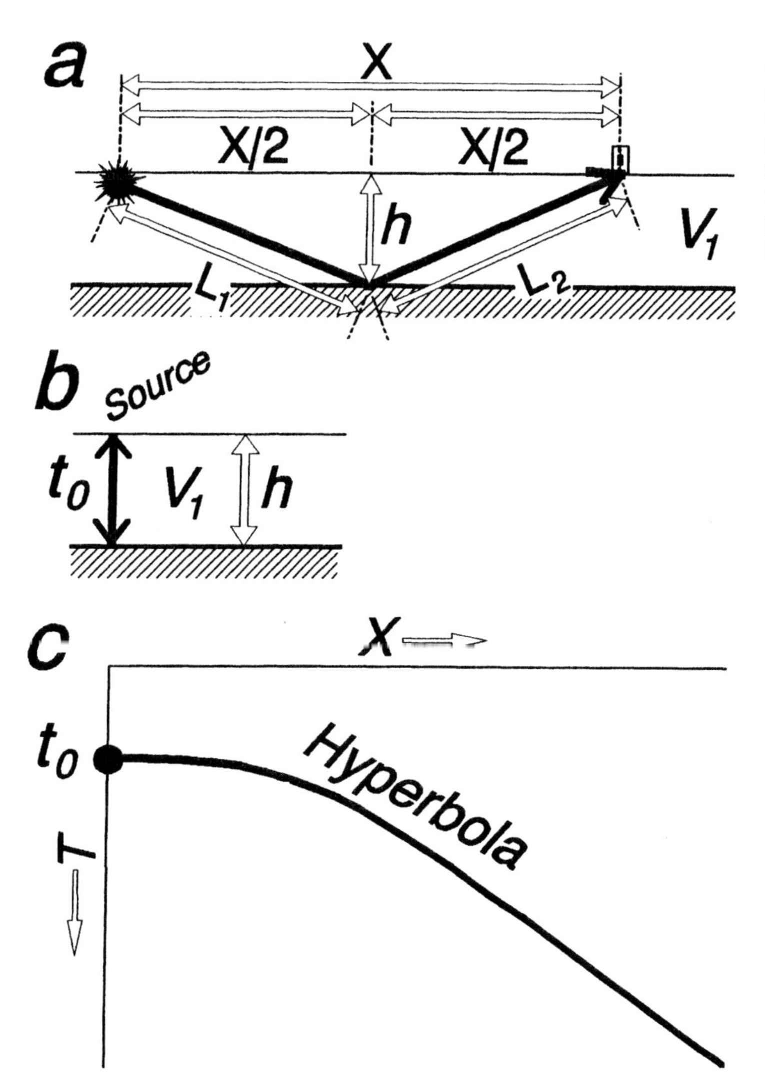

FIGURE 3.29 a) Geometry used to determine travel time of ray reflected from horizontal interface. b) Geometry of ray directly down to horizontal interface at depth (h), reflected back to the source location. The travel time (t_0) is $2h/V_1$. c) The reflected wave appears as a hyperbola on a travel-time graph, with the T-axis intercept at t_0.

Fig. 3.29b illustrates that the time (t_0) to go vertically down to the interface and straight back up to the shot location is a constant, given by:

$$t_0 = \frac{2h}{V_1}$$

The travel-time equation for a reflected wave from a horizontal interface overlain by a constant velocity medium is therefore:

$$T_f^2 = t_0^2 + \left(\frac{1}{V_1}\right)^2 X^2$$

or:

$$\boxed{\mathbf{T_f = \sqrt{t_0^2 + \frac{X^2}{V_1^2}}}}$$

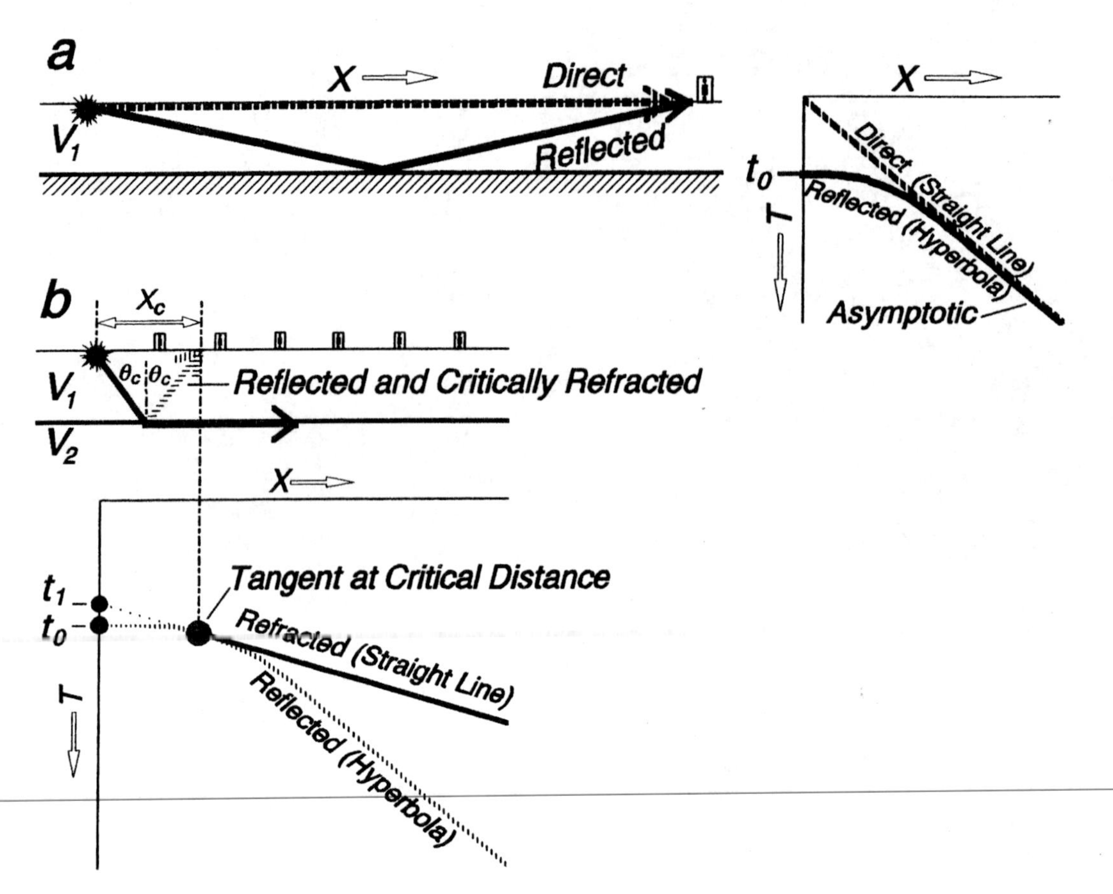

FIGURE 3.30 Selected raypaths and travel-time graphs illustrating relationships between reflected and other arrivals. a) The direct and reflected arrival curves are asymptotic at large distance from the source. b) The critically refracted arrival curve is tangent to that of the direct arrival at the critical distance (X_c).

This is the equation of a hyperbola (Fig. 3.29c), with:

$$t_0 = \text{T-axis intercept time.}$$

Two special circumstances illustrate how the travel-time graph for the reflected wave relates to the direct and critically refracted arrivals (Fig. 3.30). a) *Long distance from the source.* As X becomes very large, the T-intercept time (t_0) becomes insignificant, so that:

$$T_f = \sqrt{t_0^2 + \frac{X^2}{V_1^2}} \approx \frac{X}{V_1} = T_d$$

The travel-time curve (T_f) for reflections recorded at large distances is therefore approximately the same as for the direct wave through the upper layer (T_d); the reflected wave is *asymptotic* to the direct wave. b) *Critical distance.* At the critical distance (X_c), the reflected and critically refracted waves have the same arrival time; the straight line for the refracted wave is *tangent* to the hyperbola of the reflection.

Summary of Raypaths, Travel-Time Curves, and Equations For a horizontal, two-layer model (Fig. 3.20), arrivals of direct, critically refracted, and reflected waves show simple geometric relationships when plotted on a travel-time graph

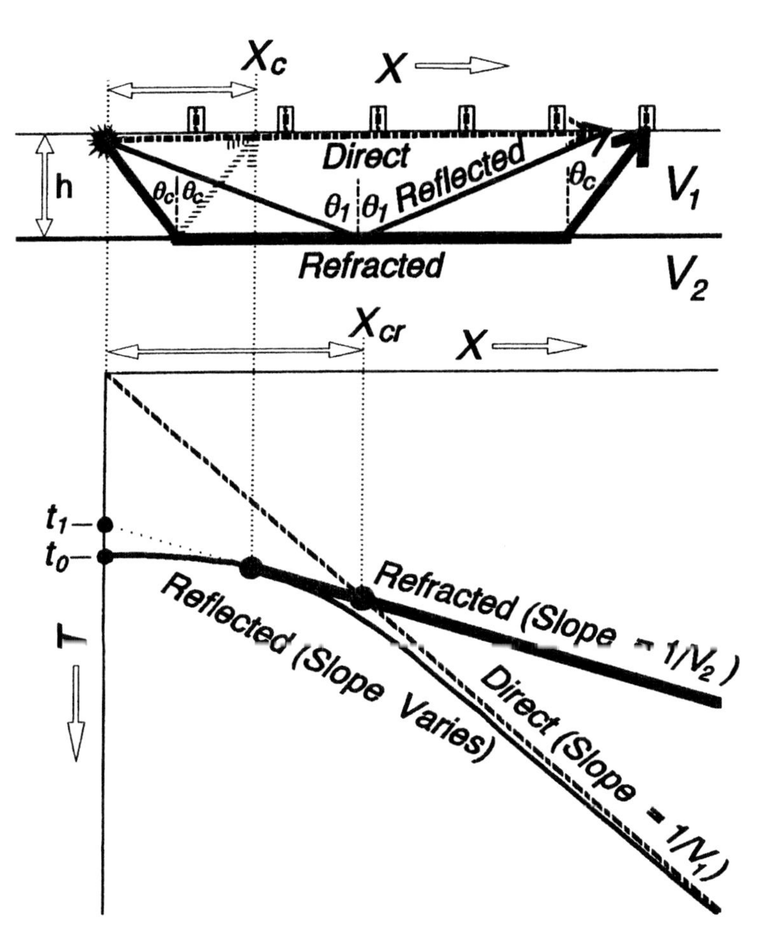

FIGURE 3.31 Selected raypaths and travel-time graph of direct, critically refracted, and reflected waves for a horizontal interface separating a higher velocity (V_2) layer from a lower velocity (V_1) surficial layer. $\mathbf{X_c}$ is the *critical distance* (closest distance from the source where the critically refracted wave is observed) and $\mathbf{X_{cr}}$ the *crossover distance* (beyond that distance the critically refracted wave arrives before the direct wave).

(Fig. 3.31). Equations used to compute travel times for the three arrivals are summarized below.

Direct Wave:

$$\mathbf{T_d = \frac{X}{V_1}}$$

T_d = travel time of direct wave
X = horizontal distance from source to receiver
V_1 = seismic velocity of upper layer.

Critically Refracted Wave:

$$\mathbf{T_r = t_1 + \frac{X}{V_2}}$$

T_r = travel time down to, along, and up from refracting interface
V_2 = seismic velocity of layer below refracting interface.
t_1 = T-axis intercept:

$$t_1 = \frac{2h\cos\theta_c}{V_1}$$

h = thickness of surficial layer.

θ_c = critical angle:

$$\theta_c = sin^{-1}\left(\frac{V_1}{V_2}\right)$$

Reflected Wave:

$$\mathbf{T_f = \sqrt{t_0^2 + \frac{X^2}{V_1^2}}}$$

T_f = travel time along V-shaped path, down to and back up from reflecting interface

t_0 = T-axis intercept (time to go vertically down to the interface and straight back up to the shot location):

$$t_0 = \frac{2h}{V_1}$$

True Velocity and Apparent Velocity

True velocity refers to the actual speed at which seismic energy (that is, a seismic wave) travels through a material. The distance (ΔY) traversed along a straight line, divided by the elapsed time (ΔT), gives the true velocity (V_t) of the material (Fig. 3.32):

$$\mathbf{V_t = \frac{\Delta Y}{\Delta T}}$$

A seismic wave emerging at Earth's surface, however, appears to travel with a certain velocity across a horizontal array of receivers (Fig. 3.33). The *apparent velocity* (V_{ap}) is the distance (ΔX) between two closely spaced receivers, divided by the difference in travel time (ΔT) to the two receivers:

$$\mathbf{V_{ap} = \frac{\Delta X}{\Delta T}}$$

A direct arrival from a surface source has vertical wave fronts that travel horizontally (Fig. 3.34), striking geophones at a velocity equal to the shallow velocity (V_1). Wave fronts from a critically refracted wave emerge from the ground at an angle, passing geophones at a rate that appears faster than V_1. The apparent velocity for such a critically refracted wave is:

$$V_{ap} = \frac{\Delta X}{\Delta T} = \frac{\Delta X}{\Delta Y / V_1}$$

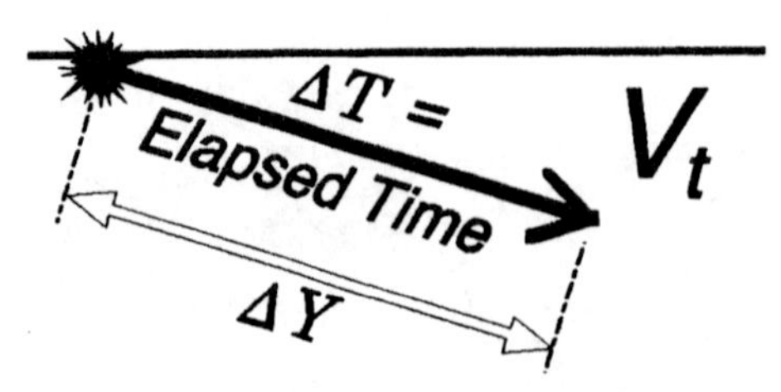

FIGURE 3.32 The *true velocity* (V_t) depends on the time (ΔT) it takes the seismic energy to travel a distance (ΔY).

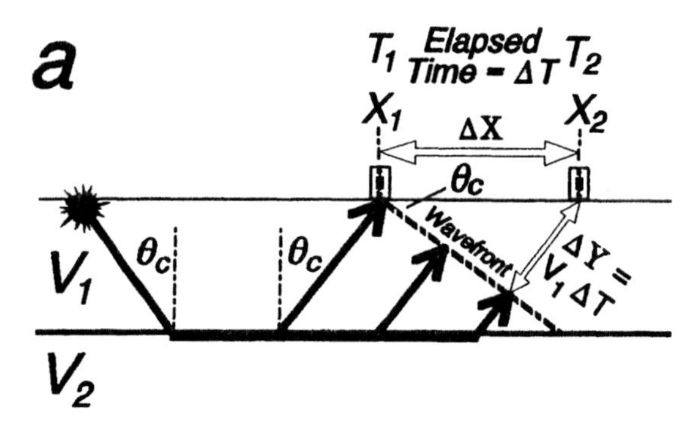

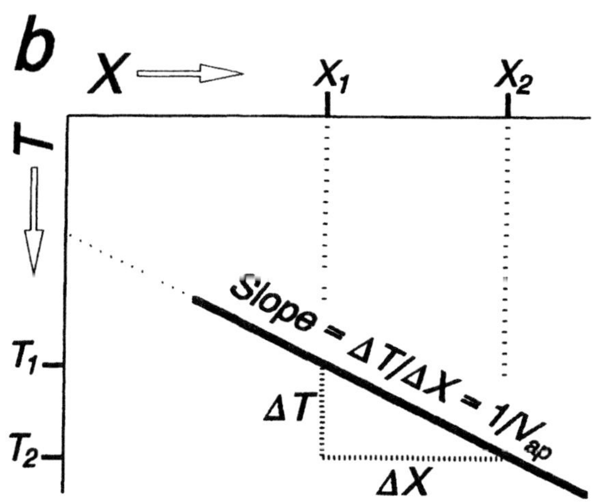

FIGURE 3.33 *Apparent velocity.* a) Selected raypaths and wavefront for an emerging, critically refracted wave. A time ΔT elapses as the wave front strikes the receiver at X_1 then at X_2. b) Travel-time graph for critically refracted arrival. The slope is the inverse of the velocity (V_{ap}) that the wave appeared to travel along the surface from X_1 to X_2.

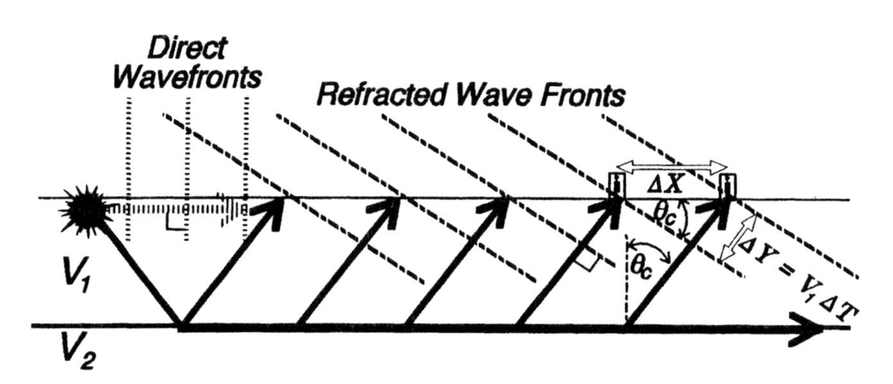

FIGURE 3.34 Refracted arrivals have higher apparent velocities than direct arrivals. Direct wave fronts are vertical; the time it takes a wave front to travel from one receiver to the next is a function of the true velocity of the material (V_1). Refracted wave fronts strike the surface at an angle equal to the critical angle (θ_c); the apparent velocity is $V_1/\sin\theta_c = V_2$.

$$= V_1 \frac{\Delta X}{\Delta Y}$$

But: $\sin\theta_c = \Delta Y/\Delta X$, meaning $\Delta Y = \Delta X \sin\theta_c$

so that:

$$V_{ap} = V_1 \frac{\Delta X}{\Delta X \sin\theta_c}$$

$$= \frac{V_1}{\sin\theta_c}$$

From Snell's Law: $\sin\theta_c = V_1/V_2$

so that:

$$V_{ap} = \frac{V_1}{V_1/V_2} = V_2$$

$$\boxed{\mathbf{V_{ap} = V_2}}$$

The apparent velocity (V_{ap}) measured at the surface for the critically refracted wave is therefore exactly the same as the seismic velocity of the refracting layer (V_2). Notice that the inverse of the slope of the line for the refracted arrival on the travel-time graph (Fig. 3.33b) yields the same apparent velocity:

$$\text{Slope} = \frac{\Delta T}{\Delta X} = \frac{1}{V_{ap}} = \frac{1}{V_2}$$

$$\frac{1}{\text{Slope}} = \frac{1}{(\Delta T/\Delta X)} = V_{ap} = V_2$$

The travel-time equation for the critically refracted wave can be used to demonstrate the same relationship:

$$T_r = t_1 + \frac{X}{V_2}$$

$$\text{Slope} = \frac{dT_r}{dX} = \frac{1}{V_2}$$

$$\frac{1}{\text{Slope}} = \frac{1}{(dT/dX)} = V_2$$

This result means that, for the simple, flat layer case, V_2 can be determined by taking the inverse of the slope ($1/V_{ap}$) for the refracted arrival. Likewise, for other arrivals *the apparent velocity is the inverse of the slope on a travel-time graph.* Three situations are apparent from the raypath diagram and travel-time graph in Fig. 3.35. 1) *A ray emerging almost vertically has high apparent velocity.* The reflected wave that travels straight down and up, arriving at time t_0, thus appears to have infinite velocity. 2) *At the critical distance raypaths for the reflected and critically refracted arrivals emerge at the same angle.* The refracted wave is therefore tangent to the reflected wave on the travel-time curve, signifying the same apparent velocity (V_2). 3) *The apparent velocity of a ray travelling almost horizontally is approximately the same as the true velocity of the near-surface material.* The direct and reflected waves observed at great distance thus have apparent velocity $\approx V_1$.

EXERCISES

3-1 Fig. 3.10 shows a curve plotted through data points from Fig. 3.9, suggesting that compressional wave velocity and density are *directly* proportional. The equation:

$$V_P = \sqrt{\frac{k + 4/3\,\mu}{\rho}}$$

implies that P-wave velocity is *inversely* proportional to density. Explain the paradox.

3-2 Using the information in Fig. 3.9, explain the anomalous positions of V_P and V_S for ice in Fig. 3.10.

3-3 Use equations relating seismic velocities (V_P and V_S) to elastic constants (k and μ) and density (ρ) to discuss the following questions about the Earth.

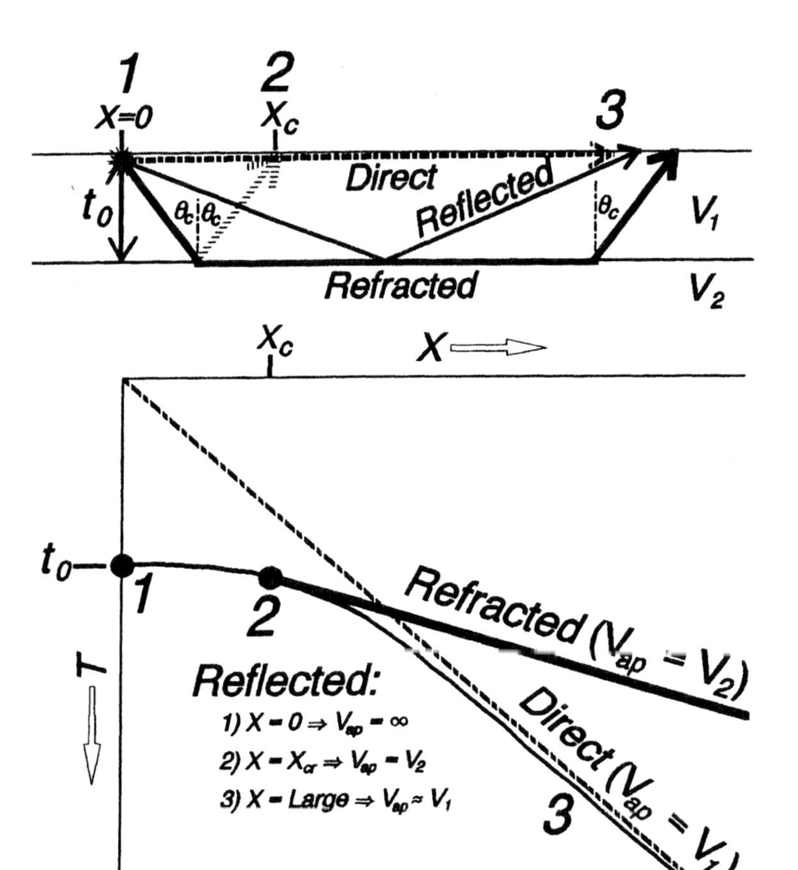

FIGURE 3.35 Selected raypaths and travel-time graph illustrating apparent velocity situations (1, 2, and 3) discussed in text.

a) How does the *crust* differ from the *mantle* in terms of chemical composition and density? Explain why seismic waves travel *faster* through the mantle than through the crust.

b) How does the *lower lithosphere* differ from the *asthenosphere* in terms of chemical composition and density? Explain why seismic waves travel *slower* through the asthenosphere than through the lower lithosphere.

c) How does the *asthenosphere* differ from the *mesosphere* in terms of chemical composition and density? Explain why seismic waves travel *faster* in the mesosphere than in the asthenosphere.

d) How does the *mesosphere* differ from the *outer core* in terms of chemical composition and density? Explain why seismic waves travel *slower* in the outer core than in the mesosphere.

e) How does the *outer core* differ from the *inner core* in terms of chemical composition and density? Explain why S-waves travel in the inner core but not in the outer core. Explain why P-waves travel *faster* in the inner core than in the outer core.

3-4 "THE WAVE" was a phenomenon of the 1980's and early 1990's. Spectators at athletic events stood up and sat down, making "THE WAVE" propagate around the stadium.

a) Comparing the people to particles within the Earth, describe and illustrate the type of seismic wave represented by "THE WAVE."

b) Describe and illustrate the motion the people could make to simulate other types of seismic waves.

3-5 **a)** For the horizontally layered model below, choose an appropriate horizontal scale and plot a *travel-time graph* for the following arrivals: i) direct compressional wave; ii) direct shear wave; iii) direct Rayleigh waves (include schematic portrayal of dispersion range); iv) air wave (compressional "sound" wave from the source that travels through the air to the receivers); v) reflected compressional wave; vi) refracted compressional wave.

b) For the *direct P, refracted P* and *reflected P,* sketch raypaths on the model and compute the *apparent velocity* that would be observed at 0 m, 2500 m, and 25000 m from the source.

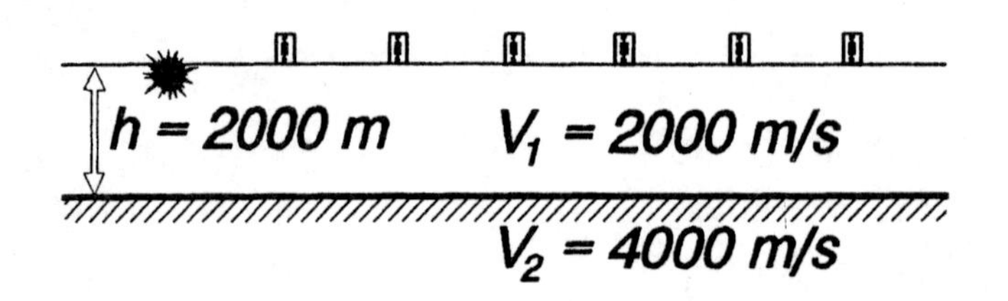

3-6 For the model below illustrate and explain (verbally and mathematically) why critical refraction cannot occur when $V_2 < V_1$.

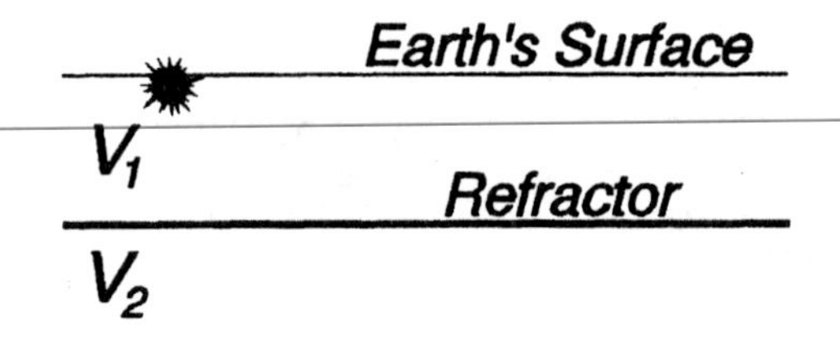

3-7 The sedimentary section off the east coast of the United States is cut by domal structures that could be either *granite intrusions* or *salt diapirs.* Based on the information in Figs. 3.9 to 3.11, explain how you might combine observations from *seismic refraction* and *gravity* data to a) find the domal structures; and b) distinguish between the two types of domes. In addition to drawing cross sections and graphs, be specific about the *physical properties* that each method is able to resolve (see Figs. 1.2 to 1.4).

SELECTED BIBLIOGRAPHY

Bolt, B. A., *Earthquakes,* 1988, New York: W. H. Freeman and Comp., 282 pp.

Griffiths, D. H., and R. F. King, 1981, *Applied Geophysics for Geologists and Engineers: The Elements of Geophysical Prospecting* (2nd ed.), New York: Pergamon Press, 230 pp.

Kinsler, L. E., A. R. Frey, A. B. Coppens, and J. V. Sanders, 1982, *Fundamentals of Acoustics* (3rd ed.), New York: John Wiley and Sons, Inc., 480 pp.

Lay, T., and T. C. Wallace, 1995, *Modern Global Seismology,* International Geophysics Series Number 58, New York: Academic Press, 521 pp.

Ludwig, W. J., J. E. Nafe, and C. L. Drake, 1971, Seismic refraction, In: *The Sea* (v. 4, part I), edited by A. E. Maxwell, New York: John Wiley and Sons, Inc., pp. 53–84.

Ramsey, J. G., and M. I. Huber, 1983, *The Techniques of Modern Structural Geology. Volume 1: Strain Analysis,* New York: Academic Press, 307 pp.

Suppe, J., *Principles of Structural Geology,* 1985, Englewood Cliffs, N. J.: Prentice Hall, Inc., 537 pp.

Telford, W. M., L. P. Geldart, R. E. Sheriff, and D. A. Keys, 1976, *Applied Geophysics,* Cambridge: Cambridge University Press, 1976, 860 pp.

Twiss, R. J., and E. M. Moores, 1992, *Structural Geology,* New York: W. H. Freeman and Comp., 532 pp.

C H A P T E R 4

Seismic Refraction Interpretation

seismic *(sīz mik′)* ***adj.,*** [< *Gr.* seismos, *an earthquake* < seiein, *to shake*], *relating to an earthquake or artificial shaking of the Earth.*
refraction *(ri frak′ shən)* ***n.,*** [< *L.* refractus, *pp. of* refringere < *re-, back,* + frangere, *to break*], *the bending of a ray or wave of light, heat, or sound as it passes obliquely from one medium to another.*
interpret *(in tur′ prit)* ***vt.,*** [< *MFr.* < *L.* interpretari < interpres, *negotiator*], *to explain the meaning of; to give one's own understanding of.*
seismic refraction interpretation *(sīz mik′ ri frak′ shən in tur′ prə tā′ shən)* ***n.,*** *an explanation of subsurface conditions that led to the bending of seismic waves in such a way that the waves were observed at Earth's surface*

The refraction of seismic waves provides constraints on the composition and structure of certain parts of the Earth. Waves refract because they encounter changes in seismic velocity. Velocity changes relate to changes in bulk modulus, rigidity and density; measuring how waves refract thus tells us something about those properties in regions traversed by the seismic waves.

The refraction method is most useful where there is an abrupt increase in velocity with depth, because critically refracted P-waves eventually arrive ahead of other waves (Fig. 3.27). Fig. 4.1 illustrates two problems addressed effectively: 1) *Crustal thickness.* Seismic waves bend abruptly when refracted from the crust into the underlying, higher velocity mantle. It was the observation of these critically refracted waves that led Mohorovičić to the discovery of the crust/mantle boundary (or "Moho") in the early part of the 20th century. 2) *Depth to Bedrock.* Hard rock underlying loose material generates a critical refraction; the "bedrock" depth is thereby mapped for engineering, groundwater, or other applications. The method is also effective at resolving seismic velocity of material below refracting interfaces; internal crustal velocities and the velocity of the uppermost mantle may thus be mapped.

Seismic refraction experiments must be long enough so that critically refracted events appear before other events on travel-time graphs (Fig. 4.2). As a "first arrival," the slope (and hence apparent velocity) of the event can be readily determined; as a secondary arrival, the event may be masked by interference from direct, reflected, or other refracted events. Receivers therefore must extend well beyond the crossover distance for the deepest refractor of interest. A general rule is that the length of the array of receivers ("spread length") should be at least twice the crossover distance. Unlike reflection experiments, where the spread length is about equal to the depth of the deepest reflector, seismic refraction spread lengths are about five to ten times the depth of the deepest refractor.

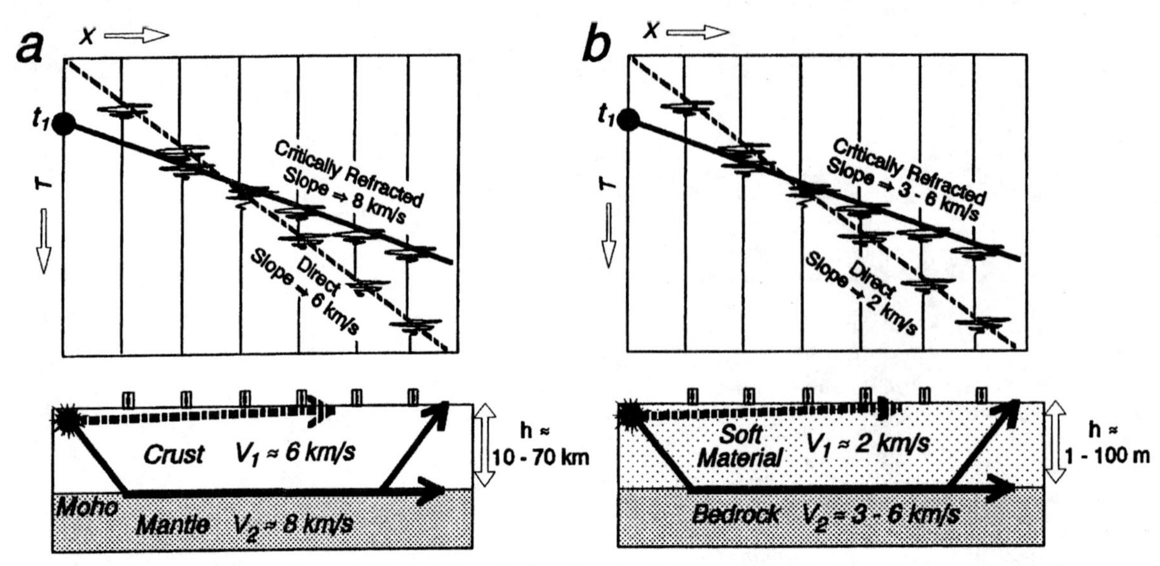

FIGURE 4.1 *Two problems addressed effectively by seismic refraction.* a) Critical refraction from the top of the mantle facilitates mapping of crustal thickness. b) The strong contrast in velocity between soft (unconsolidated or weathered) material and firm bedrock results in critical refraction. Note that travel time (T) is commonly plotted downward in seismic refraction and reflection studies, because travel time relates to depth within the Earth.

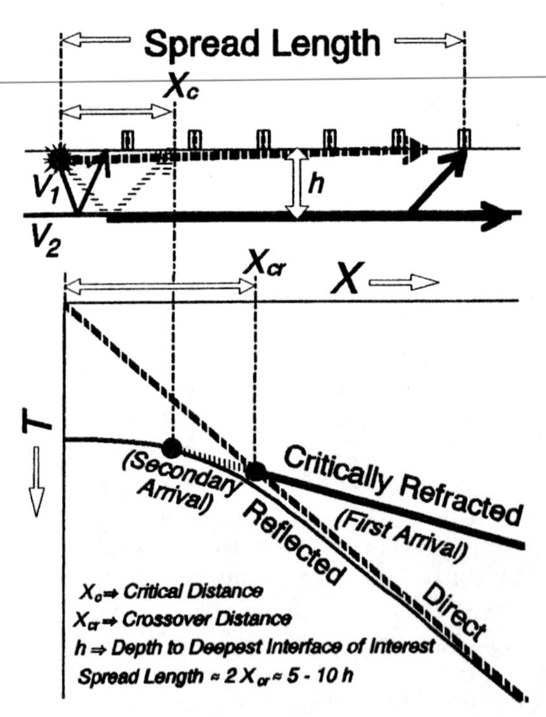

FIGURE 4.2 In order to see a critical refraction clearly as a first arrival, the spread length should be at about twice the crossover distance.

INTERPRETATION MODELS

The theory for the refraction of seismic waves is developed in Chapter 3. Interpretation of seismic refraction data involves the identification of linear arrivals on travel-time plots (Fig. 4.3). The slopes of lines drawn through arrivals relate to velocities encountered by direct and critically refracted waves. For layers with constant velocity separated by horizontal interfaces, apparent velocities determined from slope inverses correspond exactly to true layer velocities (Burger, 1992, pp. 72–74). Where interfaces dip, true layer velocities do not correspond exactly to apparent velocities; a more elaborate analysis is necessary (see, for example, pp. 281–284 of Telford et al., 1976; pp. 433–439 of Sheriff and Geldart, 1995; pp. 124–126 of Fowler, 1990).

The models below show travel-time graphs that would result from a single horizontal interface, several horizontal interfaces, and a single dipping interface. Equations developed from the models can be used in two ways (see Chapter 1): 1) *forward modeling* gives anticipated travel-time graphs; and 2) *inversion* yields velocities, thicknesses, and dips of interfaces from actual refraction profiles.

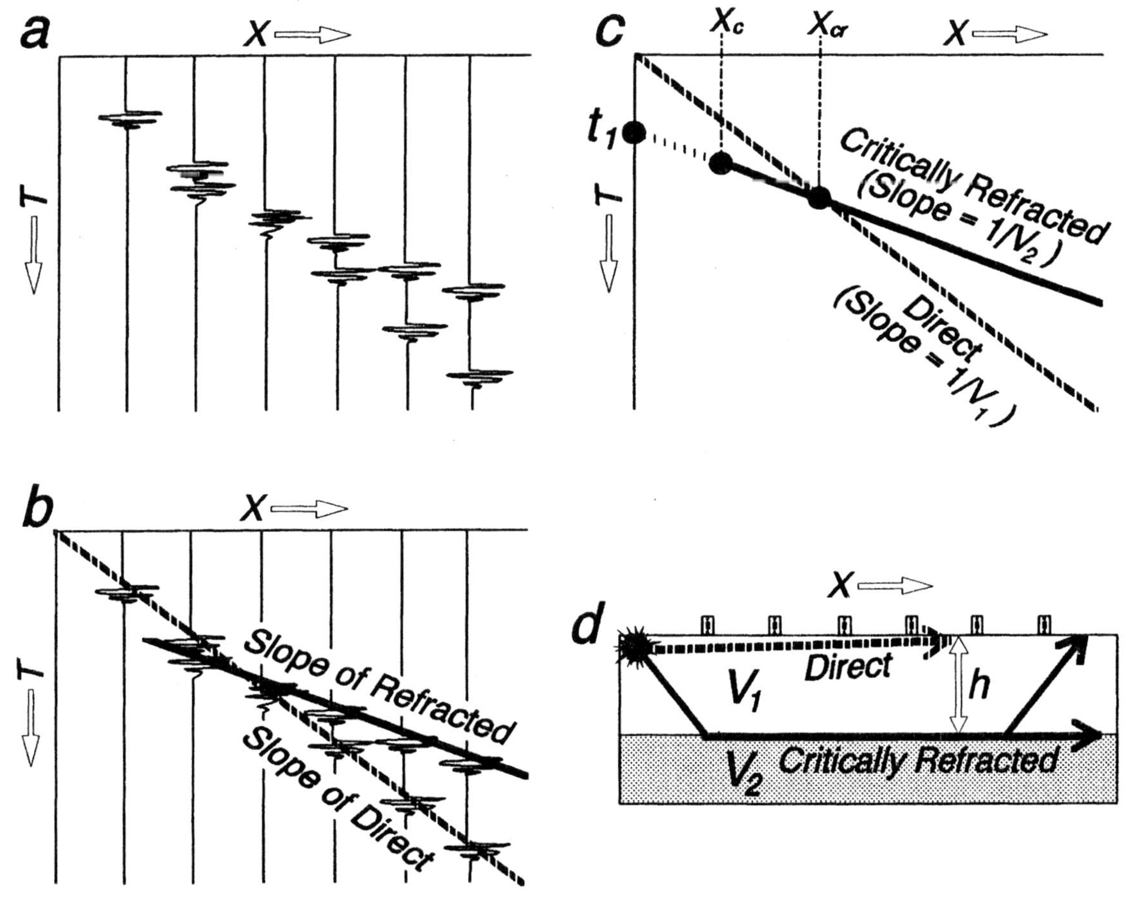

FIGURE 4.3 *Refraction interpretation from travel-time graph.* a) Seismic traces showing events on uninterpreted record. b) Straight lines drawn through events define direct and critically refracted arrivals. c) Events plotted on travel-time graph, with the T-axis intercept (t_1), slopes for direct and critically refracted arrivals, critical distance (X_c), and crossover distance (X_{cr}) identified. d) Simple horizontal interface model used to interpret arrivals.

Single Horizontal Interface

The theory behind a critical refraction from a single, horizontal interface was developed in Chapter 3. The model (Fig. 4.3d) involves an interface at depth (h), separating a lower velocity (V_1) from a higher velocity (V_2). The travel time (T) to a receiver at horizontal distance (X) from the source is:

$$\mathbf{T} = \mathbf{t_1} + \frac{\mathbf{X}}{\mathbf{V_2}}$$

where:

$$t_1 = \text{T-axis intercept} = \frac{2h \cos\theta_c}{V_1}$$

$$\theta_c = \text{critical angle} = \sin^{-1}\left(\frac{V_1}{V_2}\right)$$

$$X_c = \text{critical distance} = 2h \tan\theta_c$$

$$X_{cr} = \text{crossover distance} = 2h\sqrt{\frac{V_2 + V_1}{V_2 - V_1}}$$

The above equations can be *forward modeling equations*; when applied to a hypothetical model (Fig. 4.3d), they yield a predicted travel-time graph (Fig. 4.3c).

Inversion, on the other hand, can be used to interpret the velocity structure from an observed refraction profile (Fig. 4.3a). The intercept time (t_1) and the slopes of the direct and refracted arrivals are read directly from the travel-time plot (Fig. 4.3b,c). The observed slopes and intercept time can then be solved for the true velocities (V_1, V_2) and the depth to the interface (h), using the following *inversion equations*:

$$\text{Slope of Direct} = \frac{1}{V_1} \Rightarrow \mathbf{V_1} = \frac{\mathbf{1}}{\textbf{slope of direct}}$$

$$\text{Slope of Refracted} = \frac{1}{V_2} \Rightarrow \mathbf{V_2} = \frac{\mathbf{1}}{\textbf{slope of refracted}}$$

$$\theta_c = \sin^{-1}\left(\frac{V_1}{V_2}\right)$$

$$t_1 = \frac{2h\cos\theta_c}{V_1} \Rightarrow \mathbf{h} = \frac{\mathbf{t_1 V_1}}{\mathbf{2\cos\theta_c}}$$

Fig. 4.3d, in this case, represents the inversion model that results from the observed refraction profile (Fig. 4.3a).

Crustal Thickness The single-layer case illustrates the utility of the seismic refraction method to map changes in crustal thickness (Fig. 4.4). The T-axis intercept (t_1) can be thought of as a *"delay time"*; the critically refracted arrival from a deep Moho is delayed (Fig. 4.4c), compared to the arrival where the Moho is shallow (Fig. 4.4b). Thus, relative to areas of normal-thickness continental crust, t_1 will be large where continental crust is thick (collisional mountain ranges). Where the

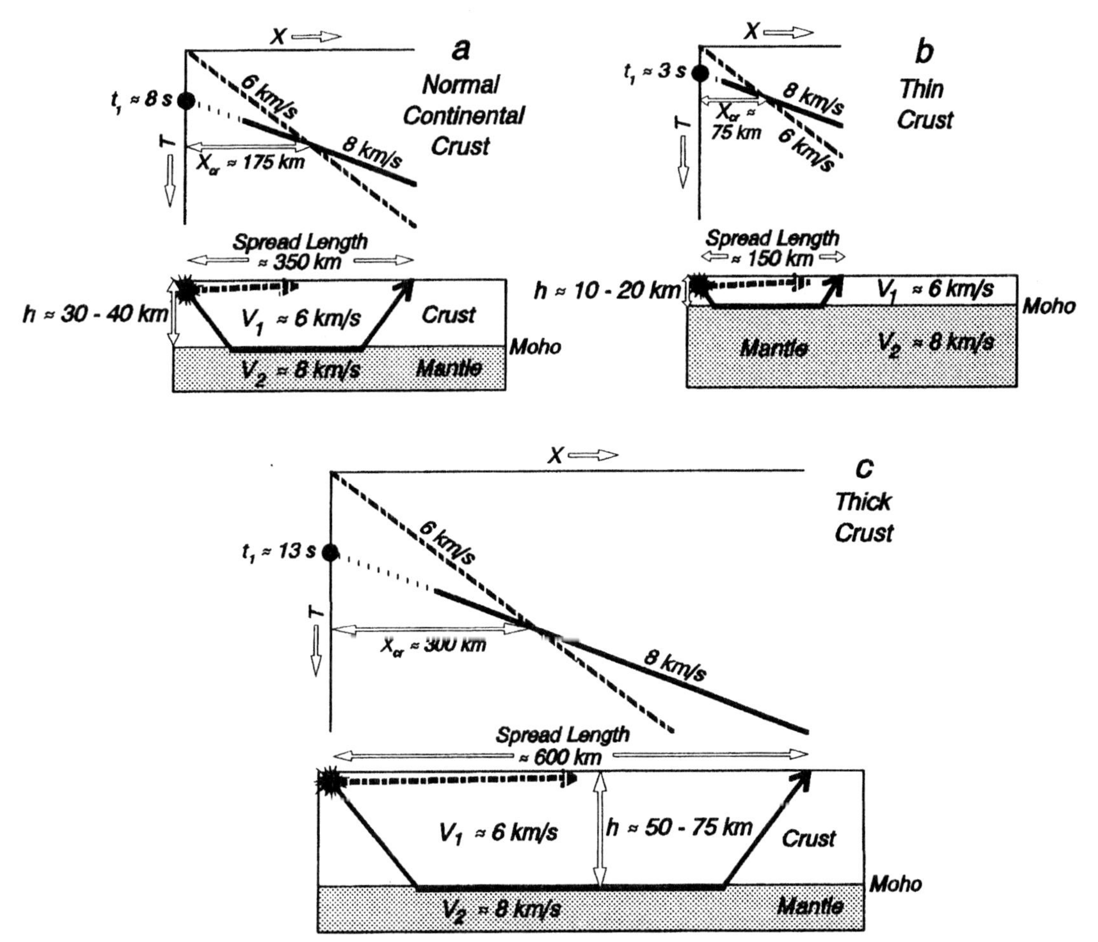

FIGURE 4.4 Comparisons of intercept times (t_1) and crossover distances (X_{cr}) for different crustal thickness. The grossly simplified models illustrate the approximate spread lengths ($2X_{cr}$) necessary to resolve the depth to Moho (h). The travel-time graphs were determined using forward modeling equations presented in text. Inversion equations can be used to interpret crustal thickness if the T-axis intercept (t_1) and apparent velocities are read from observed refraction profiles. a) The distance from the source to the farthest receiver must be about 350 km to resolve the crustal thickness in regions of typical continental crust; the T-axis intercept is about 8 s. b) Oceans and regions of very thin continental crust require about 150 km spread lengths, where a shorter T-intercept of about 3 s might be expected. c) Very deep Moho beneath some mountain ranges necessitates very long spread lengths (≈ 600 km), and results in large T-axis intercept times (≈ 13 s).

crust is thin (continental rift zones; passive continental margins; ocean basins), t_1 is small. Similarly, the crossover distance (X_{cr}) is large where the crust is thick and small for thin crust. In order to resolve crustal thickness, refraction spread lengths must be very long (≈ 300–600 km) in continental areas, much shorter (≈ 100–150 km) in the oceans.

Depth to Bedrock Hard bedrock beneath weathered material or loose sedimentary deposits results in critical refraction. Refraction surveying is therefore an effective tool for mapping changes in depth to bedrock in engineering projects (Fig. 4.5). The major difference between crustal scale (Fig. 4.4) and bedrock surveying (Fig. 4.5) is that the depth to the refractor is much shallower (≈ 1–100 m deep) for the latter. The receiver arrays only need to extend 10s to 100s of *meters* from the source, compared to 10s to 100s of *kilometers* for crustal surveys.

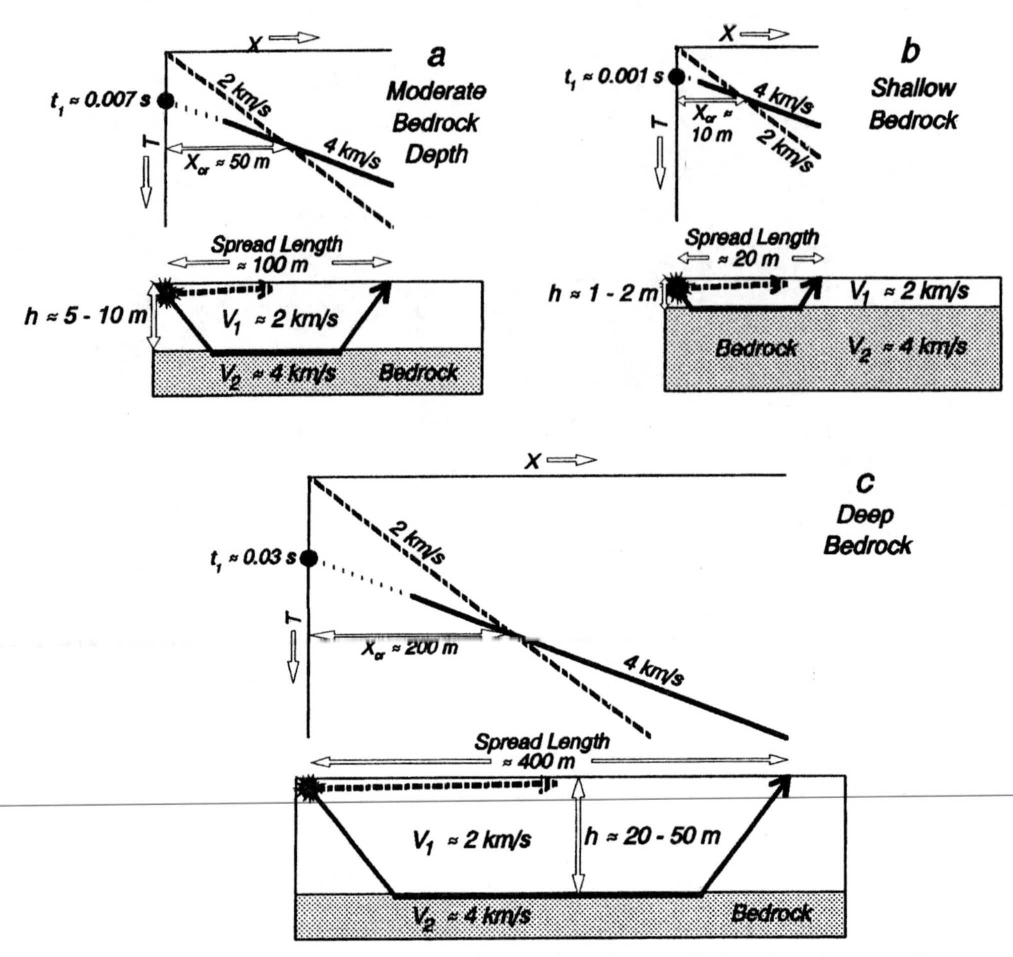

FIGURE 4.5 Approximate T-axis intercept times (t_1), crossover distances (X_{cr}), and required spreadlengths for bedrock depths (h) that are (a) moderate; (b) shallow; and (c) deep. Travel-time graphs were determined using forward modeling equations presented in text.

Several Horizontal Interfaces

Consider the case of several horizontal interfaces, where the velocity of a deeper layer is always greater than that of the layer above ($V_4 > V_3 > V_2 > V_1$ in Fig. 4.6). A ray bends more and more toward the horizontal as it crosses deeper and deeper interfaces, traveling horizontally along the critically refracting layer (Fig. 4.6a). Rays refracted along a higher-velocity interface at depth emerge at higher angles than those from shallow, lower-velocity refractors; the apparent velocities corresponding to deep refractors are thus higher than those associated with shallow interfaces.

Each critically refracted ray bends across successive interfaces, according to Snell's Law (Fig. 4.6a):

first critical refraction:

$$\sin\theta_{1,2} = V_1/V_2$$

second critical refraction:

$$\sin\theta_{1,3} = V_1/V_3$$

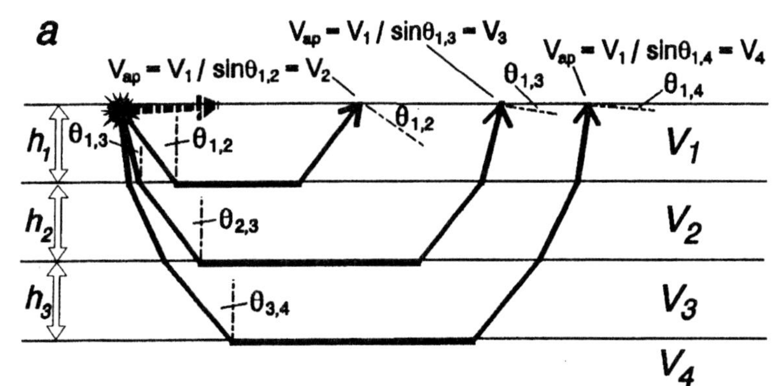

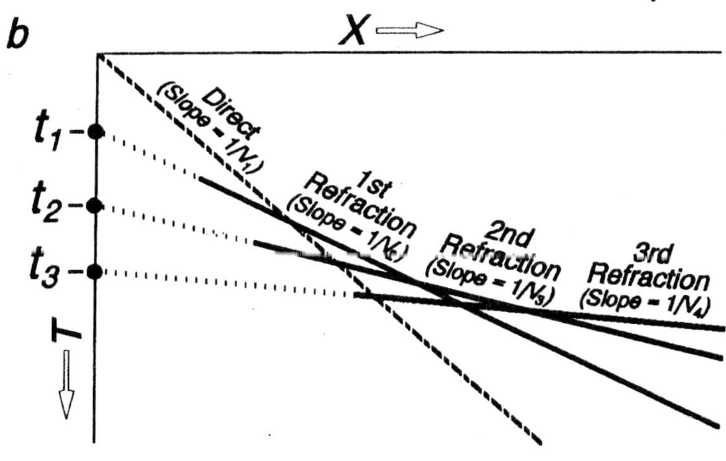

FIGURE 4.6 *Model for critical refraction from three horizontal interfaces.* a) Increasing velocity with depth ($V_4 > V_3 > V_2 > V_1$) results in a critical refraction from each of the interfaces. Each refracted ray emerges at an angle indicating the velocity of the corresponding refracting layer. b) Travel-time graph resulting from (a). The inverses of the slopes for the direct and refracted arrivals indicate the four layer velocities.

$$\sin\theta_{2,3} = V_2/V_3$$

third critical refraction:

$$\sin\theta_{1,4} = V_1/V_4$$

$$\sin\theta_{2,4} = V_2/V_4$$

$$\sin\theta_{3,4} = V_3/V_4$$

where $\theta_{1,2}$, $\theta_{2,3}$, and $\theta_{3,4}$ are the critical angles for the second, third, and fourth layers, respectively.

Apparent velocities recorded along Earth's surface (Fig. 4.6a) are inverses of slopes observed on the travel-time graph (Fig. 4.6b):

$$V_{ap2} = \frac{V_1}{\sin\theta_{1,2}} = V_2$$

$$V_{ap3} = \frac{V_1}{\sin\theta_{1,3}} = V_3$$

$$V_{ap4} = \frac{V_1}{\sin\theta_{1,4}} = V_4$$

where V_{ap2}, V_{ap3}, and V_{ap4} are the apparent velocities for refractions from the second, third, and fourth layers, respectively.

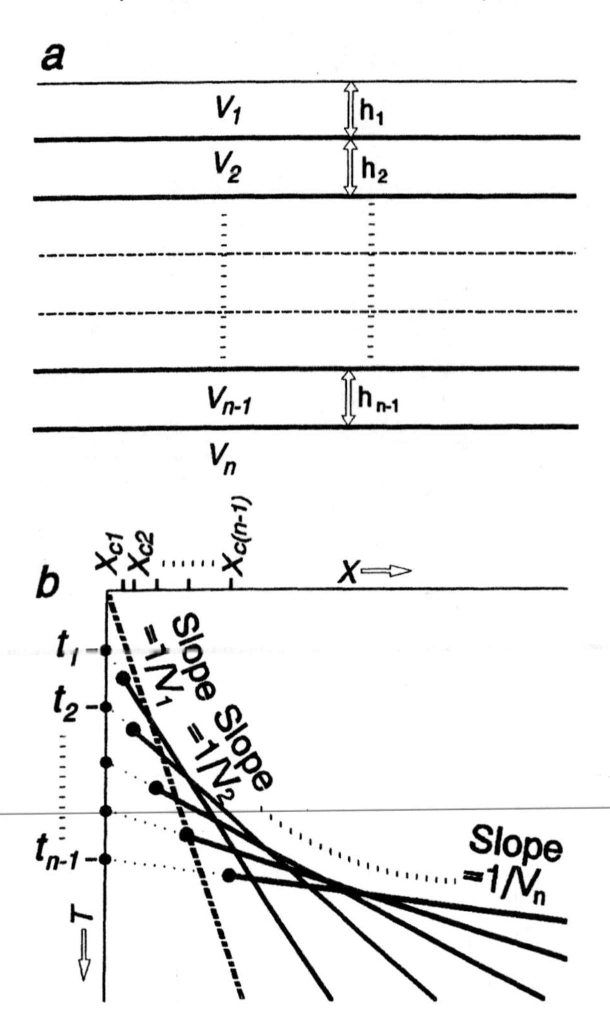

FIGURE 4.7 a) Generalized refraction model for several horizontal interfaces. There are n layers; layer velocity increases with depth ($V_n > V_{n-1} \ldots V_2 > V_1$). b) Travel-time graph showing T-axis intercept times ($t_1, t_2, \ldots t_{n-1}$) and slopes for direct and critically refracted arrivals from (a). See text for travel-time equations for each event.

For a model with several horizontal interfaces (Fig. 4.7), the generalized travel-time equation for a critical refraction from the top of layer "n" is (Telford et al., 1976, p. 280–281; Burger, 1992, p. 79–80):

$$\boxed{\mathbf{T_n = t_{n-1} + \frac{X}{V_n}}}$$

where:

T_n = travel time down from the source, horizontally along the top of layer n, and back up to a receiver at X

t_{n-1} = T-axis intercept for the refraction from layer n:

$$t_{n-1} = \sum_{i=1}^{n} \frac{2h_i \cos\theta_{i,n}}{V_i}$$

X = horizontal distance from the source to the receiver

V_n = velocity of layer n.

An example for a *4-layer case* (Fig. 4.6) is given below.

n = 1 ⇒ Direct Arrival

travel time: $T_1 = \frac{X}{V_1}$

T-Axis intercept: 0

critical distance: 0

n = 2 ⇒ 1st Refraction

travel time: $T_2 = t_1 + \frac{X}{V_2}$

T-Axis intercept: $t_1 = \frac{2h_1 \cos\theta_{1,2}}{V_1}$

critical distance: $X_{c1} = 2h_1 \tan\theta_{1,2}$

n = 3 ⇒ 2nd Refraction

travel time: $T_3 = t_2 + \frac{X}{V_3}$

T-Axis intercept: $t_2 = \frac{2h_1 \cos\theta_{1,3}}{V_1} + \frac{2h_2 \cos\theta_{2,3}}{V_2}$

critical distance: $X_{c2} = 2h_1 \tan\theta_{1,3} + 2h_2 \tan\theta_{2,3}$

n = 4 ⇒ 3rd Refraction

travel time: $T_4 = t_3 + \frac{X}{V_4}$

T-Axis intercept: $t_3 = \frac{2h_1 \cos\theta_{1,4}}{V_1} + \frac{2h_2 \cos\theta_{2,4}}{V_2} + \frac{2h_3 \cos\theta_{3,4}}{V_3}$

critical distance: $X_{c3} = 2h_1 \tan\theta_{1,4} + 2h_2 \tan\theta_{2,4} + 2h_3 \tan\theta_{3,4}$

The model (Fig. 4.7a) can be applied to any number of horizontal layers, as long as *velocity does not decrease with depth.* Inversion of observed data (travel-time graph in Fig. 4.7b) can be accomplished by: 1) reading the velocities (V_1, V_2, ..., V_n) as the inverses of the slopes of the direct and refracted arrivals; 2) reading the T-intercepts (t_1, t_2, ..., t_{n-1}); then 3) solving for the thicknesses (h_1, h_2, ..., h_{n-1}) from the above equations.

Refraction Surveys of Oceanic Crust The character of a typical ophiolite sequence (Fig. 2.24) was first recognized through seismic refraction observations in ocean basins (for example, Fig. 4.8). Based on inversion of many refraction profiles, a consistent pattern of P-wave velocity and thickness has emerged (Fig. 4.9a):

Velocity:

direct arrival in water ≈ 1.5 km/s
refraction from Layer 1 ≈ 2 km/s
refraction from Layer 2 ≈ 5.07 ± 0.63 km/s

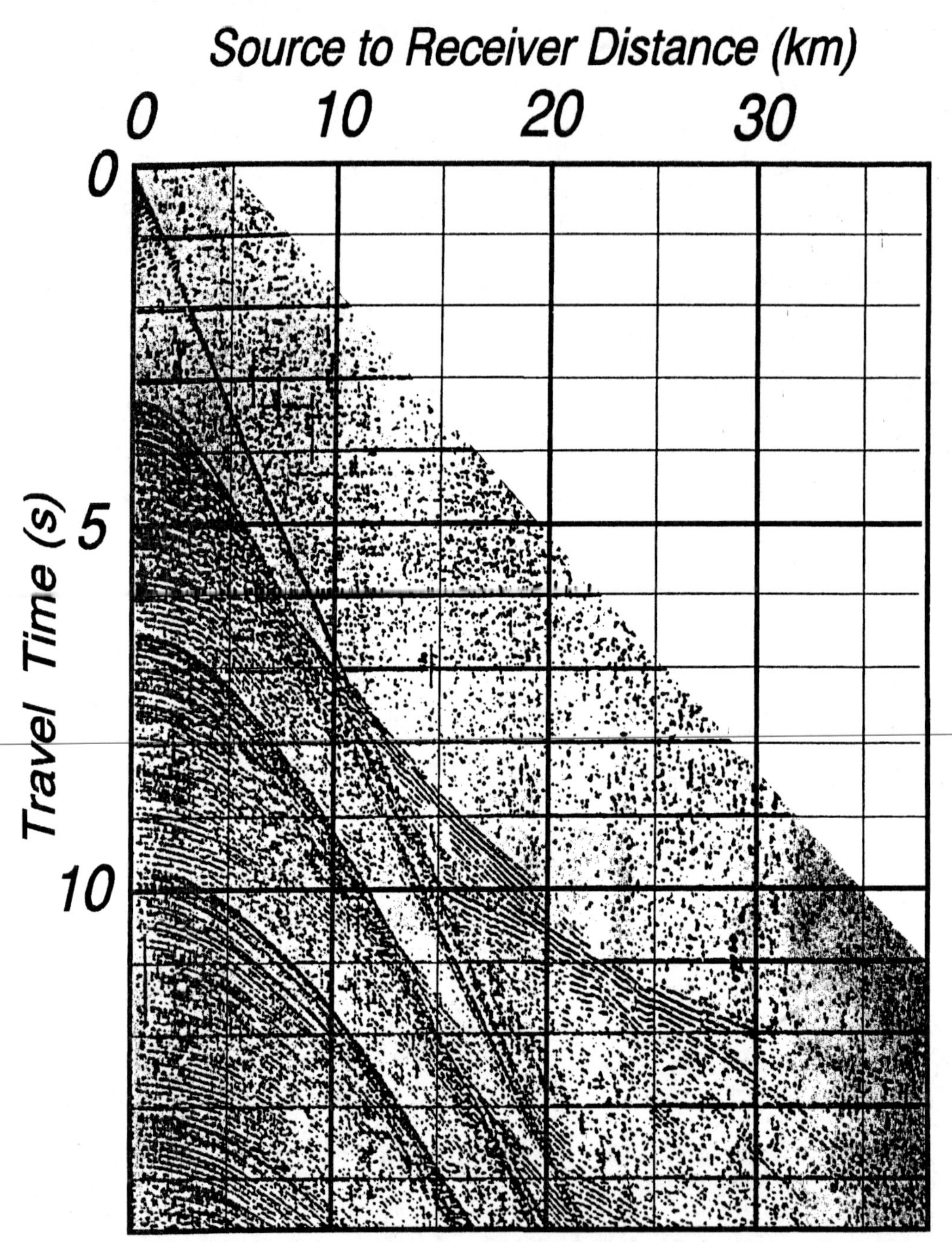

FIGURE 4.8 Seismic refraction profile recorded in an ocean basin. Modified from *Exploration Seismology*, 2nd ed., by R. E. Sheriff and L. P. Geldart, © 1995. A sonobuoy was tossed into the water; the ship then periodically fired an air gun as it moved away (to the right). The horizontal scale gives the distance from the source (ship) to the receiver (sonobuoy). The vertical scale represents the elapsed time between firing the air gun and arrival of events at the sonobuoy. The linear event extending through the origin is the direct arrival. Linear events observed at greater distances on the record, some as first arrivals, are refractions from within the crust and from the Moho. Hyperbolic events are reflections from the bottom of the water (see Chapters 5 and 6).

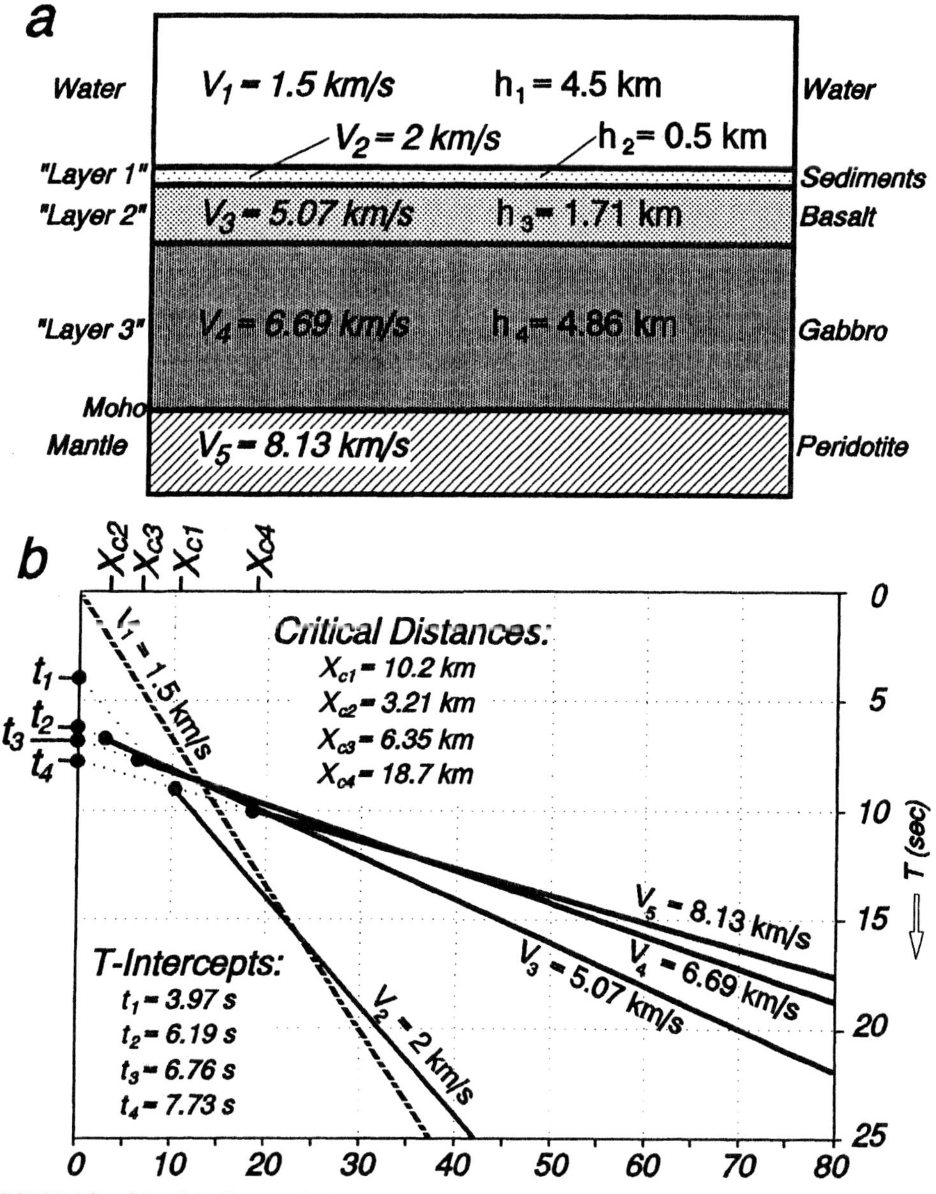

FIGURE 4.9 a) Model of layer velocities and thicknesses of an ophiolite, determined by averaging results of seismic refraction studies of oceanic crust. Modified from *The Solid Earth: An Introduction to Global Geophysics*, by C. M. R. Fowler, © 1990. b) Travel-time graph computed through forward modeling of the velocities and thicknesses in (a).

refraction from Layer 3 $\approx$ 6.69 $\pm$ 0.26 km/s
refraction from Layer 4 $\approx$ 8.13 $\pm$ 0.24 km/s

Thickness:

water $\approx$ 4.5 km
Layer 1 $\approx$ 0.5 km
Layer 2 $\approx$ 1.71 $\pm$ 0.75 km
Layer 3 $\approx$ 4.86 $\pm$ 1.42 km.

Deep-sea drilling and comparisons with ophiolite sequences exposed on land have led to the following stratigraphy:

water column
Layer 1 = deep-sea sediments
Layer 2 = basalt (volcanics and shallow intrusions)
Layer 3 = gabbro (deeper intrusions)
Layer 4 = peridotite (uppermost mantle; gives Moho refraction).

The general appearance of refraction profiles in areas of oceanic crust thus consists of a direct arrival through the water and refractions from four interfaces (Fig. 4.9b).

Problem Situations There are two common situations where the equations presented above will yield erroneous results: a low-velocity layer and a thin layer.

1. *Low-velocity layer ($V_2 < V_1 < V_3$).* There is *no critical refraction from the layer with velocity V_2* (Fig. 4.10a). It will appear from the travel-time graph that the layer with velocity V_3 lies directly below the layer with V_1 (Fig. 4.10b). The equations will yield no layer with V_2, and the layer with V_1 will be too thick (Fig. 4.10c).
2. *Thin layer ($V_3 > V_2 > V_1$, but with h_2 very small).* The layer with velocity V_2 is so thin (Fig. 4.11a) that *nowhere is the critical refraction observed as a first*

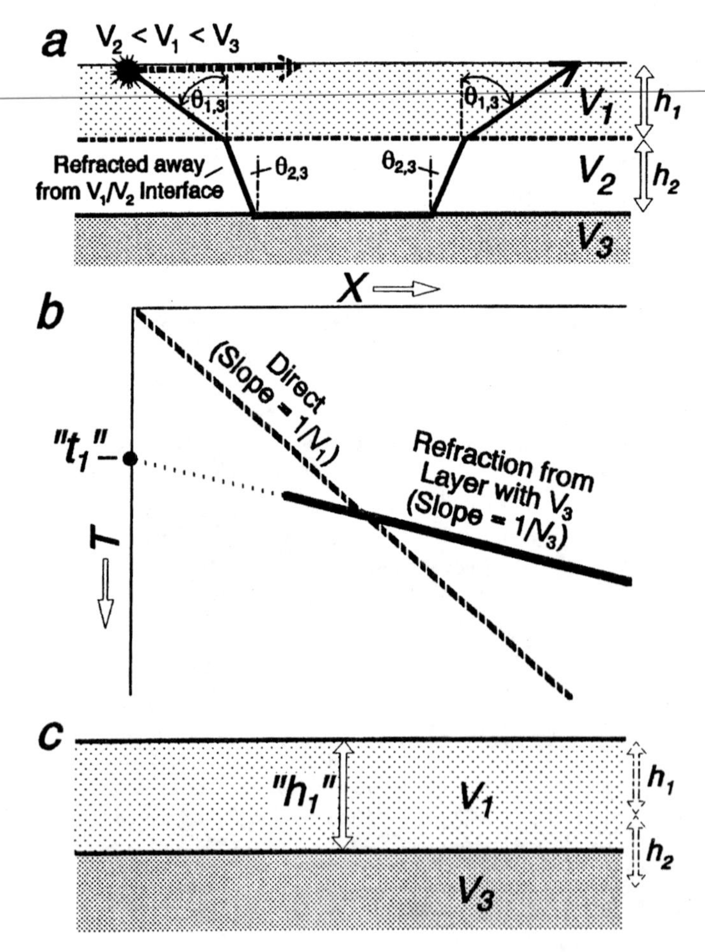

FIGURE 4.10 a) Model of low-velocity (V_2) layer, sandwiched between higher velocities (V_1, V_3). No critical refraction can occur from the low-velocity layer (Fig. 3.23a). b) Travel-time graph for (a). The only critical refraction observed is from the layer with velocity (V_3). c) Model that would result from inversion of velocities (V_1 and V_3) and intercept time ("t_1") determined from (b), using the two-layer equations presented in text. Note that "h_1" is thicker than the upper layer, and that the low-velocity layer is absent.

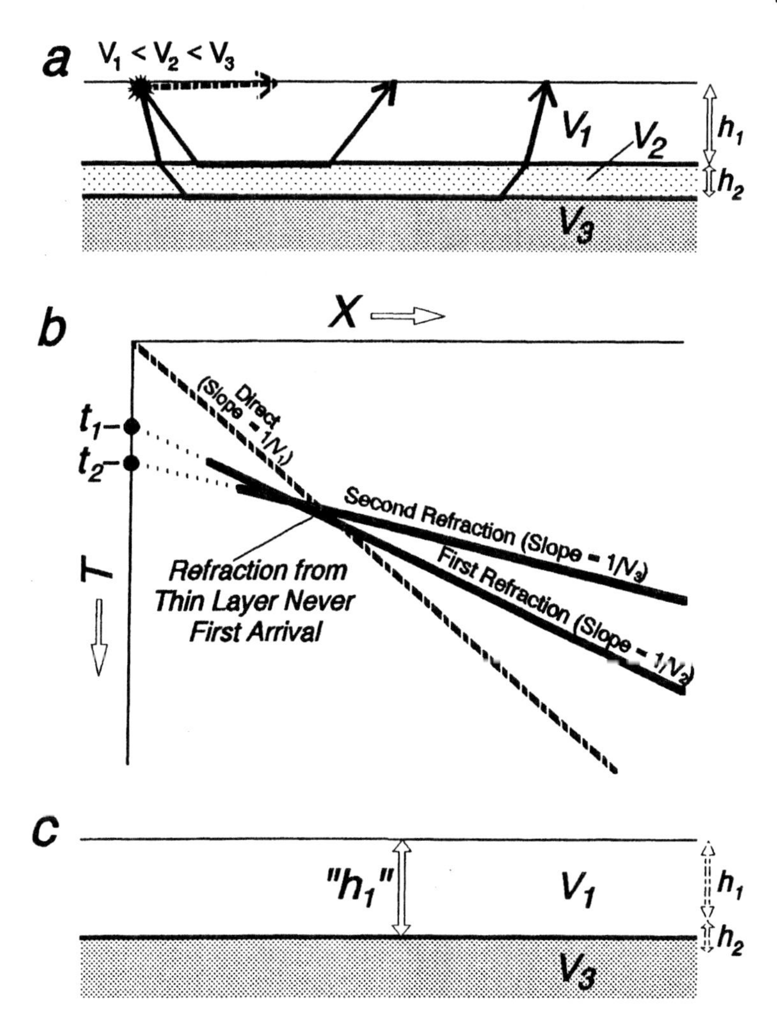

FIGURE 4.11 a) Model showing velocity increase with depth, but with a thin layer (h_2 very small). b) The first refraction, from the top of the thin layer, arrives either after the direct arrival, or after the second refraction (or after both); it may be difficult to identify the first refraction on the record. c) If the first refraction is overlooked, inversion equations yield a model with velocity (V_1) directly over the V_3 material. The thickness ("h_1") computed for the V_1 layer will be too large, and the depth (also "h_1") to the top of the V_3 layer will be too small.

arrival (Fig. 4.11b). The thin layer would have to be recognized through a secondary arrival; if it is not recognized, the equations yield an erroneous depth to the top of the layer with velocity V_3 (Fig. 4.11c). The sediment layer overlying oceanic crust (oceanic "Layer 1") is often so thin that its critical refraction nowhere appears as a first arrival (Figs. 4.8, 4.9).

Reduced Travel-Time Plots In the literature, seismic refraction records are often plotted not with the actual travel time, but with a *reduced travel time*. That is because, when seismic traces are plotted according to actual travel time (Fig. 4.12a), the slopes of refracted arrivals often appear similar; it may be difficult to see and identify the different events.

Identification of arrivals may be improved by replotting the seismic traces on a reduced travel-time graph (Fig. 4.12b). On such a display of the data, an amount of time is subtracted from the actual travel time (T) of each seismic trace according to the distance (X) from the source to the receiver and a reduction velocity (V_{red}):

$$\mathbf{T_{red} = T - \frac{X}{V_{red}}}$$

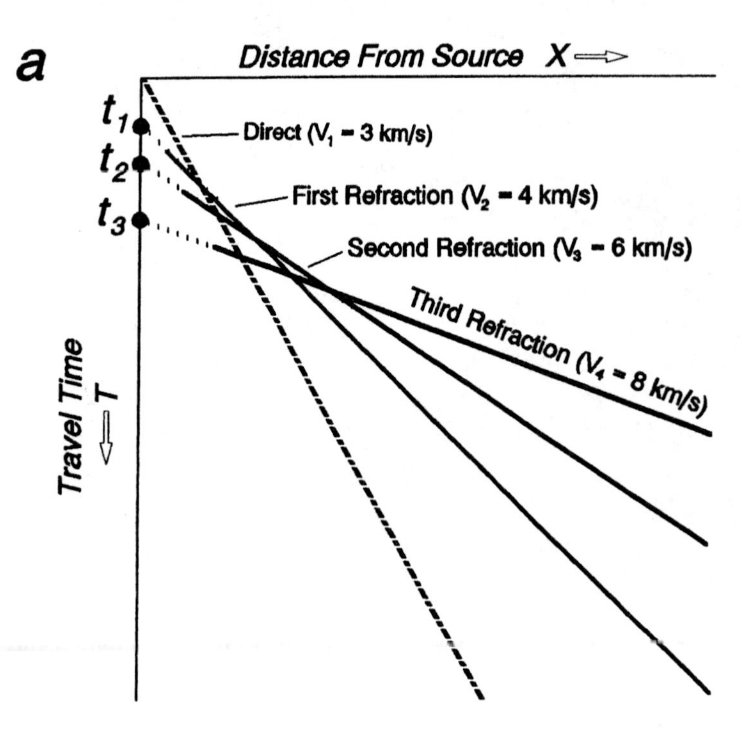

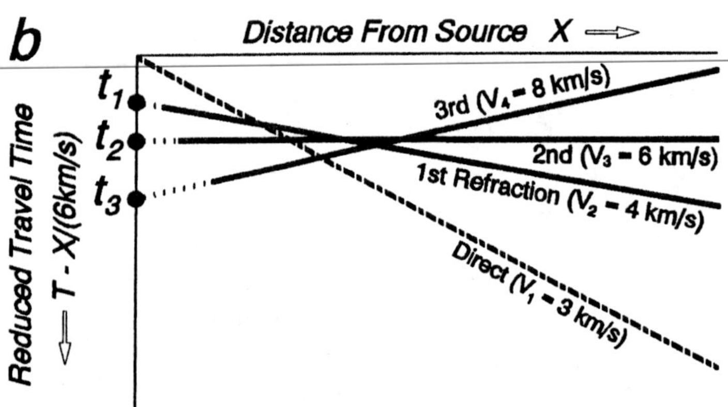

FIGURE 4.12 a) Positions of direct and critically refracted arrivals on seismic refraction profile. b) Same profile as (a), but plotted with a *reduced travel time* of 6 km/s. Arrivals with apparent velocity of 6 km/s are horizontal on the graph. Velocities < 6 km/s have positive slopes (toward the right); velocities > 6 km/s slope in a negative direction (toward the left).

where:

T_{red} = reduced travel time.

The same information for the travel-time graph in Fig. 4.12a is plotted in Fig. 4.12b with a *reduced travel time of 6 km/s*. Instead of plotting actual travel time (T), the reduced travel time (T_{red}) is plotted, where:

$$T_{red} = T - \frac{X}{6 \text{ km/s}}$$

The reduced travel-time display makes it easier to identify events relative to an apparent velocity (V_{ap}) of 6 km/s: 1) events with $V_{ap} < 6\ km/s$ have *positive slopes*; 2) events with $V_{ap} = 6\ km/s$ are *horizontal*; and 3) events with $V_{ap} > 6\ km/s$ have *negative slopes*. The positive slopes thus show that the direct arrival and first refraction have velocities of less than 6 km/s (perhaps indicating sedimentary strata). The

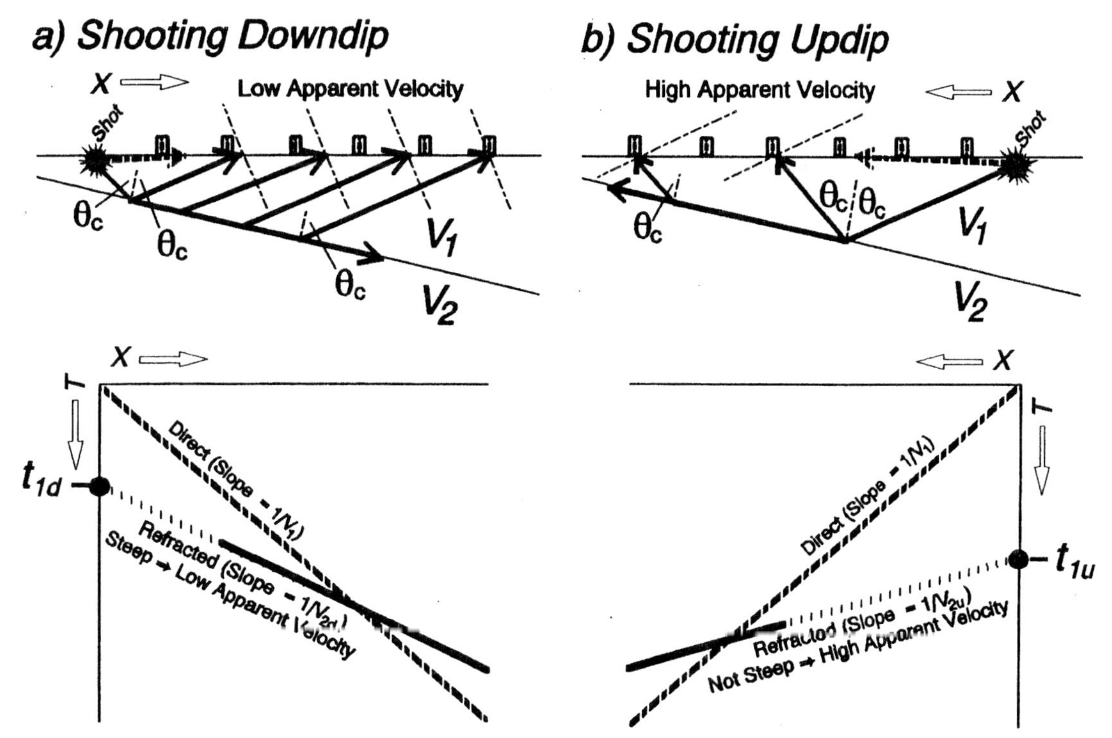

FIGURE 4.13 *Seismic survey of a dipping refractor.* a) Selected raypaths shooting toward receivers in a downdip direction. Raypaths for the critical refraction emerge at a shallow angle, resulting in an apparent velocity (V_{2d}) lower than the true velocity (V_2). b) When the receivers are updip from the source, rays emerge at a steeper angle; an apparent velocity (V_{2u}) higher than V_2 results.

6 km/s refraction might be from crystalline basement. The higher velocity arrival (8 km/s; probably from the Moho) stands out because of its negative slope.

Single Dipping Interface

For a dipping interface (Fig. 4.13), apparent velocities observed at the surface are not equal to the true velocity of the refracting layer. When the source shoots downdip toward the receivers, the apparent velocity is lower than the true velocity (Fig. 4.13a); a velocity higher than the true velocity results from shooting updip (Fig. 4.13b).

The dipping interface can be resolved by recording a *reversed refraction profile.* A profile is shot in one direction (as from a shotpoint at A to receivers extending to B), then in the other direction (from B to A; Fig. 4.14a). The seismic travel-time records (Fig. 4.14b) are superimposed with the same horizontal and vertical scales, then analyzed according to the equations presented below (see also Burger, 1992, p. 80–85; Telford et al., 1976, p. 281–284).

For a dipping interface, the *intercept times* shooting in the downdip and updip directions are not equal:

$$t_{1d} \neq t_{1u}$$

where:

t_{1d} = T-axis intercept when shooting downdip (from A to B)
t_{1u} = T-axis intercept when shooting updip (from B to A).

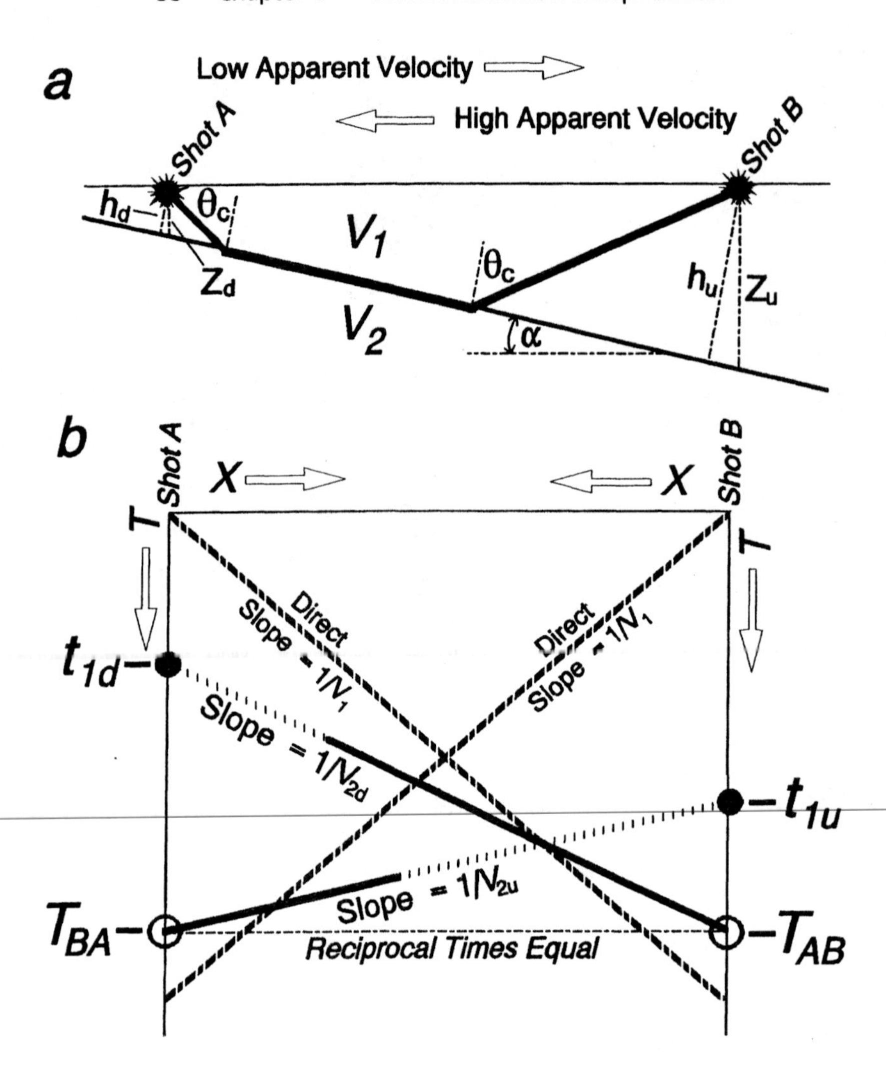

FIGURE 4.14 A reversed refraction profile is the combination of surveys shooting in downdip and updip directions (Fig. 4.13). a) A ray from the source at A, shooting to a receiver at B, traverses the same course as a ray from the shot at B shooting to a receiver at A. The reciprocal times T_{AB} and T_{BA} in (b) are therefore exactly the same. Apparent velocities V_{2d} and V_{2u} in (b) are different, because the rays emerge at different angles. b) Superposition of travel-time curves for the downdip (Fig. 4.13a) and updip (Fig. 4.13b) surveys. Note that, because the interface is deeper beneath Shot B, the intercept time (t_{1u}) for updip shooting is greater than when shooting downdip (t_{1d}).

The travel time from the shot at A to a receiver at B, however, has to be the same as the travel time from the shot at B to a receiver at A, because the exact raypath is utilized. Thus the *reciprocal times* must be equal:

$$T_{AB} = T_{BA}$$

where:

T_{AB} = travel time from shot at A to receiver at B
T_{BA} = travel time from shot at B to receiver at A.

In picking events on a reversed refraction plot (Fig. 4.14b), one can verify if refractions are from the same interface by determining if $T_{AB} = T_{BA}$. If reciprocal times are not the same, the analysis below will be erroneous.

The *apparent velocities* for the refracted arrival when shooting in the downdip and updip directions are:

$$V_{2d} = \frac{V_1}{\sin(\theta_c + \alpha)}$$

$$V_{2u} = \frac{V_1}{sin(\theta_c - \alpha)}$$

where:

V_{2d} = apparent velocity shooting downdip
V_{2u} = apparent velocity shooting updip
V_1 = velocity of the overlying layer
θ_c = critical angle
α = dip of the interface.

The *dip of the interface* and the *critical angle* can be determined by solving for V_{2d} and V_{2u} simultaneously:

$$\alpha = \frac{sin^{-1}(V_1/V_{2d}) - sin^{-1}(V_1/V_{2u})}{2}$$

$$\theta_c = \frac{sin^{-1}(V_1/V_{2d}) + sin^{-1}(V_1/V_{2u})}{2}$$

The *true velocities* can be determined from the travel-time graph (Fig. 4.14b) by reading the velocity of the overlying layer (V_1) as the inverse of the slope of the direct arrival, and then solving for the velocity of the refracting layer (V_2) from Snell's Law:

$$V_1 = \frac{1}{\text{slope of direct arrival}}$$

$$V_2 = \frac{V_1}{sin\theta_c}$$

The *travel times* to a receiver at horizontal distance (X), shooting downdip (T_d) and updip (T_u), are:

$$T_d = t_{1d} + \frac{X}{V_{2d}}$$

$$T_u = t_{1u} + \frac{X}{V_{2u}}$$

where:

$$t_{1d} = \frac{2h_d cos\theta_c}{V_1}$$

$$t_{1u} = \frac{2h_u cos\theta_c}{V_1}$$

h_d = *perpendicular* distance to interface when shooting downdip:

$$h_d = \frac{V_1 t_{1d}}{2cos\theta_c}$$

h_u = *perpendicular* distance to interface when shooting updip:

$$h_u = \frac{V_1 t_{1u}}{2cos\theta_c}$$

The *vertical depths to the interface* below points A and B, respectively, are (Fig. 4.14a):

$$Z_d = \frac{h_d}{\cos\alpha}$$

$$Z_u = \frac{h_u}{\cos\alpha}$$

where:

Z_d = *vertical* distance to interface when shooting downdip.
Z_u = *vertical* distance to interface when shooting updip.

(Note that Z_d and Z_u are sensitive to error in intercept times; t_{1d} and t_{1u} should be determined carefully from the records).

The method can be expanded for multiple dipping interfaces, equations becoming complex with more layers (Sheriff and Geldart, 1995, p. 433–439). The utility of a multiple dipping-layer model should be examined carefully before application; assumptions about constant velocity within layers and uniform directions of strike and dip of interfaces may not be valid.

TECTONIC INTERPRETATION OF SEISMIC REFRACTION PROFILES

The seismic refraction method is a useful tool to resolve crustal thickness (Fig. 4.4) and, in places, gross velocity layering within the crust (Fig. 4.9). For some regions, refraction velocity determined for the uppermost mantle can be an indicator of thermal state.

Maps of Crustal Thickness and Uppermost Mantle Velocity

The depth to the crust/mantle boundary (Moho; Fig. 2.8) often relates to tectonic history. Regions of plate divergence commonly have shallow Moho depths (continental rift zones, Figs. 2.13b, 2.14a; passive continental margins, Fig. 2.14c; mid-ocean ridges, Figs. 2.13c, 2.14b,c). Low mantle refraction velocities can indicate zones where hot asthenosphere is shallow at mid-ocean ridges and continental rifts (Fig. 2.24).

The Moho geometry is altered considerably where lithospheric plates converge. At subduction zones, the oceanic Moho deepens on the subducting plate, while oceanic to continental depth Moho occurs on the overriding plate (Figs. 2.16, 2.17). In continental collision zones, the crust may thicken to twice its normal value for continental areas (Fig. 2.18).

Uppermost Mantle Velocity, United States Fig. 4.15 reveals changes in uppermost mantle velocity based on critical refraction. Compressional-wave velocity is affected by changes in rigidity (Figs. 3.9, 3.10), which relate to temperature. Normal upper mantle velocity of about 8.1 to 8.2 km/s occurs in areas where lithospheric plates are cold (continental craton). Where hot mantle is shallow (Basin and Range continental rift zone; Columbia Plateau/Yellowstone hotspot trace), velocity drops to as low as 7.8 km/s. Velocity is also low in the magmatic arc region of the

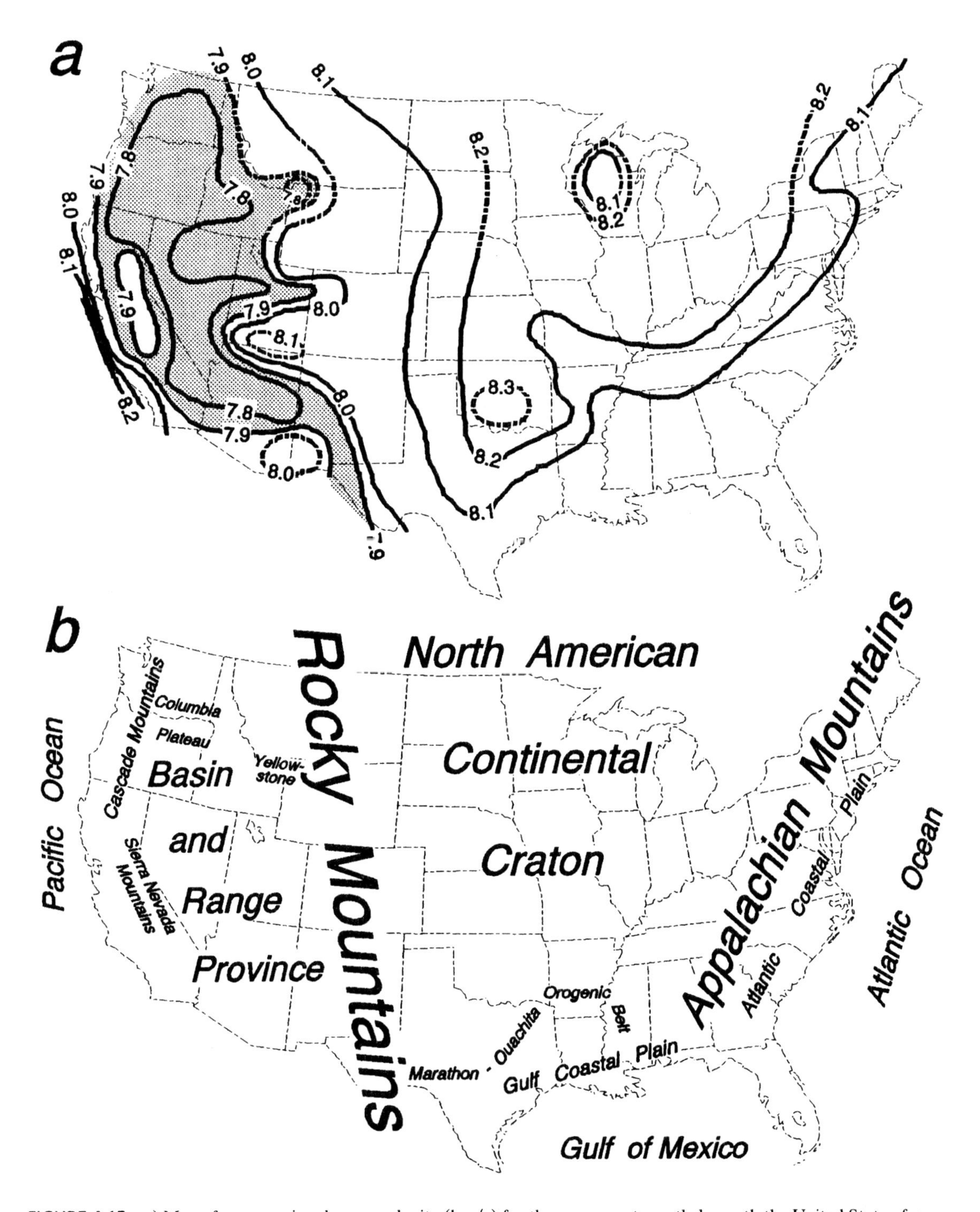

FIGURE 4.15 a) Map of compressional-wave velocity (km/s) for the uppermost mantle beneath the United States, from seismic refraction observations. From *Introduction to Geophysics*, by G. D. Garland, © 1979. Modified with permission of W. B. Saunders Comp., Toronto. Shaded region corresponds to anomalously low velocity (< 7.9 km/s). b) Major tectonic features of the United States that relate to observations in (a) and in Fig. 4.16.

Cascade Mountains. Relatively high velocity occurs in the Sierra Nevada Mountains, due to a cold slab of lithosphere that remains from past subduction.

Moho Depth, North America The map of crustal thickness for North America (Fig. 4.16) is based largely on seismic refraction observations (Mooney and Braile, 1989). The normal cratonic thickness is about 35 to 45 km. Regions that have undergone continental rifting (Basin and Range Province; Gulf and Atlantic coastal plains) have thin crust. Thick crust occurs beneath the region of high topography in the Rocky Mountains. Subdued crustal roots are associated with the Paleozoic continental collision zone in the Appalachian Mountains, and with the remnant subduction zone of the Sierra Nevada Mountains.

Moho Depth, Central Europe Seismic refraction studies in central Europe reveal changes in the depth to the Moho that relate to tectonic history (Fig. 4.17). In the older parts of the continent (Bohemian Massif; European Platform), average crustal thickness is about 35 km. A thick crustal root occurs in the continental collision zone of the Eastern Alps; a smaller depression of the Moho rims the softer zone of collision in the Carpathian Mountains. Thin crust corresponds to young continental rifting in the Pannonian Basin.

Crustal-Scale Cross Sections

Oceanic Crust The three cross sections in Fig. 4.18 reveal geometry and compressional-wave velocity of oceanic crust and upper mantle at plate boundaries and continental margins. The Mid-Atlantic Ridge profile (Fig. 4.18a) shows results of the classic study by Talwani et al. (1965). The composite of many refraction lines indicates thin crust and low uppermost mantle velocity in the vicinity of the ridge (compare with Figs. 2.13, 2.14, and 2.24a).

The right side of Fig. 4.18b, based on refraction lines off the east coast of South America, reveals seismic velocity and layer thickness typical of oceanic crust (ophiolite, Figs. 2.24a and 4.9). Approaching the passive continental margin on the west, the crust thickens as the water and sediment layers thin and the Moho deepens (Fig. 2.14c).

Fig. 4.18c, based on refraction profiles recorded in the Pacific Ocean, shows oceanic crust at both diverging and converging plate boundaries (Fig. 2.17a). At the axis of the Juan de Fuca Ridge, shallow Moho and low upper mantle velocity is indicative of an active spreading center (Fig. 2.17b). Approaching the Cascadia Subduction zone on the east, the Moho of the Juan de Fuca plate is seen descending beneath the accretionary wedge.

Continental Crust Fig. 4.19 shows velocity and crustal thickness associated with plate boundaries cutting continental lithosphere. The western part of Fig. 4.19a reveals thin crust and relatively flat Moho associated with the San Andreas Transform boundary. The crust thickens beneath the Sierra Nevada mountains, depicting a preserved root of the former convergent plate boundary. On the east, relatively thin crust ($\approx$ 30 km) and low upper mantle velocity (7.8 km/s) coincide with continental rifting in the Basin and Range Province.

In the Alps (Fig. 4.19b), crust thickens due to plate convergence that led to continental collision. Moho refractions (V_p = 8.1 to 8.2 km/s) can be traced southward to depths below 50 km, as crust of the European continental margin underthrusts rocks of the collision zone. On the south, Moho depths of 30 to 35 km represent "exotic" crust of the overriding continent.

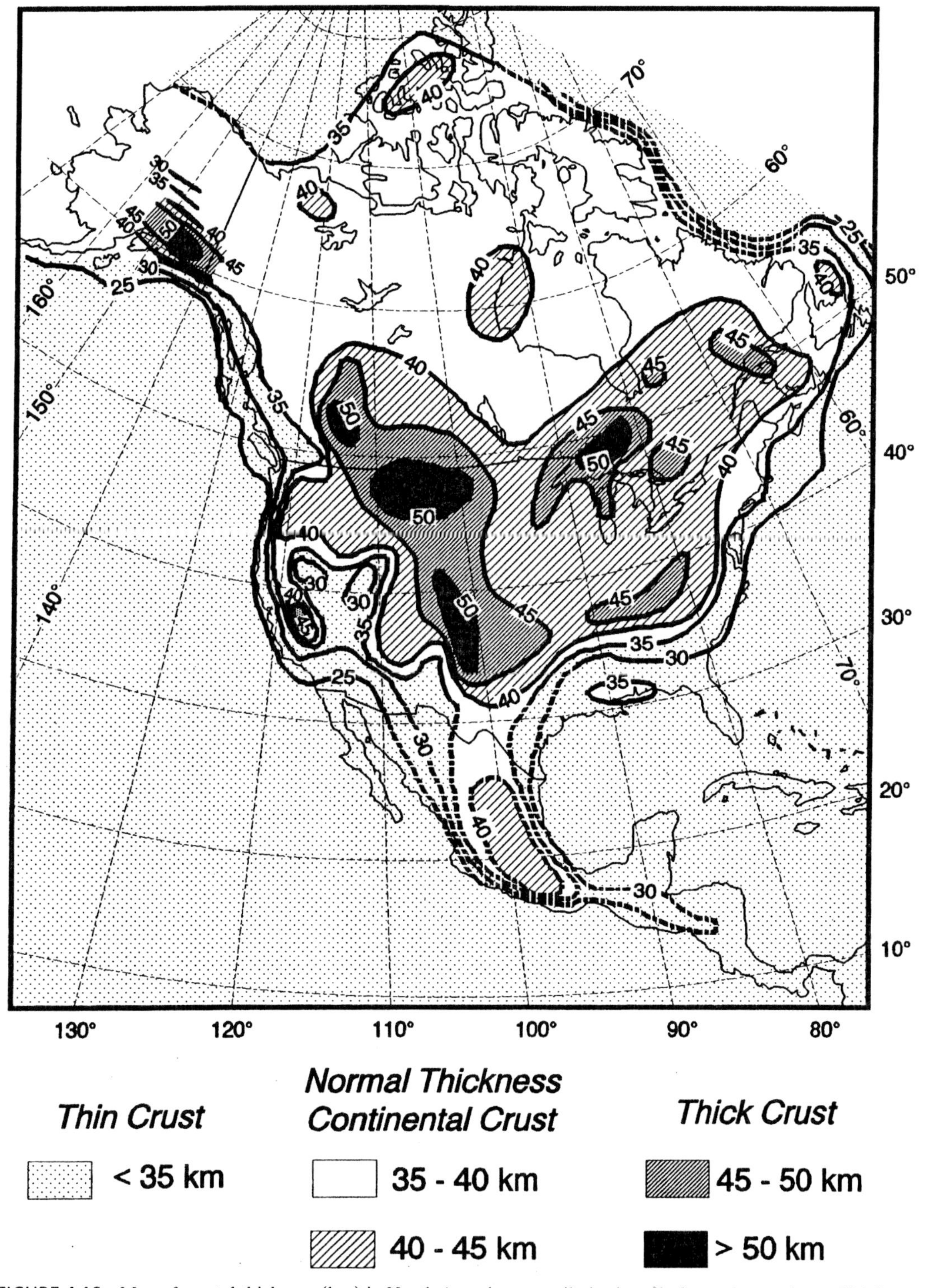

FIGURE 4.16 Map of crustal thickness (km) in North America, compiled primarily from observations of Moho refractions. From The Seismic structure of the continental crust and upper mantle of North America, by W. D. Mooney and L. W. Braile, in *The Geology of North America—An Overview,* ed. by A. W. Bally and A. R. Palmer, © 1989. Modified with permission of the Geological Society of America, Boulder, Colorado. Compare with map of major tectonic features in the United States (Fig. 4.15b).

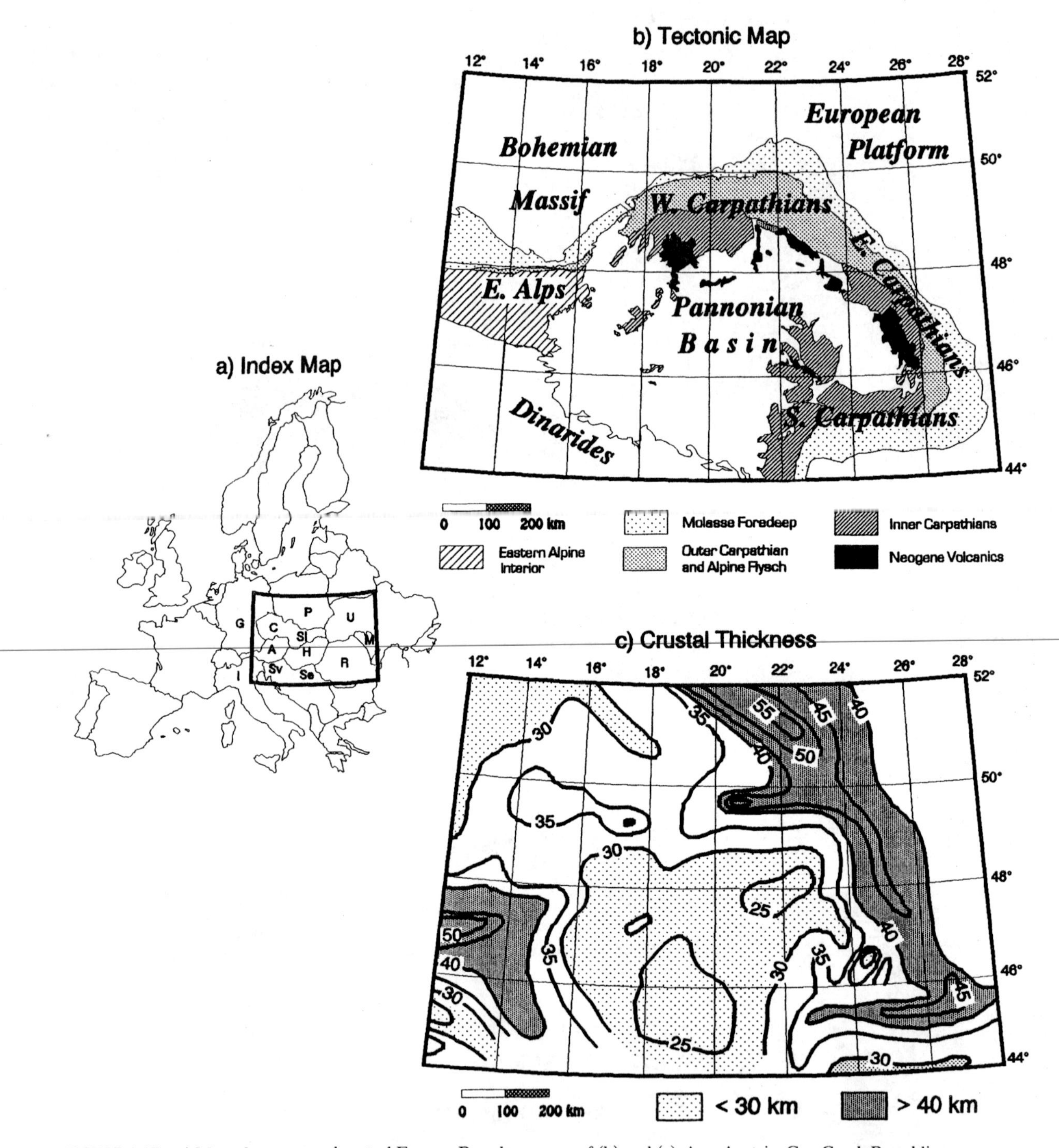

FIGURE 4.17 a) Map of western and central Europe. Box shows area of (b) and (c). A = Austria; C = Czech Republic; G = Germany; H = Hungary; I = Italy; M = Moldavia; P = Poland; R = Romania; Se = Serbia; Sl = Slovak Republic; Sv = Slovenia. b) Tectonic map of central Europe, showing Eastern Alpine and Carpathian collisional mountain ranges and Pannonian Basin continental rift zone (after Salters et al., 1988). c) Map of crustal thickness (km) in central Europe, based principally on observation of critical refractions from the top of the mantle. (Compiled from maps of Beránek, 1978; Mayerová et al., 1985; and Tomek, 1988). Maps (b) and (c) in Lillie et al. (1994).

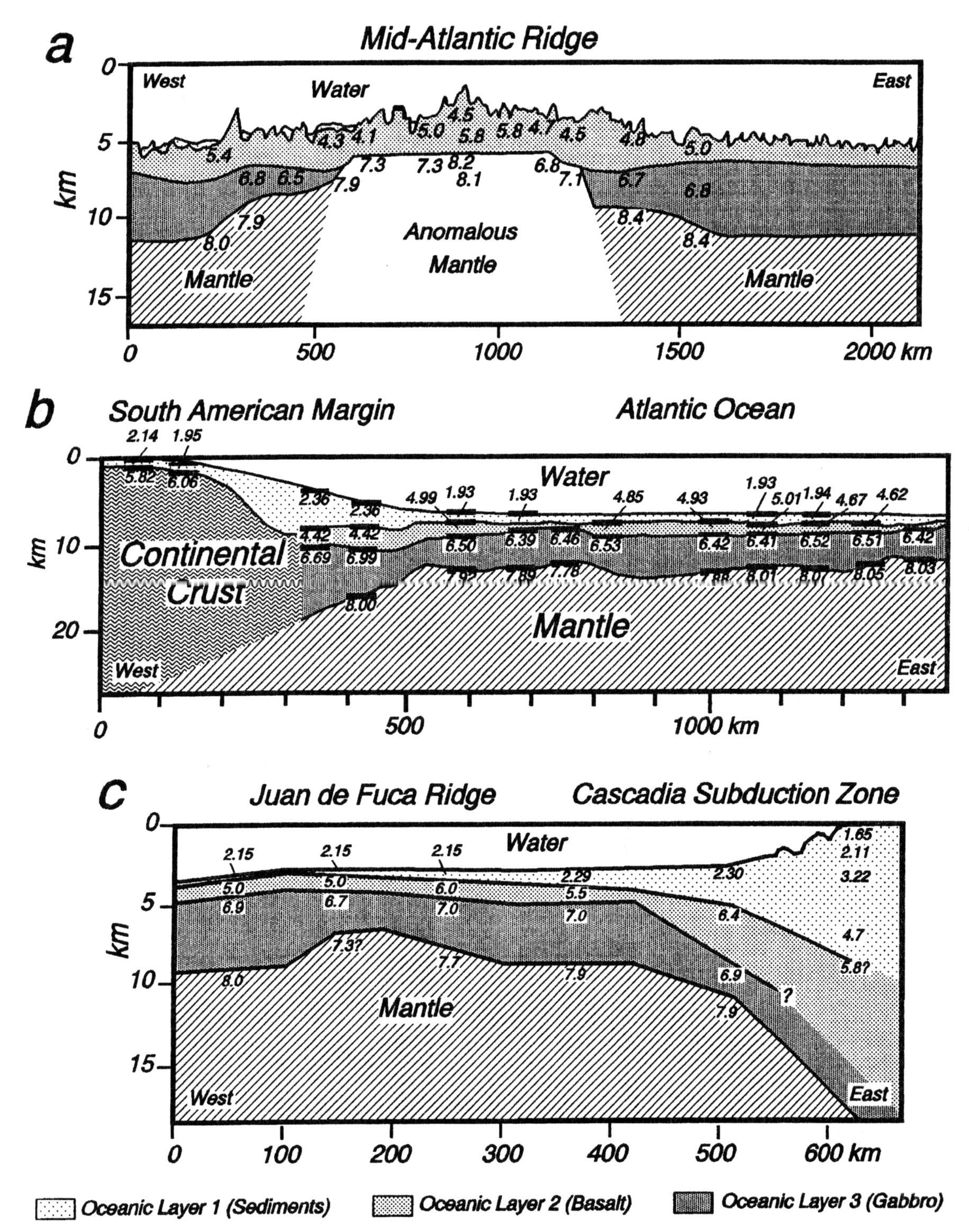

FIGURE 4.18 Composite cross sections based on seismic refraction studies in oceanic regions. Numbers are compressional-wave velocities (km/s) from refracting layers. a) Cross section of velocity and thickness, based on nine refraction profiles on and near the Mid-Atlantic Ridge (Talwani et al., 1965). b) Results of refraction profiling of the passive continental margin off Argentina (Ewing, 1965). Short, bold lines represent the tops of critically refracting layers. c) Section based on refraction studies of the Juan de Fuca Ridge and Cascadia Subduction Zone off the Oregon coast (Shor et al., 1968). (a) and (b) from *The Interior of the Earth: Its Structure, Constitution and Evolution,* 2nd ed., © 1982. Redrawn with permission of Elsevier Science Pub. Co., New York. (c) from Seismic refraction studies off Oregon and northern California, *Journal of Geophysical Research,* v. 73. pp. 2175–2194, © 1968. Redrawn with permission of the American Geophysical Union, Washington, D. C.

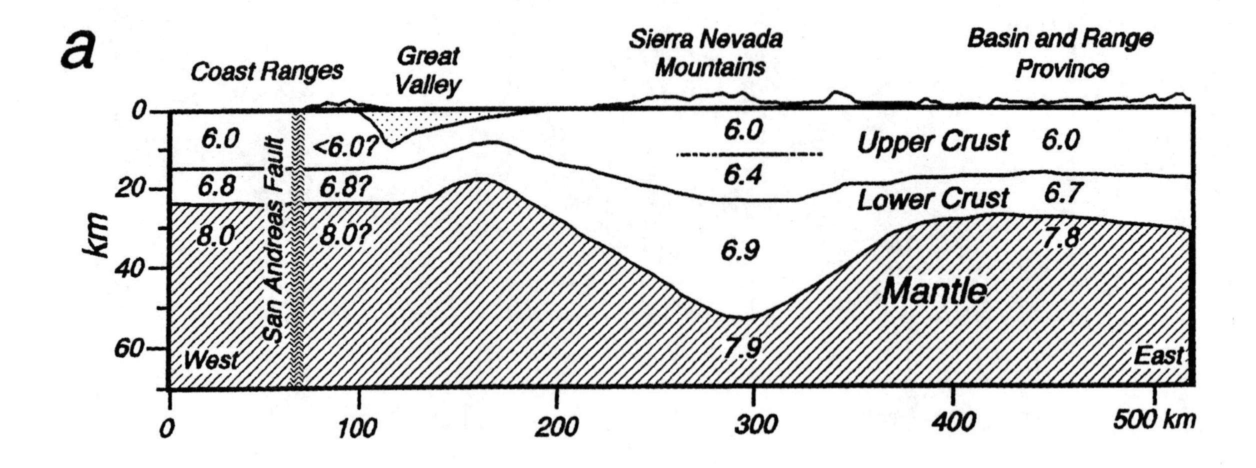

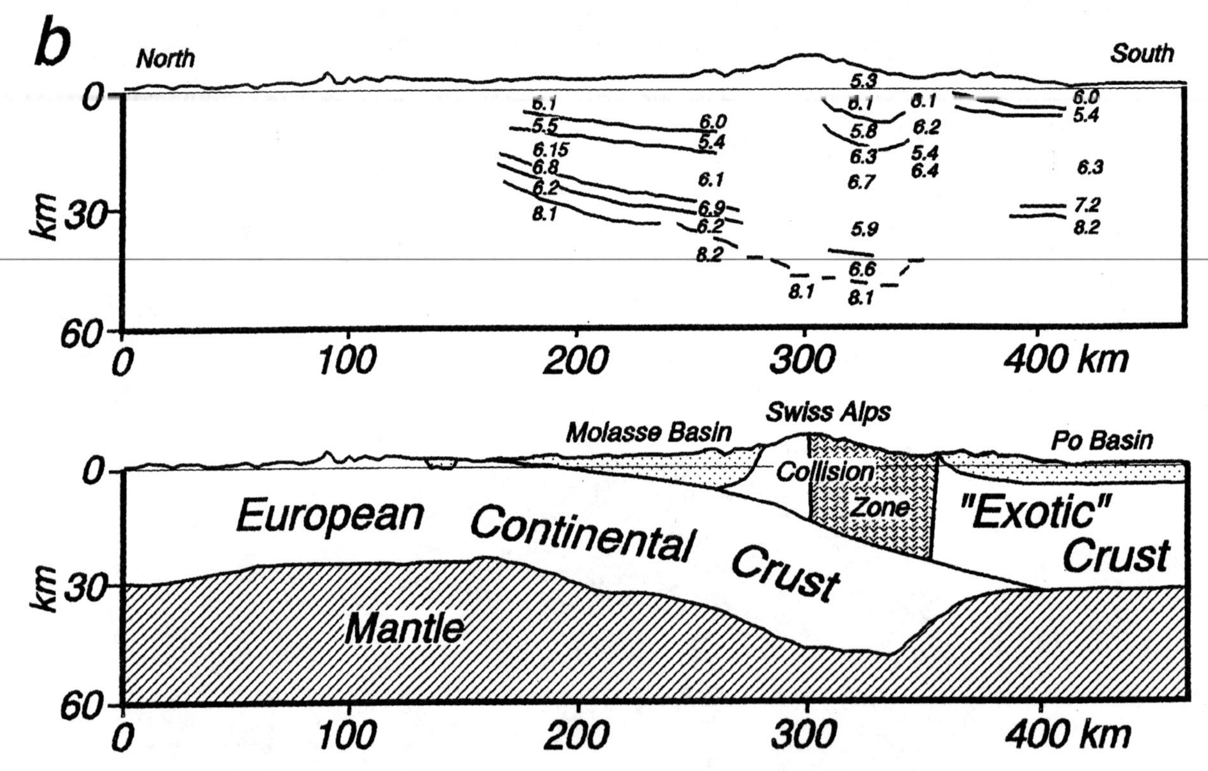

FIGURE 4.19 Refraction studies from regions of continental lithosphere. Numbers are compressional wave velocity (km/s). a) Composite section from California coast across the San Andreas Fault and Sierra Nevada Mountains, extending into the Basin and Range Province (Eaton, 1963, 1966; Bateman and Eaton, 1967). From Seismic evidence for the structure and Cenozoic tectonics of the Pacific coast states, by D. P. Hill, in *Cenozoic Tectonics and Regional Geophysics of the Western Cordillera,* ed. by R. B. Smith and G. P. Eaton, © 1978. Redrawn by permission of The Geological Society of America, Boulder, Colorado. b) Section across the western Alps, along the Swiss geotraverse (Rybach et al., 1980). Geological model below velocity profile from Trümpy (1973). From Gravity anomalies and flexure of the lithosphere at mountain ranges, by G. D. Karner and A. B. Watts, in *Journal of Geophysical Research,* v. 88, pp. 10449–10477, © 1983. Redrawn by permission of the American Geophysical Union, Washington, D. C.

EXERCISES

4-1 A hypothetical refraction line is run across a continental craton. Sketch a single, horizontal interface model showing: **a)** approximate thickness and P-wave velocity of the crust; **b)** approximate P-wave velocity of the uppermost mantle. On the model sketch: **c)** raypaths for direct, critically refracted, and reflected P-waves; **d)** wave fronts for direct and critically refracted P-waves. Compute travel times and plot the following arrivals on a travel time vs. distance graph: **e)** direct P-wave; **f)** critically refracted P-wave; **g)** reflected P-wave. Calculate and illustrate on the graph: **h)** critical distance; **i)** T-axis intercept time; **j)** crossover distance; **k)** approximate length of refraction line necessary to resolve the Moho depth.

4-2 A hypothetical refraction line is run along strike, on the coastal plain of a passive continental margin. Crustal thickness is between "normal" continental and "normal" oceanic. The sedimentary section is approximately 5 km thick. Sketch a multiple, horizontal interface model showing:

a) approximate thickness and P-wave velocity of the sedimentary strata and crystalline crust;

b) approximate P-wave velocity of the uppermost mantle.

Compute travel times and plot P-wave arrivals on a travel time vs. distance graph for:

c) direct wave through the sedimentary section;

d) critically refracted wave from the top of the crystalline basement;

e) critically refracted wave from the top of the mantle.

Calculate and illustrate on the graph:

f) crossover distance for crystalline basement refraction, relative to direct arrival;

g) crossover distance for Moho refraction, relative to basement refraction;

h) length of refraction line necessary to resolve basement depth;

i) length of refraction line necessary to resolve Moho depth.

4-3 The seismic refraction profile in Fig. 4.8 was recorded in an ocean basin. Determine seismic velocities and thicknesses, assuming a layered model with several horizontal interfaces.

a) Identify and draw straight lines through the *direct, three refracted,* and *a few reflected* arrivals on the section.

b) Give the *apparent velocity* for each of the *direct* and *refracted* arrivals.

c) Calculate *thickness* for each of the three layers overlying the highest-velocity material; that is, knowing the velocities (V_1, V_2, V_3, V_4) and T-intercepts (t_1, t_2, t_3), solve for the thicknesses (h_1, h_2, h_3).

d) Draw a cross-section illustrating the depths to the interfaces and the layer velocities and thicknesses. Speculate on materials that might comprise each layer (Figs. 2.24a, 4.9, 4.18).

4-4 Fig. 4.9a shows average results of inversion of refraction profiles recorded in deep ocean basins. Fig. 4.9b shows arrival times for direct and refracted arrivals predicted for such a model.

a) Compare the hypothetical travel-time graph in Fig. 4.9b with the actual seismic profile in Fig. 4.8. How are the records similar and how are they different? Explain the similarities and differences.

b) Compare the model of an ophiolite sequence (Fig. 4.9a) and your results of inverting the actual seismic profile in Exercise 4-3. How are the models similar and how are they different? Explain the similarities and the differences.

c) The record in Fig. 4.8 was recorded in Baffin Bay (Canadian Arctic). Discuss how and why the water depth you compute differs from that observed across a normal ocean basin.

d) Discuss how differences in crustal layer thicknesses and velocities for the models in Fig. 4.9a and Exercise 4-3 might be due to actual differences and how they might be due to interpretation error. List the differences and possible sources of error.

4-5 The travel-time graph below shows results of a reversed refraction profile recorded with the shotpoint at A, then at B. Assuming a single dipping interface model, interpret through inversion: **a)** true velocity for the layers above and below the interface; **b)** dip of the interface; **c)** depths to the interface vertically below points A and B.

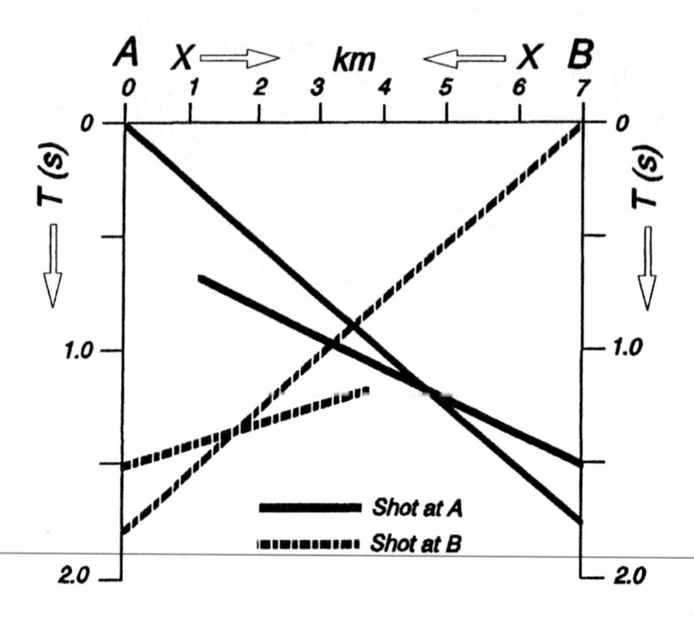

4-6 **a)** Based on the map of North America (Fig. 4.16), draw a cross section of changing Moho depth along 40° N latitude. **b)** From Fig. 4.15, add P-wave velocities for the uppermost mantle to your cross section. **c)** Discuss reasons for the crustal thicknesses and mantle velocities you illustrated on your cross section.

4-7 **a)** Based on the central European maps in Fig. 4.17, draw a cross section along 47° N latitude. **b)** Discuss reasons for the crustal thickness changes you illustrated in (a).

SELECTED BIBLIOGRAPHY

Bateman, P. C., and J. P. Eaton, 1967, Sierra Nevada batholith, *Science*, v. 158, pp. 1407–1417.

Beránek, B., 1978, Results of deep seismic sounding measurements in Czechoslovakia and their contribution to solution of contact of Alpine system with European platform, *Czechoslovak Geology and Global Tectonics*, Bratislava: Slovak Academy of Sciences, pp. 248–256.

Bott, M. H. P., 1982, *The Interior of the Earth: Its Structure, Constitution and Evolution* (2nd ed.), New York: Elsevier Science Pub. Co., 403 pp.

Burger, H. R., 1992, *Exploration geophysics of the shallow subsurface*, Englewood Cliffs, N.J.: Prentice Hall, Inc., 489 pp.

Couch, R. W., and R. P. Riddihough, 1989, The crustal structure of the western continental margin of North America, in: *Geophysical Framework of the Continental United States*, edited by L. C. Pakiser and W. D. Mooney, Boulder, Colorado: Geological Society of America, Memoir 172, pp. 103–128.

Dobrin, M. B., 1976, *Introduction to Geophysical Prospecting* (3rd ed.), New York: McGraw-Hill, 446 pp.

Eaton, J. P., 1963, Crustal structure from San Francisco, California, to Eureka, Nevada, from seismic-refraction measurements, *Journal of Geophysical Research*, pp. 5789–5806.

Eaton, J. P., 1966, Crustal structure in northern and central California from seismic evidence, in: *Geology of Northern California, Bulletin California Division of Mines and Geology*, v. 190, pp. 419–426.

Ewing, M., 1965, The sediments of the Argentine basin, *Quarterly Journal of the Royal Astronomical Society*, v. 6, pp. 10–27.

Ewing, J. I., and M. Ewing, 1959, Seismic refraction measurements in the Atlantic Ocean basins, in the Mediterranean Sea, on the mid-Atlantic ridge, and in the Norwegian Sea, *Geological Society of America Bulletin*, v. 70, pp. 291–318.

Fowler, C. M. R., 1990, *The Solid Earth: An Introduction to Global Geophysics*, Cambridge: Cambridge University Press, 472 pp.

Garland, G. D., 1979, *Introduction to Geophysics* (2nd ed.), Toronto: W. B. Saunders Comp., 494 pp.

Harding, A. J., J. A. Orcutt, M. E. Kappus, E. E. Vera, J. C. Mutter, P. Buhl, R. S. Detrick, and T. M. Brocher, 1989, Structure of young oceanic crust at 13° N on the East Pacific Rise from expanding spread profiles, *Journal of Geophysical Research*, v. 94, pp. 12163–12196.

Healy, J. H., 1963, Crustal structure along the coast of California from seismic refraction measurements, *Journal of Geophysical Research*, v. 68, pp. 5777–5787.

Healy, J. H., and L. G. Peake, 1975, Seismic velocity structure along a section of the San Andreas Fault near Bear Valley, California, *Bulletin Seismological Society of America*, v. 65, pp. 1177–1197.

Heirtzler, J. R. and X. Le Pichon, 1965, Crustal structure of the mid-ocean ridges: 1. Seismic refraction measurements, *Journal of Geophysical Research*, v. 70, pp. 318–339.

Hill, D. P., 1978, Seismic evidence for the structure and Cenozoic tectonics of the Pacific coast states, in *Cenozoic Tectonics and Regional Geophysics of the Western Cordillera*, edited by R. B. Smith and G. P. Eaton, Boulder, Colorado: Geological Society of America, Memoir 152, pp. 145–174.

Ibrahim, A. K., G. Latham, and J. Ladd, 1986, Seismic refraction and reflection measurements in the Middle America Trench offshore Guatemala, *Journal of Geophysical Research*, v. 84, pp. 5643–5649.

Jarchow, C. M., G. A. Thompson, R. D. Catchings, and W. D. Mooney, 1993, Seismic evidence for active magmatic underplating beneath the Basin and Range province, western United States, *Journal of Geophysical Research*, v. 98, pp. 22095–22122.

Karner, G. D., and A. B. Watts, 1983, Gravity anomalies and flexure of the lithosphere at mountain ranges, *Journal of Geophysical Research*, v. 88, pp. 10449–10477.

Kearey, P. and M. Brooks, 1984, *An Introduction to Geophysical Exploration*, Boston: Blackwell Scientific Publications, 296 pp.

Keller, B., W. A. Prothero, A. M. Trehu, and D. J. Stierman, 1983, Ray trace model of the Santa Barbara Channel, California, land-sea seismic refraction experiment, *Geophysical Research Letters*, v. 10, pp. 933–936.

Koch, K., and B. W. Stump, 1996, Constraints for upper mantle shear-wave models of the Basin and Range from surface wave inversion, *Bulletin Seismological Society of America*, v. 86, pp. 1591–1607.

Lillie, R. J., M. Bielik, V. Babuška and J. Plomerová, 1994, Gravity modelling of the lithosphere in the Eastern Alpine-Western Carpathian-Pannonian Basin region, *Tectonophysics*, v. 23, pp. 215–235.

Mayerová, M., Z. Nakládalová, I. Ibrmajer and H. Herrmann, 1985, Plošne rozložení Moho-plochy v ČSSR sestavené z výsledků profilových měření HSS a technickych odpalů, 8 celostátni konference geofyziků, Geofyzika Brno, pp. 44–55.

Mooney, W. D., and R. H. Colburn, 1985, A seismic-refraction profile across the San Andreas, Sargent, and Calaveras faults, west-central California, *Bulletin Seismological Society of America*, v. 75, pp. 175–191.

Mooney, W. D., and L. W. Braile, 1989, The seismic structure of the continental crust and upper mantle of North America, in *The Geology of North America—An Overview*, edited by A. W. Bally and A. R. Palmer, Boulder, Colorado: Geological Society of America, Decade of North American Geology, v. 1, pp. 39–52.

Rybach, L., S. Mueller, A. Milnes, J. Ansorge, D. Bernoulli, and M. Frey, 1980, The Swiss Geotraverse Basel-Chiasso—A review, *Ecologae Geol. Helv.*, v. 73, pp. 437–462.

Salters, V., S. Hart, and G. Pantó, 1988, Origin of late Cenozoic volcanic rocks of the Carpathian Arc, Hungary, in: *The Pannonian basin: A study in Basin Evolution*, edited by L. H. Royden and F. Horváth, Tulsa: American Association of Petroleum Geologists, Memoir 45, pp. 279–292.

Sheriff, R. E., and L. P. Geldart, 1995, *Exploration Seismology* (2nd ed.), Cambridge: Cambridge University Press, 592 pp.

Shor, G. G., Jr., P. Dehlinger, H. K. Kirk, and W. S. French, 1968, "Seismic refraction studies off Oregon and northern California," *Journal of Geophysical Research*, v. 73, pp. 2175–2194.

Steinhart, J. S., and T. J. Smith (editors), 1966, *The Earth Beneath the Continents*, Washington: American Geophysical Union, Monograph 10, 663 pp.

Talwani, M., X. Le Pichon, and M. Ewing, 1965, Crustal structure of the mid-ocean ridges: 2. Computed model from gravity and seismic refraction data, *Journal of Geophysical Research*, v. 70, pp. 341–352.

Telford, W. M., L. P. Geldart, R. E. Sheriff and D. A. Keys, 1976, *Applied Geophysics*, Cambridge: Cambridge University Press, 860 pp.

Tomek, C., 1988, Geophysical investigation of the Alpine-Carpathian Arc, in *Evolution of the Northern Margin of Tethys, Volume 1*, edited by M. Rakus, J. Dercourt and A. E. M. Nairn, Paris: Mem. Soc. Geol., Nouvelle Serie No. 154, Paris, pp. 167–199.

Trümpy, R. 1973, The timing of orogenic events in the central Alps, in *Geology and Tectonics*, edited by K. A. deJong and R. Scholten, New York: John Wiley and Sons, pp. 229–252.

Warren, D. H., and J. H. Healy, 1973, Structure of the crust in the conterminous United States, *Tectonophysics*, v. 20, pp. 203–213.

C H A P T E R 5

Seismic Reflection: Acquisition, Processing, And Waveform Analysis

seismic *(sīz mik')* ***adj.,*** [< *Gr.* seismos, *an earthquake* < seiein, *to shake*], *relating to an earthquake or artificial shaking of the Earth.*
reflect *(ri flekt')* ***vi.,*** [< *Mfr.* < *L.* < *re-, back* + flectere, *to bend*], *to bend back.*
reflection *(ri flek'shən)* ***n.,*** *the bending back of a ray or wave of light, heat, or sound into the same medium, as it encounters a medium with different properties.*
acquire *(ə kwīr')* ***vt.,*** [*L.* acquirere < *ad-, to* + quaerere, *to seek*], *to get or gain by one's own efforts or actions.*
process *(präs' es)* ***n.,*** [< *Ofr.* < *L.* procedere < *pro-, forward* + cedere, *to go*], *a particular method of making or doing something, in which there are a number of steps.*
wave *(wāv)* ***n.,*** [< *OE.* wafian], *motion that periodically advances and retreats as it is transmitted progressively from one particle in a medium to the next.*
form *(fôrm)* ***n.,*** [< *Of.* < *L.* forma], *the shape, outline or configuration of anything.*
seismic waveform *(sīz mik' wāv fôrm)* ***n.,*** *the configuration of a seismic trace, due to the interference pattern of reflected waves and various types of noise.*

Horizontal strata commonly contain abrupt changes in seismic velocity and density, resulting in the reflection of sound waves back to the surface. The seismic reflection method was thus developed in the 1920s and 30s as a tool for oil and gas exploration in sedimentary basins (Petty, 1976). Financial incentive led the petroleum industry to refine the technique to sophisticated levels by the 1970s; many advances in computer equipment and data processing can be attributed to this effort.

Compared to sedimentary strata, igneous and metamorphic rocks commonly are acoustically homogenous and structurally complex; stratification, if present, is generally not horizontal. The reflection method initially was thought of little value in studying areas of crystalline rocks. Since the 1970s, however, the technique has provided valuable constraints on the composition and structure of the mid-to-lower crust (Oliver, 1986, 1996).

Seismic reflection and refraction studies differ in their field geometries and in the types of problems they can resolve (Fig. 5.1). Refraction requires that receivers extend to well beyond the crossover distance, so that refraction spread lengths need to be five to ten times the depth to the deepest refractor of interest. At large distances, direct and critically refracted waves often interfere with reflections. The most useful reflections are usually those arriving nearly vertically, at less than the critical distance; reflection spread lengths therefore approximate the depth of the deepest reflector of interest.

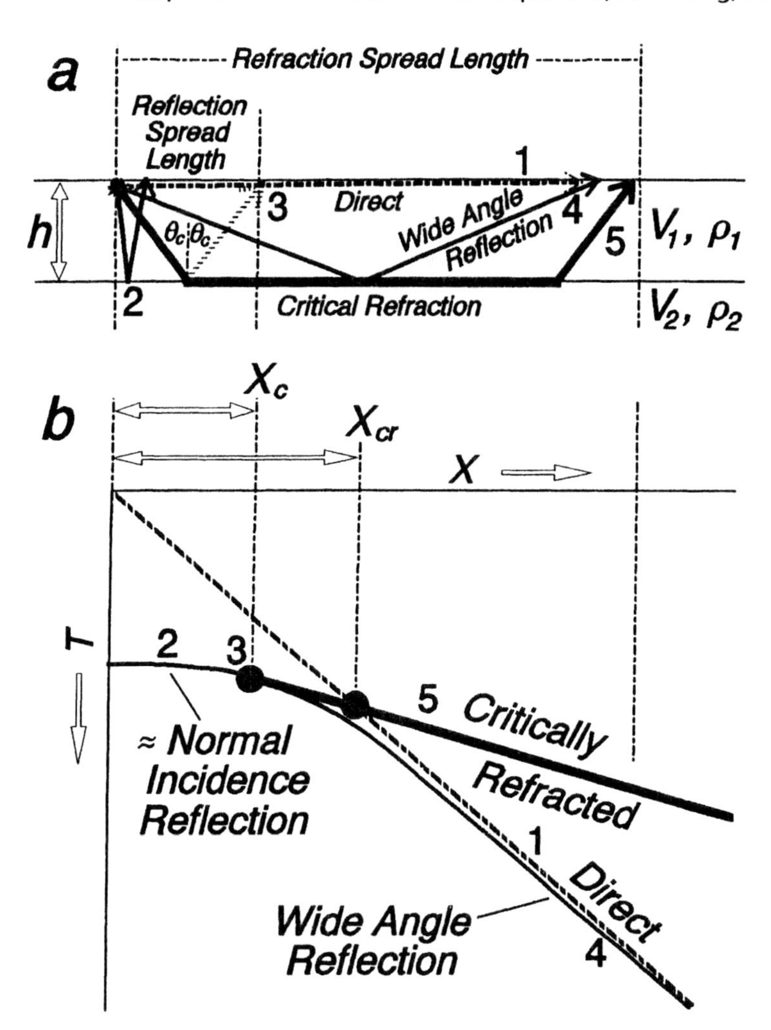

FIGURE 5.1 a) Model of horizontal interface at depth (h), separating layers of contrasting velocity (V) and density (ρ). Selected raypaths illustrated for: 1. direct arrival; 2. reflection closer than the critical distance (≈ normal incidence reflection); 3. reflection/refraction at the critical distance; 4. reflection beyond the critical distance (wide angle reflection); 5. critical refraction. b) Travel-time graph for events illustrated in (a). The reflection method commonly utilizes events recorded at distances closer than the critical distance (X_c), away from interference from the other events. Refraction typically utilizes events recorded beyond the crossover distance (X_{cr}). Spread lengths are thus approximately equal to h for reflection studies, and 5h to 10h for refraction.

Refraction and reflection studies are complimentary, refraction providing gross crustal velocities and thicknesses, reflection showing finer details of structure and stratigraphy. Critically refracted waves require abrupt increases in velocity; slopes of refracted waves thus provide velocity estimates for thick intervals of the crust. Reflections occur when there are changes in seismic velocity and/or density; boundaries or layers that are nearly flat and continuous are especially resolvable.

The seismic reflection method is popular with geologists for two main reasons: 1. reflection data are commonly portrayed as profiles resembling *geologic cross sections*; 2. under certain circumstances (for example, flat-lying strata in sedimentary basins) reflection profiles offer *high resolution of subsurface detail.* When one or both of these reasons are valid, reflection data are a powerful tool to constrain interpretations of parts of the Earth. Wise use of reflection profiles, however, requires appreciation of both the utility and the pitfalls of the method. This chapter presents the standard methods used to acquire and process such profiles, along with an overview of the factors that influence the form of the waves recorded.

WHAT IS A SEISMIC SECTION?

Seismic reflection data are normally recorded with several receivers offset from each source. A seismic section used for interpretation, however, has undergone processing that condenses the results of many shots. The standard portrayal of the data, an *unmigrated time section*, presents each seismic trace *as if a source and a receiver were in the same position* (Fig. 5.2). Reflections from the common source/receiver positions result from raypaths striking reflectors at 90°, or *normal incidence*.

Seismic waves reflect when they encounter changes in the acoustic impedance (product of velocity and density) of the Earth. There are two general categories of problems addressed through seismic profiling (Fig. 5.3): 1. *Seismic Stratigraphy*. Relatively small changes in lithology, porosity, and/or thickness of layers cause changes in acoustic impedance. Those changes are analyzed by observing lateral changes in the *seismic waveform*. 2. *Geologic Structure*. The relatively large-scale *geometry of reflected events* can be used to map structures in the subsurface.

Seismic acquisition and processing procedures are designed to attenuate unwanted signals and noise, and to present the data in an interpretable format. Parameters determined in attempts to optimize acquisition and processing sometimes yield information on rock properties (for example, seismic velocities). Ideally, a seismic interpreter is involved in acquisition and processing; at the very least, the interpreter should be knowledgeable of acquisition and processing parameters used in the survey.

ACQUISITION

Field geometries vary considerably according to the environment (land vs. marine), the nature of the geologic problem, and the accessibility of the area. Geophones are designed to record only frequencies near that of the input signal (source), discriminating against noise outside source frequencies. Most geophones are designed to respond to motions that are vertical (Fig. 5.4); reflected compressional waves (vertical particle motion) are enhanced, at the expense of events that produce horizontal motions at the surface (direct compressional waves; reflected shear waves).

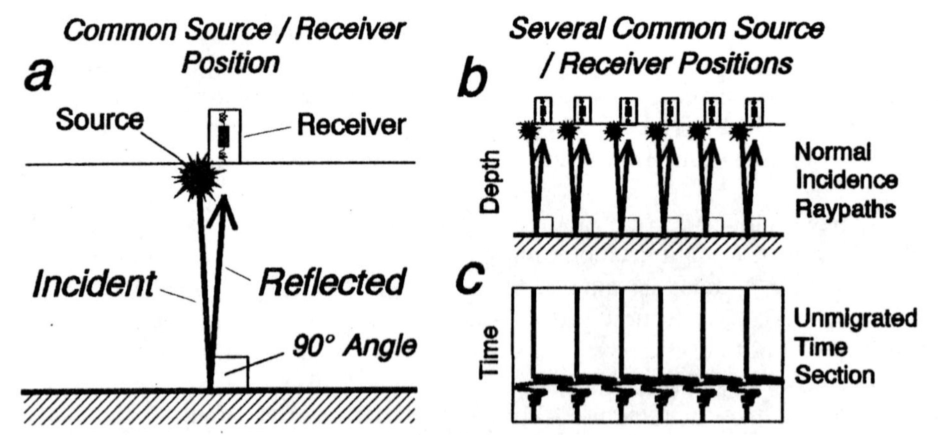

FIGURE 5.2 *Normal incidence seismic section.* a) A reflected ray must strike a reflector at normal incidence (90°) in order to return to a receiver at the same position as the source (common source/receiver position). b) A seismic section (profile) mimics normal incidence reflections to several common source/receiver positions. c) A common portrayal of reflection data, the unmigrated time section, mimics responses to the normal incidence configuration depicted in (b).

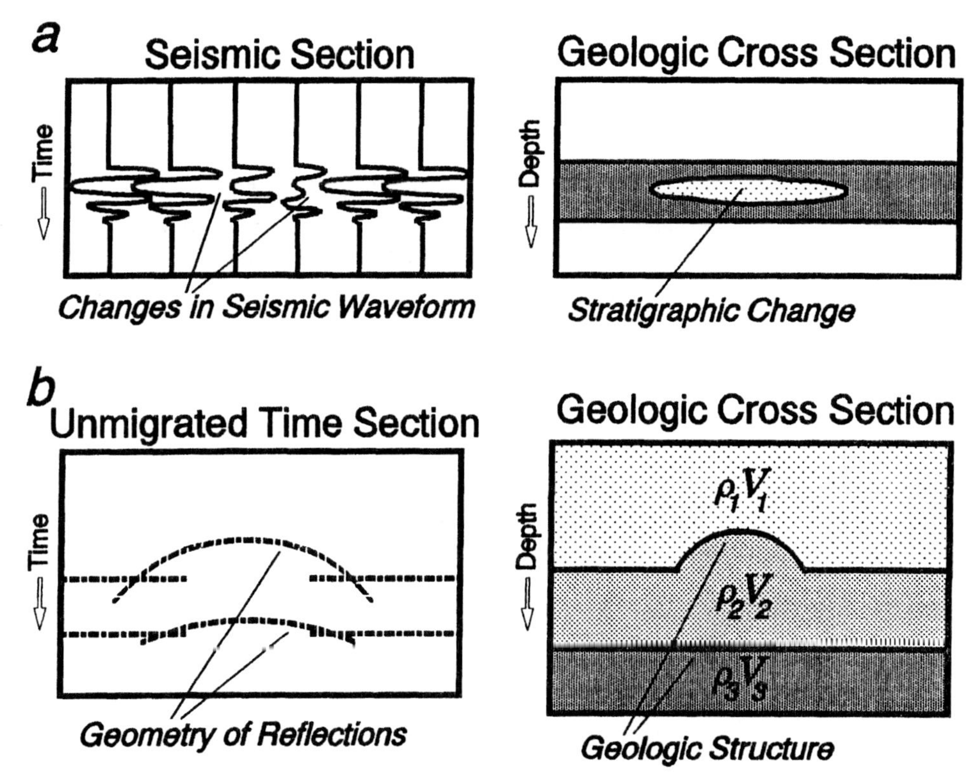

FIGURE 5.3 Two types of problems addressed by seismic reflection profiling. a) *Seismic stratigraphy,* based on analysis of seismic waveform. b) *Geologic structure,* according to the overall geometry of reflections.

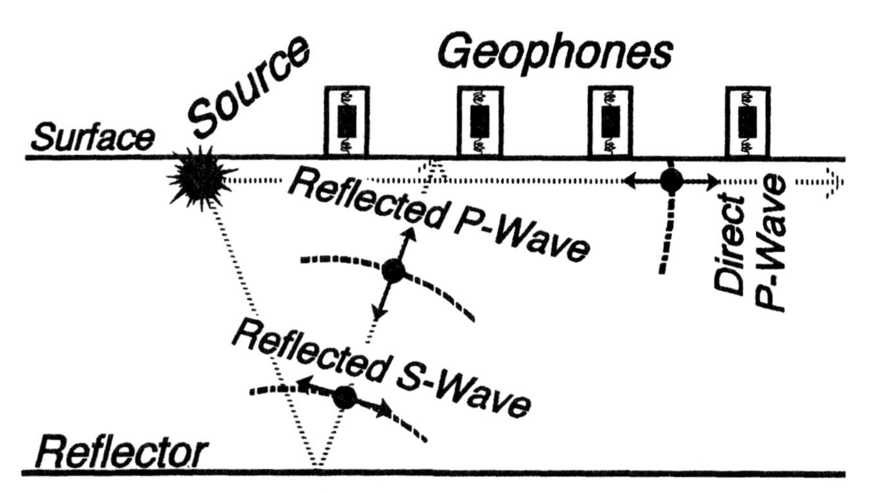

FIGURE 5.4 *Enhancement of reflected compressional waves compared to other arrivals.* Reflected P-waves have particle motions that are nearly vertical, exciting vertically oriented geophones. Direct P-waves and reflected S-waves have nearly horizontal particle motions, resulting in little response from the geophones.

Receiver Arrays

Most surveys are designed so that the source (explosive; Vibroseis; airgun) is "fired" into stations of receivers (geophones; hydrophones). Each receiver station commonly is not a single instrument, but rather an *array* of several receivers, connected electronically and centered around a point (Fig. 5.5a). The geometry of an array (or *geophone group*) is designed to cancel certain unwanted signals, while enhancing reflected events. Rayleigh waves are an example of unwanted noise; they are also called *groundroll,* because the waves produce an up-and-down, rolling motion of the ground

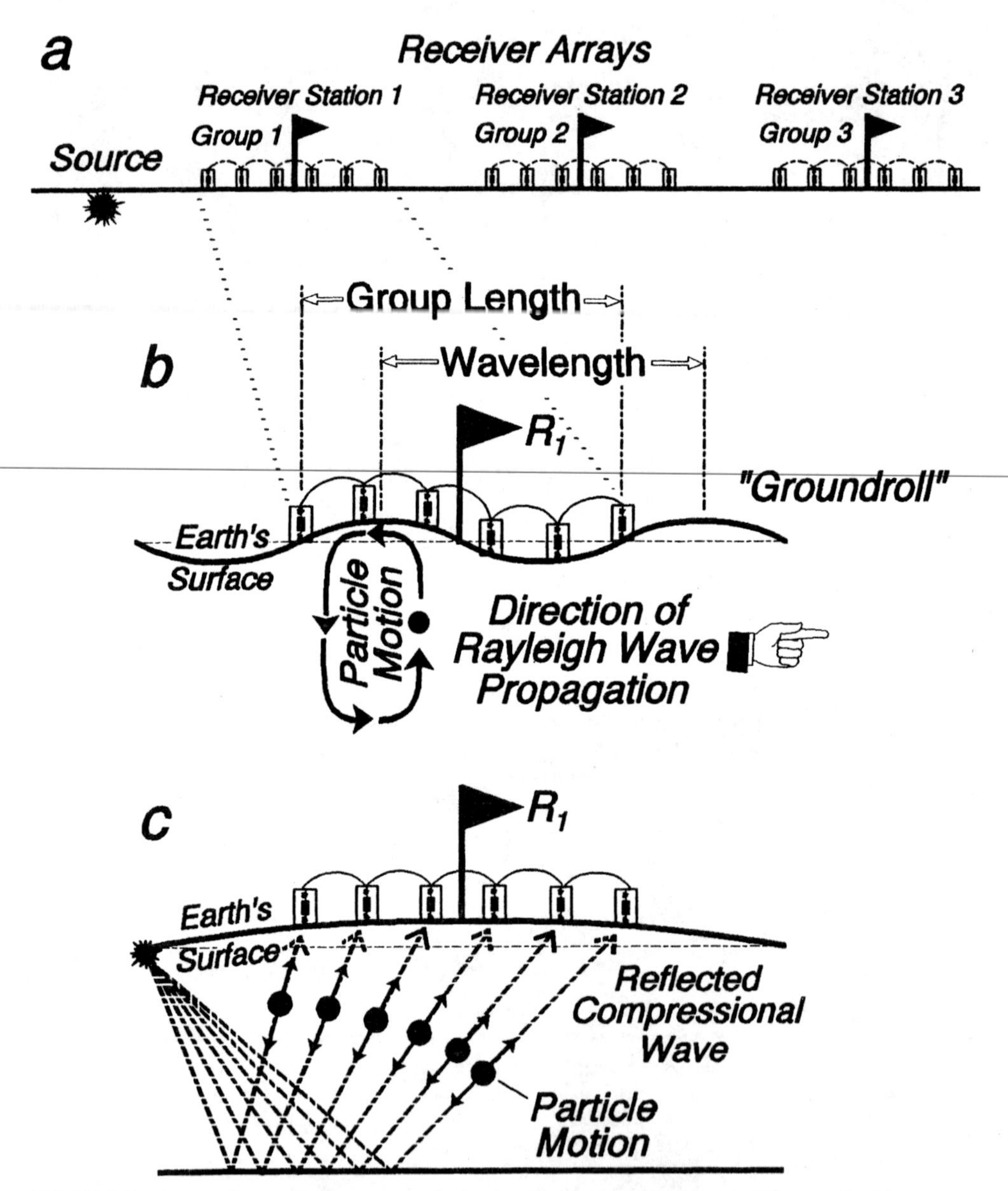

FIGURE 5.5 *Attenuation and enhancement of seismic waves by geophone array.* a) Source fired into receiver stations of connected geophones ("arrays" or "groups"). The geophone responses for a group add together as a single seismic trace. b) Response of one geophone group. If the group length is equal to the wavelength of a Rayleigh wave, positive signals from geophones moving up cancel negative signals from those moving down; the Rayleigh wave is thus attenuated. c) A reflected P-wave at close distance to the source moves the group of geophones up or down in unison, enhancing the arrival.

surface (Fig. 3.12). The array can be proportioned to the wavelength of the groundroll, so that half the geophones in the array are moving up, while half are moving down (Fig. 5.5b). The corresponding positive and negative electrical signals sent by the geophones cancel, so that the groundroll is attenuated. Reflections commonly arrive at high angles to the ground surface, (that is, with high apparent velocity; see Chapter 3). The result is that geophones in the array move up and down in unison; their electrical signals add in phase, enhancing reflected signals at the expense of other arrivals (Fig. 5.5c).

Common Midpoint Method

A seismic reflection survey is commonly conducted in the *common midpoint* (*CMP*) mode, whereby seismic traces from different shots have a common point on the surface, midway between source and receiver pairs (Fig. 5.6). (The term *common depth point,* or *CDP*, is used to express the same concept). The CMP method provides redundancy of information that enhances signal-to-noise ratio (see "Seismic Waveform" discussion below) and facilitates the determination of seismic velocities used in processing and interpretation.

In Fig. 5.6 a source is shot into arrays of six receiver groups. During the first shot, the source and receivers are toward the right. As the survey progresses, the source and receiver positions are "pulled" in steps of one interval toward the left. During the first shot, the midpoint between the source (S_1) and the first receiver group (R_1) is the same midpoint as S_2/R_3 and S_3/R_5 for shots 2 and 3, respectively. Likewise, there are common midpoints for other combinations: (S_1/R_2, S_2/R_4, S_3/R_6); (S_2/R_1, S_3/R_3, S_4/R_5); and (S_2/R_2, S_3/R_4, S_4/R_6).

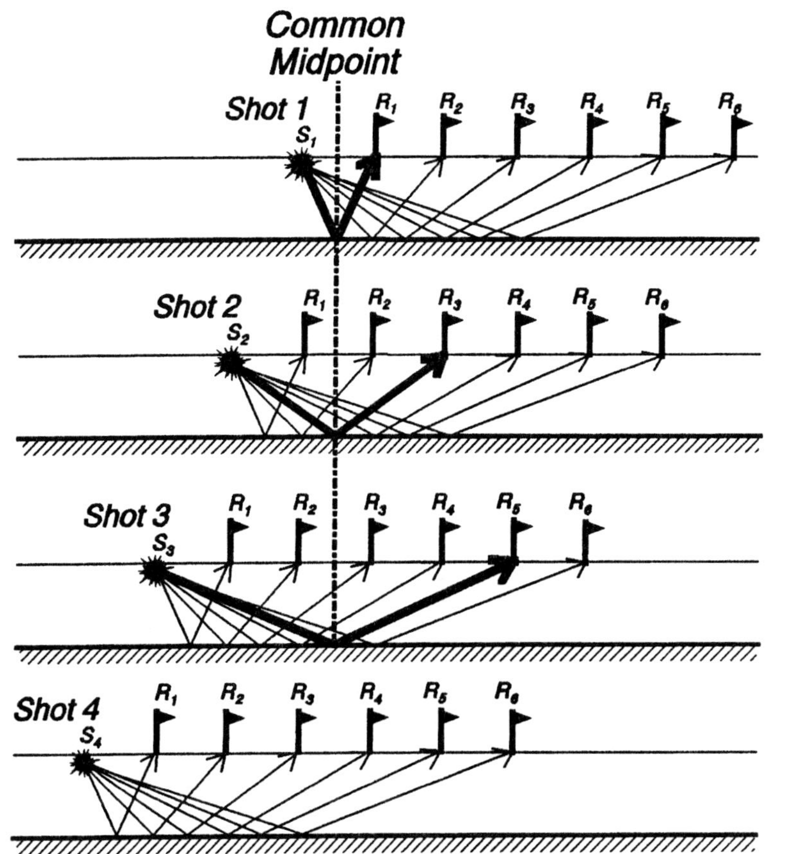

FIGURE 5.6 *Common midpoint (CMP) method.* The six seismic traces for Shot 1 are illustrated in Fig. 5.7a. Fig 5.7b shows the three traces for a common midpoint, highlighted by the bold raypaths.

For the shots illustrated in Fig. 5.6, each record of six seismic traces is known as a *shot gather* (Fig. 5.7a). A display of traces corresponding to reflections around a common midpoint, plotted side-by-side according to horizontal distance (X) from each source, is known as a *CMP* (or *CDP*) *gather* (Fig. 5.7b). A reflection from a horizontal interface, overlain by constant velocity, appears as a hyperbola on both the shot and CMP gathers. Before the three seismic traces on the CMP gather are added together ("stacked"), the event must be corrected for differences in travel time, according to the hyperbola. The differences in travel time are called *normal moveout* (*NMO*). The travel time for a reflection from a horizontal interface is (Chapter 3):

$$\mathbf{T_f = \sqrt{t_0^2 + \frac{X^2}{V_1^2}}}$$

where:

T_f = travel time from source, to interface, to receiver
X = horizontal distance from source to receiver
V_1 = seismic velocity of layer above interface
t_0 = T-axis intercept (time directly down to interface, straight back up to shot location):

$$t_0 = \frac{2h}{V_1}$$

h = thickness of layer above interface
V_1 = seismic velocity of layer above interface.

The normal moveout (T_{NMO}; Fig. 5.7b) is the difference between the travel time along the hyperbola (T_f) and the intercept time (t_0):

$$\mathbf{T_{NMO} = T_f - t_0}$$

A single trace on a seismic section is a composite of the traces from a CMP gather, corrected for T_{NMO}, then added together as a *stacked trace* (Fig. 5.7c). The *fold of stack* is the number of traces from the CMP gather comprising a stacked trace. A resulting profile, comprising numerous stacked CMP traces, is a *stacked seismic section*.

PROCESSING

Seismic reflection data are subjected to various processing steps to enhance reflected signals, attenuate noise not dealt with during acquisition, and to present the data in a more interpretable format. In most cases, signals to be enhanced are primary reflections that occur within (or very nearly within) the plane of the seismic profile.

A *primary reflection* results from energy that undergoes only one reflection before returning to the surface (Fig. 5.8a). Energy that reflects more than once is a multiple reflection (or *multiple*). Multiples can be of many forms, depending on the geometry of the reflected raypaths. A common multiple is a reflection that reflects back down at the surface, then reflects back up again (Fig. 5.8b). Multiples as well as other noise must be attenuated during processing so that primary reflections can be seen.

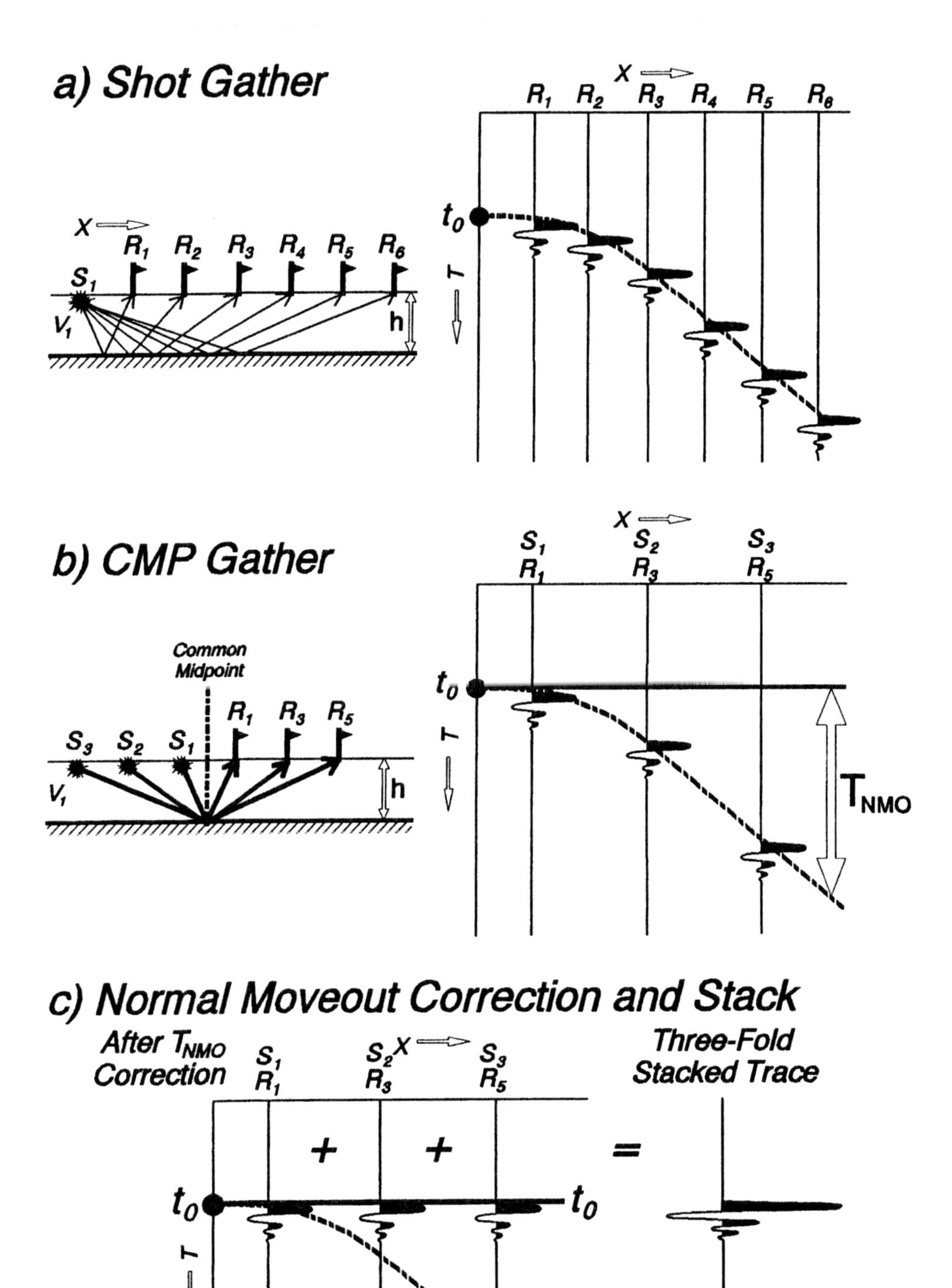

FIGURE 5.7 *Raypath diagrams and travel-time graphs for shot and CMP gathers.* a) Left: Raypaths for one shot with six receiver positions (Shot 1 in Fig. 5.6). Right: A *shot gather* is a travel-time graph of the resulting seismic traces, plotted according to horizontal distance (X) from the source. b) Left: Raypaths centered around a common midpoint, selected from shots 1, 2, and 3 in Fig. 5.6. Right: A *common midpoint (CMP) gather* is the travel-time graph, plotted according to horizontal distance (X) from the respective sources. c) Left: CMP gather after normal moveout (T_{NMO}) correction. The events from the horizontal interface are in phase; peaks and troughs align so that they constructively interfere when the traces are added together. Right: The reflected event on the resulting (3-fold) seismic trace shows enhanced amplitude.

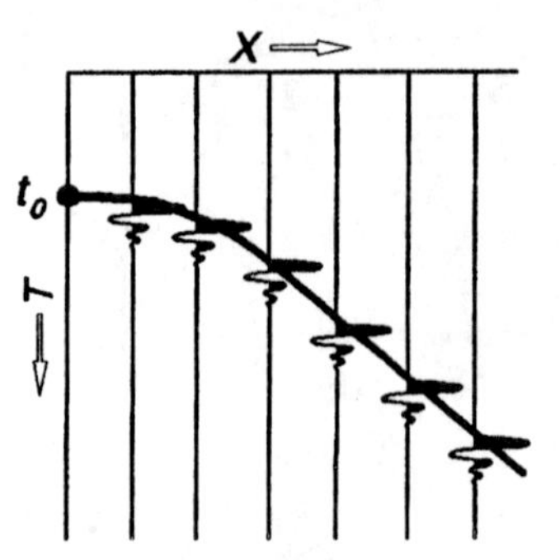

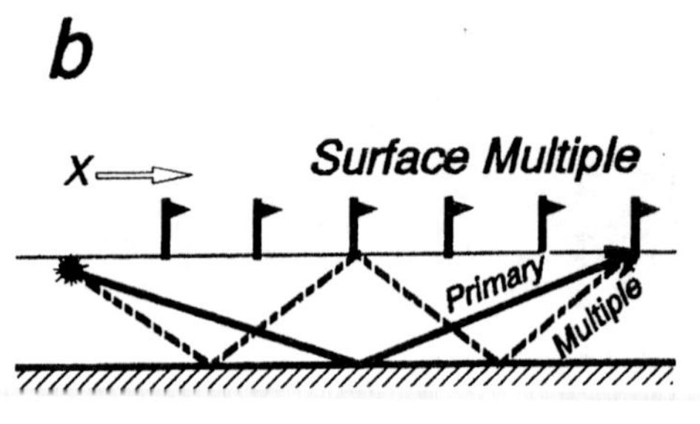

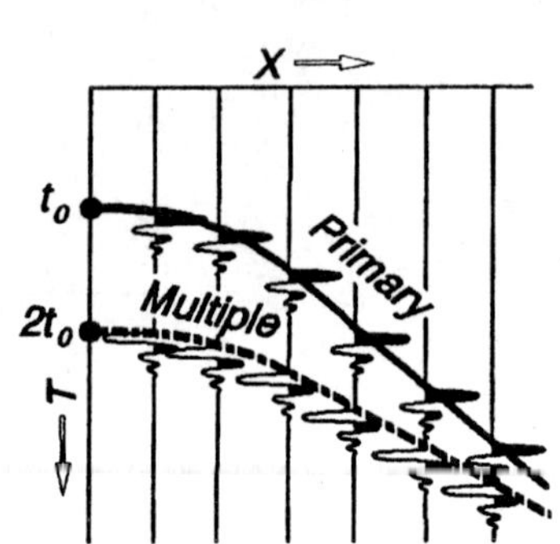

FIGURE 5.8 a) Selected raypath and shotpoint gather for a primary reflection. b) Comparison of primary reflection and surface multiple on raypath diagram and shotpoint gather.

One appealing aspect of the seismic reflection method is that the data may be plotted with the time axis downward, mimicking depth within the Earth. Certain processing steps are thus aimed at enhancing the image created, so that the profile is an abstract representation of a geological cross section. The final product must be interpreted wisely, however, as there are many pitfalls to literal interpretation of seismic reflection profiles (Chapter 6).

General Processing Sequence

Objectives and descriptions of the general steps taken to process seismic reflection data are presented below. More thorough and detailed treatment of processing can be found elsewhere (Robinson and Treitel, 1980; Chap. 9 of Sheriff and Geldart, 1995; pp. 370–395 of Telford et al., 1976).

1. Gather In the gather process, traces from different shot gathers are rearranged as common midpoint (CMP) gathers. The three traces highlighted in bold in Fig. 5.9a comprise the CMP gather in Fig. 5.9b; the same CMP gather, taken through the processing steps in subsequent illustrations, comprises one 3-fold trace on the stacked seismic profile.

2. Velocity Analysis A reflection from a horizontal interface follows a hyperbola, according to:

$$\mathbf{T_f = t_0 + T_{NMO}}$$

where:

T_f = travel time from source, to interface, to receiver

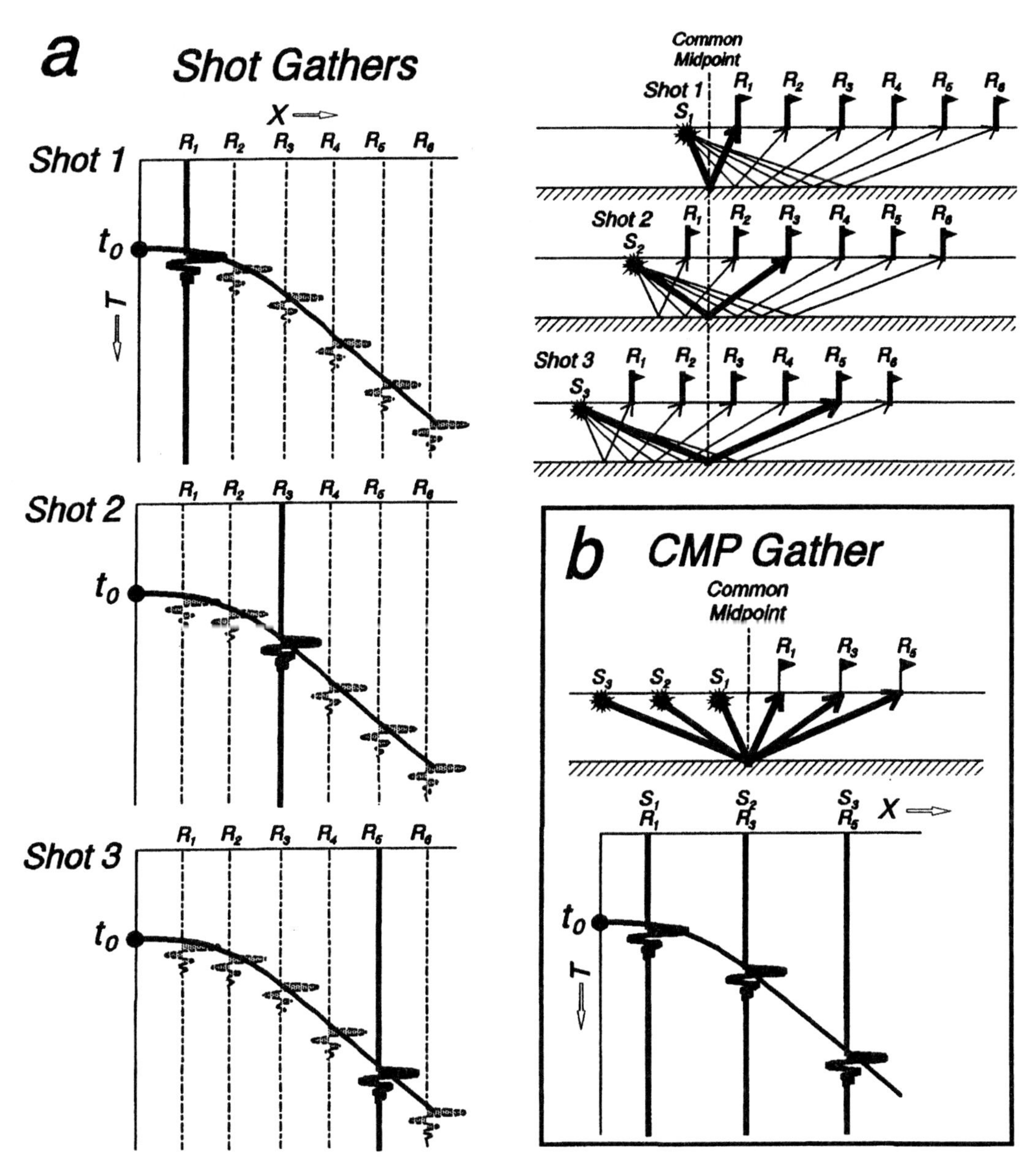

FIGURE 5.9 *Gather.* a) Raypath diagram and shot gathers for three shot records. Highlighted raypaths represent three common midpoint (CMP) traces. b) Raypath diagram and gather for the three CMP traces highlighted in (a).

t_0 = T-axis intercept (time directly down to interface, straight back up to shot location)

T_{NMO} = hyperbolic increase in time with increasing distance from the source.

Velocity analysis determines velocities that best "tune" primary reflections when traces are stacked (Fig. 5.10). The process is commonly trial and error, whereby different normal moveout time (T_{NMO}) corrections are applied to traces in CMP gathers. A desired velocity removes T_{NMO} so that events that originally followed

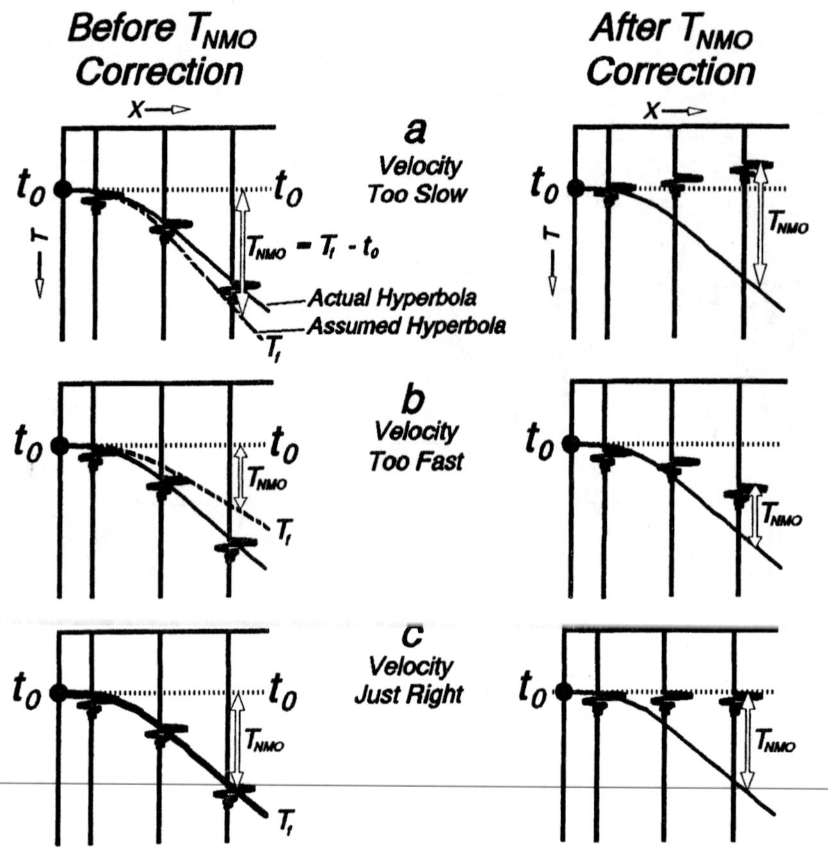

FIGURE 5.10 *Velocity analysis.* A reflection from a horizontal interface follows a hyperbola (left). Corrections according to different velocities attempt to remove T_{NMO}, the increase in travel time with distance from the source (right). a) A slow velocity employs a hyperbola that is too steep, removing too much T_{NMO}. The events are over corrected, lining up out of phase. b) A high velocity does not employ enough T_{NMO} correction; the reflection hyperbola is under corrected, with events out of phase. c) The correct velocity assumes the same hyperbola as the actual events; the T_{NMO} correction aligns the events in phase.

hyperbolas align horizontally. Upon stacking, primary reflections add in phase (constructively), while different types of noise add out of phase (destructively); the analysis thus yields *stacking velocities.*

3. Normal Moveout (T_{NMO}) Correction Normal moveout corrections are determined not just for one event, but for several prominent reflections in a CMP gather. For a given t_0 and velocity, events originally falling along hyperbolas (Fig. 5.11a) align after T_{NMO} corrections (Fig. 5.11b). Commonly, a deeper reflection is corrected for T_{NMO} with a higher velocity [$V_{RMS}(2)$] than that used for a shallower event [$V_{RMS}(1)$].

4. Mute At distance from the source, the tops of records commonly have unwanted noise (direct P and S waves, refractions, events distorted by T_{NMO} corrections), with few or no reflections (Fig. 5.12a,b). Rather than stack this noise together with reflections recorded near the source, a (commonly triangular) region of CMP gather traces are set to zero amplitude, or muted (Fig. 5.12c). On the tops of actual seismic reflection profiles, including many of those in Chapter 6, artifacts of muting can be seen as V-shaped "valleys" in regions where only far offsets comprise the stack traces.

5. Statics Corrections On land, seismic source and receiver stations follow the topography. Corrections must be made for time delays at stations of high eleva-

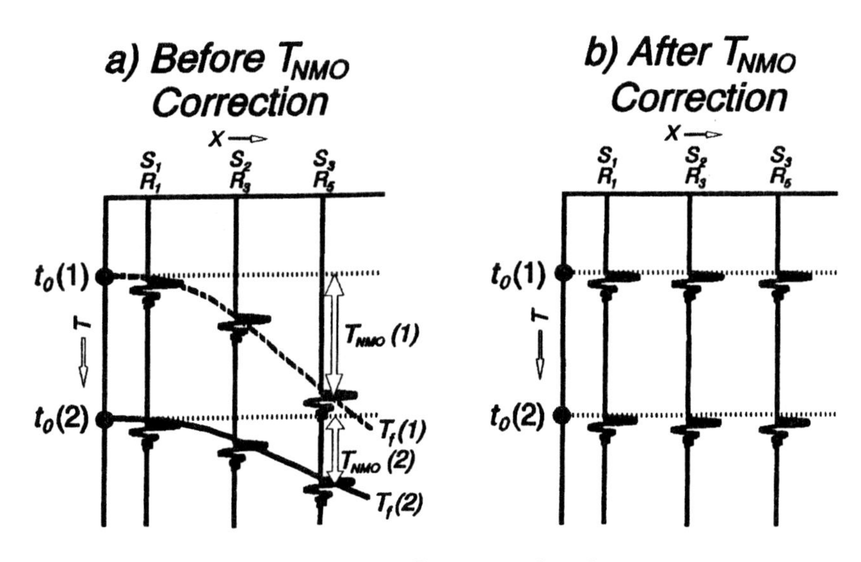

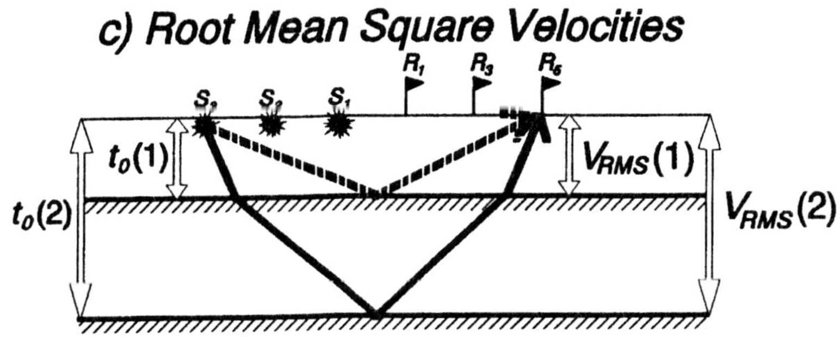

FIGURE 5.11 *Normal moveout corrections*. a) Shotpoint gather showing two hyperbolic events, with T-axis intercepts at $t_0(1)$ and $t_0(2)$. b) T_{NMO} corrections applied according to velocities that best move the hyperbolic events to horizontal alignment at $t_0(1)$ and $t_0(2)$. c) Model of the two-way travel times to the events [$t_0(1)$ and $t_0(2)$] and the velocities that best align the events [$V_{RMS}(1)$ and $V_{RMS}(2)$]. See discussion below on "Types of Velocities" for explanation of root mean square velocities (V_{RMS}) and their correspondence to normal moveout correction ("stacking") velocities.

tion, relative to lower stations. Time is added or subtracted to traces within CMP gathers, according to 1) the estimated near-surface velocity; and 2) the source and receiver elevations relative to a horizontal datum (Fig. 5.13). After these statics corrections are made, T_{NMO}-corrected reflections should be more in phase.

6. Stack After undergoing T_{NMO}, mute, and statics corrections (Fig. 5.14a), CMP traces are added (stacked) together (Fig. 5.14b). The *fold of stack* refers to the number of seismic traces that are combined to make one trace. Stacked traces are displayed side-by-side to make an unmigrated time section (Fig. 5.14c). The display mimics the situation that would result if each seismic trace were recorded at a common source/receiver position; thus, an unmigrated time section is also referred to as a "normal incidence" section (Fig. 5.2).

7. Migration In two dimensions, a reflected event could have come from any position along a semicircle through the event, centered about the common source/receiver location (Chapter 6). Migration spreads events along the potential locations on the semicircle; events on adjacent traces will add in phase at the true position of the reflector, out of phase away from the true position. A migrated time section (Fig. 5.14d) thus attempts to move events to their true horizontal positions, relative to common source/receiver positions on the surface.

8. Depth Conversion The relative positions of events are distorted on time sections because of vertical and lateral changes in velocity (Chapter 6). If velocities

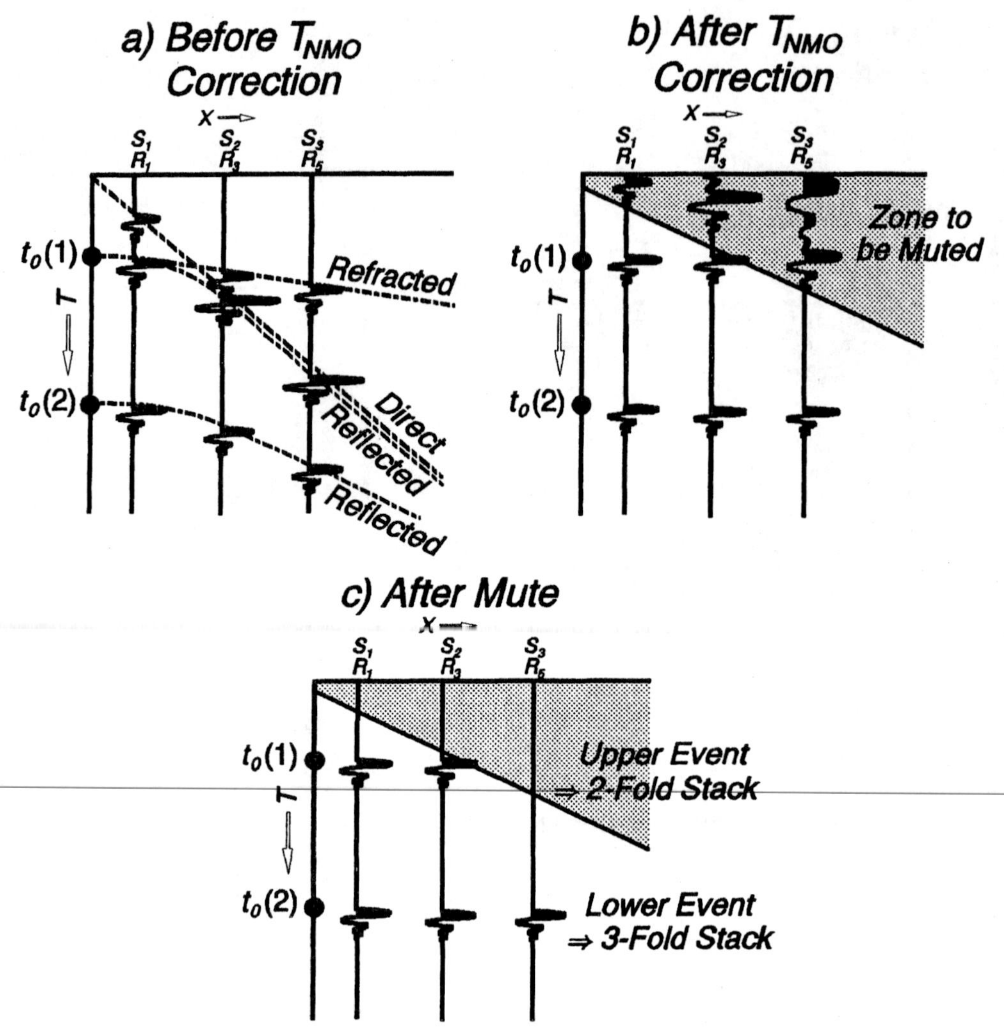

FIGURE 5.12 *Mute.* a) CMP gather before normal moveout corrections, showing a direct and refracted arrival, in addition to two reflections. b) After normal moveout corrections, the direct and refracted events align so as to interfere destructively with the upper reflection. Waveforms in the shaded region have also been distorted ("stretched") as a result of the T_{NMO} corrections. c) Setting the amplitude of traces in the shaded region to zero (muting) removes misaligned events from the records. Though comprised of only two (non-zeroed) traces (2-fold), the upper reflection will tune better than it would have without muting.

are known well enough, the section can be converted from a vertical scale in two-way travel time to a depth scale (Fig. 5.14e).

Other Processing Procedures

At various stages, other types of processing may be done on seismic traces. *Frequency filtering* keeps only a selected band of frequencies, deleting frequencies outside that range. If the input seismic signal is 10–50 Hz, it would not be possible to have reflected signal outside that range; removing frequencies below 10 Hz and above 50 Hz would therefore improve signal-to-noise ratio.

The signal that gets into the ground may not be a simple pulse, but rather short-term multiples (or "reverberation") from near-surface layering. In other words, the input pulse is "convolved" into a closely-spaced series of overlapping,

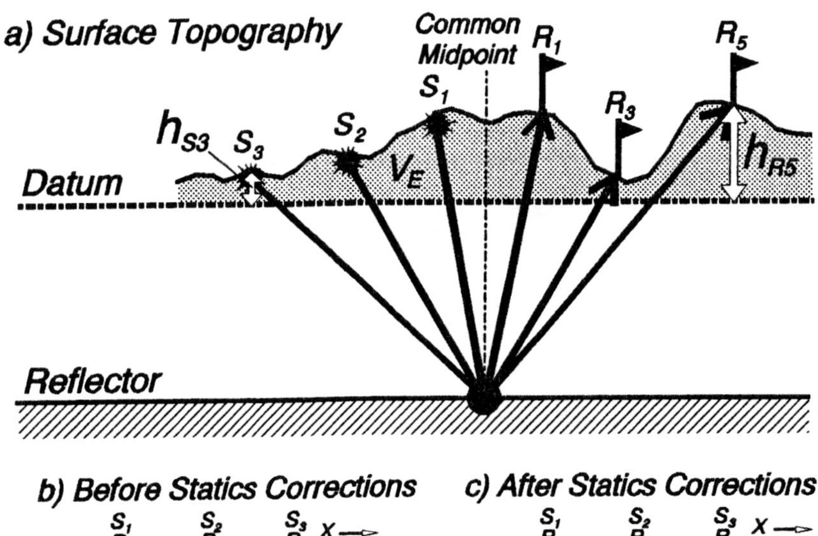

FIGURE 5.13 *Elevation statics corrections.* a) Raypaths for a common midpoint gather, with sources and receivers laid out along the topography. To correct to the datum for the trace associated with S_3 and R_5, the statics correction is $(h_{S3} + h_{R5})V_E$. Relative to the datum, h_{S3} is the elevation of shot 3 and h_{R5} is the elevation of receiver 5; V_E is the assumed velocity of the material above the datum (elevation statics velocity). b) CMP gather after T_{NMO} corrections and mute (Fig. 5.12c), but before statics corrections. Notice that, on each trace, the upper and lower events deviate from horizontal lines by approximately the same amounts. c) Elevation statics corrections applied to the record in (b). Note that events line up in phase, improving amplitudes when stacked.

very shallow reflections. *Deconvolution* attempts to remove the reverberation, sharpening the reflected events.

Seismic signals are attenuated as they penetrate the Earth for three reasons: 1) *Spherical Divergence.* Amplitude of seismic waves decrease with distance from the source. Like the thinning skin of an inflating balloon, the finite amount of seismic energy is spread out over a wavefront of increasing surface area. 2) *Absorption.* As seismic particles vibrate, some of their energy is lost through friction. The amplitudes of seismic waves decrease, as seismic energy changes to heat. 3) *Reflection.* Energy that is reflected back is no longer available to be transmitted deeper into the Earth as the primary wave front. The net effect of these three factors is that reflections from depth have amplitudes several orders of magnitude less than shallow reflections. *Automatic Gain Control (AGC)* balances amplitudes along the length of a seismic trace, generally amplifying the trace as time increases.

Types of Velocities

Various types of seismic velocities are reported on seismic profile legends and in reports of seismic processing and interpretation (Fig. 5.15).

Average Velocity (V_{av}) The distance to an interface, divided by the one-way travel time to that interface, is the *average velocity* for the material above the interface (Fig. 5.15a).

$$V_{av} = z/t$$

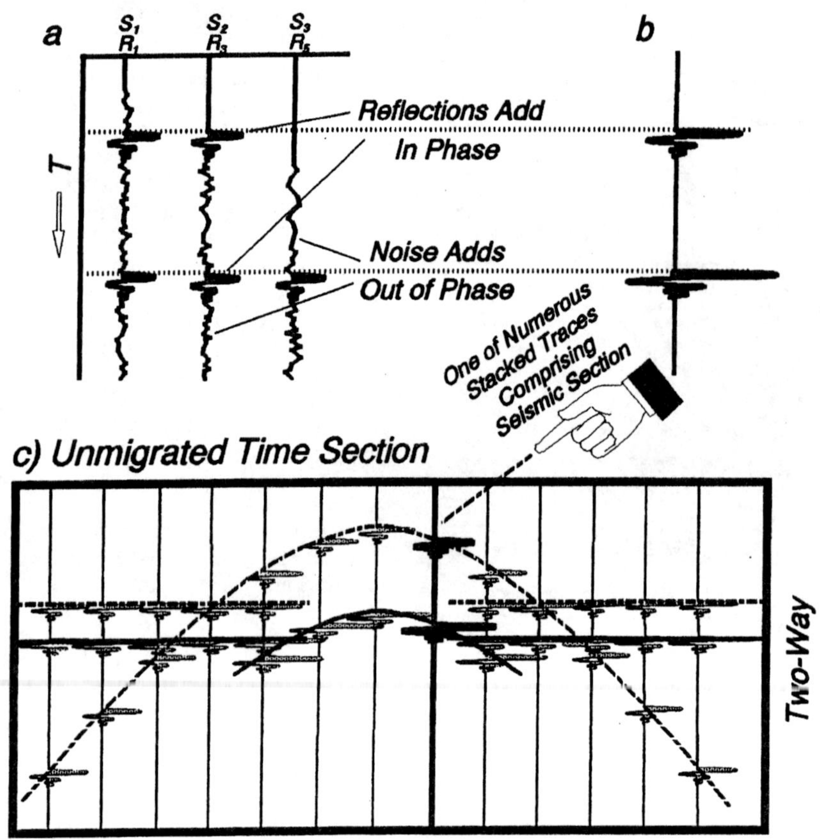

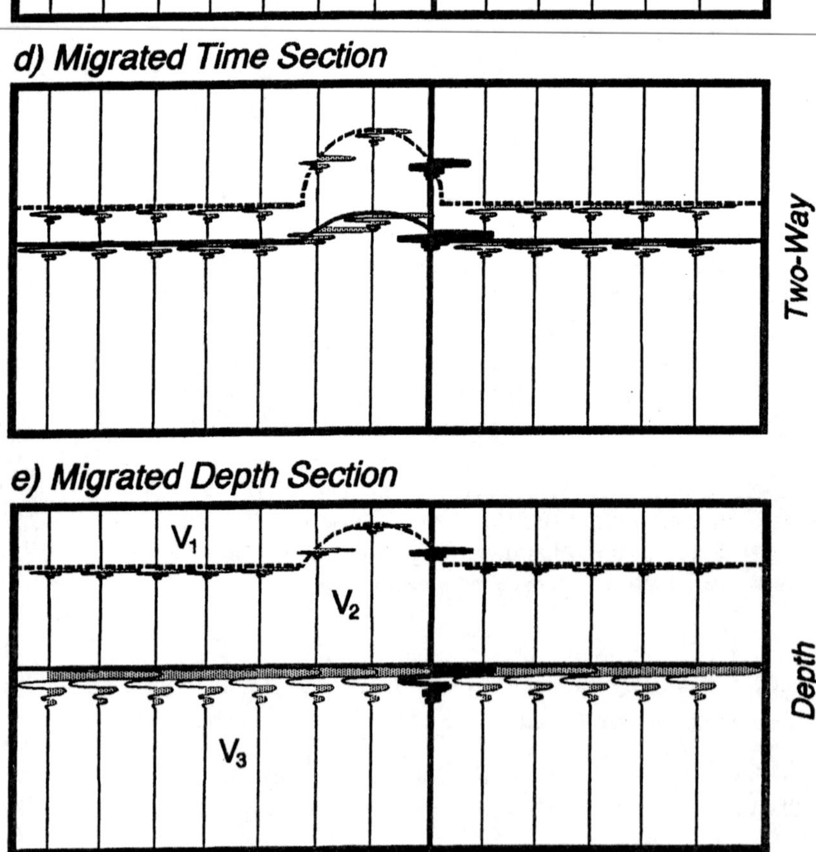

FIGURE 5.14 *Stack, migration, and depth conversion.* a) CMP gather (including random noise), after velocity analysis, T_{NMO}, mute, and elevation statics corrections. b) *Three-fold, stacked seismic trace.* The reflected events from (a) add constructively (in phase), while the noise adds with destructive interference (out of phase). c) The trace from (b) is one of numerous stacked traces that comprise the *unmigrated time section.* d) *Migration.* Events from (c) are moved to their true horizontal positions, relative to the surface. e) *Depth Conversion.* The section in (d) is converted to depth according to velocities of materials above reflectors. The result can be a truer perspective of depths to reflectors within the Earth, and of thicknesses of materials between reflectors.

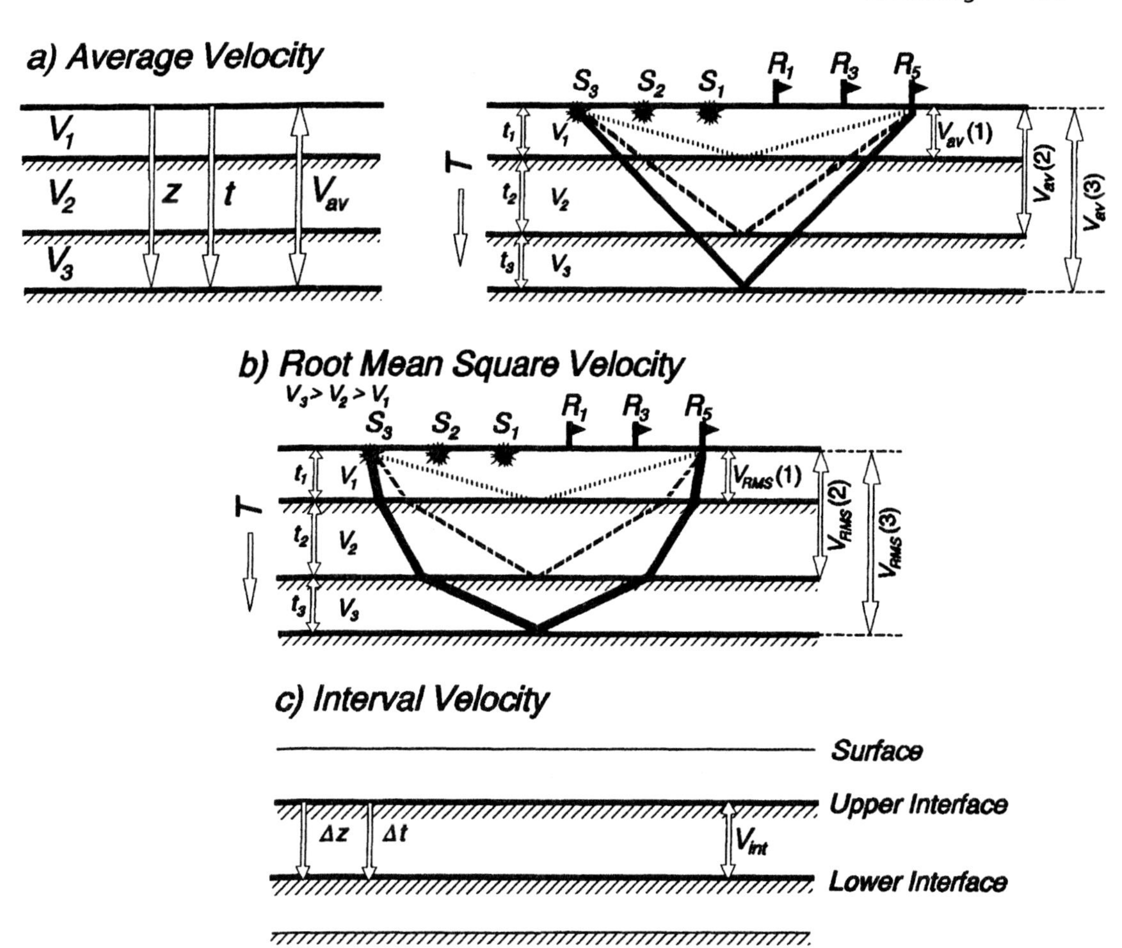

FIGURE 5.15 *Types of velocity*. a) Left: The *average velocity* of the material above an interface (V_{av}) is the depth to the interface (z) divided by the time (t) it takes a wave to travel vertically from the surface to the interface. Right: Hypothetical reflection raypaths, assuming no bending due to Snell's Law. The average velocity is the same as if each ray traveled vertically through each layer. b) *Root mean square velocity*. Raypath bending leads to disproportionally long paths in high-velocity layers. For example, for the bold raypath, near-vertical travel in the low-velocity (V_1) layer leads to a very short ray segment; in the high-velocity (V_3) layer, more horizontal travel leads to a long ray segment. The average velocity along such a U-shaped raypath is approximated by the root mean square velocity (V_{RMS}). c) The *interval velocity* (V_{int}) for a layer is the thickness of the layer (Δz) divided by the travel time spent in the layer (Δt).

where:

V_{av} = average velocity above the interface
z = depth from the surface to the interface
t = one-way travel time from surface to interface.

For a layered sequence above a reflector, the average velocity can be envisioned according to "V-shaped" rays that do not bend across interfaces; the average is time-weighted according to:

$$V_{av} = \frac{\sum_{i=1}^{n} V_i t_i}{\sum_{i=1}^{n} t_i}$$

where:

V_i = seismic velocity of the ith layer
t_i = vertical, two-way travel time within the ith layer.

Root Mean Square (RMS) Velocity (V_{RMS}) Snell's Law describes bending as rays refract across an interface separating different velocities (Fig. 3.22). Rays are refracted so that they travel more horizontally along high-velocity layers, compared to more vertical travel through low-velocity layers (Fig. 5.15b). Consequently, rays spend proportionally more travel time in higher-velocity layers. The *root mean square (RMS) velocity* is a weighted average; it accounts for the disproportionate travel time in high-velocity layers by squaring the velocities in the $V_i t_i$ term, then taking the square root of the averaged sum:

$$V_{RMS} = \sqrt{\frac{\sum_{i=1}^{n} V_i^2 t_i}{\sum_{i=1}^{n} t_i}}$$

where:

V_{RMS} = root mean square velocity
V_i = seismic velocity of the ith layer
t_i = vertical, two-way travel time within the ith layer.

Stacking Velocity The *stacking velocity* is the velocity that best corrects an event on a CMP gather for normal moveout (Fig. 5.10). A stacking velocity is a type of average for the material above the reflecting interface. Due to bending as velocities change across interfaces, raypaths follow "U-shaped" paths (Fig. 5.15b), as opposed to straight, "V-shaped" paths (Fig. 5.15a). A stacking velocity is therefore an approximation of the RMS velocity for the material above the reflecting interface.

Interval Velocity (V_{int}) The *interval velocity* is the average velocity of the material between two interfaces (Fig. 5.15c):

$$V_{int} = \Delta z / \Delta t$$

where:

V_{int} = average velocity between two interfaces
Δz = thickness of layer between the two interfaces
Δt = one-way travel time between the two interfaces.

Velocity information from seismic reflection surveys is in the form of stacking velocities determined during processing. The Dix Equation (p. 130 of Sheriff and Geldart, 1995) computes the interval velocity between reflecting interfaces by assuming that the stacking velocities are RMS velocities (Fig. 5.16):

$$V_{int} = \sqrt{\frac{V_2^2 t_2 - V_1^2 t_1}{t_2 - t_1}}$$

where:

V_{int} = interval velocity between upper and lower interface
V_1 = stacking velocity for reflection from upper interface [$V_{RMS}(1)$]
V_2 = stacking velocity for reflection from lower interface [$V_{RMS}(2)$]

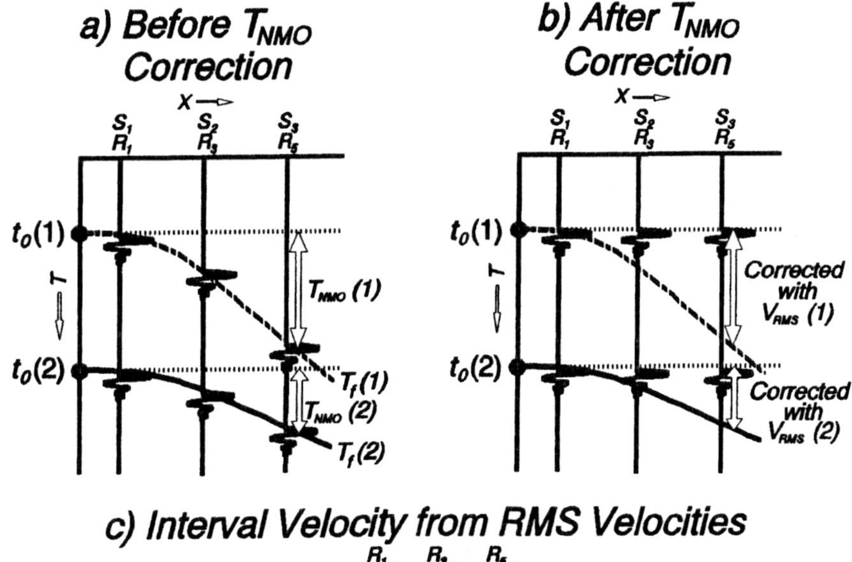

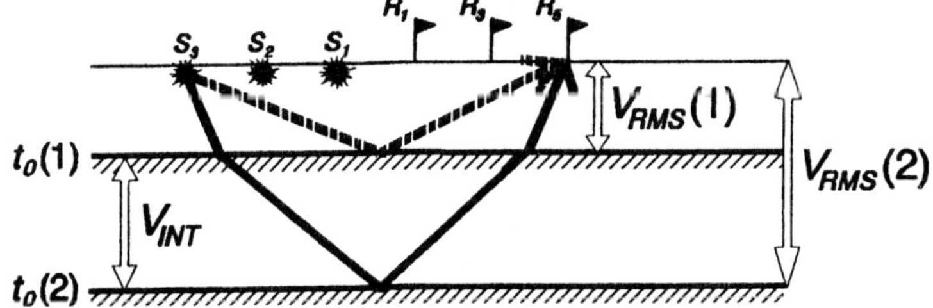

FIGURE 5.16 Determining interval velocities from seismic reflection surveys. T_{NMO} corrections applied to a CMP gather (a) are used to estimate stacking velocities down to each of the two reflectors (b). Assuming the stacking velocities are root mean square velocities (V_{RMS}), the interval velocity between the two reflectors (V_{INT})is computed from the Dix Equation (c).

t_1 = vertical, two-way travel time to upper interface [$t_0(1)$]
t_2 = vertical, two-way travel time to lower interface [$t_0(2)$].

Internal velocities determined from stacking velocities may be used to infer rock velocities, but only under ideal conditions of flat layers with no lateral velocity changes.

Multiples

Multiples (Fig. 5.8) are a common type of noise on seismic reflection profiles. A *long-path multiple* arrives as a distinct event, some time after the primary event. An example would be the reflection from an interface that reflects downward from the surface, then reflects upward again from the same interface (Fig. 5.17a). *Short-path multiples* arrive soon after the primary event, so that they interfere with the primary event. An example is a "peg-leg" multiple that reverberates in a near-surface layer, then reflects from deeper (Fig. 5.17b).

The processing sequence of normal moveout corrections then stacking can attenuate multiples, particularly the long-path variety (Fig. 5.18). A multiple may arrive near the time of a primary reflection from a deeper interface (Fig. 5.18a). The primary reflection, however, will lie along a reflection hyperbola representing higher RMS-velocity than the steeper, lower-velocity hyperbola of the multiple (Fig. 5.18b). Careful selection of stacking velocities will line up the two primary reflections, while the multiple will be out of phase (Fig. 5.18c).

If multiples are not adequately dealt with in processing, they will overprint primary reflections on stacked sections. A surface multiple will have twice the travel

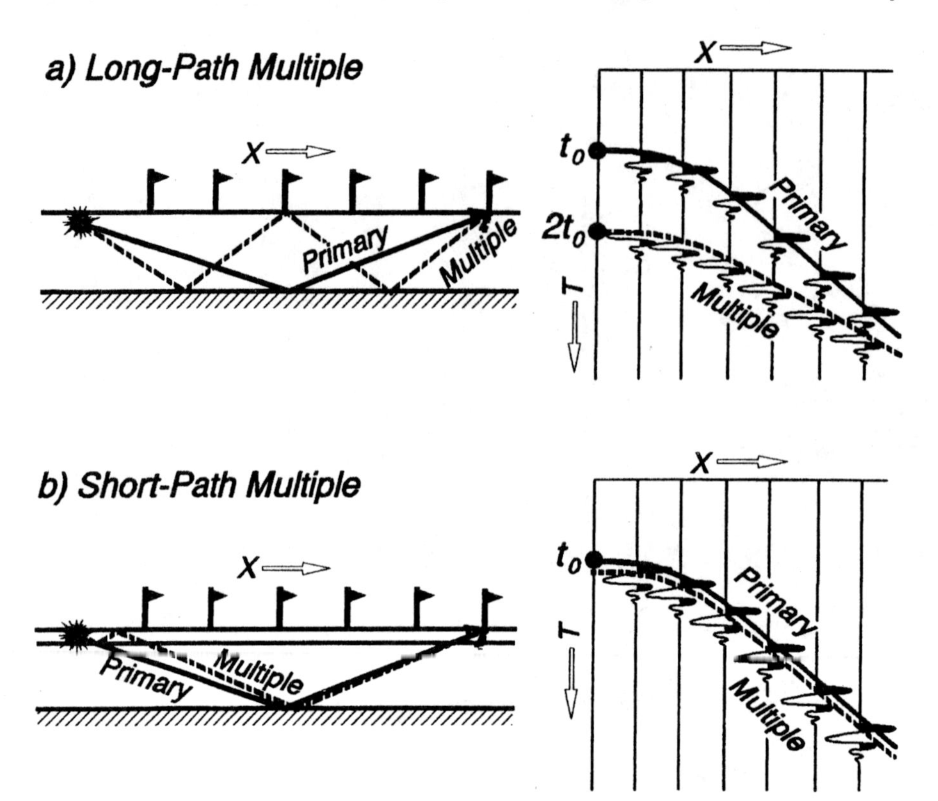

FIGURE 5.17 Raypath diagrams and shotpoint gathers for primary and multiple reflections. a) Long-path (surface) multiple, resulting in an event distinct from the primary reflection. b) Short-path ("peg-leg") multiple, resulting in interference with the primary reflection.

time and twice the apparent dip as the primary event. An example is the *water-bottom multiple* commonly observed on marine profiles (Fig. 5.19).

SEISMIC WAVEFORM

Amplitudes and interference patterns (waveforms) observed on seismic traces provide clues to compositions and thicknesses of Earth materials. Seismic energy is reflected when a material of differing acoustical properties is encountered; a seismic trace resulting from a homogeneous medium would have no reflections (Fig. 5.20a), while an interface separating vastly different materials would generate a strong reflection (Fig. 5.20b).

Factors that determine the appearance of an individual seismic trace are: 1) the *input seismic signal*, dependent on the frequency and phase spectra of the source; 2) *changes in acoustic impedance* within the Earth that are encountered by the seismic signal; and 3) *noise* of many varieties, introduced through the acquisition and processing procedures as well as from cultural and other sources.

Input seismic signal

Although there are many kinds of artificial sources used to generate seismic energy, two basic categories are represented on the final seismic section: 1) those that begin as an abrupt pulse at the time of the reflected event (minimum phase); and 2) those that depict half of the pulse at the time of the event and half after (zero phase).

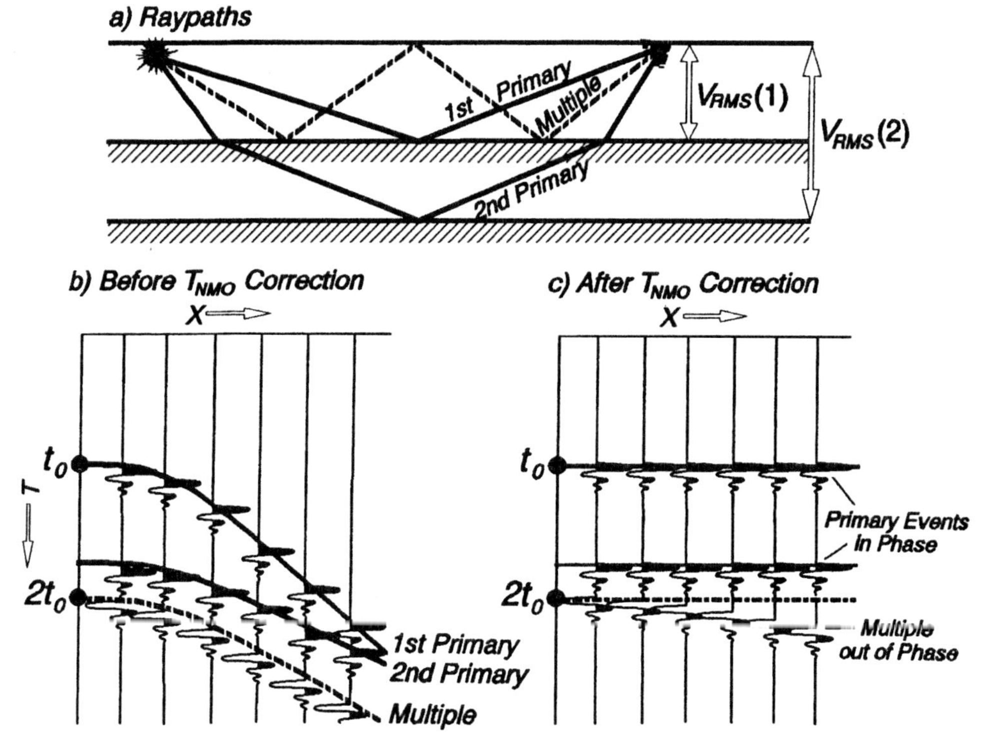

FIGURE 5.18 *Discrimination of primary vs. multiple reflections.* a) Raypath diagram of two primary reflections and a surface multiple reflection from the shallower interface. b) On the gathered traces, the multiple has more normal moveout than the deep primary reflection arriving at about the same time. c) Proper stacking velocity selection will line up the two primary reflections in phase, while miss-aligning the multiple.

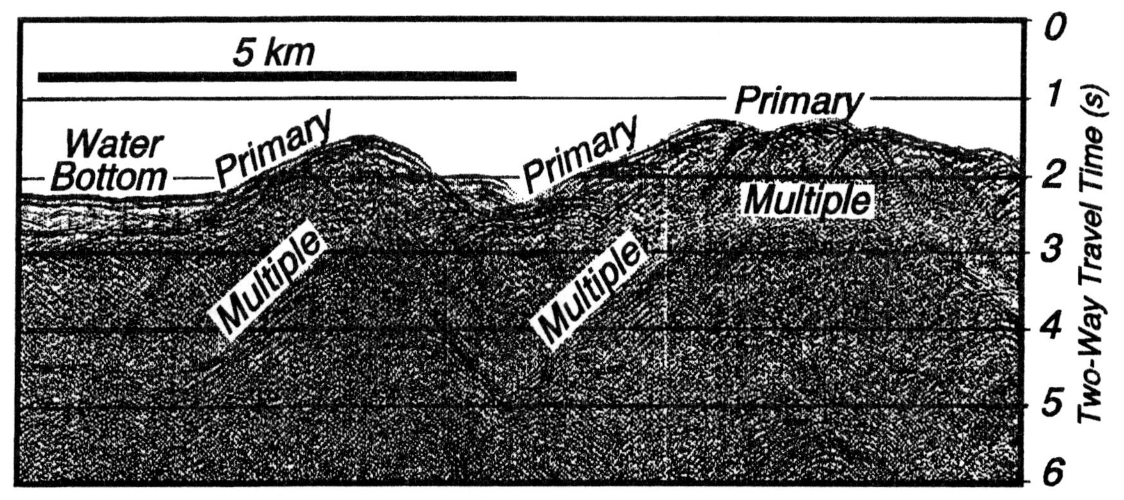

FIGURE 5.19 *Water bottom multiple on stacked seismic section.* The multiple was not completely removed during processing, so that it overprints deeper primary events. The water bottom reflection occurs at 2.5 s in the middle of the section, with the multiple at twice the travel time, or 5.0 s. Similarly, on the left side of the section the water bottom is at 2.25 s, the multiple at 4.5 s. Unmigrated time section from Middle America Trench. From The Middle American Trench, by B. P. Collins and J. S. Watkins, p. 31, in *Seismic Images of Modern Convergent Margin Tectonic Structure,* Studies in Geology No. 26., ed. by R. von Huene, © 1986. Reprinted by permission of the American Association of Petroleum Geologists, Tulsa, Oklahoma.

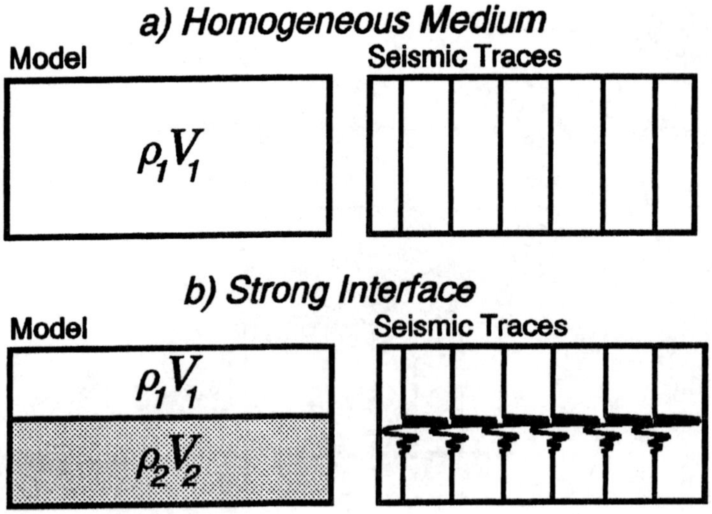

a) Homogeneous Medium

Model

Seismic Traces

$\rho_1 V_1$

b) Strong Interface

Model

Seismic Traces

$\rho_1 V_1$

$\rho_2 V_2$

FIGURE 5.20 Models and resulting seismic traces for: (a) a medium with constant acoustic impedance ($\rho_1 V_1$); and (b) a medium with an interface separating a layer of acoustic impedance ($\rho_1 V_1$) from a layer with different acoustic impedance ($\rho_2 V_2$). ρ = density; V = seismic velocity.

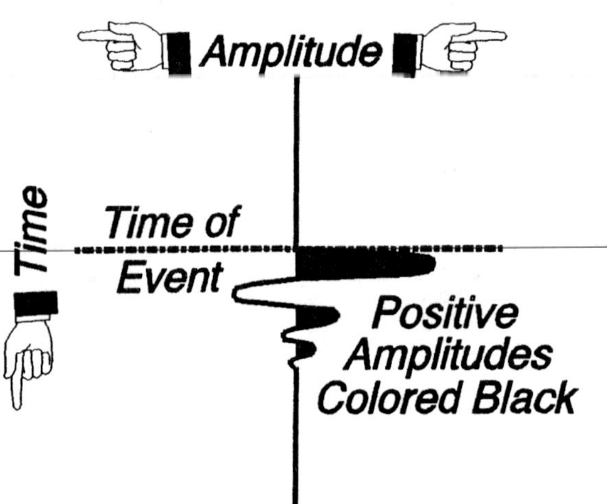

FIGURE 5.21 *Minimum phase seismic pulse.*

Minimum Phase Pulse An explosive source, like *dynamite* on land or *air guns* at sea, results in a burst of energy near the time of the event, followed by reverberation that diminishes with time (Fig. 5.21).

Zero Phase Pulse Instead of an abrupt explosion, vibrations can be sent into the ground as a *"sweep" of continuously varying frequencies.* In one technique, known as *Vibroseis* (trademark CONOCO, Inc.), a massive plate beneath a truck (Fig. 5.22a) vibrates for a few seconds, first with low frequencies, then gradually changing to higher ones (or vice versa; Fig. 5.22b). Computer processing (Vibroseis Correlation) of the resulting seismic trace (Fig. 5.22c) results in a waveform symmetrical about the time of the event (Fig 5.22d).

Width of Seismic Pulse The scale of geologic layering that can be resolved by the seismic reflection method depends on the frequency spectrum of the input pulse. The width (or "wavelength") of pulse of a given frequency can be found through the relationship:

$$V = f\lambda$$

where:

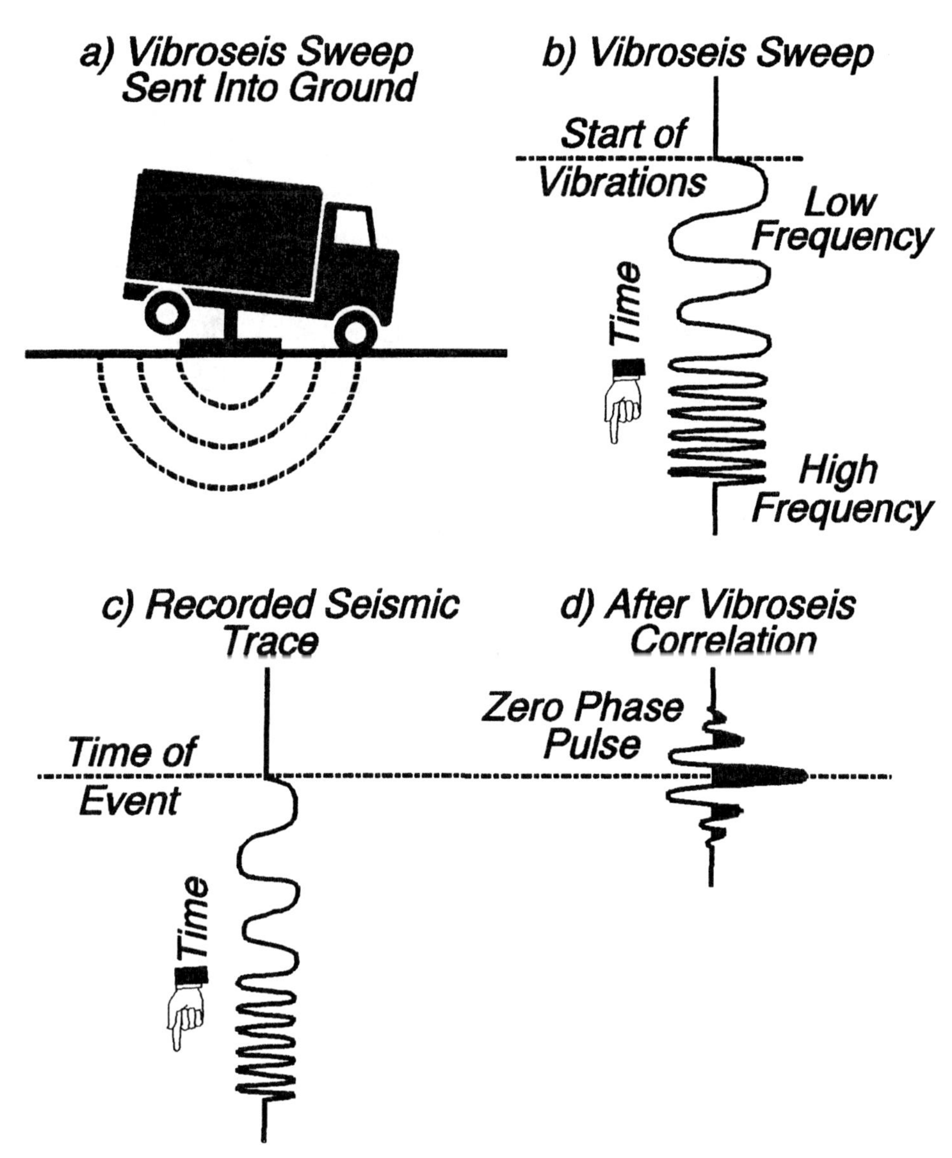

FIGURE 5.22 *Zero phase seismic pulse.* Metal plate beneath truck (a) sends a sweep of continuously changing frequencies (b) into the ground. Beginning at the time of an event, the recorded seismic trace (c) mimics the sweep. After a correlation is made to find the input signal within the recorded trace (Vibroseis correlation), a zero phase pulse (d) represents the event.

V = seismic velocity
f = frequency
λ = wavelength.

The wavelength is thus:

$$\lambda = V/f.$$

A typical reflection survey contains frequencies in the range of about 10 to 50 Hz (1 Hz = 1 cycle/s). For a frequency of 30 Hz and seismic velocity of 3000 m/s, the wavelength is:

$$\lambda = \frac{3000 \text{ m/s}}{30 \text{ ~/s}} = 100 \text{ m}$$

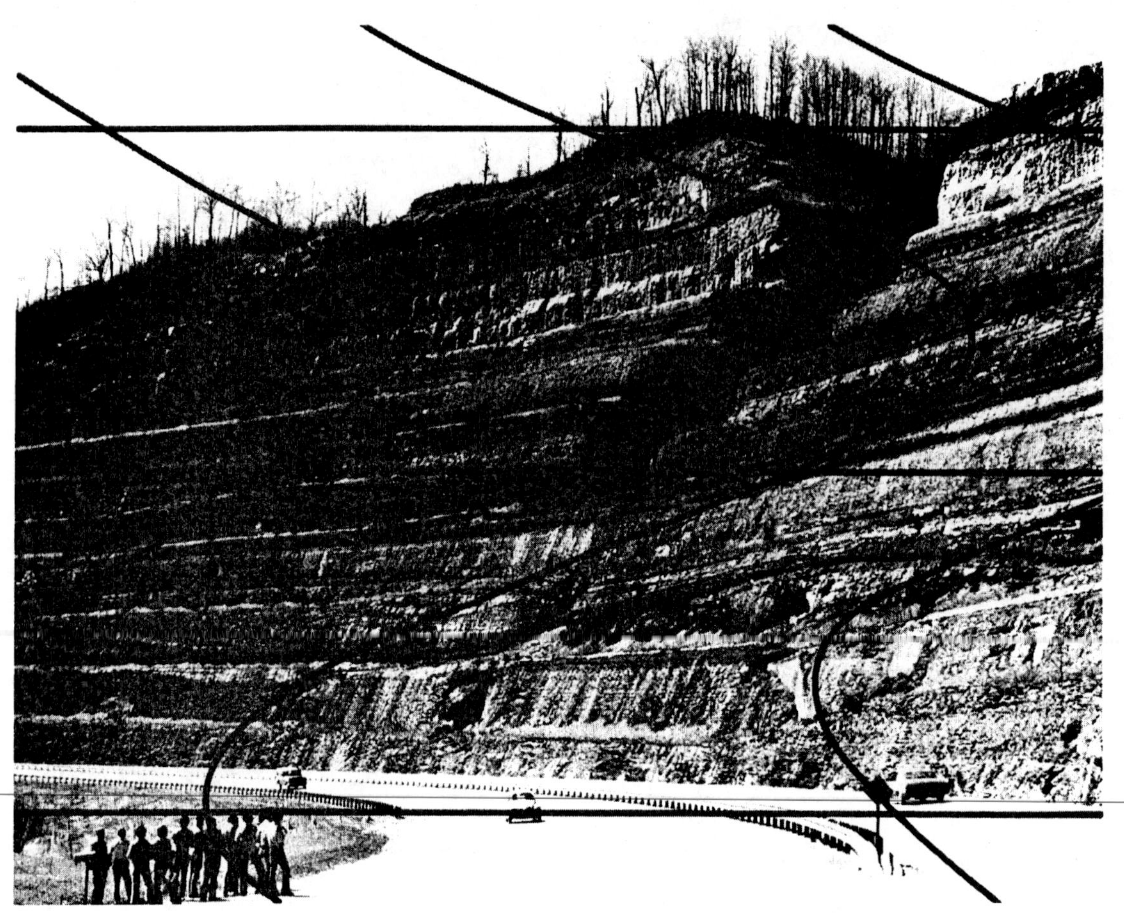

FIGURE 5.23 Portion of typical seismic trace, superimposed on outcrop of sedimentary strata. From *Principles of Sedimentary Basin Analysis,* by A. D. Miall, © 1984. Reprinted by permission of Springer-Verlag, New York.

Fig. 5.23 shows how layering in outcrop scale might compare to a portion of a seismic trace. Note the numerous thin layers that together encompass approximately one half of a seismic wavelength; the reflection method thus responds to geologic changes on a much larger scale than the layers in the photograph.

Acoustic Impedance and Reflection Coefficient

The amount of energy reflected back from an interface between two layers depends on the difference in the *acoustic impedance* of the layers:

$$I = \rho V$$

where:

I = acoustic impedance
ρ = density
V = seismic velocity.

The *reflection coefficient* (Fig. 5.24) expresses the amplitude and polarity of the wave reflected from an interface, relative to the incident wave. It depends on the acoustic impedance of the two materials:

$$RC = \frac{(I_2 - I_1)}{(I_2 + I_1)} = \frac{(\rho_2 V_2 - \rho_1 V_1)}{(\rho_2 V_2 + \rho_1 V_1)}$$

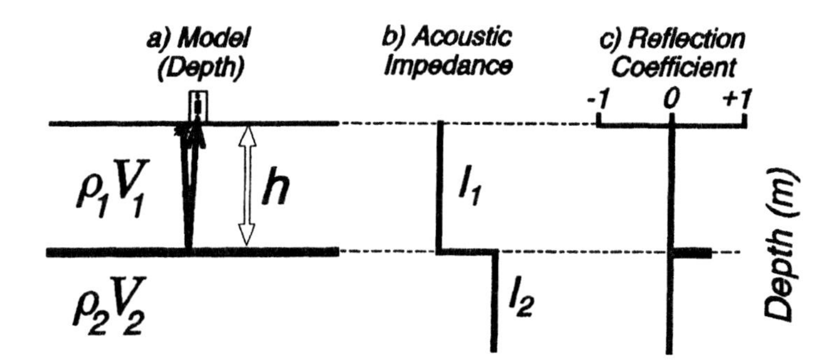

FIGURE 5.24 a) Model of interface at depth (h), separating material of density (ρ_1) and seismic velocity (V_1) from material of density (ρ_2) and velocity (V_2). b) Graph of change in acoustic impedance, from $I_1 = \rho_1 V_1$ to $I_2 = \rho_2 V_2$, across the interface in (a). c) Reflection coefficient resulting from acoustic impedance change in (b).

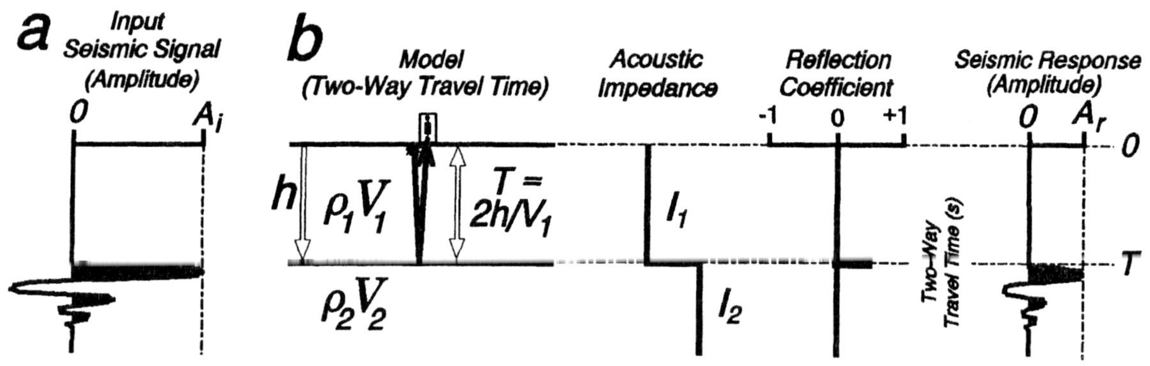

FIGURE 5.25 a) The input seismic signal is a specific waveform with a certain amplitude (A_i). b) The seismic response to an interface has the same waveform as the input signal, scaled to the amplitude of the reflection coefficient (A_r); it appears at two-way travel time (T) dependent on the thickness (h) and velocity (V_1) of the material above the interface.

where:

RC = reflection coefficient of the interface between the two layers
I_1, I_2 = acoustic impedance of the layers above and below the interface
ρ_1, ρ_2 = density of the layers above and below the interface
V_1, V_2 = seismic velocity of the layers above and below the interface.

Reflection from Single Interface In the absence of noise, attenuation, or other seismic arrivals, a *seismic trace* can be viewed as the result of convolving the *input seismic signal* with a series of *reflection coefficients*. For a single interface buried at depth h (Fig. 5.25), the resulting seismic trace has:

$$T = 2h/V_1$$

and

$$A_r = (A_i)(RC)$$

where:

T = travel time from the surface to the interface and back (two-way travel time)
A_i = amplitude of the incident (input) seismic wave
A_r = amplitude of the reflected seismic wave.

The time of the reflected wave is thus a function of the velocity above the interface and the depth of the layer. The amplitude of the reflected wave depends on the

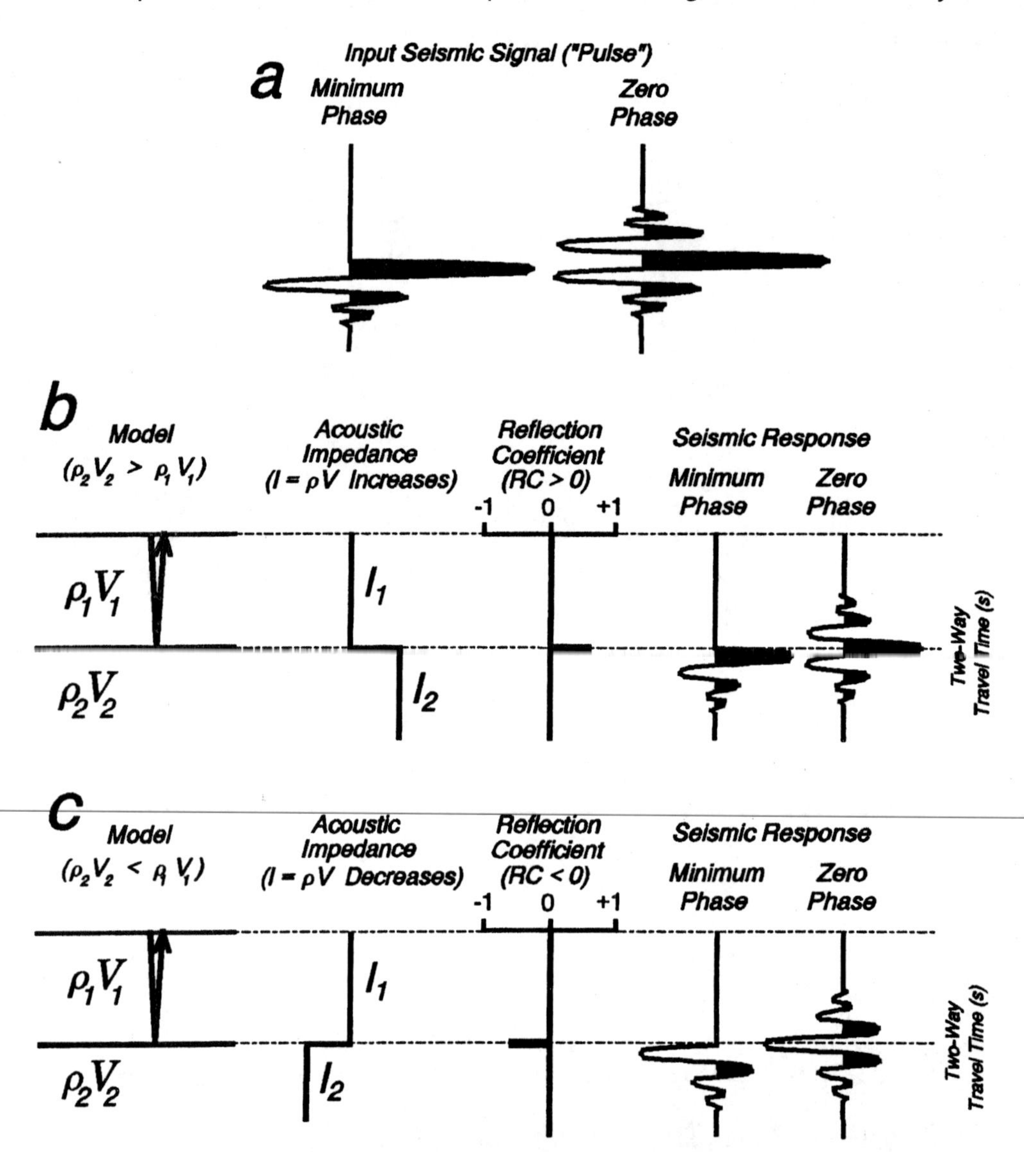

FIGURE 5.26 a) Input minimum phase and zero phase seismic signals. b) Seismic response for reflection at interface with positive reflection coefficient. c) Seismic response for negative reflection coefficient.

amplitude of the incident wave and the strength of the reflection coefficient, which depends on the difference in the acoustic impedance of the two layers.

Whether the reflected wave has the same or opposite polarity as the input seismic signal (Fig. 5.26a) depends on whether the reflection coefficient is positive or negative. If the acoustic impedance of the reflecting layer is greater than that of the layer above ($I_2 > I_1$), the reflection coefficient will be positive ($RC > 0$). The resulting reflection has the *same polarity* as the input signal (Fig. 5.26b). If the acoustic impedance of the reflecting layer is smaller ($I_2 < I_1$), the reflection coefficient is negative ($RC < 0$). The reflection is the *opposite polarity* of the input signal (Fig. 5.26c).

Reflections from Several Interfaces When there is more than one reflecting interface the seismic trace is the sum of the responses for each interface. If the interfaces are sufficiently far apart (compared to the width of the input signal), three distinct events will appear on the seismic trace (Fig. 5.27a). If, however, the input signal

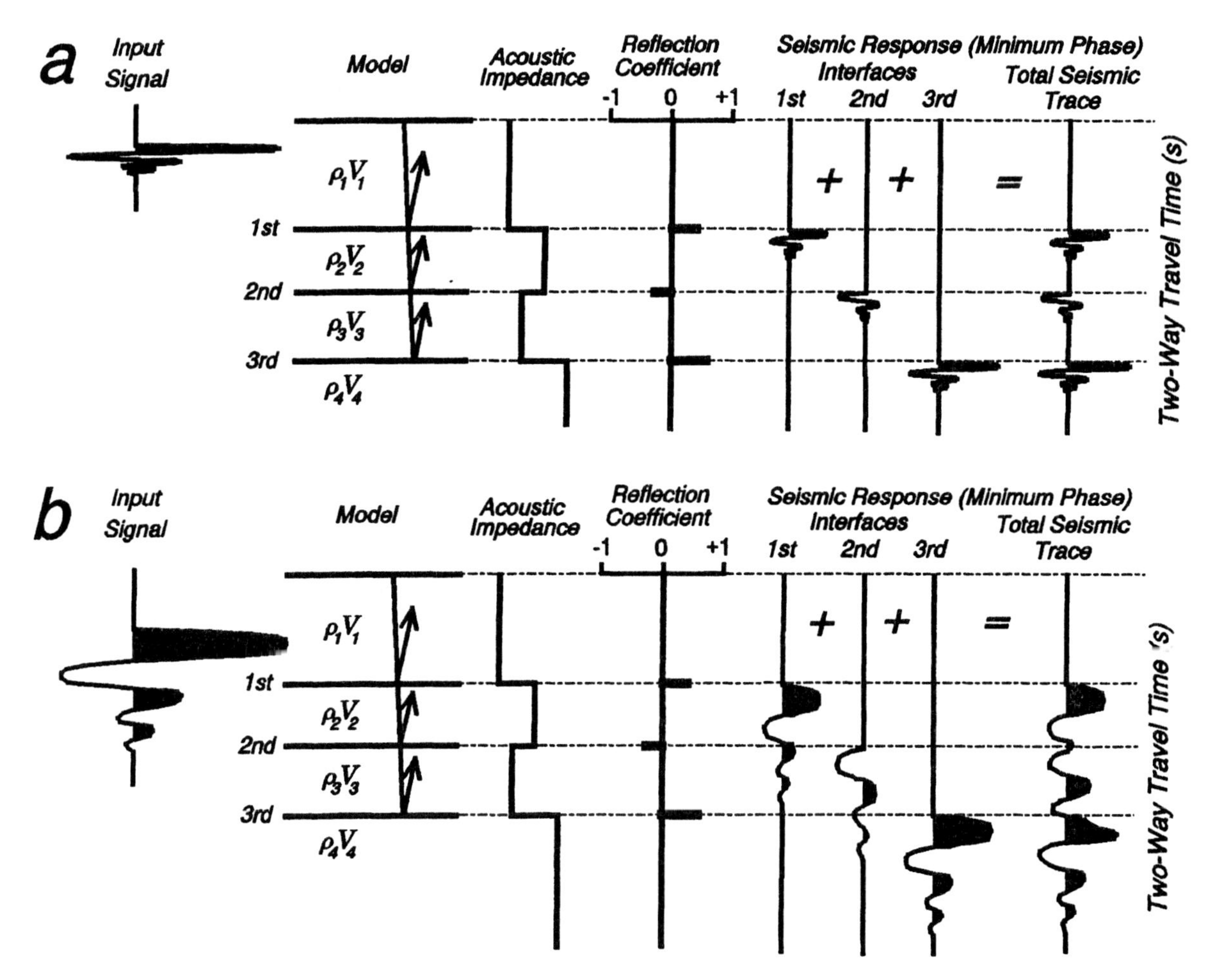

FIGURE 5.27 Response of reflections from several interfaces with (a) *high frequency* and (b) *low frequency* input seismic signals.

is broad relative to the time spacing of the reflection coefficients, the seismic trace shows the interference pattern caused by the overlapping events (Fig. 5.27b).

Noise

Noise on a seismic reflection profile can be defined as *the deflection of a seismic trace caused by anything other than energy reflected once from an interface.* Noise can be natural, cultural, or induced by the seismic method itself.

Natural Noise Seismic receivers (geophones) can be shaken as the result of natural phenomena. The *wind* shakes trees, bushes, and other objects, causing vibrations of the ground that are recorded by the geophones. Likewise, *animals* walking near geophones, or *water* flowing in nearby streams can cause ground vibrations. Wind, rain, and animals can also shake the cables connecting geophones, causing the geophones to vibrate.

Cultural Noise Activities of people can contribute to noise. *Cars, trucks, trains,* as well as *people* walking near the survey line, cause the ground to rattle. *Electrical power lines* cause a magnetic field which can interfere with electrical systems of a seismic survey.

Noise Induced by Seismic Acquisition and Processing The seismic reflection method attempts to image *primary reflections* (Figs. 5.5c, 5.8, 5.17). Other seismic events, or phenomena introduced by acquisition or processing, interfere with the primary reflections. *Rayleigh waves* (also referred to as groundroll, Fig. 5.5b) appear as low-frequency events cutting across the reflected arrivals. *Direct waves, critically refracted waves,* and *diffractions* (energy radiating from a point source) can also arrive at the same time as reflections (Fig. 5.12a).

Multiple reflections (energy reflected more than once from the same interface) also interfere (Figs. 5.8b, 5.17, 5.18, 5.19). Seismic surveys best attenuate noise when survey lines are straight and continuous. Surveys may have to skip or go around areas with buildings, highways, rivers or other obstructions, causing gaps or bends in the recorded data. Noise can also be introduced in the processing of data, through factors like over-migration or poor normal moveout corrections (Fig. 5.10).

Total Seismic Trace

Convolution A *seismic trace* can be viewed as 1) the input seismic signal, convolved with 2) a reflection coefficient time series (resulting from acoustic impedance changes), with 3) noise superimposed (Fig. 5.28). Convolution means replacing each reflection coefficient with the input seismic signal, scaled to the amplitude and polarity of the reflection coefficient (Fig. 5.28a). Mathematically, convolution of an input signal with a reflection coefficient series is (see p. 280 of Sheriff and Geldart, 1995):

$$I(t) * RC(t) = A(t)$$

where:

$I(t)$ = amplitude of the input signal at time t
$*$ = convolution operator
$RC(t)$ = amplitude of the reflection coefficient at time t
$A(t)$ = amplitude of the seismic trace at time t.

Noise adds algebraically to the convolved seismic trace (Fig. 5.28b):

$$[I(t) * RC(t)] + N(t) = A(t)$$

where:

$N(t)$ = amplitude of the noise at time t.

Signal to Noise Ratio The ability to see stratigraphic changes on seismic profiles depends on the amplitude of reflections, relative to the level of noise (*signal-to-noise ratio*, Fig. 5.29). Acquisition and processing techniques are designed to filter out noise while keeping signals resulting from primary reflections, thus enhancing the signal-to-noise ratio.

Frequency and Phase Components of Seismic Trace

A continuous function, like a seismic trace represented as a time series, may be expressed as the sum of sine wave frequency components; each component has a certain *amplitude* and degree of *phase shift*, relative to other frequency components. The components can be expressed through a *Fourier series* (see Telford et al., 1976, p. 371; Sheriff and Geldart, 1995, p. 531). For the portion of a seismic trace which

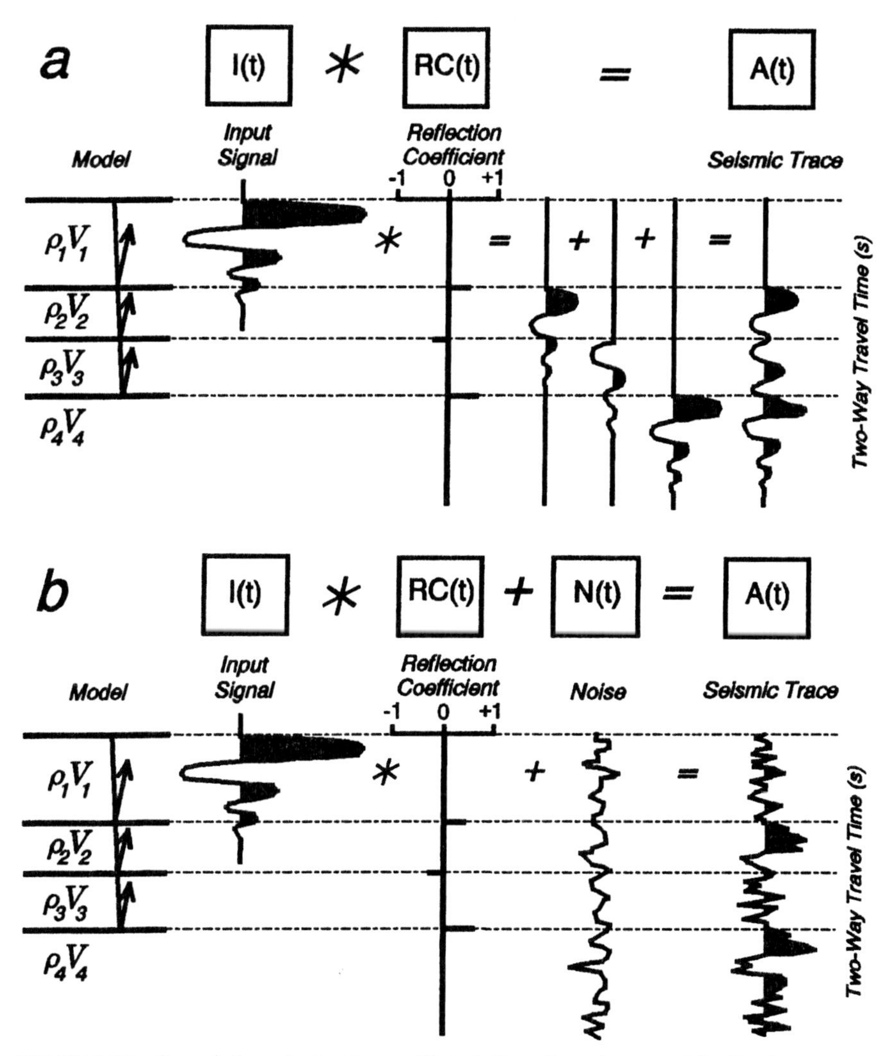

FIGURE 5.28 Convolution of reflection coefficients from three-layer model with a low-frequency input signal (Fig. 5.27b). a) Without noise. b) With noise.

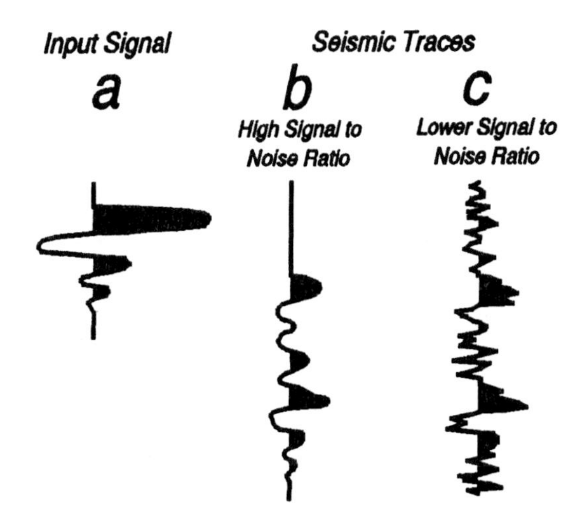

FIGURE 5.29 *Signal-to-noise ratio.* a) Input seismic signal. b) Events mimicking the input signal can be recognized on seismic trace with high signal-to-noise ratio. c) On a trace with low signal-to-noise ratio, seismic events are hidden within the noise.

extends from time 0 to time T, a Fourier series expansion of the trace can be expressed as:

$$A(t) = \sum_{n=0}^{\infty} a_n \cos(2\pi nt/T) + \sum_{n=1}^{\infty} b_n \sin(2\pi nt/T)$$

where:

$A(t)$ = amplitude of the seismic trace at time t
a_n, b_n = coefficients that express the amplitude and phase shift of a given frequency component.

Visualization of the sine wave components of the input signal can help us understand the appearance of reflected events on seismic traces. For example, assume that the zero-phase pulse depicted in Fig. 5.22d has frequencies from 10 to 50 Hz. The Fourier series simply expresses that the sine wave components between 10 and 50 Hz (Fig. 5.30a) can be added with equal amplitudes, in phase exactly at the time of the event (Fig. 5.30b). Near the time of the event ("1" in Fig. 5.30b), there is a central peak or trough, depending on the polarity of the reflection coefficient (Fig. 5.26). Away from this central "lobe" there are alternating troughs and peaks, or "sidelobes," resulting from the different frequency components being in and out of phase at various times ("2" and "3" in Fig. 5.30b).

Fig. 5.31 illustrates two 10–50 Hz pulses. The *frequency spectra* show that only frequencies of 10 to 50 cycles per second (or Hz) are represented, and that their amplitudes are equal. The *phase spectra* reveal how the central peaks of the frequency components for each pulse line up relative to one another. Fig. 5.31a shows that, for a zero-phase pulse, each of the frequency components are exactly at their peak amplitude at the time of the event, thus adding together constructively to give the central lobe. For the minimum-phase pulse (Fig. 5.31b), the components are

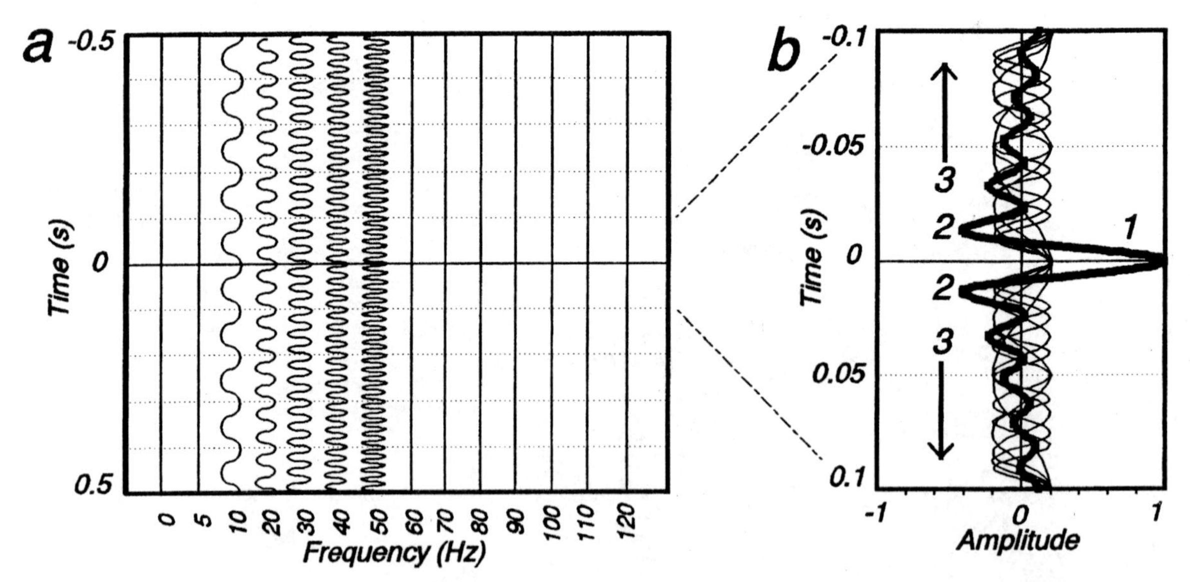

FIGURE 5.30 a) Sine waves with frequencies from 10 to 50 cycles per second constitute a 10–50 Hz pulse; frequencies below 10 Hz and above 50 Hz have zero amplitude. b) Frequency components (thin lines) of a 10–50 Hz, zero-phase pulse (bold line). 1. At the time of the event (t = 0), the frequency components add together to give a central peak. 2. At about t = +0.015 and −0.015 s, the individual frequency components add to minima, giving the negative sidelobes. 3. At larger positive and negative times, the components add to varying degrees, producing other (but smaller) sidelobes.

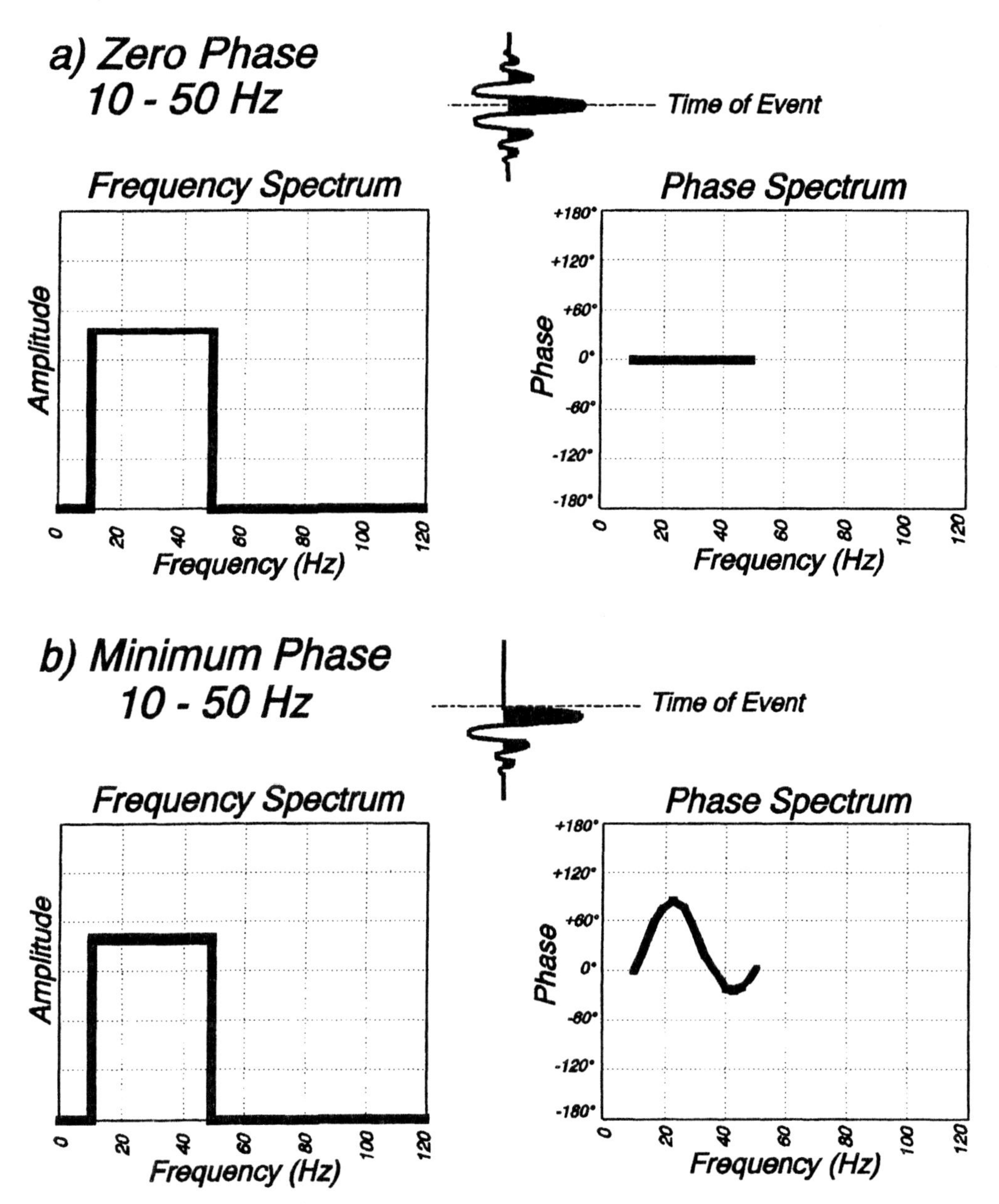

FIGURE 5.31 *Frequency and phase spectra for 10–50 Hz pulse.* For both pulses illustrated, the frequency spectra show equal amplitudes for the 10 Hz to 50 Hz components, zero amplitude for frequencies outside that range. a) *Zero-phase pulse.* The phase spectrum shows that the 10 to 50 Hz frequency components are in phase, so that the pulse is symmetrical about the time of the event. b) *Minimum-phase pulse.* Frequency components are phase-shifted relative to one another, in such a way that the bulk of the pulse arrives just after the time of the event.

shifted in phase in such a way that they sum to zero before the time of the event, reach a maximum value soon after the time of the event, then gradually approach zero amplitude (see Sheriff, 1978, 1984).

Extreme cases are illustrated in Fig. 5.32. A 40 Hz pulse has only one frequency component, producing a pure sine wave with no way to tell when the event occurred (Fig. 5.32a). In contrast, a wave that has all frequencies of equal amplitude

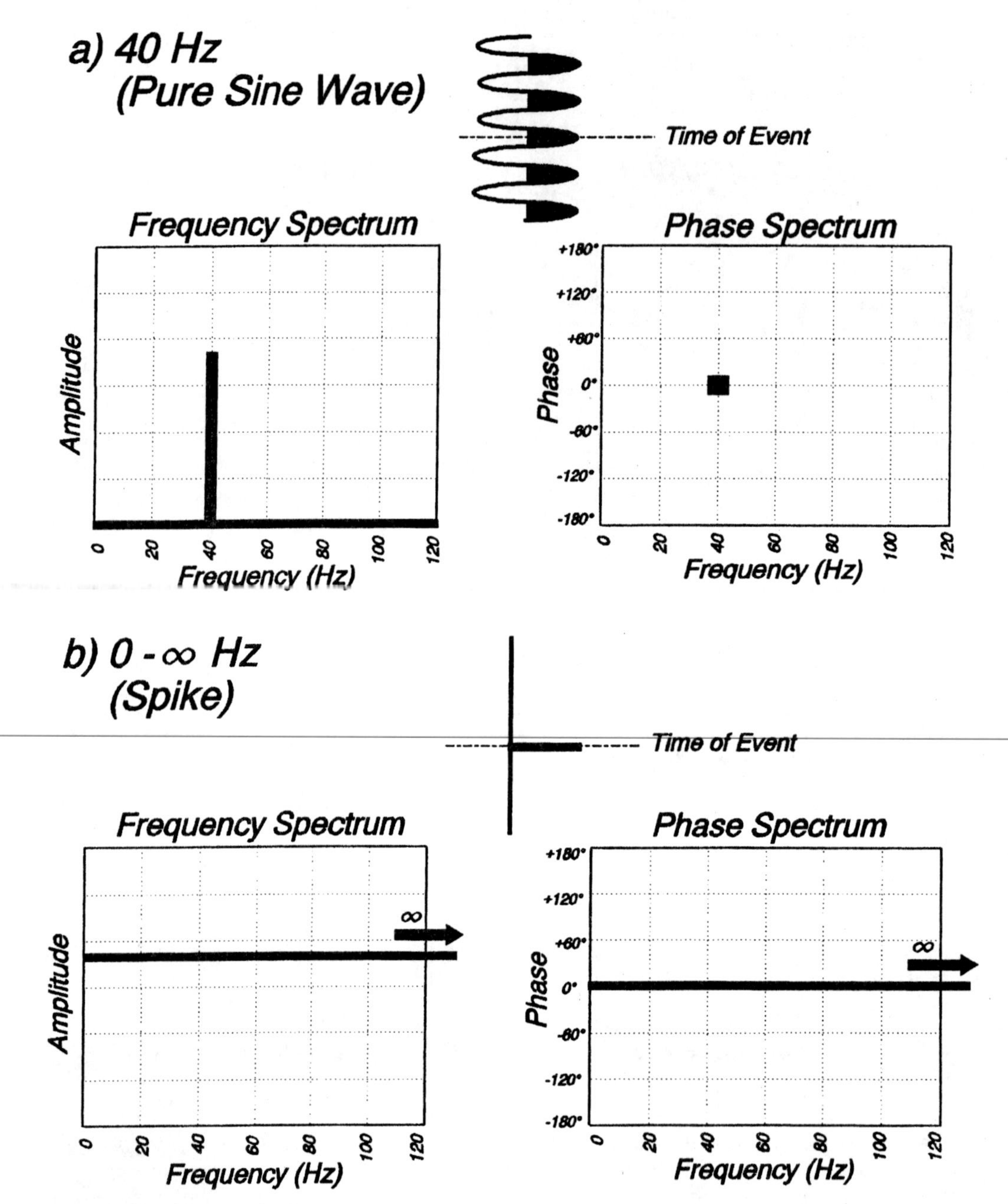

FIGURE 5.32 Seismic pulse, frequency spectrum, and phase spectrum for: a) only one frequency component (40 Hz); and b) all frequency components (0 – ∞ Hz) represented with equal amplitude.

(zero to infinite Hz), results in zero amplitude everywhere, except at the time of the event, where a discrete event ("spike") occurs (Fig. 5.32b).

Seismic Resolution and Definition

The ability to see stratigraphic changes on seismic profiles depends on two factors: 1) a *high signal-to-noise ratio*; and 2) a *seismic pulse with a broad bandwidth encompassing both low and high frequencies*.

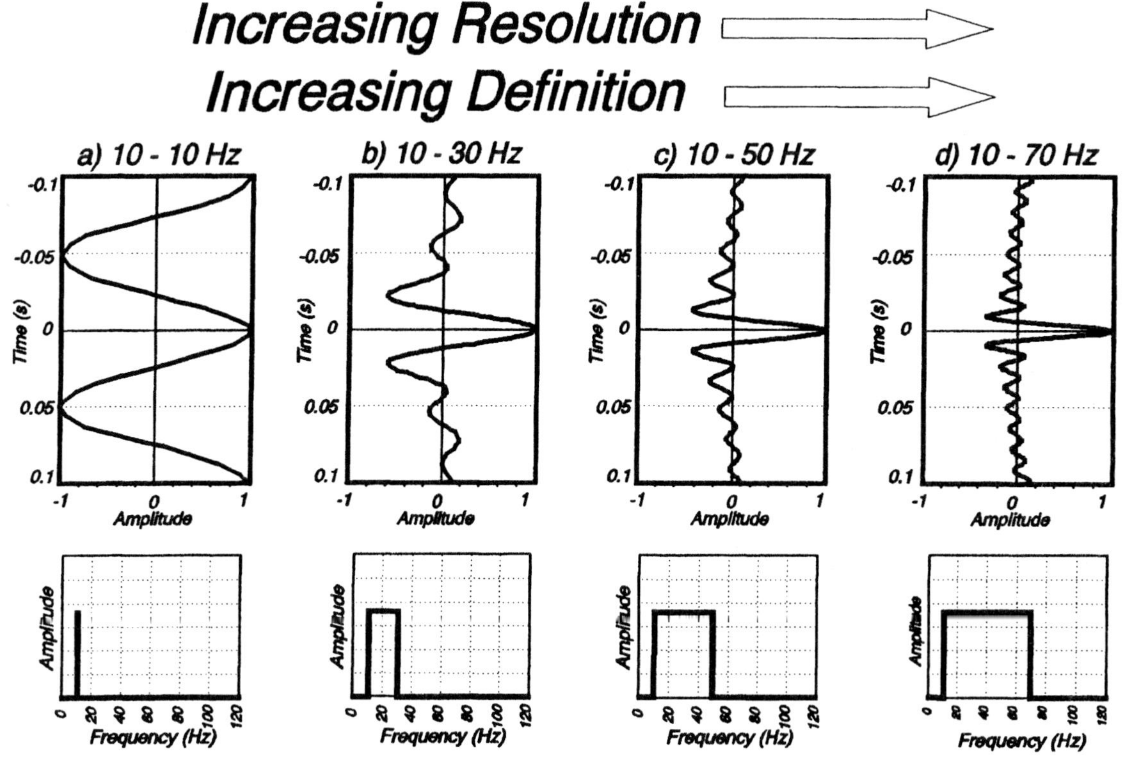

FIGURE 5.33 Change in form of zero phase seismic pulse as more and more frequencies are added to the high end of the spectrum. Both *seismic resolution* (content of high frequencies) and *seismic definition* (bandwidth) improve from (a) to (d).

The thickness of layering that can be distinguished on seismic profiles depends on how well a seismic pulse can *resolve* those layers. An equally important quality, however, is the ability of the pulse to *define* the presence of interfaces. Both resolution and definition are dependant on the frequency content of the pulse.

Fig. 5.33 illustrates two things that happen to a pulse as more and more frequency components are added: 1) the time between the main lobe ("peak") and the adjacent sidelobes ("troughs") becomes less and less as higher and higher frequencies are added (*resolution* increases); and 2) the ratio of the amplitude of the main lobe to the adjacent sidelobes becomes greater as the bandwidth of frequencies increases (*definition* increases).

Seismic Resolution Seismic resolution depends on having *high enough frequencies* to recognize closely spaced reflection coefficients. A pulse with only low frequencies is very broad, with sidelobes far apart from the main pulse; it is not possible to resolve thin layering with such a pulse (Fig. 5.27b). When high frequencies are present, the pulse is narrower, with sidelobes closer to the time of the main pulse; better resolution results (Fig. 5.27a). Higher resolution is possible with the pulses in Figs. 5.33c and 5.33d, compared to the pulses in Figs 5.33a and 5.33b.

Seismic Definition Even if high frequencies are present, it may not be possible to determine that layers are present. Seismic definition requires that there be a *broad enough bandwidth* of frequencies to detect that layering exists (Fig. 5.34). The

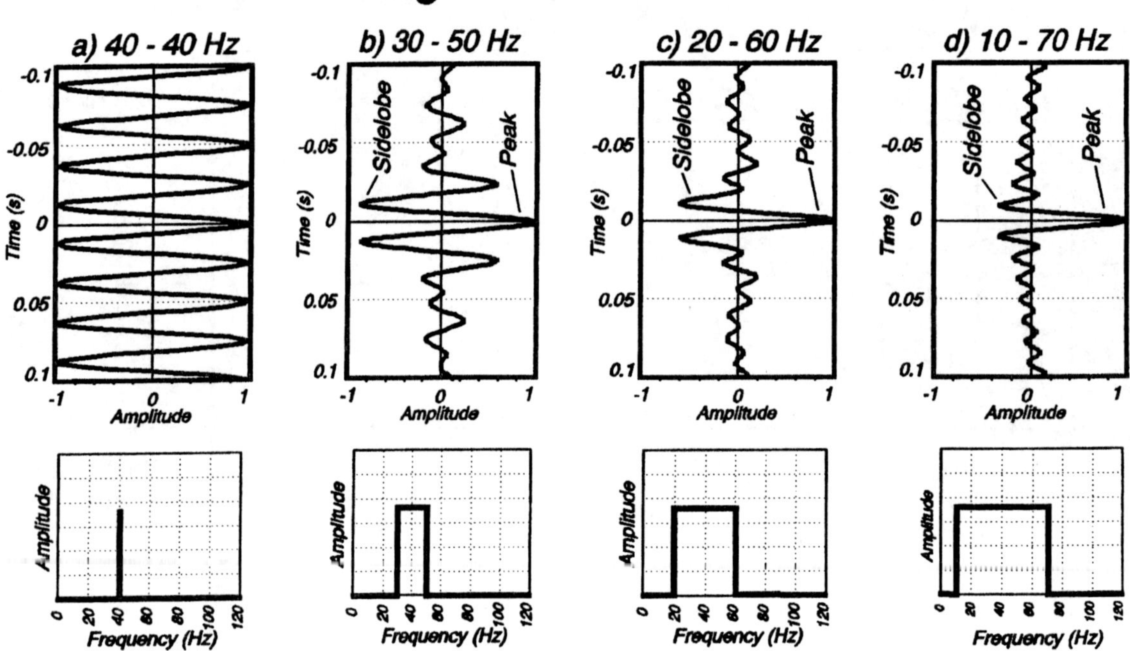

FIGURE 5.34 *Increasing seismic definition.* The pulses have the same central frequency (40 Hz) and, hence, the same power of resolution. Bandwidth increases from (a) to (d), improving the potential to define interfaces. a) Single, 40 Hz pulse is a pure sine wave; side lobes have the same amplitude as the central peak (at Time = 0 s), so that defining power is zero. b) Compared to the pulse in (a), a 30–50 Hz pulse spans the same time ($\approx$ 0.0125 s) between the main peak and adjacent sidelobes (troughs). The central peak has higher amplitude than adjacent troughs, so that the pulse is better able to define an interface at Time = 0. c,d) Peak-to-sidelobe amplitude ratio increases as bandwidth broadens, improving seismic definition.

bandwidth of a seismic pulse can be thought of in the same sense as octaves in music (Fig. 5.35). Doubling the frequency steps up by one octave; four times the frequency expands by two octaves. A pulse spanning only one octave will have sidelobes that are comparable in size to those of the main pulse (Fig. 5.35a). Increasing the number of octaves attenuates the sidelobes relative to the amplitude of the main pulse, making it easier to define where an interface is present (Fig. 5.35b).

Importance of High Frequencies and Broad Bandwidth Interpretation of fine detail of stratigraphy requires that the input pulse have *both* high frequencies and broad bandwidth. Fig. 5.34 shows four pulses, each with central frequency of 40 Hz, and thus each with the same power of resolution (compare the widths of the central pulses). The pulses portray different definition, however, with the broad bandwidth, 10–70 Hz signal approaching a spike, and the no bandwidth (40–40 Hz) pulse revealing a pure sine wave, with no ability to define that an interface might be present.

Fig. 5.35a shows a comparison of pulses with poor definition, encompassing only one octave (10–20 Hz; 20–40 Hz; 40–80 Hz). Note that resolution improves with the higher frequency pulse, but definition remains the same. Fig. 5.35b illustrates much better definition, because two octaves of bandwidth are represented (5–20 Hz; 10–40 Hz; 20–80 Hz). Note that the 20–80 Hz signal approaches a spike, because it includes both high frequencies and broad bandwidth.

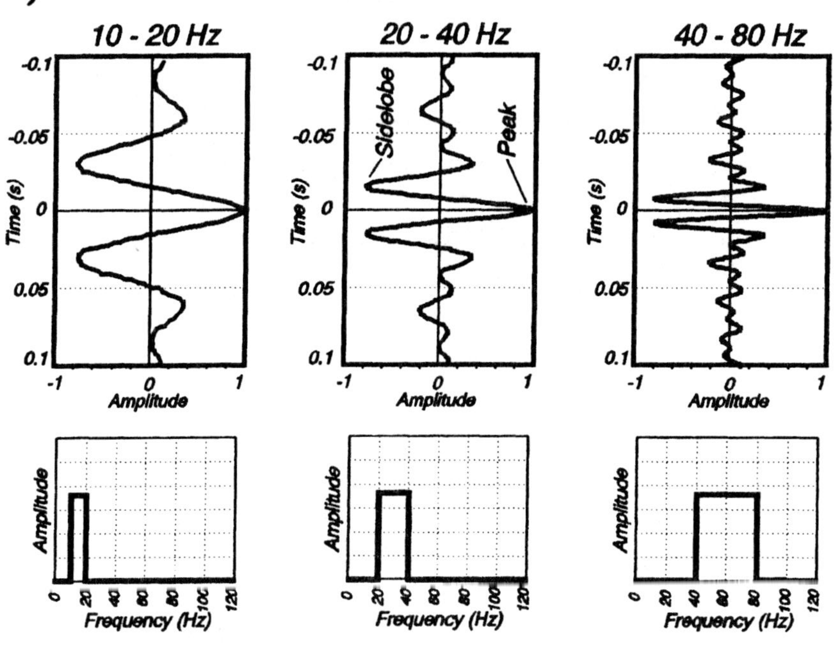

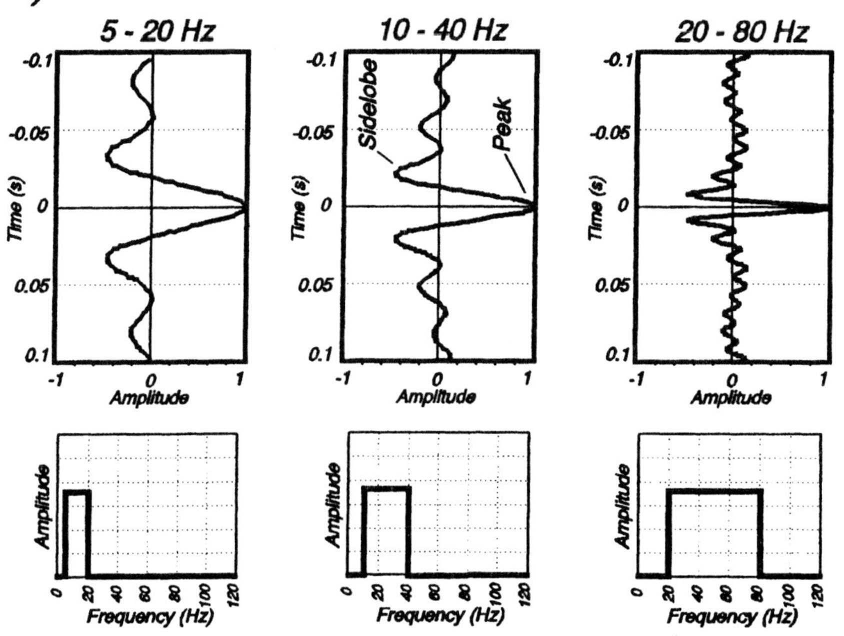

FIGURE 5.35 Improving seismic definition and resolution. *Definition*. Pulses in (a) span one octave, because their bandwidths represent doubling of frequency. Ratios of sidelobes to central peaks for each of the three pulses are the same. In (b) pulses span two octaves of bandwidth, because the highest frequencies are four times the lowest. Compared to the pulses in (a) the pulses in (b) have higher peak-to-sidelobe ratios and, hence, higher definition. *Resolution*. In (a) each of the pulses span one octave, and thus have the same defining power; resolution is higher going from left to right, however, because higher frequencies are involved. The same is true for the 2-octave pulses in (b).

Examples of Waveforms on Seismic Profile

Fig. 5.36 shows a seismic reflection profile recorded across the Middle America Trench off Mexico. The profile reveals packages of varying seismic waveform that can be used to interpret gross stratigraphy. The overlying layer of water has constant acoustic impedance and, hence, zero amplitude reflection coefficients; a region of

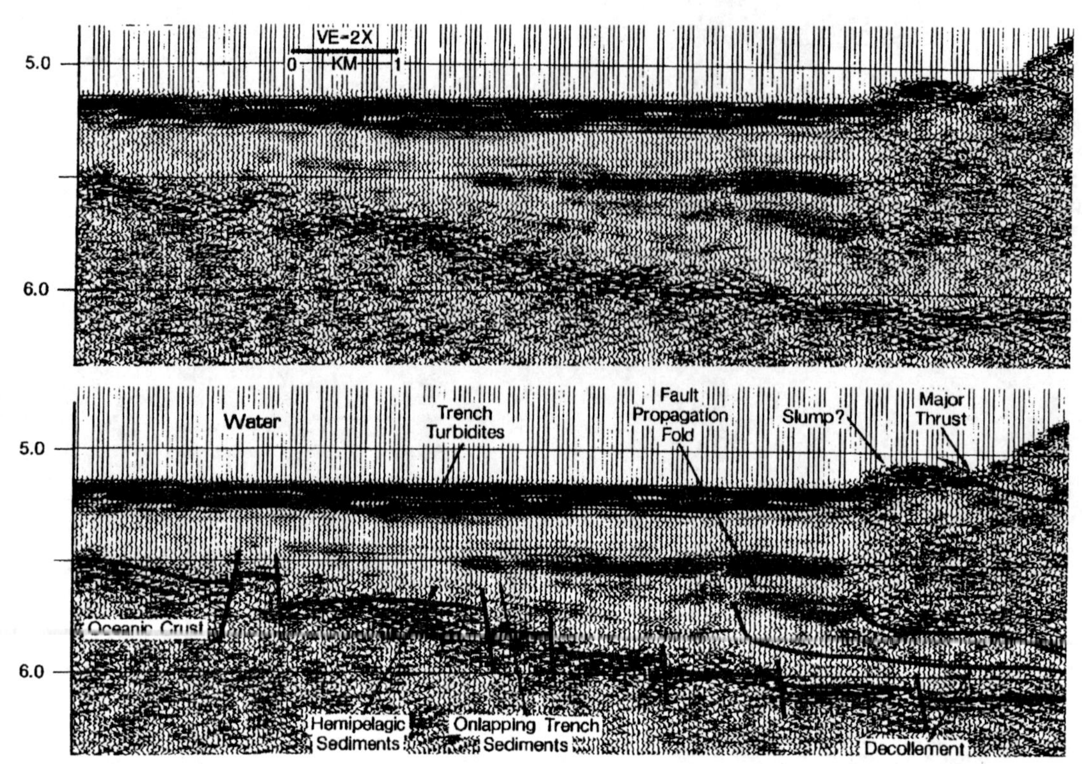

FIGURE 5.36 Seismic waveform packages observed on a migrated depth section from the Middle America Trench off Mexico (Line Mex-2; Moore and Shipley, 1988). The top figure shows the seismic traces with no interpretation; the same section is shown on the bottom, but with interpretations superimposed. Note that the uppermost 4.8 km of the water column is not shown. Prominent reflection packages include the water layer (zero amplitude); trench turbidites (high amplitude); hemipelagic sediments (low amplitude, "transparent" zone); deformed turbidites on right side of section ("chaotic" seismic expression); and oceanic crust ("chaotic"). From G. F. Moore and T. H. Shipley, Mechanisms of sediment accretion in the Middle America Trench off Mexico, in *Journal of Geophysical Research,* v. 93, pp. 8911–8927, © 1988. Redrawn with permission of the American Geophysical Union, Washington, D. C.

blank seismic traces results. On the left side of the section, trench turbidites have much higher density and seismic velocity than the water, leading to a strong water bottom reflection. The trench turbidites have layers of contrasting acoustic impedance, leading to a sequence of high-amplitude reflections. Hemipelagic sediments deposited on top of the oceanic crust are much more homogeneous, leading to very low amplitude reflections (a seismically "transparent" zone). Two regions of complex layering include the deformed trench turbidites on the right side of the section, and the crystalline basement of the oceanic crust; those regions have a more random or "chaotic" seismic appearance.

EXERCISES

5-1 The model on the following page shows compressional wave velocities (V_p), densities (ρ) and thicknesses (h) for layers of horizontal sedimentary strata.

a) Calculate the *acoustic impedance* for each layer.

b) Pretend a well is drilled to the bottom of the model. Draw well logs of *seismic velocity, density, acoustic impedance,* and *reflection coefficient*, with *depth* as the vertical scale.

Surface

Shale	$V_p = 2500$; m/s	$\rho = 2.2$; g/cm³	$h = 1000$ m
Tight Sandstone	$V_p = 3800$;	$\rho = 2.4$;	$h = 50$
Porous Sandstone	$V_p = 3000$;	$\rho = 2.1$;	$h = 50$
Shale	$V_p = 2500$;	$\rho = 2.2$;	$h = 1000$

Model for Exercise 5-1

c) Draw well logs of seismic velocity, density, acoustic impedance, and reflection coefficient, with *two-way travel time* as the vertical scale.

d) Convolve the reflection coefficient series with a *zero-phase wavelet of 20 to 80 Hz frequency ($\approx$ 50 Hz center frequency)* and plot the resulting seismic trace.

e) Convolve the reflection coefficient series with a *zero-phase wavelet of 10 to 40 Hz frequency ($\approx$ 25 Hz center frequency)* and plot the resulting seismic trace.

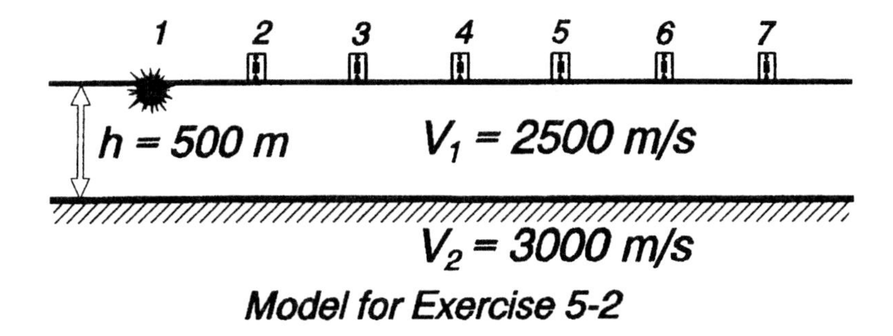

Model for Exercise 5-2

5-2 Given the simple, horizontally layered model above.

a) Plot a *travel time vs. distance graph* for the: i) direct P; ii) direct Rayleigh; iii) critically refracted P; iv) P primary reflection; v) first P multiple reflection.

b) What is the *apparent velocity* of each of these arrivals at a distance of *zero m, 500 m, 2000 m, 6000 m, and 20,000 m* from the source?

c) Repeat (a) and (b), but with the interface at *5000 m* depth instead of 500 m.

5-3 For the Exercise 5-2 model:

a) Draw a *stacking diagram* for a series of shots at each of the locations 1 through 7, moving the entire spread one shot interval to the right each time. What is the *fold of stack* for each resulting *CMP point*?

b) Calculate a *normal moveout correction* for each of the events plotted in Exercise 5-2(a), assuming a *stacking velocity of 2500 m/s*.

c) Replot the travel-time curve at the same scale as in 5-2(a). Which events would stack *in phase*, and which *out of phase*?

5-4 a) For the (shotpoint gather) record in Fig. 4.8, identify the primary and the first two multiple reflections from the water bottom.

b) Compute the water depth based on the primary and two multiple reflections.

c) Draw raypaths illustrating the primary and the two multiples, to a receiver 15 km from the source.

SELECTED BIBLIOGRAPHY

Anstey, N. A., 1977, *Seismic Interpretation: The Physical Aspects*, Boston: International Human Resources Development Corporation, 625 pp.

Badley, M. E., 1985, *Practical Seismic Interpretation*, Boston: International Human Resources Development Corporation, 266 pp.

Bally, A. W. (editor), 1983, *Seismic Expression of Structural Styles*, American Association of Petroleum Geologists, Studies in Geology Series 15, (3 volume set).

Burger, H. R., *Exploration Geophysics of the Shallow Subsurface*, 1992, Englewood Cliffs, N.J.: Prentice Hall, Inc., 489 pp.

Collins, B. P., and J. S. Watkins, 1986, The Middle America Trench, in: *Seismic Images of Modern Convergent Margin Tectonic Structure*, R. von Huene (editor), Tulsa: American Association of Petroleum Geologists, Studies in Geology 26, pp. 30–32.

Dobrin, M. B., 1976, *Introduction to Geophysical Prospecting*, (3rd ed.), New York: McGraw-Hill, 446 pp.

Dohr, G., 1981, *Applied Geophysics: Introduction to Geophysical Prospecting*, Geology of Petroleum, v. 1, New York: Halsted Press, 231 pp.

Halbouty, M. T. (editor), 1982, *The Deliberate Search for the Subtle Trap*, Tulsa: American Association of Petroleum Geologists, Memoir 32, 351 pp.

Miall, A. D., 1984, *Principles of Sedimentary Basin Analysis*, New York: Springer-Verlag, 490 pp.

Moore, G. F., and T. H. Shipley, 1988, Mechanisms of sediment accretion in the Middle America Trench off Mexico, *Journal of Geophysical Research*, v. 93, pp. 8911–8927.

Oliver, J., 1986, A global perspective on seismic reflection profiling of the continental crust, in: *Reflection Seismology: The Continental Crust*, M. Barazangi and L. Brown (editors), Tulsa: American Association of Petroleum Geologists, Geodynamics Series 14, pp. 1–3.

Oliver, J., 1996, *Shocks and Rocks: Seismology in the Plate Tectonics Revolution. The Story of Earthquakes and the Great Science Revolution of the 1960s*, Washington: American Geophysical Union, 189 pp.

Payton, C. E. (editor), 1977, *Seismic Stratigraphy: Applications to Hydrocarbon Exploration*, Tulsa: American Association of Petroleum Geologists, Memoir 26, 516 pp.

Petty, O. S., 1976, *Seismic Reflections*, Houston: Geosource, Inc., 81 pp.

Robinson, E. S., and C. Çoruh, 1988, *Basic Exploration Geophysics*, New York: John Wiley and Sons, 562 pp.

Robinson, E. A., and S. Treitel, 1980, *Geophysical Signal Analysis*, Englewood Cliffs, N.J., Prentice Hall, Inc., 466 pp.

Ryder, R. T., M. W. Lee, and G. N. Smith, 1981, *Seismic Models of Sandstone Stratigraphic Traps in Rocky Mountain Basins*, Tulsa: American Association of Petroleum Geologists, Methods in Exploration Series, 77 pp.

Sengbush, R. L., 1983, *Seismic Exploration Methods*, Boston: International Human Resources Development Corp., 296 pp.

Sheriff, R. E., 1978, *A First Course in Geophysical Exploration and Interpretation*, Boston: International Human Resources Development Corp., 313 pp.

Sheriff, R. E., 1984, *Encyclopedic Dictionary of Exploration Geophysics* (2nd ed.), Tulsa: Society of Exploration Geophysicists, 323 pp.

Sheriff, R. E., and L. P. Geldart, 1995, *Exploration Seismology* (2nd ed.), Cambridge: Cambridge University Press, 592 pp.

Telford, W. M., L. P. Geldart, R. E. Sheriff, and D. A. Keys, 1976, *Applied Geophysics*, Cambridge: Cambridge University Press, 860 pp.

C H A P T E R 6

Structural And Tectonic Interpretation Of Seismic Reflection Profiles

seismic *(sīz mik') **adj.**,* [< *Gr.* seismos, *an earthquake* < seiein, *to shake*], *relating to an earthquake or artificial shaking of the Earth.*
reflect *(ri flekt') **vi.**,* [< *Mfr.* < *L.* < *re-, back* + flectere, *to bend*], *to bend back.*
reflection *(ri flek' shən) **n.**, the bending back of a ray or wave of light, heat, or sound into the same medium, as it encounters a medium with different properties.*
interpret *(in tur' prit) **vt.**,* [< *MFr.* < *L.* interpretari < interpres, *negotiator*], *to explain the meaning of; to give one's own understanding of.*
seismic reflection interpretation *(sīz mik' ri flek'shən in tur'prə tā' shən) **n.**, an explanation of subsurface conditions that led to the reflection of seismic waves back to Earth's surface.*

Seismic reflection profiles are used extensively to map geological structures in the subsurface. They are especially useful when reflections can be tied to firm constraints provided by surface outcrop and drilling (Fig. 1.8). Even when not strictly constrained, reflection geometries provide information on gross structural form within the crust. Seismic profiles are thus valuable tools to unravel the tectonic history of a region, and to compare and contrast features associated with continental margins and plate boundaries.

Though resembling geological cross sections, seismic reflection profiles must be interpreted wisely, with appreciation for both the utility and pitfalls of the method. This chapter examines the seismic expression of different structural geometries, then shows reflection patterns characteristic of tectonic settings.

APPEARANCE OF STRUCTURES ON REFLECTION PROFILES

Patterns of events on seismic reflection profiles provide constraints on subsurface structure. Structural interpretation first requires appreciation of how simple geometries would appear on different types of seismic section displays. A common display is the unmigrated time section; the data may be processed further to yield migrated time and migrated depth sections.

An *unmigrated time section* is an attempt to display reflection data *as if the source and receiver for each seismic trace were at the same surface position* (Fig. 6.1). This situation requires that each seismic ray recorded at the surface be reflected normal (at a 90° angle) to an interface below the surface. An unmigrated seismic section is thus referred to as a *normal incidence section.* A reflection (event) appears

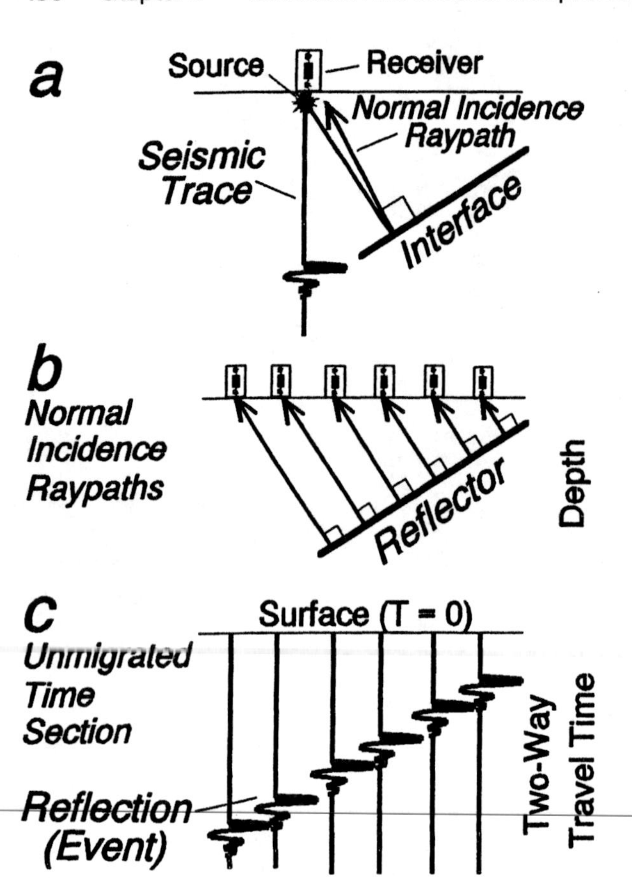

FIGURE 6.1 a) On a seismic trace, a reflection from a dipping interface is plotted vertically below the common source/receiver position. b) Normal incidence raypaths for dipping reflector, to several common source/receiver positions. c) The seismic reflection profile resulting from (b) is an *unmigrated time section*.

directly below the position of the common source/receiver position, even if the reflection was not from directly below. The event appears at the time it takes the ray to travel to the reflecting interface, then back to the surface (two-way travel time).

Migration moves seismic events back to their true positions, *as if each ray went straight down then back up* (Fig. 6.2). A *migrated time section* is thus referred to as a *vertical incidence section.* After migration the vertical scale can be *converted from two-way travel time to depth,* yielding a *migrated depth section* (Fig. 6.3). For an average velocity (V) above the interface, the conversion of a point from two-way travel time (T) to depth (d) is:

$$\mathbf{d = (T/2) \times V.}$$

Fig. 6.4 shows the three section displays for a seismic profile of the Nankai Trough, off Japan (Nasu et al., 1982; see also Moore et al., 1990). The top of oceanic basement is imaged just above 8 s, two-way travel time, on the unmigrated and migrated sections. Overlying the basement and deep-marine sedimentary cover is undeformed, trench fill material on the right side of the profiles. The left side of the section shows trench-fill that is deformed into an accretionary wedge through thrusting and folding. Notice that the distorted image of the water bottom on the unmigrated section is clarified through migration. Likewise, faults that are not in their proper positions before migration move to apparent offsets of strata within the accretionary wedge on the migrated profile. The depth section shows that the water layer, which is low-velocity material, encompasses far less thickness than appears on

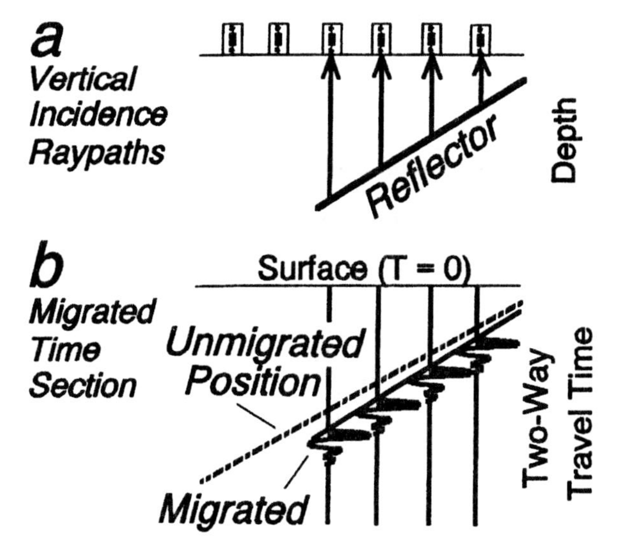

FIGURE 6.2 a) A migrated time section mimics the hypothetical situation of vertical incidence raypaths. b) *Migrated time section*, illustrating how the event in Fig. 6.1c moves from its normal incidence (unmigrated) position to a vertical incidence (migrated) position.

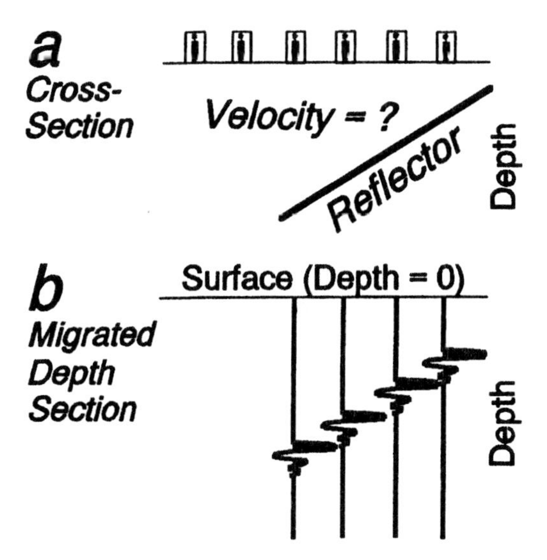

FIGURE 6.3 The average velocity of material above a reflector (a) is used to convert from a migrated time section (Fig. 6.2b) to a *migrated depth section* (b).

the time sections. The top of the basement, which looks horizontal on the time sections, dips gently to the left on the depth section, revealing a truer orientation of the subducting plate.

Migration and depth conversion require accurate knowledge of the velocity and travel paths of seismic energy. Thorough interpretation requires viewing not just migrated time and depth sections, but also seeing how events appear before migration and depth conversion.

Four factors must be taken into account when viewing seismic profiles. 1. *Geometric effects are due to reflections from interfaces that are not horizontal. Migration* of the data attempts to put the events back into their proper orientations. 2. *Velocity effects cause time shifts of some reflectors relative to others.* A common example is "velocity pullup," whereby events below high-velocity material appear shallow on time sections. Proper *depth conversion* removes velocity effects. 3. *Raypath bending distortions occur as seismic waves penetrate complex structures.*

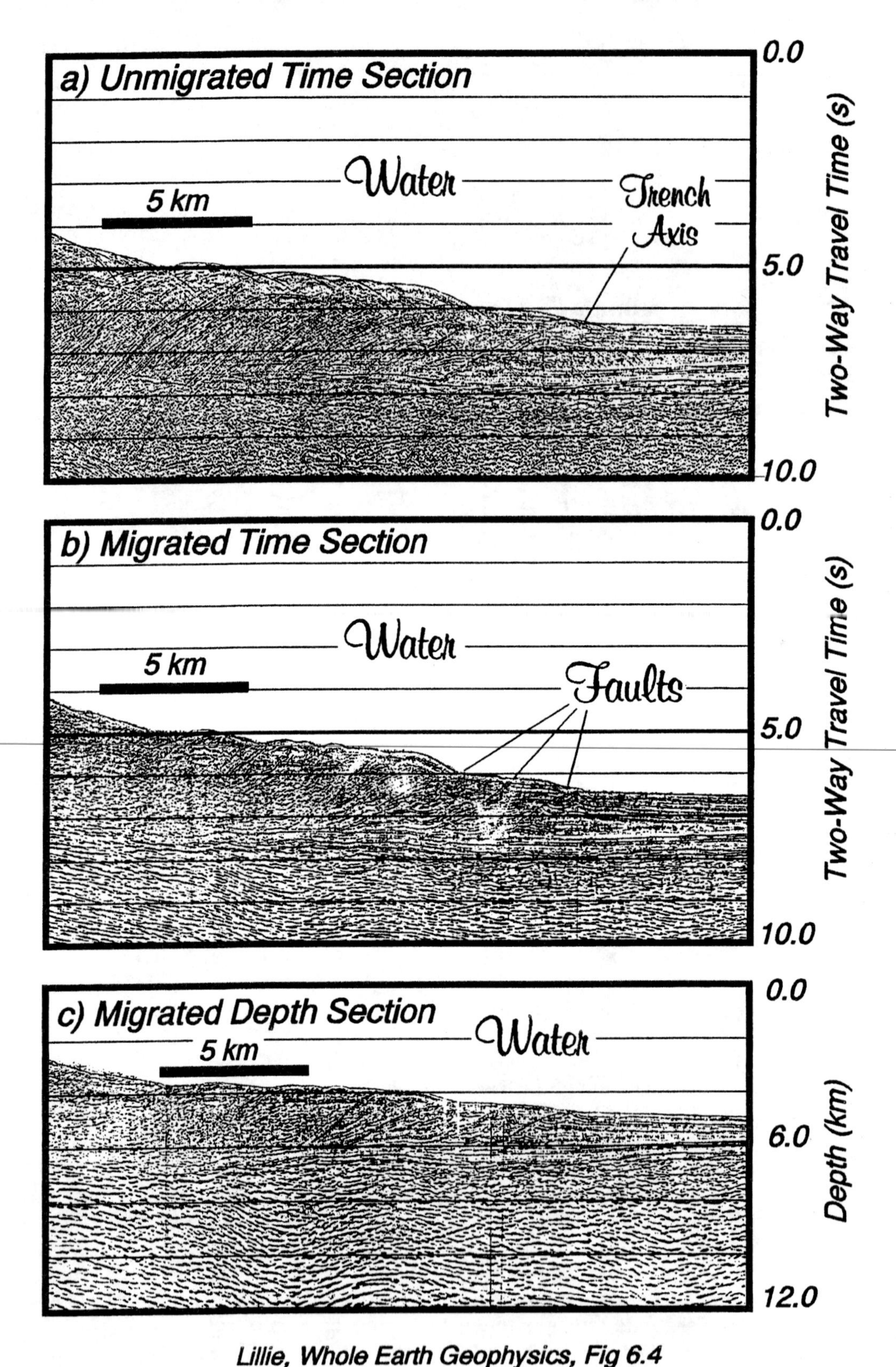

Lillie, Whole Earth Geophysics, Fig 6.4

FIGURE 6.4 *Migration and depth conversion of a seismic profile from the Nankai Trough, Japan.* From N. Nasu and others, Multi-channel seismic reflection data across Nanki Trough, *IPOD, Japan Basic Data Series #4,* © 1982, Ocean Research Institute, University of Tokyo, Tokyo, Japan.

The bending is caused by changes in velocity across interfaces, according to Snell's Law. 4. *Three-dimensional effects produce distortions in geometry and travel time as events are reflected from out of the plane of the section.* Such "sideswipe" makes events appear deeper on seismic sections than they would appear if they originated from directly beneath the survey line. The four effects are illustrated below, for simple then more complex geometries. For effects 1, 2, and 3, the discussions assume two-dimensional structures, striking and extending infinitely in and out of the page.

1. Geometric "Migration" Effects

An unmigrated section has events plotted directly below the common source/receiver position, even though the actual reflecting point might not have been directly below (Fig. 6.1). The actual reflecting point could have been any point below the surface that is the same travel time away from the receiver. Thus, the event appearing

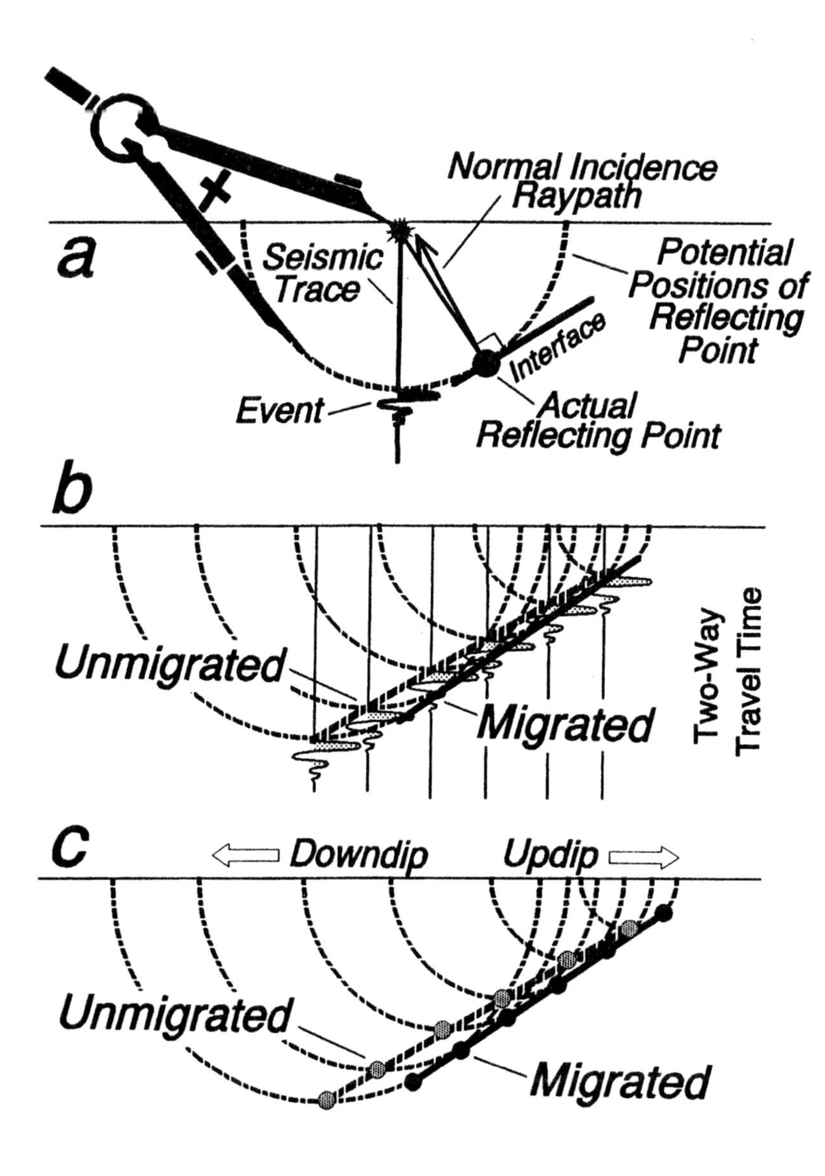

FIGURE 6.5 a) Assuming constant velocity material and two-dimensional structure, an event on an unmigrated seismic trace could have come from anywhere along a semicircle drawn through the event. b) Migration is equivalent to swinging circular arcs through the event. The migrated position of the event is the line segment tangent to each of the arcs. c) Migration shallows, shortens, and steepens an unmigrated event, moving points in an updip direction.

on the seismic trace could have been from anywhere on a hemisphere (semicircle for two-dimensional structure) drawn through the event (Fig. 6.5a).

Migration of a linear reflection segment can be envisioned as swinging arcs through points along the event. The migrated position of the segment is defined by the line segment running tangent to each of the arcs (Fig. 6.5b). Notice relationships of the unmigrated and migrated positions of the event (Fig. 6.5c): 1) each point of the unmigrated event is *downdip* from the corresponding point of the migrated event; 2) unmigrated points are *deeper* than corresponding migrated points; 3) the unmigrated event is *less steep* than the migrated event; 4) the unmigrated event is *longer* than the migrated event. Migrating a seismic profile thus tends to make reflecting segments move updip, shallow, steepen, and become shorter.

Dipping Interface Connecting Two Horizontal Interfaces The normal incidence raypaths in Fig. 6.6 illustrate that a dipping event appears downdip from its actual position. The unmigrated event is deeper in time than it should be, and it cuts across horizontal events. In the updip direction, there is a gap where the event has moved away from the horizontal event it should touch. Migration shortens, steepens, and moves the event updip to where it belongs, connecting the horizontal interfaces.

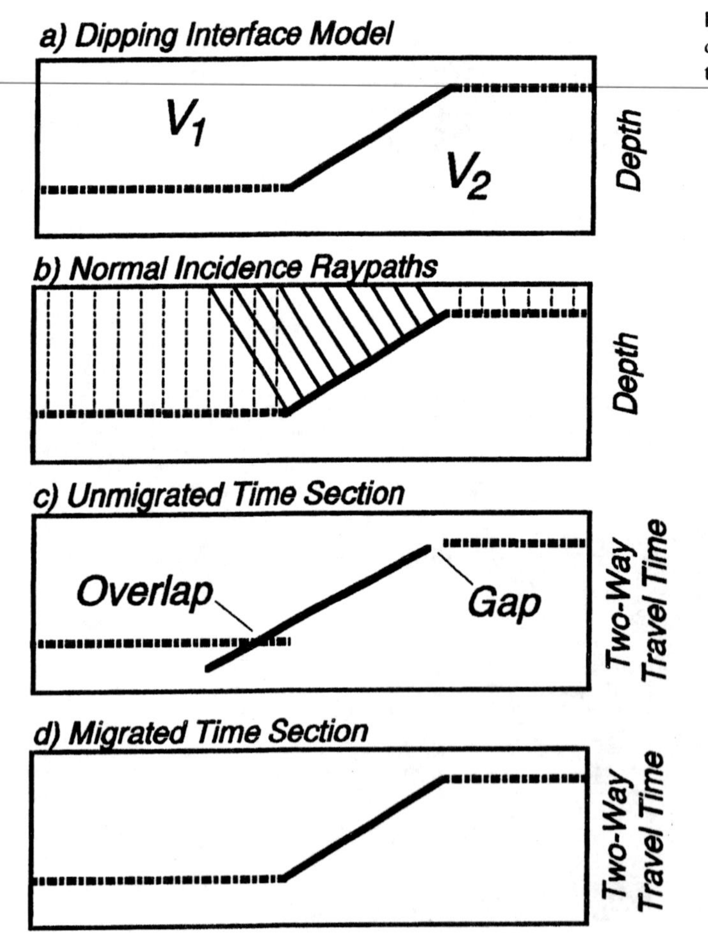

FIGURE 6.6 Seismic expression of dipping interface (solid line) connecting two horizontal interfaces (dashed lines).

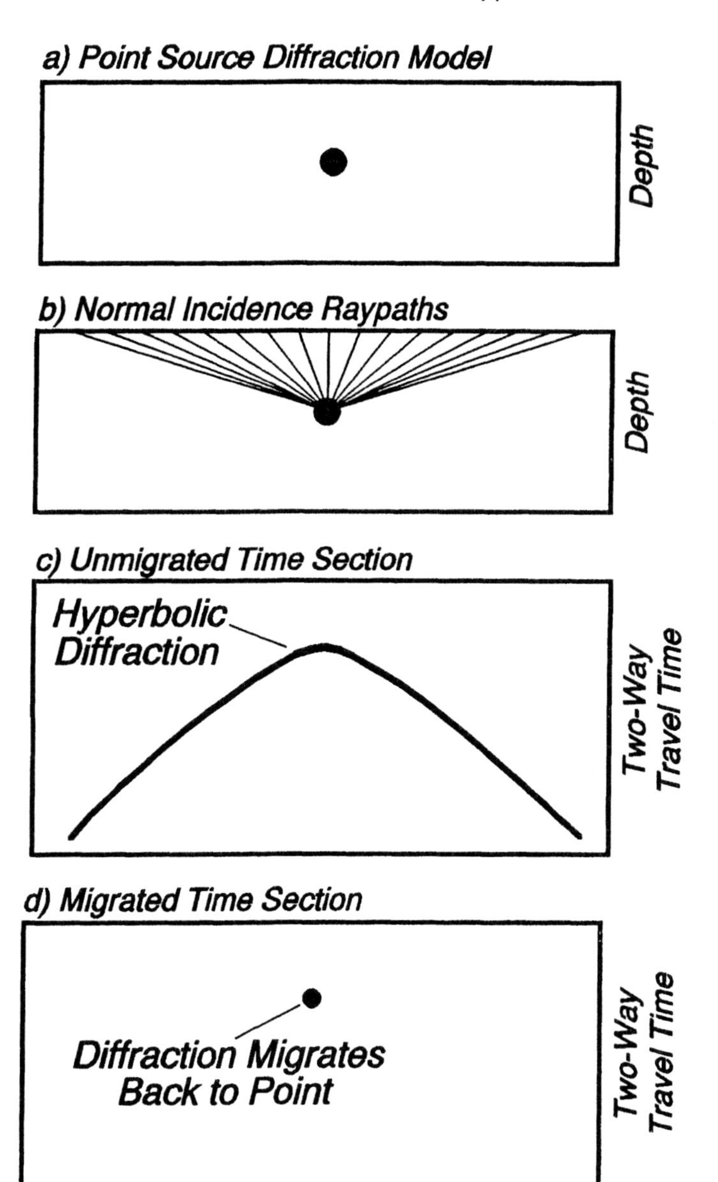

FIGURE 6.7 Seismic expression of the top of a very small sphere is equivalent to a point source diffraction, producing a hyperbolic pattern.

Very Small Sphere (Point-Source Diffraction) A point in the subsurface can be thought of as a very small sphere, leading to normal incidence reflections from its upper surface (Fig. 6.7). Reflection of energy from such a point is known as *diffraction*. The unmigrated event has a hyperbolic pattern, depending on the depth of the point and the seismic velocity of the material above the point. Migration collapses the hyperbola back to a point.

Anticline Reflection from the top of an anticline is similar to that of a point-source diffraction, but with the sphere too large to be considered a point (Fig. 6.8). Normal incidence raypaths encounter the anticline from large distances, so that the unmigrated event is spread out, cutting across horizontal events and extending to large travel times. Migration collapses the anticline to its proper size and orientation relative to other events (Fig. 6.9).

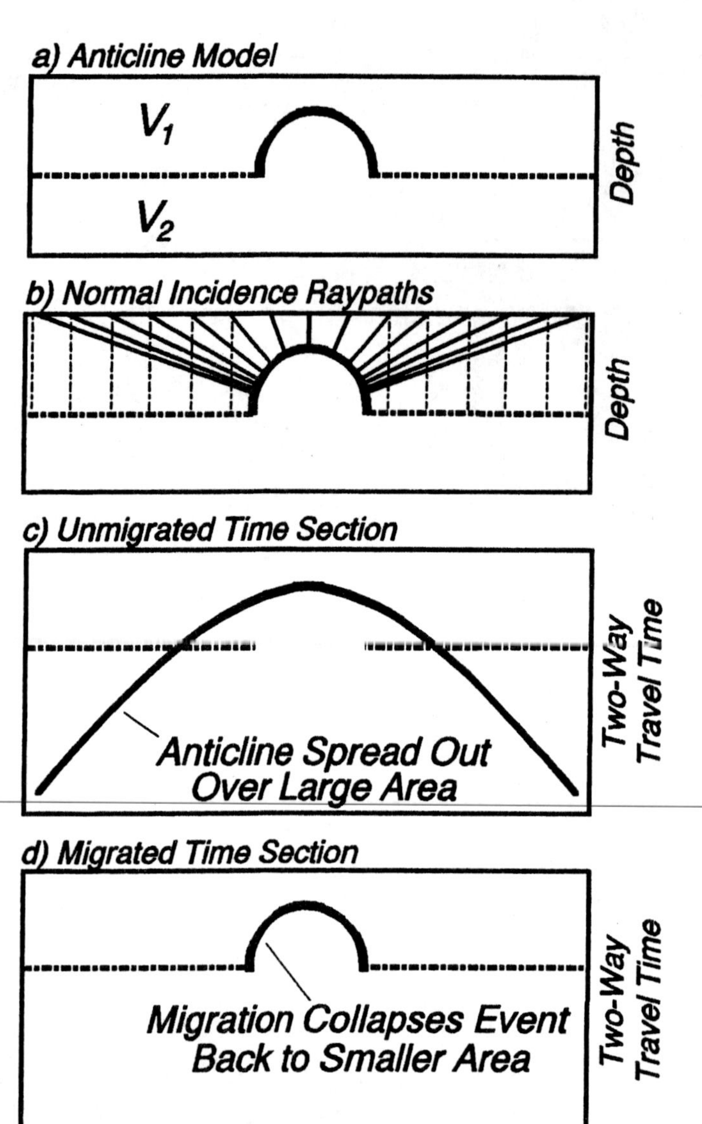

FIGURE 6.8 Seismic expression of an anticline (solid line) between two horizontal reflectors (dashed lines).

Syncline ("Buried Focus," or "Bow Tie" Effect) On an unmigrated section, events cross above the axis of a syncline (1′ and 2′ in Fig. 6.10c). If the structure is deep enough, normal incidence reflections from the bottom (concave-upward) portion of the syncline will cross at a point below the surface (Fig. 6.10b). The effect is a diffraction that radiates from the crossing point, or *buried focus*; the diffraction is a hyperbolic event beneath the syncline (3′ in Fig. 6.10c). The two crossing events and diffraction may resemble a bow tie. Migration inverts the diffraction back to concave upward and moves the crossing events to their proper positions on the flanks of the syncline (Figs. 6.10d; 6.11).

Series of Anticlines and Synclines An undulating, or "hummocky," surface shows effects similar to a series of adjacent anticlines and synclines (Fig. 6.12a). The spread-out events from the structural highs cross above buried-focus diffractions from the lows (Fig. 6.12b), resulting in numerous bow ties (Fig. 6.12c). Migration

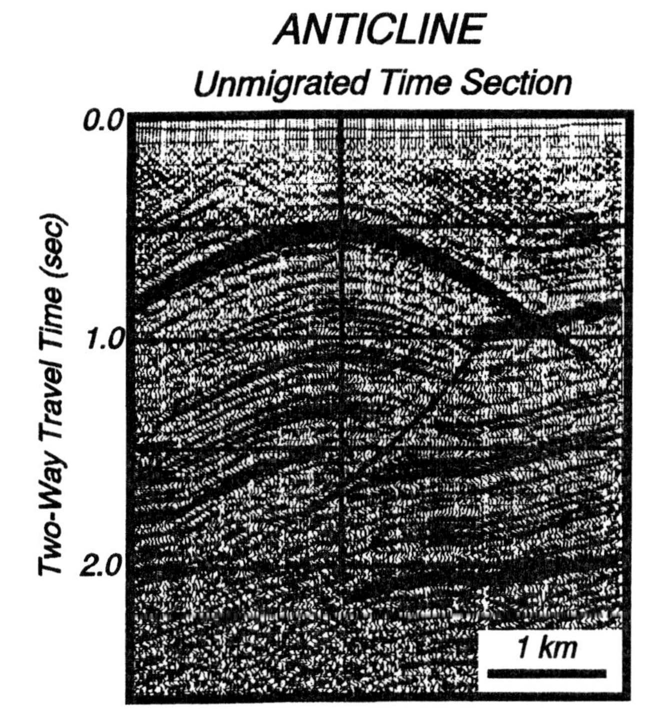

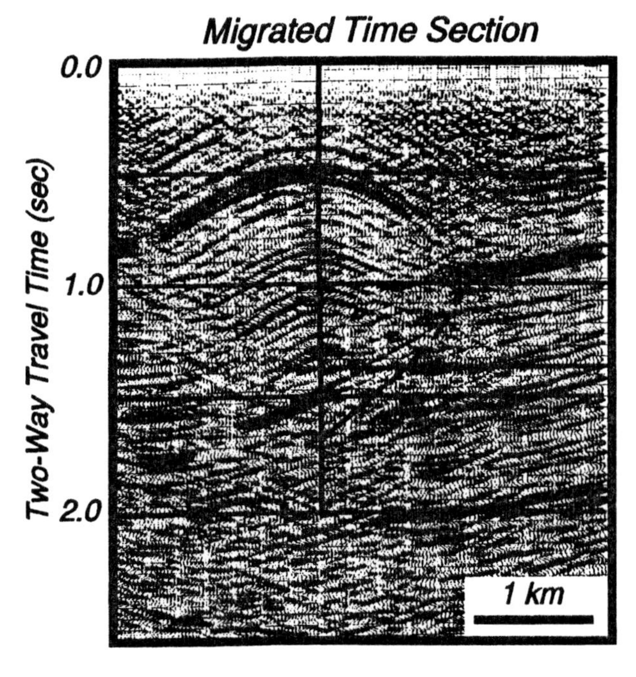

FIGURE 6.9 Unmigrated and migrated seismic profiles showing an anticline.

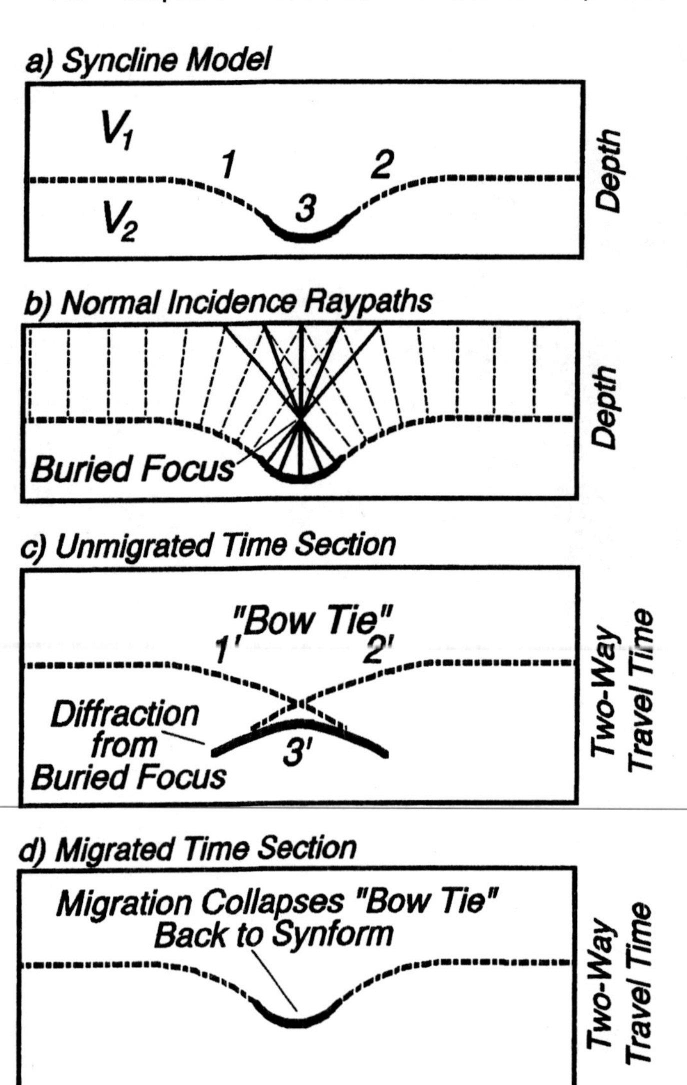

FIGURE 6.10 *Seismic expression of buried focus syncline.* Solid line (3) shows the concave-upward part of the structure that results in raypaths focused on a buried point; a hyperbolic diffraction results (3′). Dashed line portions of the structure (1, 2) produce crossing events (1′, 2′) that, together with the diffraction, appear as a "bow tie."

cures the bow ties so that structural highs and lows can be seen (Fig. 6.12d). Seismic profiles from deep-sea trenches often reveal hummocky surfaces resulting from: 1) thrust faults offsetting the surface of the accretionary wedge; and 2) normal faults that formed soon after oceanic basement was created at a mid-ocean ridge (Fig. 6.13).

2. Velocity Effects

Changes in velocities cause changes in the times that reflections appear on seismic sections. Vertical changes in velocity cause differences in the two-way travel times of layers of equal thickness, while lateral velocity changes make events appear to move up or down on time sections.

Vertical Velocity Changes For the model in Fig. 6.14a, the middle layer has higher velocity (V_2) than that of the upper layer (V_1). Even though the layers are of equal thickness, the middle layer occupies a shorter interval on a time section

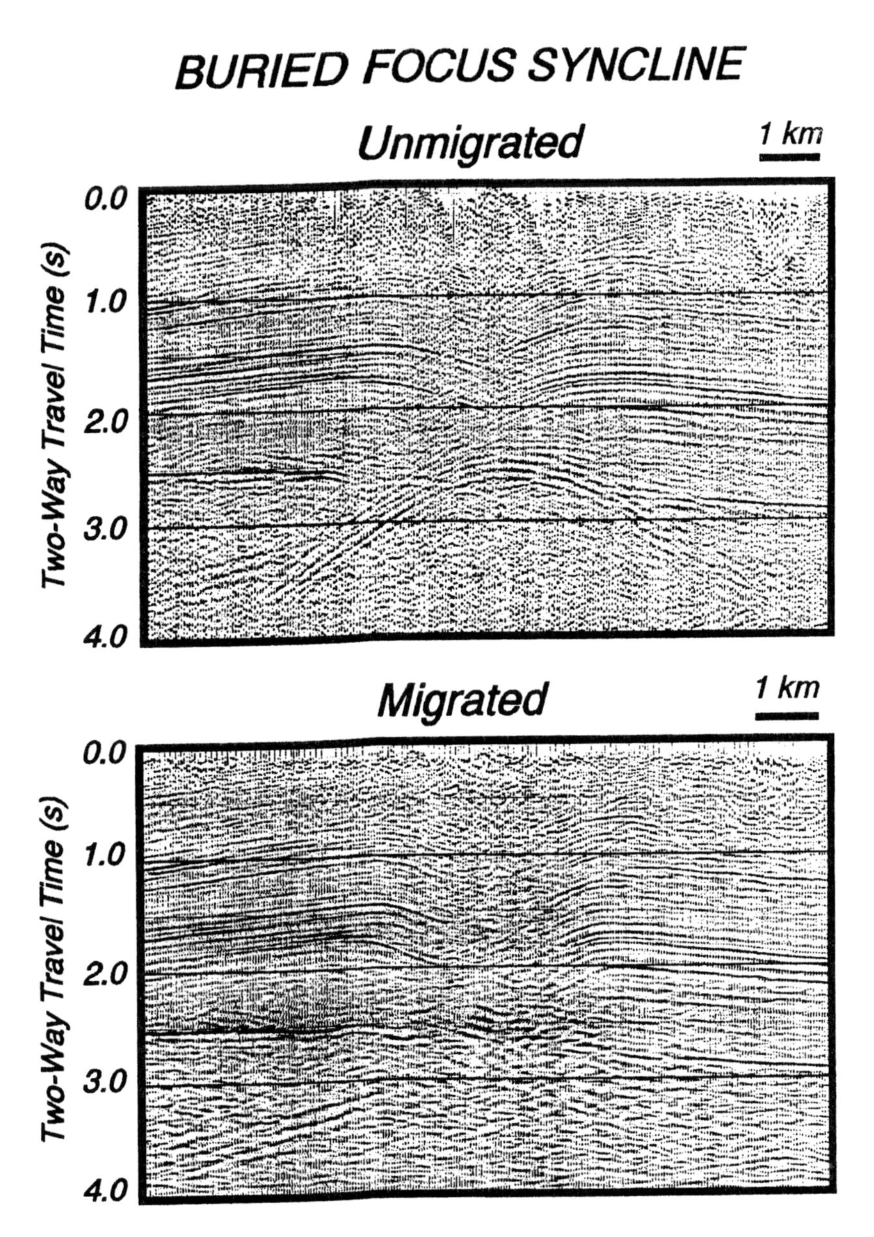

FIGURE 6.11 *Unmigrated and migrated seismic sections recorded across a syncline in the Potwar Plateau of Pakistan.* Portion of line CW-13 (Lillie et al., 1987). Published with permission of the Oil and Gas Development Corporation, Pakistan.

(Fig. 6.14b). If velocities are known accurately, depth conversion restores thicknesses to their proper perspective (Fig. 6.14c). On the time section in Fig. 6.15, the middle layer appears about $1\frac{1}{2}$ times the thickness of the upper layer; the depth section suggests the middle layer is actually three times the thickness of the upper.

Lateral Velocity Changes The model in Fig. 6.16 shows horizontal interfaces below layers of lower (V_1) and higher (V_2) velocity. Seismic waves take less time to travel to the interfaces beneath the left side of the diagram; the corresponding time shift of reflections is referred to as *velocity pullup*. Seismic profiles from the Nankai Trough (Fig. 6.17a) show velocity pullup. Notice that the reflection from the top of oceanic basement appears nearly horizontal on the time section. The event is pulled up to the left, however, because the accretionary wedge is higher velocity than the

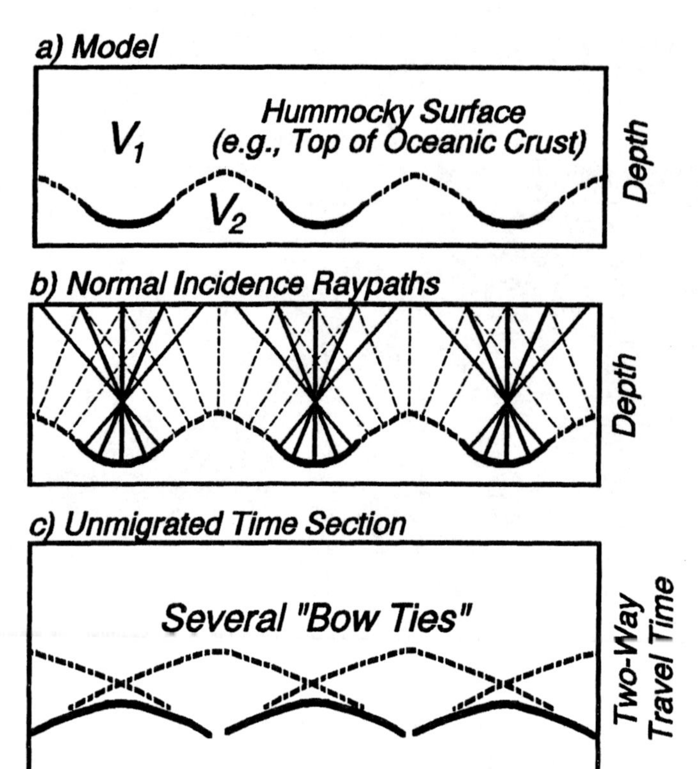

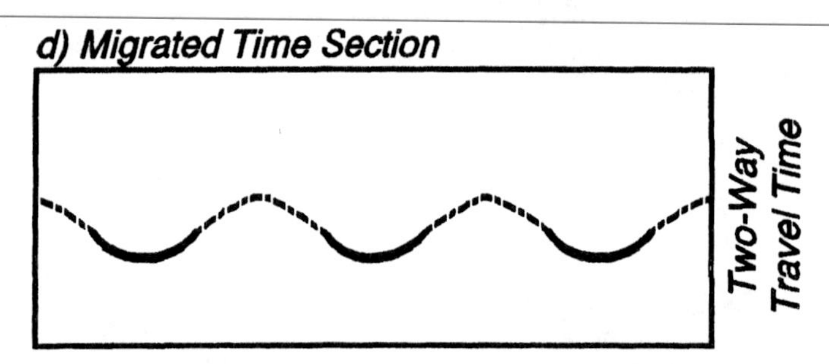

FIGURE 6.12 *Seismic expression of an undulating, or hummocky surface.* Solid lines = synclinal portions; dashed lines = anticlinal portions.

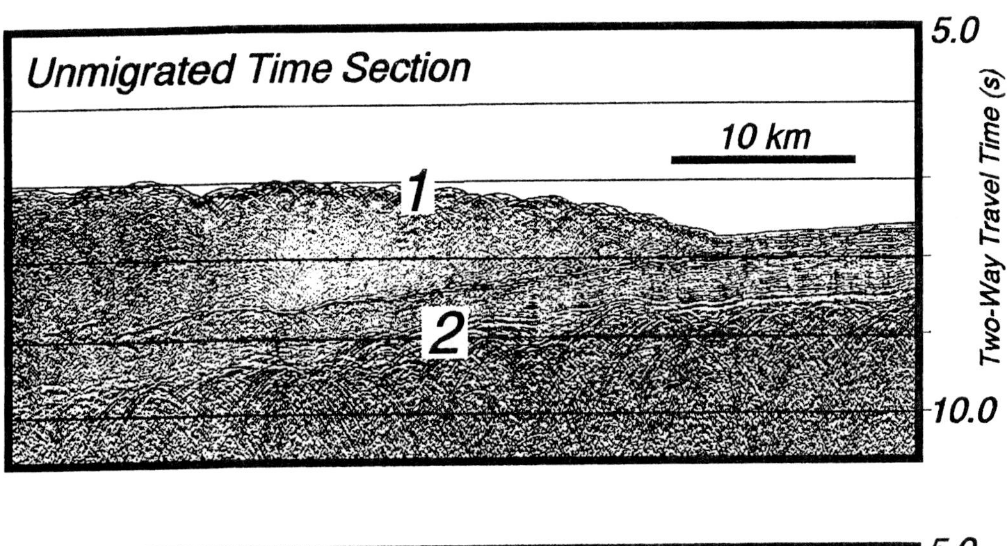

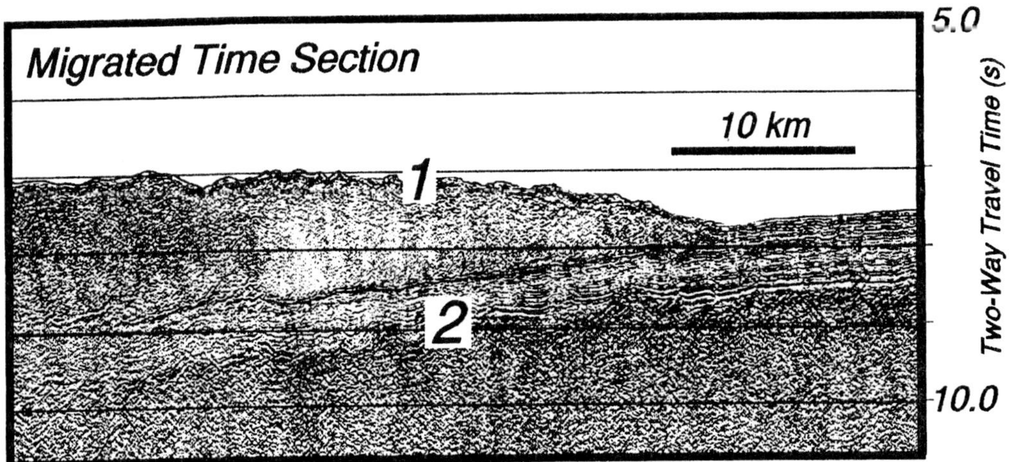

FIGURE 6.13 Unmigrated and migrated seismic reflection profiles recorded across the deep sea trench (right) and accretionary wedge (left) of the Barbados Ridge Complex. Hummocky surfaces are: 1) the water bottom at the top of the accretionary wedge; and 2) the top of the basement (oceanic layer 2; Figs. 4.9, 4.18). From N. L. Bangs and G. K. Westbrook, *Journal of Geophysical Research*, vol. 96, pp. 3853–3866, © 1991. Redrawn with permission of the American Geophysical Union, Washington, D.C.

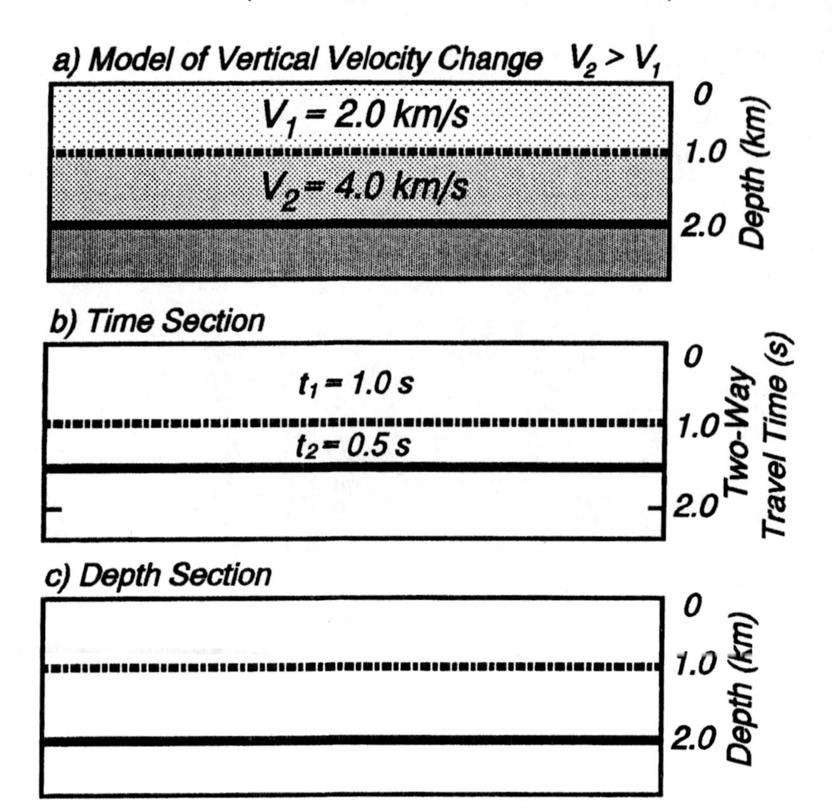

FIGURE 6.14 *Effects of vertical changes in velocity.* The middle layer in (a), having twice the velocity of the upper layer, occupies only half the two-way travel time in (b). Proper depth conversion (c) restores the true thickness.

water filling the trench. Depth conversion shows that the basement actually dips from the trench region beneath the accretionary wedge (Fig. 6.17b).

3. Raypath Bending

Rays change direction (refract), according to Snell's Law, when they encounter interfaces obliquely (Fig. 3.22). A horizontal interface placed below the anticline of Fig. 6.8 shows overlapping events (raypath bending effect) on an unmigrated time section, in addition to velocity pullup (Fig. 6.18). Rays originating from outside the bounds of the anticline refract downward when they cross the interface separating the layer with lower velocity (V_1) from the higher velocity layer (V_2; Fig. 6.18b). The resulting, normal incidence reflections from the horizontal interface below the anticline are thus spread over a larger area, crossing horizontal events (Fig. 6.18c).

Most migration routines will not sufficiently account for the raypath-bending effect (Fig. 6.18d), sometimes leading to erroneous interpretation of faulting of horizons that are continuous. Fig. 6.19b shows the results of a standard migration, with the assumption that raypaths are not bent. A depth conversion still shows distortion of the lower event (Fig. 6.19c). A more sophisticated migration, which takes raypath bending into account, produces the section in Fig. 6.19d; notice that the lower event is more continuous, suggesting it may not be faulted.

Combination of Effects 1, 2, and 3

A model that produces geometric, velocity, and raypath bending effects is shown in Fig. 6.20. The unmigrated and migrated time sections have the following advantages for understanding interpretation techniques and potential problems. 1) Only primary

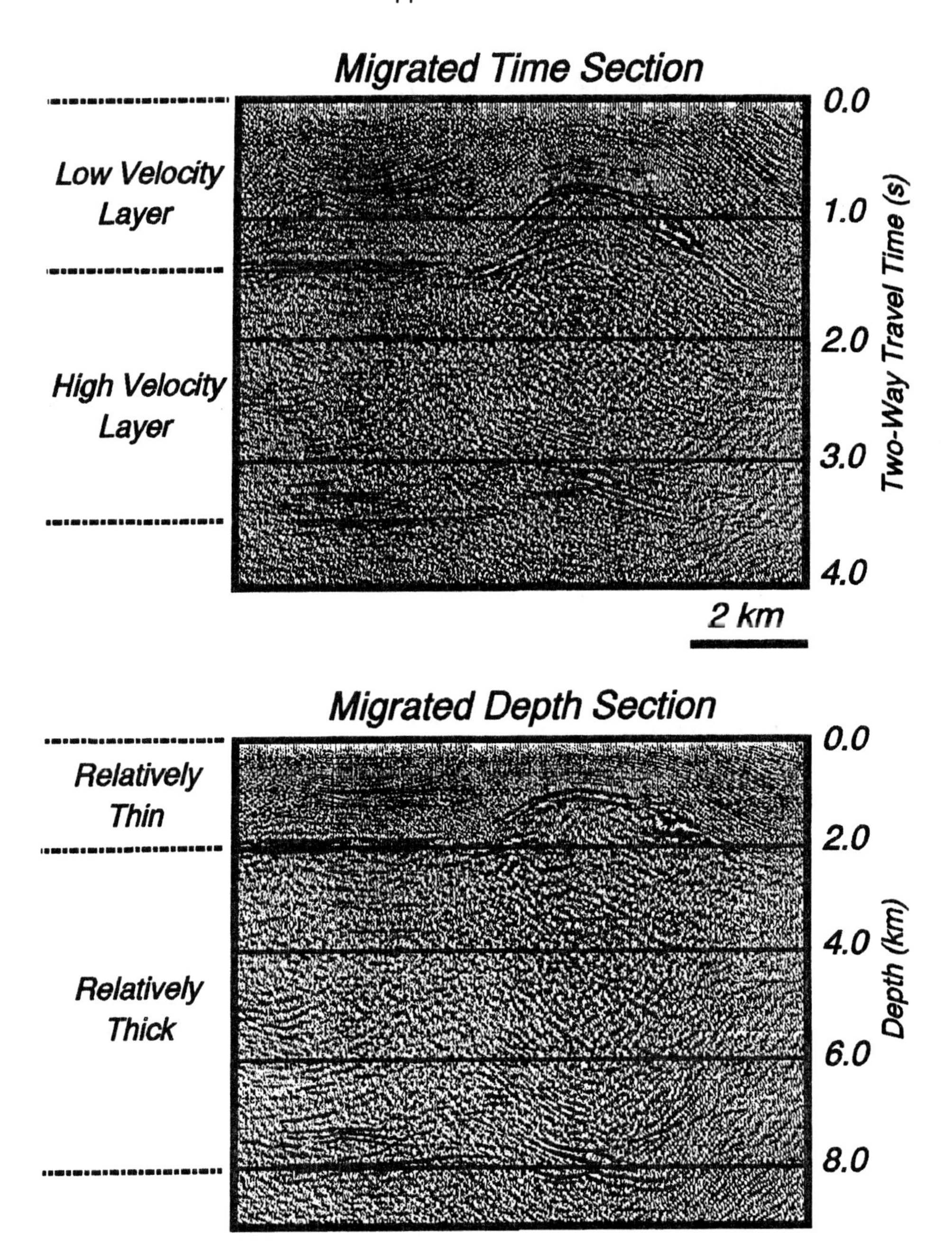

FIGURE 6.15 *Migrated time and migrated depth sections in a region of drastic difference in layer velocity.* From K. Larner, B. Gibson, and R. Chambers, Imaging beneath complex structure, *AAPG Studies in Geology Series,* no. 15, vol. 1, pp. 26–39, © 1983. Redrawn with permission of the American Association of Petroleum Geologists, Tulsa, Oklahoma, USA.

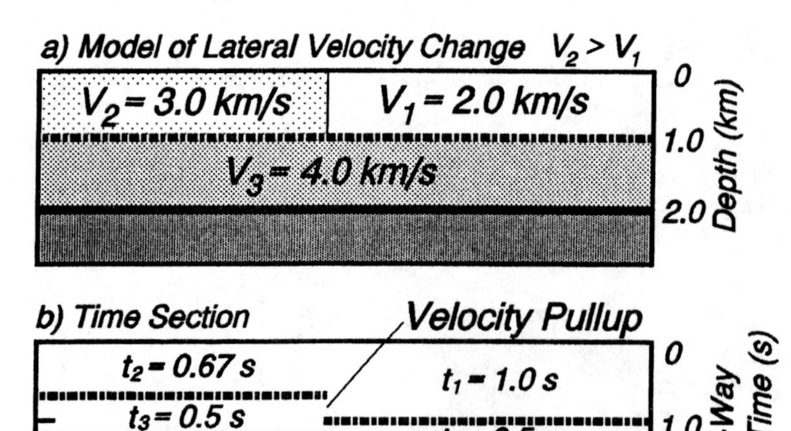

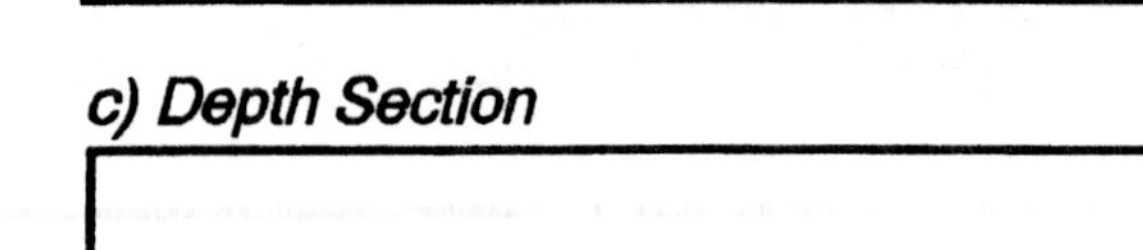

FIGURE 6.16 *Effects of lateral change in velocity*. a) Model of horizontal interfaces beneath layers of different velocity. b) Horizontal reflections on the left side of the time section arrive early, relative to reflections on the right (they are "pulled up"). c) Proper depth conversion restores the continuity and horizontal nature of the interfaces.

reflections are incorporated into the synthetically generated sections. 2) The unmigrated section was developed as if an actual seismic survey were recorded, including generating reflections for numerous shotpoint gathers, convolving with a zero-phase, 10–50 Hz pulse, and applying standard processing techniques (Figs. 5.9–5.14c); a 12-fold section was so generated. 3) No random or other noise was put on the original, shotpoint gather traces. 4) Migration is precise, because velocity at each part of the model is perfectly known.

Model The model (Fig. 6.20a) is a simple fault ramp, resulting in an anticline ("snakehead" thrust fault). Selected raypaths are highlighted for events that result in geometric (g, f_1, f_2), velocity (v_1, v_2) and raypath bending (r_1, r_2) effects.

Unmigrated Time Section *Geometric effects* on the unmigrated time section (Fig. 6.20b) include crossing events on the left and right flanks of the anticline (g), and downdip displacement, shortening and lessening of dip of the reflection from the dipping fault ramp (f_1, f_2). *Lateral velocity changes* result in pullup of horizontal interfaces beneath the anticline (v_1, compared to v_2). Discontinuity and crossing of events from continuous horizontal interfaces is due to *raypath bending* (r_1, r_2).

Migrated Time Section Migration of the time section (Fig. 6.20c) shows a more proper perspective of the structure. The flanks of the anticline are, for the most part, shown to be continuous (g'). The dipping fault ramp reflections have moved to a position of offset strata (f_1', f_2'). There are, however, still some problems in interpretation, even under these ideal conditions of known velocities and no

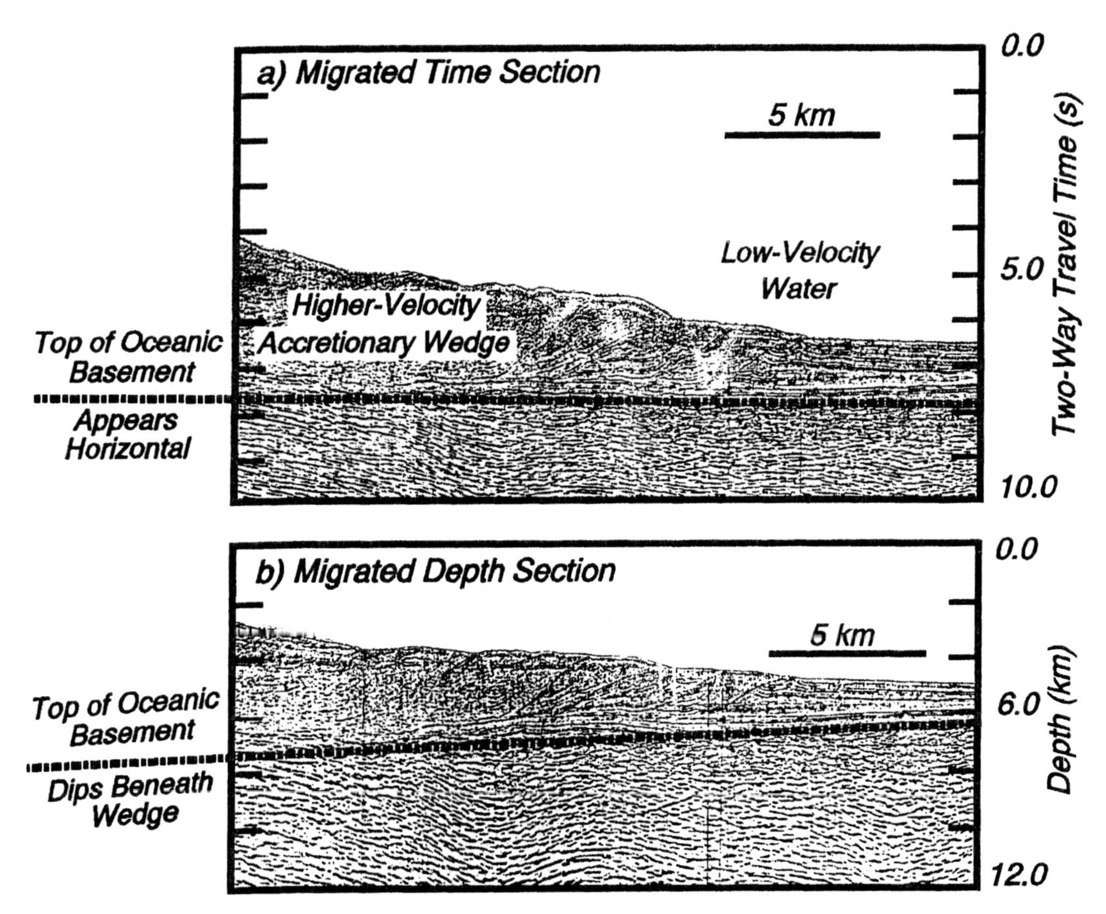

FIGURE 6.17 Seismic reflection profile of the trench (right) and toe of the accretionary wedge (left) in the Nankai Trough, Japan. From N. Nasu and others, Multi-channel seismic reflection data across Nanki Trough, *IPOD, Japan Basic Data Series #4,* © 1982, Ocean Research Institute, University of Tokyo, Tokyo, Japan. a) The top of oceanic basement (layer 2) appears horizontal on the time section. b) The relatively low-velocity water layer gradually thins from the right to the left side of the section; depth conversion thus reveals that the top of basement dips downward from right to left.

noise. It may be ambiguous to determine how far the fault has propagated up through the right flank of the anticline (p_1). Below the structure, lateral velocity changes and raypath bending effects make it difficult to realize that the lowermost interface is completely flat and continuous (p_2).

4. Three-Dimensional ("Sideswipe") Effects

Seismic lines are commonly run perpendicular to the strike of geologic structures, so that reflections are from within the plane of the section. If structures are oblique to the line of section, reflections can come from outside the plane; such events are termed *sideswipe*.

Herman et al. (1982) illustrate three-dimensional effects encountered in seismic profiling. Their first model (Fig. 6.21a) shows a buried dome above two horizontal interfaces. The bold lines are the surface locations of synthetic seismic profiles shown in Fig. 6.21b. Line 101, over the center of the dome, shows the true, unmigrated position for the dome relative to reflections from the upper and lower interfaces. Line 91 crosses the edge of the dome; it shows the domal event lower in time,

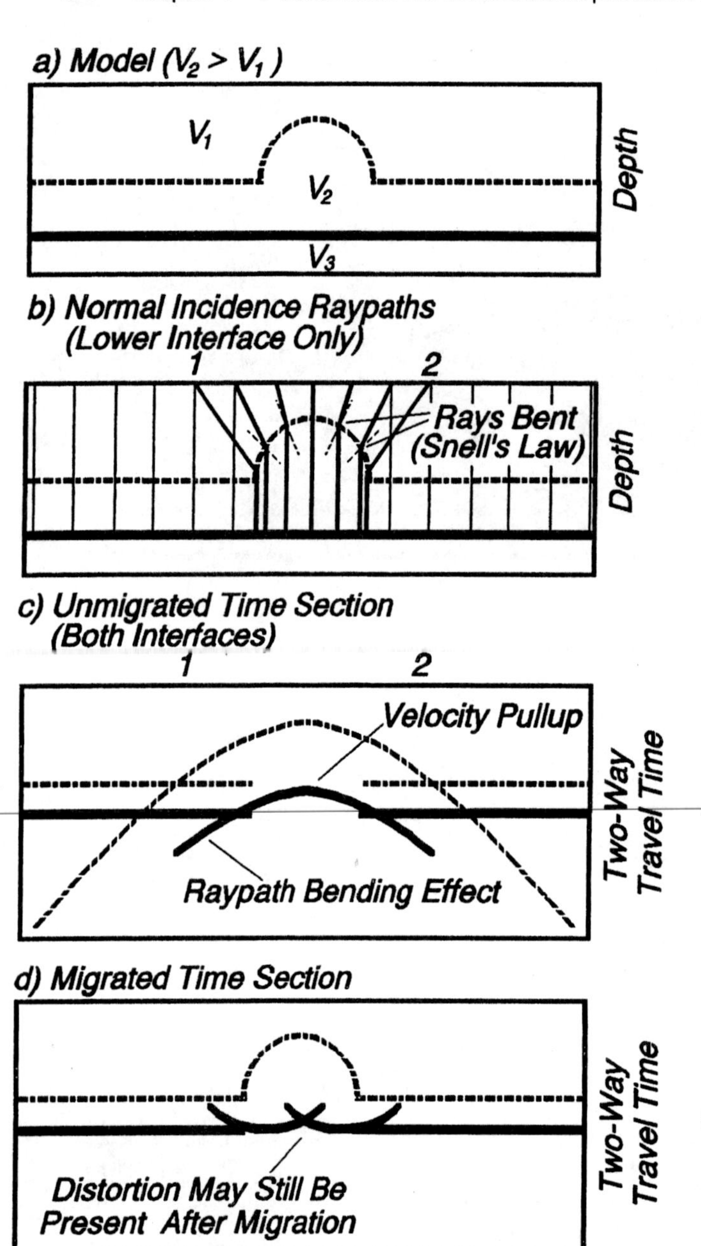

FIGURE 6.18 *Raypath bending.* a) Deep, horizontal interface added to anticline model (Fig. 6.8). b) Rays refract toward the vertical when crossing the interface on top of the anticline, because V_2 is greater than V_1. The portion of the horizontal interface directly below the anticline is thus imaged over a broader region (from point 1 to point 2). c) Unmigrated time section shows overlap in events from the horizontal interface, in addition to velocity pullup. d) Migration may not accurately remove raypath bending effects, leaving artifacts (distortion) that might be misinterpreted. (For this same model, see profiles with seismic traces in Fig. 5.14).

because waves were reflected from the dome slightly off to the side. Lines 81 and 61 show the event even deeper on the time section; sideswipe from the dome was from longer horizontal distances.

A more complicated model (Fig. 6.22a) illustrates a seismic profile run in the vicinity of two domes and a fault escarpment. The bold line 108 represents the vertical "shadow" of a seismic profile run on the surface, some distance above the structures. Note that the line crosses the side of the right dome, just misses the edge of the left dome, and crosses the fault escarpment at an oblique angle. The synthetic seismic profile (Fig. 6.22b) shows how complex and potentially erroneous interpretations could be, if the three-dimensional nature of the problem is not recognized. Fig. 6.22c is a standard, two-dimensional migration of that profile, while Fig. 6.22d was

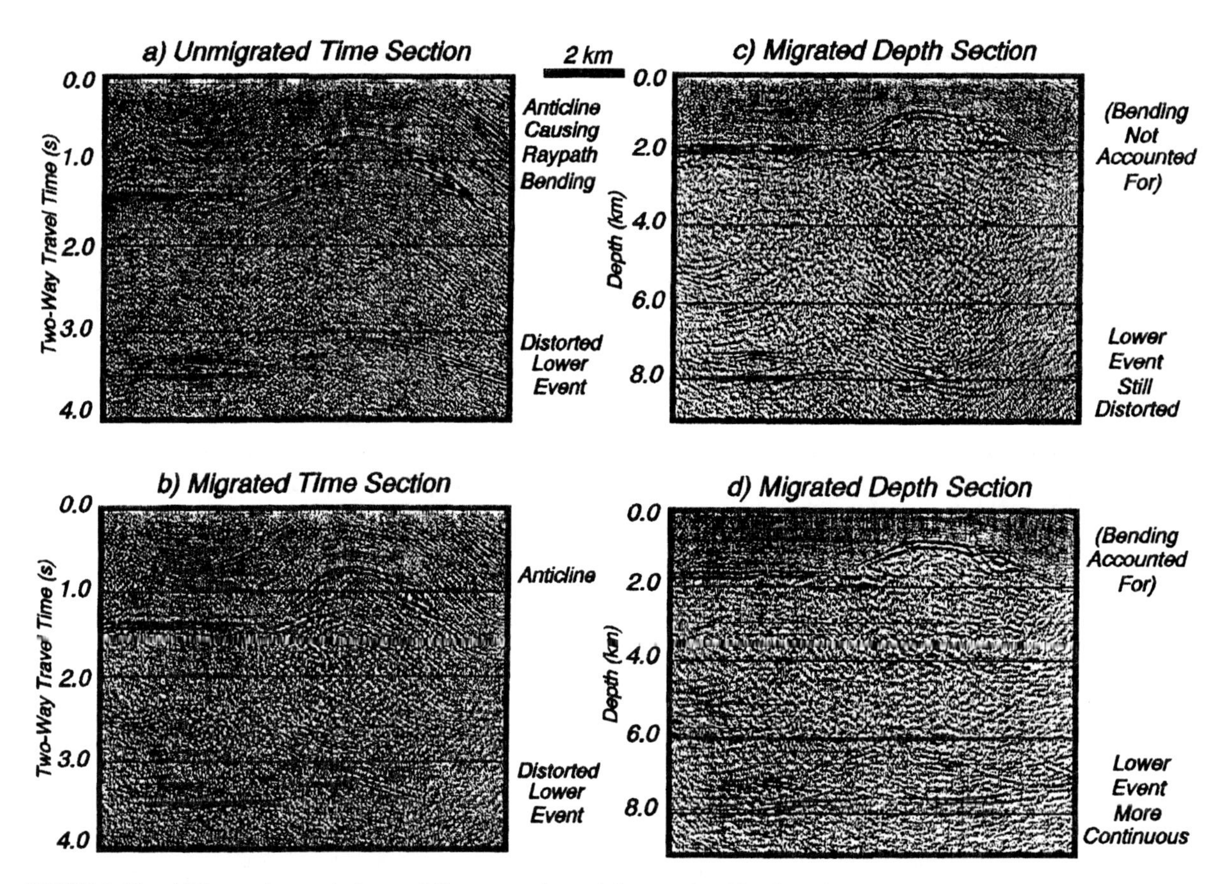

FIGURE 6.19 a) Distorted event below anticline on unmigrated time section. The distortion may be due to a combination of velocity pullup and raypath bending (Fig. 6.18b,c). b) Standard migration of (a). The continued distortion of the lower event may be an artifact of the raypath bending distortion. c) Depth conversion after standard migration (b) still shows distortion of the lower event. d) Migration accounting for raypath bending shows far less distortion, revealing a more continuous event after depth conversion. From K. Larner, B. Gibson, and R. Chambers, Imaging beneath complex structure, *AAPG Studies In Geology Series,* no. 15, vol. 1, pp. 26–39, © 1983. Redrawn with permission of the American Association of Petroleum Geologists, Tulsa, Oklahoma, USA.

achieved by computing 61 synthetic seismic profiles (lines 71–131 in Fig. 6.22a), then migrating all the information in three dimensions. The section in Fig. 6.22d, achieved at great expense of effort and computer time, is a close approximation of a time section for the vertical plane beneath line 108. Comparison of the two migrated profiles illustrates problems that arise if three-dimensional effects are not recognized and accommodated. The fault escarpment reflection in Fig. 6.22c has not migrated to its proper position, connecting segments of the upper reflector. The left dome still appears in the plane of the section; it should not be there, as illustrated by the three-dimensional migration (Fig. 6.22d).

SEISMIC EXPRESSION OF TECTONIC SETTINGS

Structure and stratigraphy developed in different tectonic settings can result in characteristic seismic reflection patterns. The illustrations and seismic sections presented below are a rough visualization of constraints on internal crustal structure offered by reflection data; reflection patterns that are prominent and common are shown, rather than the complete array of features that are observed. The diagrams

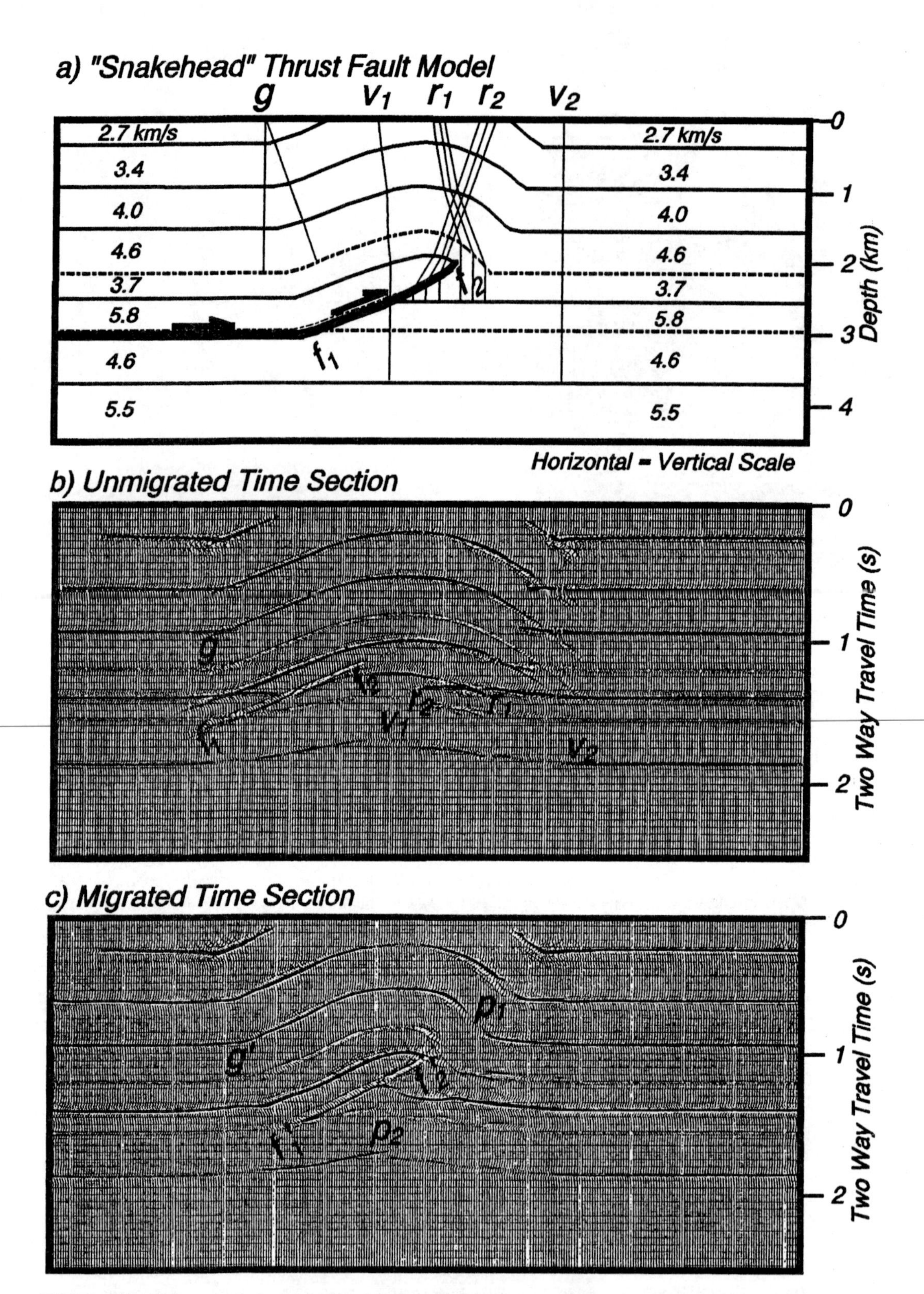

FIGURE 6.20 a) Hypothetical model with selected, normal incidence raypaths, showing geometric (g), velocity (v), and raypath bending (r) effects. Dashed lines are interfaces representing drops in velocity (negative reflection coefficients). The bold line with arrows above is the thrust fault; note the dipping segment (ramp) extending from f_1 to f_2. b) Computer generated (synthetic), unmigrated time section from (a). Letters refer to seismic expressions of the effects in (a). Vertical scale is two-way travel time (s); horizontal scale is the same as in (a). c) Time migration of (b); g′ illustrates some of the geometric effects from (b) that are resolved; f_1' and f_2' are more proper positions of ends of the fault ramp; p_1 and p_2 are potential interpretation problems discussed in text.

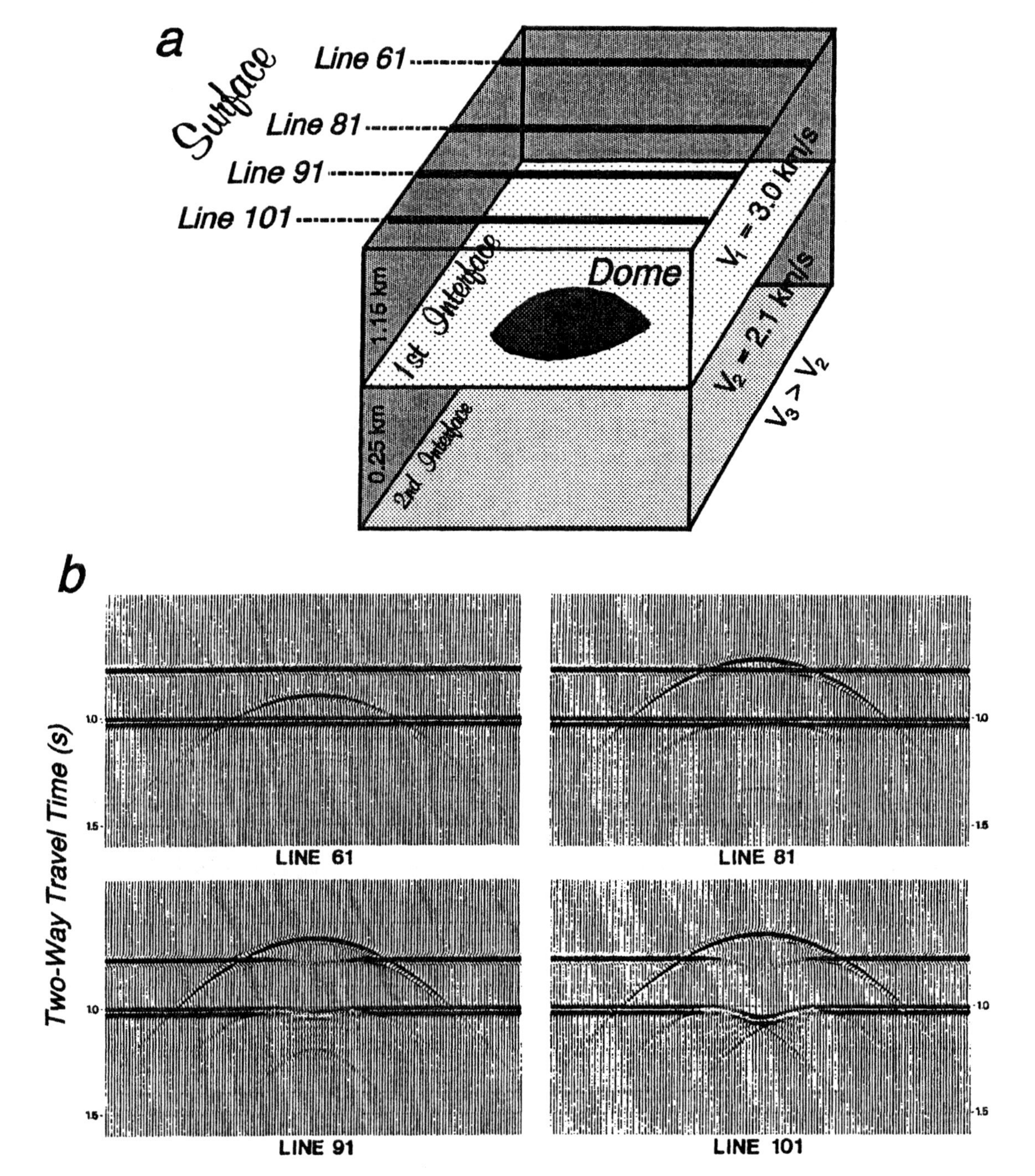

FIGURE 6.21 a) Three-dimensional model of buried dome above two horizontal interfaces. b) Synthetic, unmigrated seismic profiles from lines highlighted in (a). Vertical scales are two-way travel times in seconds. (Note: For the zero-phase pulse in these sections, positive amplitudes are left white and negative amplitudes shaded black). Modified from A fast three-dimensional modelling technique and fundamentals of 3-D frequency domain migration, by A. Herman, R. Anania, J. Chun, C. Jacewitz, and R. Pepper, *Geophysics,* vol. 47, pp. 1627–1644, © 1982. Redrawn with permission of the Society of Exploration Geophysicists, Tulsa, Oklahoma, USA.

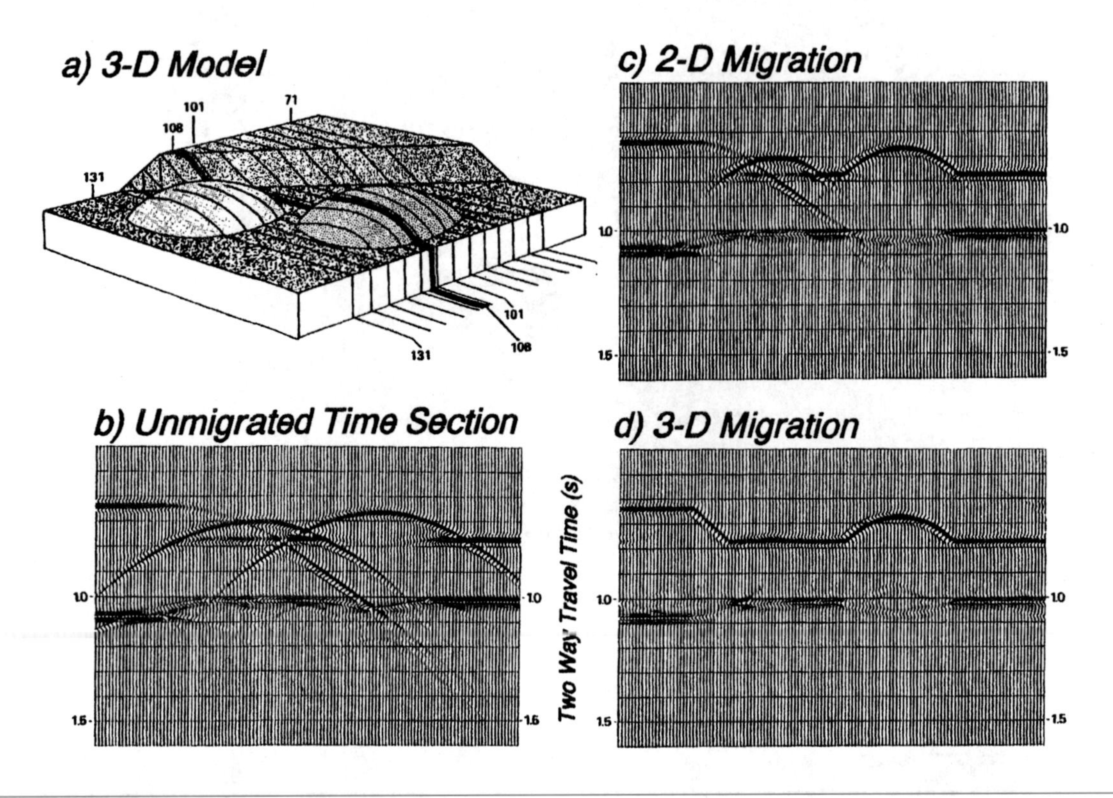

FIGURE 6.22 a) Model of two domes and two horizontal interfaces; the upper interface is offset by a fault escarpment. Interval velocities are the same as in Fig. 6.21. Lines represent vertical "shadows" of seismic profiles run along Earth's surface, some distance above the model. b) Unmigrated time section resulting from line 108 in (a). Vertical scale is two-way travel time in seconds. As in Fig. 6.21b, black shading is negative amplitude. c) Standard two-dimensional migration of (b). d) Three-dimensional migration involving 61 synthetic seismic profiles (lines 71–131 in a). Modified from A fast three-dimensional modelling technique and fundamentals of 3-D frequency domain migration, by A. Herman, R. Anania, J. Chun, C. Jacewitz, and R. Pepper, *Geophysics*, vol. 47, pp. 1627–1644, © 1982. Redrawn with permission of the Society of Exploration Geophysicists, Tulsa, Oklahoma, USA.

are presented in the sequence of the "Wilson Cycle," whereby an ocean basin is opened then closed (Wilson, 1966). As in Figs. 2.14 to 2.18, the sequence begins with continental rifting, forming an ocean basin as passive continental margins drift away from a mid-ocean ridge. The ocean closes through subduction, eventually leading to a collisional mountain range.

The cross sections are diagrammatic, in that characteristic reflection patterns at different tectonic settings are highlighted. Note that some patterns carry from one setting to another: wedge-shaped sequences from strata in continental rift grabens are seen beneath post-rift sediments on passive margins; hyperbolic diffractions from the top of oceanic layer 2, developed at mid-ocean ridges, are observed adjacent to passive margins and on the plate subducting beneath accretionary wedges. Appreciation of the entire sequence of cross sections and seismic profiles is therefore useful in overall tectonic interpretation.

Continental Rift Zone

The rifting of a continental craton thins the crust as well as the entire lithospheric plate (Figs. 2.13, 2.14, 6.23a). Patterns of reflections (reflection signatures) relate to features formed during rifting (Fig. 6.23b): strata deposited in grabens bounded by normal faults; ductile extension and underplating of the lower crust; formation of a new Moho.

1, 2. Nonreflective Upper Crust; Reflective Lower Crust As a continent rifts apart, the upper and lower crusts thin, but through different mechanisms (brittle and ductile failure, respectively). Reflection profiles from the Basin and Range Province of the western United States (Fig. 2.15) reveal that the crystalline basement of the upper crust is commonly transparent, due to structural complexity and lack of large acoustic impedance contrasts (Fig. 6.24a). The lower crust is more reflective (6.24b), exhibiting short, discontinuous events, thought to be associated with magmatism, ductile extension, or to a combination of those processes (Allmendinger et al., 1987; Potter et al., 1987).

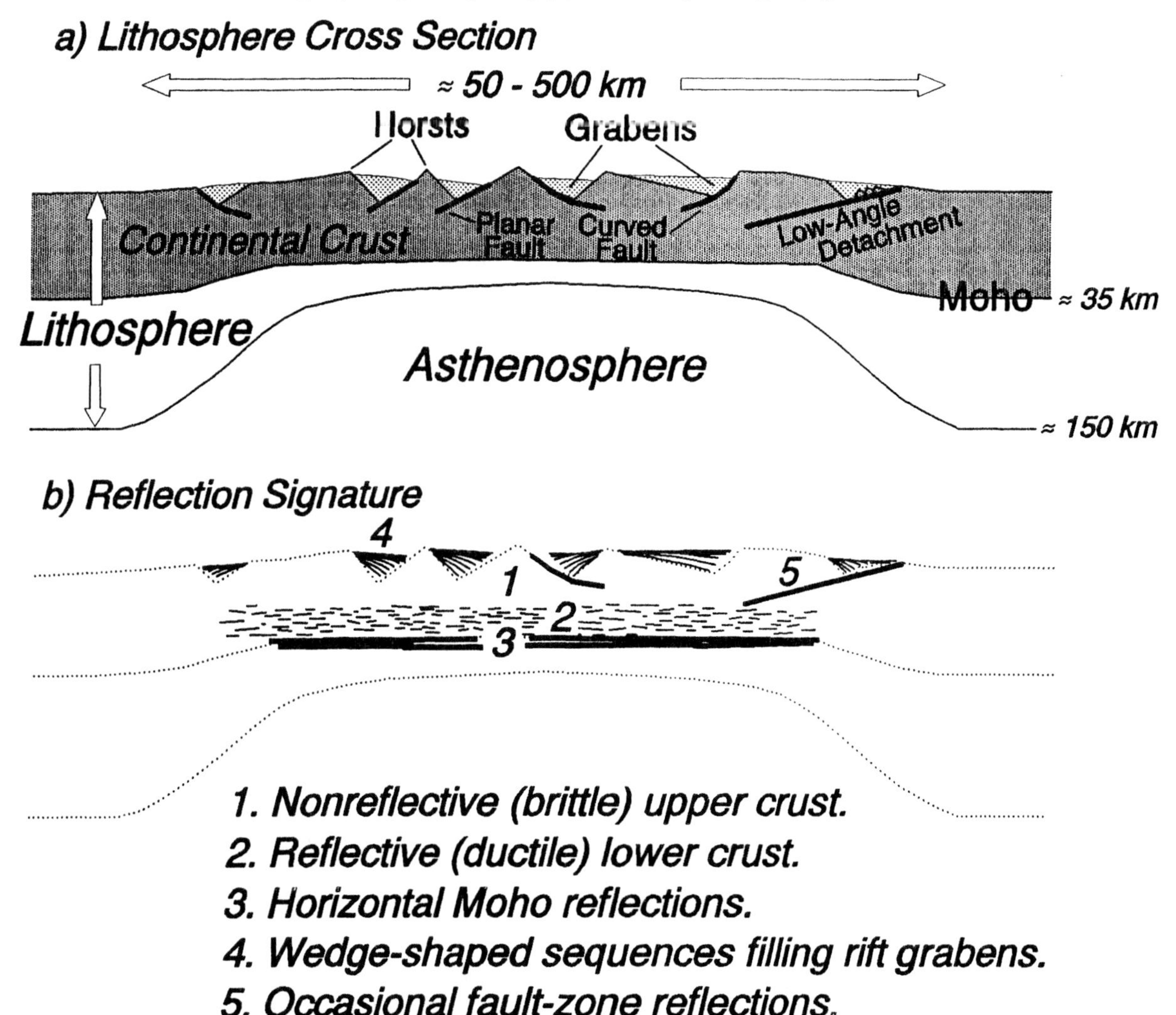

FIGURE 6.23 a) Hypothetical cross section of the lithosphere at a continental rift zone. b) Seismic reflection patterns commonly observed in continental rift zones.

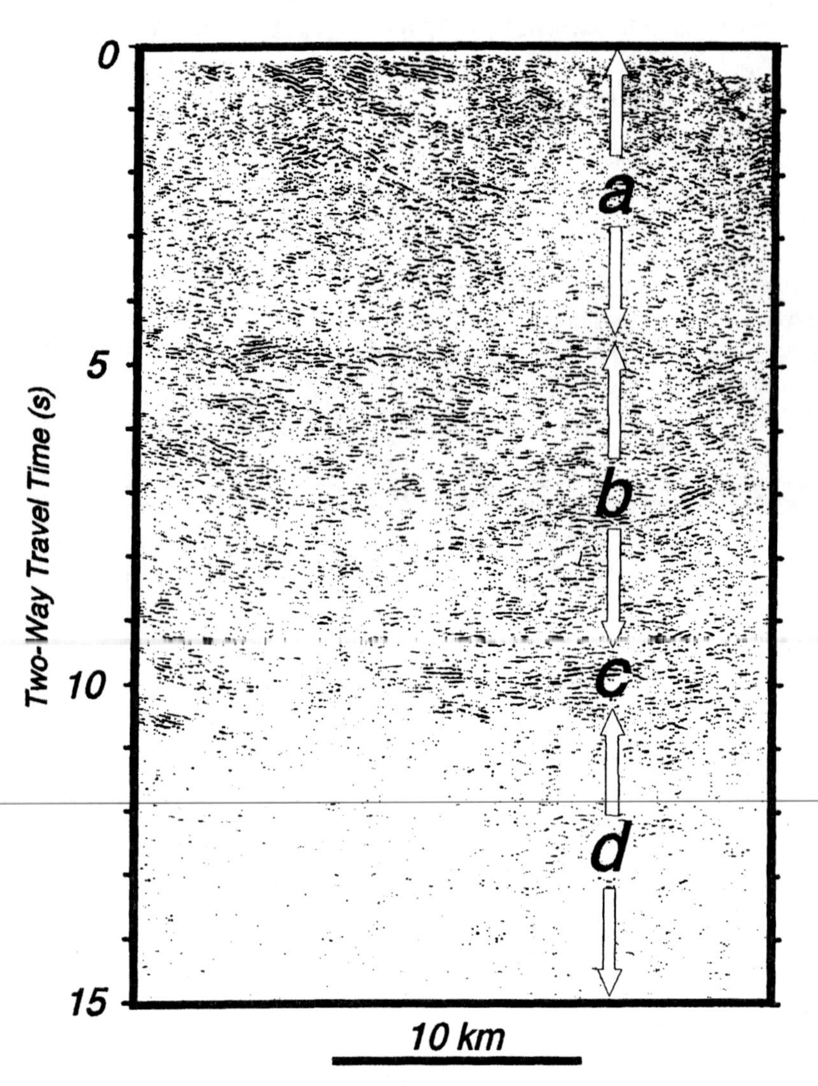

FIGURE 6.24 Seismic reflection profile from the Basin and Range Province (Consortium for Continental Reflection Profiling [COCORP] Nevada line 1). From Crustal structure of western Nevada from COCORP deep seismic-reflection data, by T. Hague, R. Allmendinger, C. Caruso, E. Hauser, S. Klemperer, S. Updyke, C. Potter, W. Sanford, L. Brown, S. Kaufman and J. Oliver, *GSA Bulletin,* vol. 98, pp. 320–329, © 1987. Redrawn with permission of the Geological Society of America, Boulder, Colorado, USA. a) Below dipping reflections from sedimentary and volcanic strata, the crystalline basement of the upper crust ($\approx$ 1 to 4.5 s) lacks prominent reflections (seismically "transparent"). b) From 4.5 to 10 s, prominent, horizontal reflections occur from the lower crust. c) Prominent reflections between about 9.5 and 10.5 s are interpreted as the zone of transition from the crust to the mantle (Moho). d) The transparent zone below 10 s is the uppermost mantle.

3. Horizontal Moho Reflections The boundary between the reflective lower crust and the nonreflective mantle in the Basin and Range Province is revealed by a zone of relatively continuous and high-amplitude reflections (Klemperer et al., 1986; Hague et al., 1987; Potter et al., 1987). The zone comprises a band of events about 0.15 to 0.3 s thick, lying between about 9.5 and 10.5 s, two-way travel time (Fig. 6.24c). Similar reflections suggest that the Moho throughout most of the Basin and Range Province is remarkably flat, lying between 29 and 32 km depth. The flatness and constant depth of the Moho suggest that it is a new boundary, formed in association with rifting.

4. Wedge-Shaped Sequences Filling Rift Grabens Seismic profiles of the upper crust in continental rift zones reveal wedge-shaped sequences of synrift sediments and volcanics filling asymmetric grabens (Anderson et al., 1983; Serpa et al., 1984). This "half-graben" geometry (Fig. 6.25a) is due to the bounding fault zone on one side having much larger offset than the normal faults on the other side; the wedge-shaped sequences thus exhibit growth, thickening and dipping toward the major fault zone. The synrift fill is commonly a few, and can be up to eight, km thick.

High amplitude reflections observed from the graben fill are due to high acoustic impedance lava flows, intercalated with lower impedance, fluvial and lacustrian sediments. A reflection-free zone is often observed in the region of the major bounding fault; that zone is interpreted as course, alluvial fan material deposited across the fault scarp, thickening as the fault grows (Anderson et al., 1983).

5. Occasional Fault-Zone Reflections The nature of faulting in rift grabens is commonly interpreted from the wedge geometry of synrift sedimentary and volcanic fill. In some places, however, reflections from actual fault zones are observed. Fault reflections could result from: 1) acoustic impedance contrast of low-velocity,

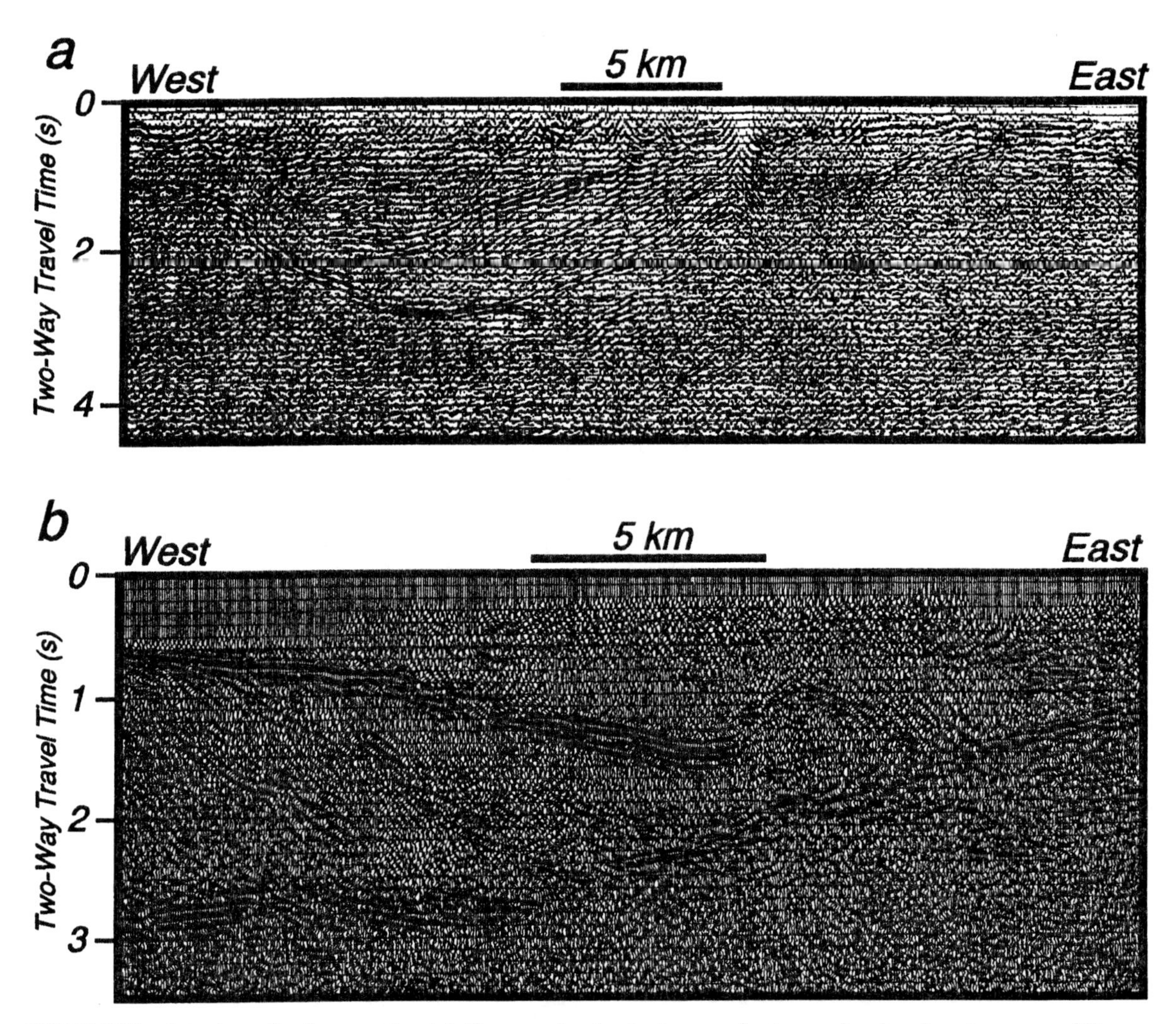

FIGURE 6.25 *Seismic profiles from continental rift zones, showing fault zone reflections and wedge-shaped sequences filling half-grabens.* a) COCORP Socorro line 1A, from the Rio Grande Rift of New Mexico (migrated time section). From Cenozoic normal faulting and the shallow structure of the Rio Grande rift near Socorro, New Mexico, by C. Cape, S. McGeary, and G. Thompson, *GSA Bulletin,* vol. 94, pp. 3–14, © 1983. Redrawn with permission of the Geological Society of America, Boulder, Colorado, USA. The rift graben wedge is terminated on its west side by a prominent, east-dipping reflection from the major fault zone. b) COCORP seismic line from the Sevier Desert, Basin and Range Province, Utah. From Cenozoic and Mesozoic structure of the eastern Basin and Range province, Utah, from COCORP seismic reflection data, by R. Allmendinger, J. Sharp, D. Von Tish, L. Serpa, L. Brown, S. Kaufman, J. Oliver, and R. Smithson, *Geology,* vol. 11, pp. 532–536, © 1983. Redrawn with permission of the Geological Society of America, Boulder, Colorado, USA. The section shows a low-angle normal fault detachment extending from 1.0 s on the east to 3.0 s on the west.

low-density graben fill, abutted against crystalline basement; or 2) the development of a sufficiently thick zone of acoustic impedance contrast within crystalline basement (for example, from a mylonitic fault zone). Reflections illustrate that some major faults are curved (or lystric, meaning spoon-shaped), as in the Socorro region of the Rio Grande Rift (Fig. 6.25a; Brown et al., 1980; Cape et al., 1983). In other places, fault zones are planar; a remarkable example from the Sevier Desert area of the Basin and Range Province shows a low-angle detachment fault extending from the surface to about 12 km depth (Fig. 6.25b).

Mid-Ocean Ridge

As a continent completely rifts apart, the fragments drift away as passive continental margins (Figs. 2.14, 6.26). An ocean basin grows as new lithosphere is created from partial melting of asthenosphere at a mid-ocean ridge (Figs. 2.7, 2.24). Prominent mid-ocean ridge reflections result from sedimentary deposits, block-faulted oceanic crust, the Moho, and (occasionally) magma chambers (Fig. 6.27).

1, 2. Transparent Water Layer; Sediments Covering Fault Blocks of Oceanic Crust For the broad wavelengths used in seismic reflection surveys, ocean water is homogenous, with no acoustic impedance contrast; the water is thus a reflection-free zone. Below the water, a blocky topography occurs on top of oceanic crust, due to normal faulting of layer 2 (Fig. 6.27a). Seismic profiles show that down-dropped valleys are gradually filled with more and more sediments; at tens to hundreds of km from the ridge axis, sediments completely cover the layer 2 topography (Fig. 6.28a).

3. Hyperbolic Events from Top of Oceanic Layer 2 On unmigrated seismic sections, the top of oceanic crust is prominent (Fig. 6.28b). The fault-block topography of layer 2 results in a series of buried focus ("bow-tie") effects (Figs. 6.10–6.12). This characteristic signature allows mapping of the extent of undeformed oceanic crust beneath passive continental margins and accretionary wedges (Fig 6.13a).

4, 5. Mid-Crustal (Axial Magma Chamber) Reflections; Horizontal Moho Reflections At the axes of some mid-ocean ridge segments, high amplitude reflections occur from a zone about 1.5 km beneath the seafloor ("AMC" in Fig. 6.28c). Those events correspond to the top of a zone of low seismic velocity (Vera et al., 1990). The high amplitude and low velocity suggest molten material ponded within the developing oceanic crust.

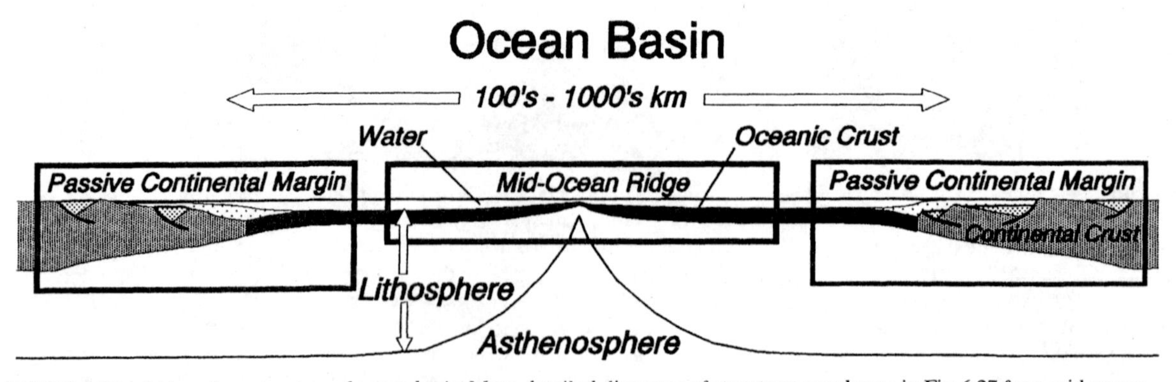

FIGURE 6.26 *Lithosphere structure of ocean basin.* More detailed diagrams of structure are shown in Fig. 6.27 for a mid-ocean ridge, and Fig. 6.29 for a passive continental margin.

Mid-Ocean Ridge

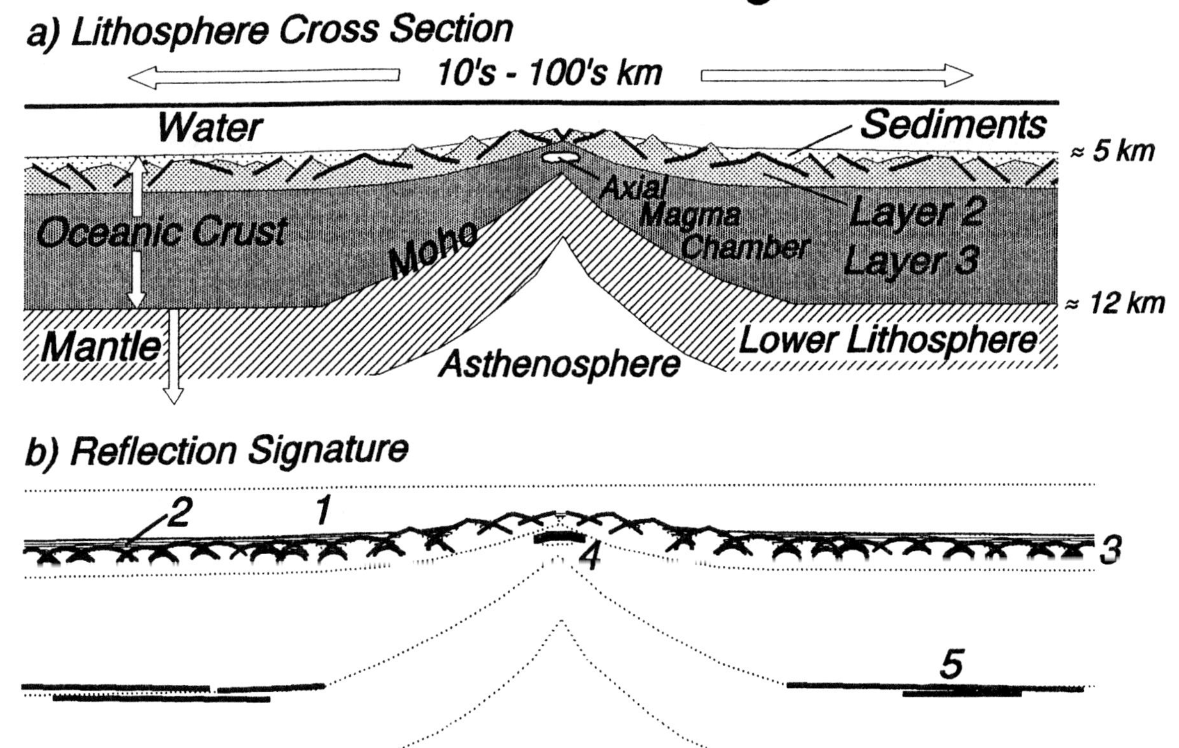

1. *Transparent water layer.*
2. *Sediments covering fault blocks of oceanic crust.*
3. *Hyperbolic events (unmigrated) from top of oceanic layer 2.*
4. *Mid-crustal (axial magma chamber) reflections.*
5. *Horizontal Moho reflections.*

FIGURE 6.27 a) Hypothetical cross section of the lithosphere at a mid-ocean ridge. b) Seismic reflection signatures commonly observed at mid-ocean ridges.

Away from mid-ocean ridge axes, reflection profiles sometimes reveal horizontal events about 2 s (two-way travel time) beneath the top of oceanic layer 2 ("M" in Fig. 6.28c). The events suggest a Moho discontinuity (or ≈ 1 km thick crust/mantle transition zone), beneath about 6 km of crystalline crust.

Passive Continental Margin

After a continent rifts apart, passive margins develop on the edges of the drifting fragments (Fig. 6.26). Postrift sediments accumulate on the margins, covering features formed during continental rifting and early mid-ocean ridge development (Fig. 6.29a). Seismic reflection patterns suggest the type of crust beneath the sediments (3, 4, and 5 in Fig. 6.29b): rift graben "wedges" indicate continental crust extended during rifting; "seaward-dipping wedges" represent volcanic material erupted in the zone separating continental from oceanic crust; hyperbolic events on unmigrated sections indicate the top of oceanic crust. Signatures of upper and lower

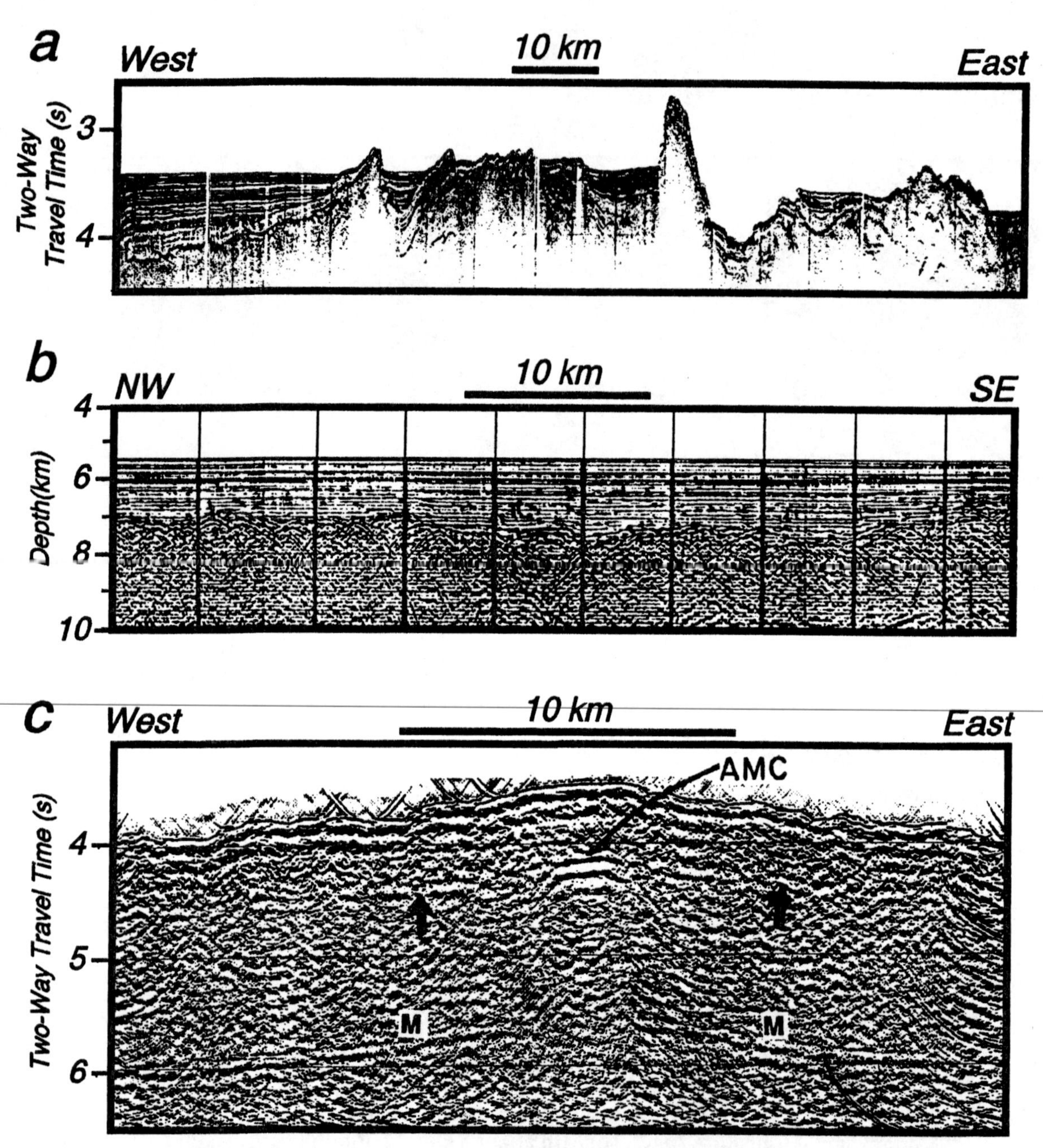

FIGURE 6.28 a) Seismic reflection profile of the Juan de Fuca Ridge. From R. Riddihough, M. Beck, R. Chase, E. Davis, R. Hyndman, S. Johnson, and G. Rogers, *Geodynamics of the Eastern Pacific Region, Caribbean and Scotia Arcs,* AGU Geodynamics Series, no. 9, pp. 5–21, © 1983. Redrawn with permission of the American Geophysical Union, Washington, D.C. The water is a reflection-free zone above layered sediments and basement. The basement (oceanic layer 2) is cut by normal faults; downdropped blocks fill with deep-marine sediments. b) Unmigrated depth section from the north Atlantic Ocean. The transparent water layer is about 5.3 km deep, covered by about 1.5 km of flat-lying sediments. Hyperbolic ("bow tie effect") reflections, about 7 km deep, reveal the hummocky top of oceanic basement. c) Migrated time section recorded across the axis of the East Pacific Rise. From E. Vera, J. Mutter, P. Buhl, J. Orcutt, A. Harding, M. Kappus, R. Detrick, and T. Brocher, *Journal of Geophysical Research,* vol. 95, pp. 15529–15556, © 1991. Redrawn with permission of the American Geophysical Union, Washington, D.C. AMC = axial magma chamber reflection; M = Moho reflections.

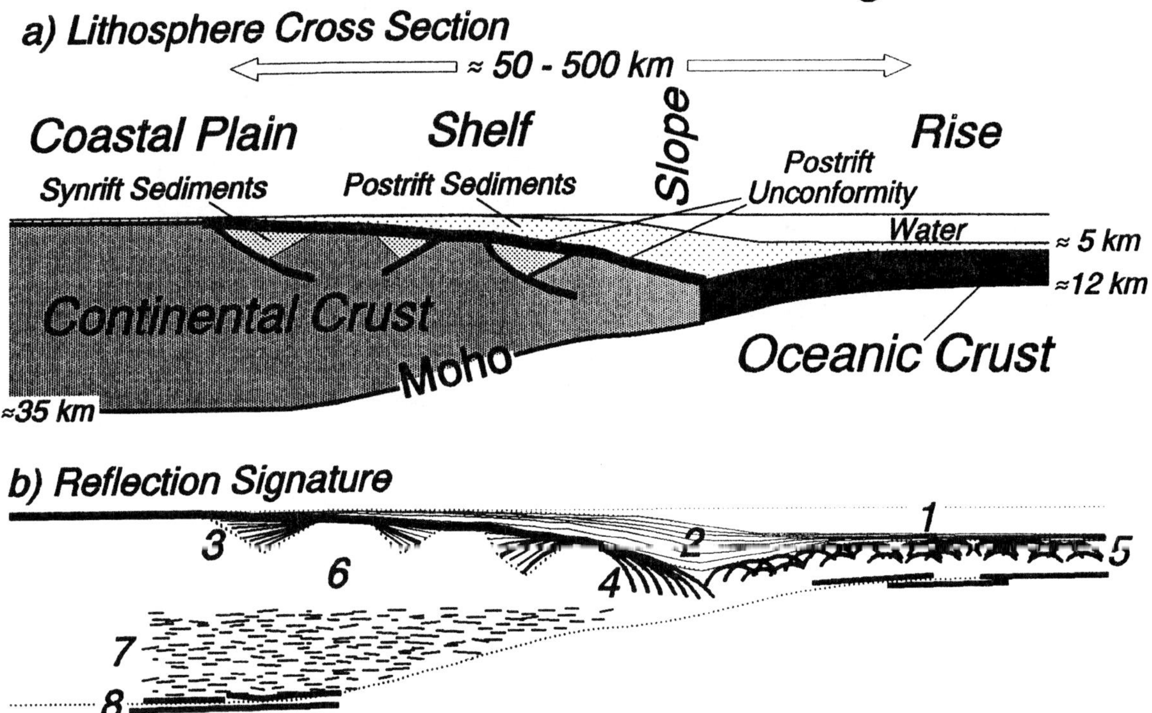

FIGURE 6.29 a) Hypothetical cross section of the crust at a passive continental margin. b) Reflection signatures characteristic of certain structure and stratigraphy of a passive margin.

continental crust, and of the Moho, may be preserved from the continental rifting stage (Fig. 6.23).

1., 2. Transparent Water Layer; Reflections from Postrift Sediments The most prominent events on seismic profiles of passive continental margins are nearly horizontal reflections from the postrift sediments, deposited after continental breakup and during the drifting and subsidence of the margin (Fig. 6.30). The continental margin may have been above sea level for some time after rifting, so that a

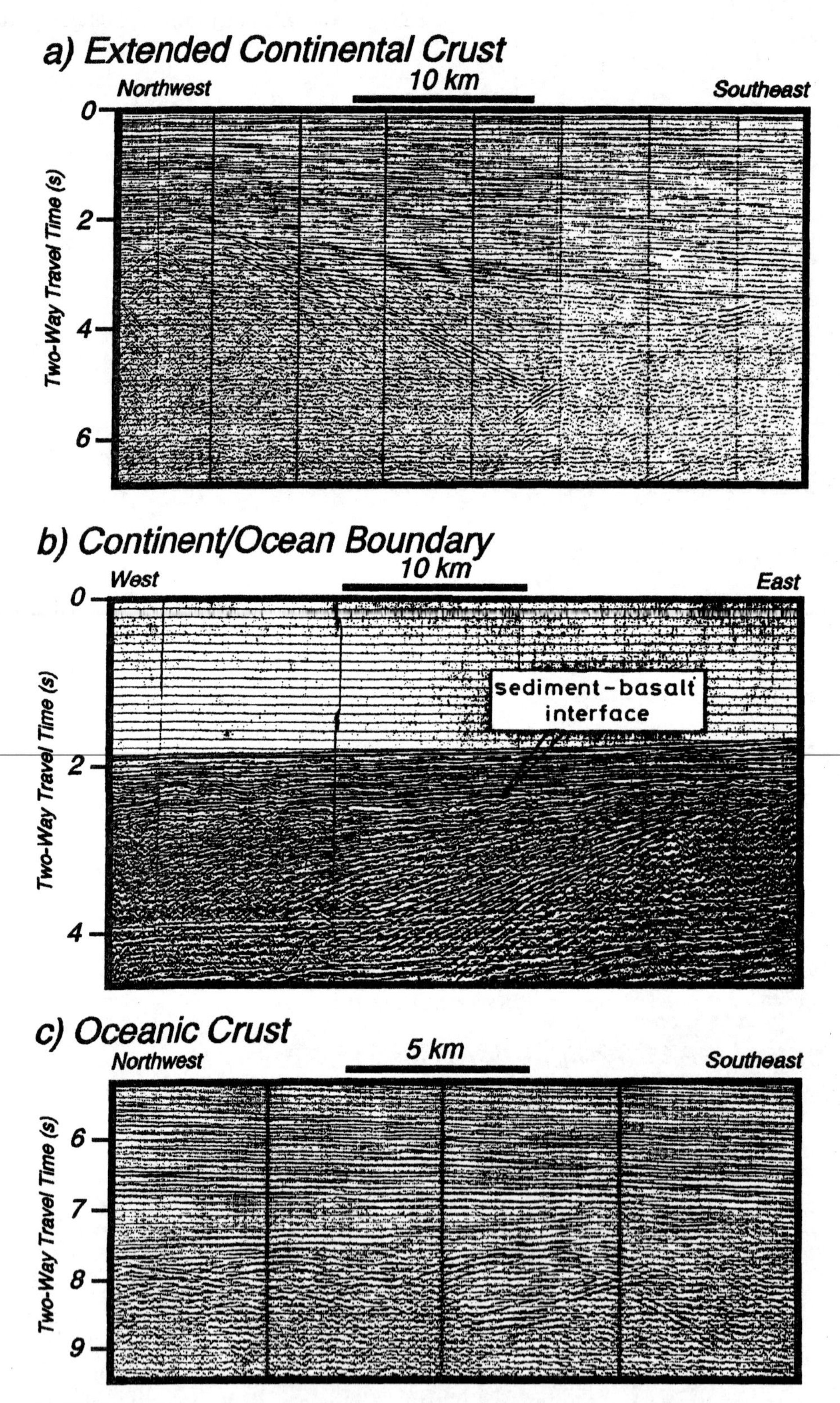

FIGURE 6.30 *Seismic reflection profiles from passive continental margins.* Each profile shows prominent horizontal reflections from the postrift sedimentary section. Deeper reflection patterns suggest the type of crystalline basement. a) *Rift graben wedge indicating continental crust extended during rifting.* U.S.G.S. line 25 from Offshore Cape Cod, part of the continental shelf off the east coast of the United States. The transparent water layer is above 0.2 s. Postrift sediments lie from

postrift (or breakup) unconformity separates the postrift sediments from the underlying rocks (prerift and synrift sediments; crystalline basement). The postrift unconformity is often flat, suggesting that horst blocks were eroded prior to subsidence of the margin below sea level.

3. Rift Graben Wedges Defining Extended Continental Crust The strata that fill rift grabens (sediments intercalated with volcanics) display strong contrasts in acoustic impedance. Wedge-shaped events observed beneath postrift sediments in coastal plain and continental shelf areas thus indicate continental basement that was extended during rifting (Fig. 6.30a; Schlee et al., 1976; Grow et al., 1979; Costain and Çoruh, 1989). Passive margins are often portrayed with normal faults dipping predominately seaward; reflection profiles, however, show half-grabens consistent with major normal faults dipping either toward or away from the ocean basin, with no dominate direction.

4. "Seaward-Dipping Wedges" at Continent/Ocean Boundary On some passive margins, the zone between extended continental crust and true oceanic crust displays prominent reflections from beneath the postrift sediments (Fig. 6.30b). Those reflections always dip toward the ocean basin and are overall wedge-shaped, spanning about two s of two way travel time; individual reflections have a convex upward appearance (Mutter et al., 1982). Drilling of the upper parts of these "seaward-dipping wedges," off the coasts of Norway and east Greenland, reveals that the wedges are almost entirely basalt that was erupted on land. The sequences are interpreted to have formed during the initial development of oceanic layer 2; fluid basalts spread laterally over tens of kilometers, when the "mid-ocean ridge" was above sea level (Mutter et al., 1984).

5. Hyperbolic Events Defining Top of Oceanic Crust The top of basement beneath continental rise areas commonly reveals the same seismic expression as the block-faulted, oceanic layer 2 formed at mid-ocean ridges (Figs. 6.27; 6.28b). The characteristic, hyperbolic events on unmigrated sections can therefore be used to map the extent of true oceanic crust beneath continental slope and shelf regions (Fig. 6.30c).

6, 7, 8. Nonreflective Upper Crust; Reflective Lower Crust, Moho Reflections Seismic reflection profiles from some passive margins show crustal reflectivity inherited from the continental rifting stage (Figs. 6.23, 6.24). Around the British Isles, a nonreflective upper crust, reflective lower crust, and Moho reflections are especially

FIGURE 6.30 (continued) 0.2 to about 2 s on the northwest, down to 3.5 s on the southeast. The postrift unconformity separates those sediments from older rocks, including crystalline basement and a prominent rift graben. The major bounding fault zone dips northwestward, toward the continent. b) *"Seaward-dipping wedge" indicating continental/oceanic crustal boundary*. Seismic reflection profile from the Vøring Plateau off the Atlantic coast of Norway. From K. Hinz, Line BFB (24-fold stack) from the Norwegian continental margin/outer Vøring Plateau, *AAPG Studies In Geology Series,* no. 15, vol. 2, pp. 2.2.3.39–2.2.3.40, © 1983. Redrawn with permission of the American Association of Petroleum Geologists, Tulsa, Oklahoma, USA. Seaward-dipping wedge lies below the sediment-basalt interface, from 2.5 to about 4.5 s. Note the upward convexity and increase in dip with depth of individual reflectors within the wedge. (Ignore nearly horizontal multiples beginning at about 4.0 s). c) *Hyperbolic diffractions indicating oceanic crust.* Unmigrated time section from the east coast of the United States (U.S.G.S. line 25, offshore Cape Cod). Hyperbolic events at about 8 s represent the top of oceanic basement (layer 2).

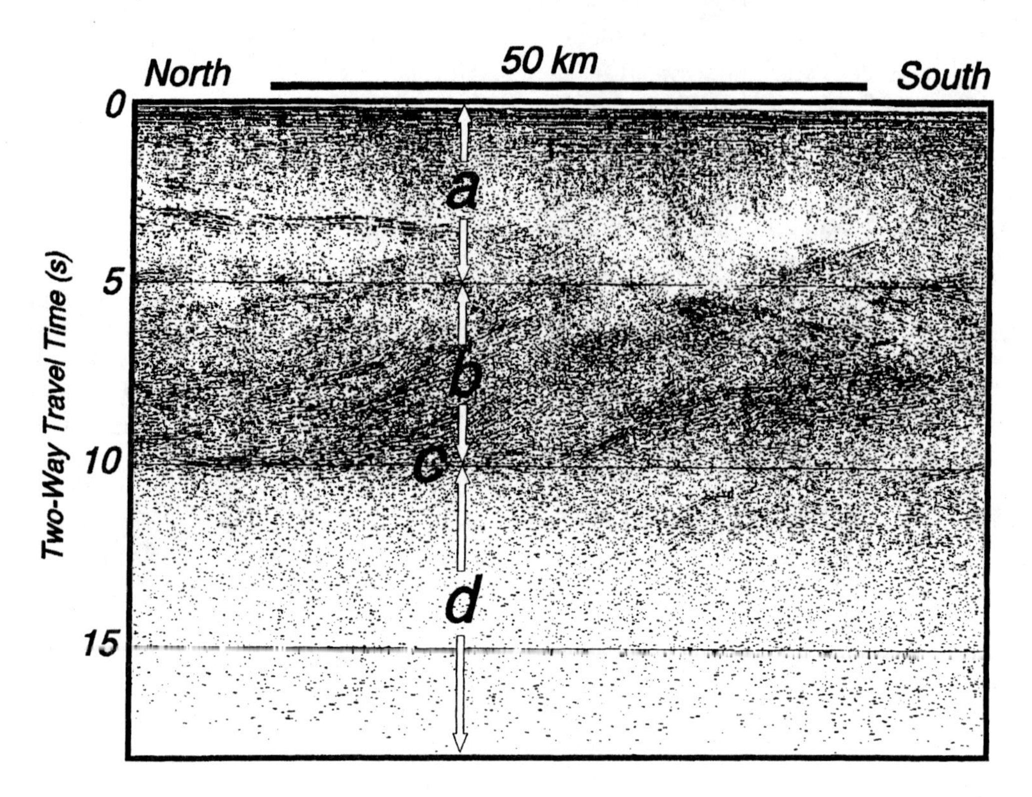

FIGURE 6.31 *Seismic reflection profile from the west coast of Ireland.* Modified from A deep seismic reflection transect across the Irish Caledonides, by S. Klemperer, P. Ryan, and D. Snyder, *Journal of the Geological Society, London,* vol. 148, pp. 149–164, © 1991. Redrawn with permission of the Geological Society of London. a) Nonreflective upper crust extends to about 5 s. b) Reflective lower crust from about 5 to 10 s. c) Moho reflections occur at about 10 s. d) Transparent upper mantle. Compare with continental rift profile, Fig. 6.24.

well displayed (Fig. 6.31; Petty and Hobbs, 1987; Reston, 1990; Klemperer et al., 1991).

Subduction Zone

At a subduction zone, a lithospheric plate capped with thin oceanic crust extends beneath a plate with thicker, island arc or continental crust (Figs. 2.16, 6.32). The seismic expression of features developed at and away from a mid-ocean ridge (hemipelagic sediments; hummocky top of oceanic crust; horizontal Moho) is often imaged beneath trench-fill sediments at the toe of the accretionary wedge (Fig. 6.33). Farther toward the volcanic arc, deformed accretionary wedge sediments and basement display a more chaotic seismic expression; flat-lying or deformed sediments in the forearc basin may also be imaged.

1, 2. Transparent Water Layer; Trench-Fill Sediments As oceanic crust approaches a subduction zone, it is covered by terrigenous sediments (turbidites) deposited in a trench (Fig. 6.34a); these undeformed sediments result in nearly horizontal reflections. At the toe of the accretionary wedge, where deformation is slight, it is often possible to image thrust structures.

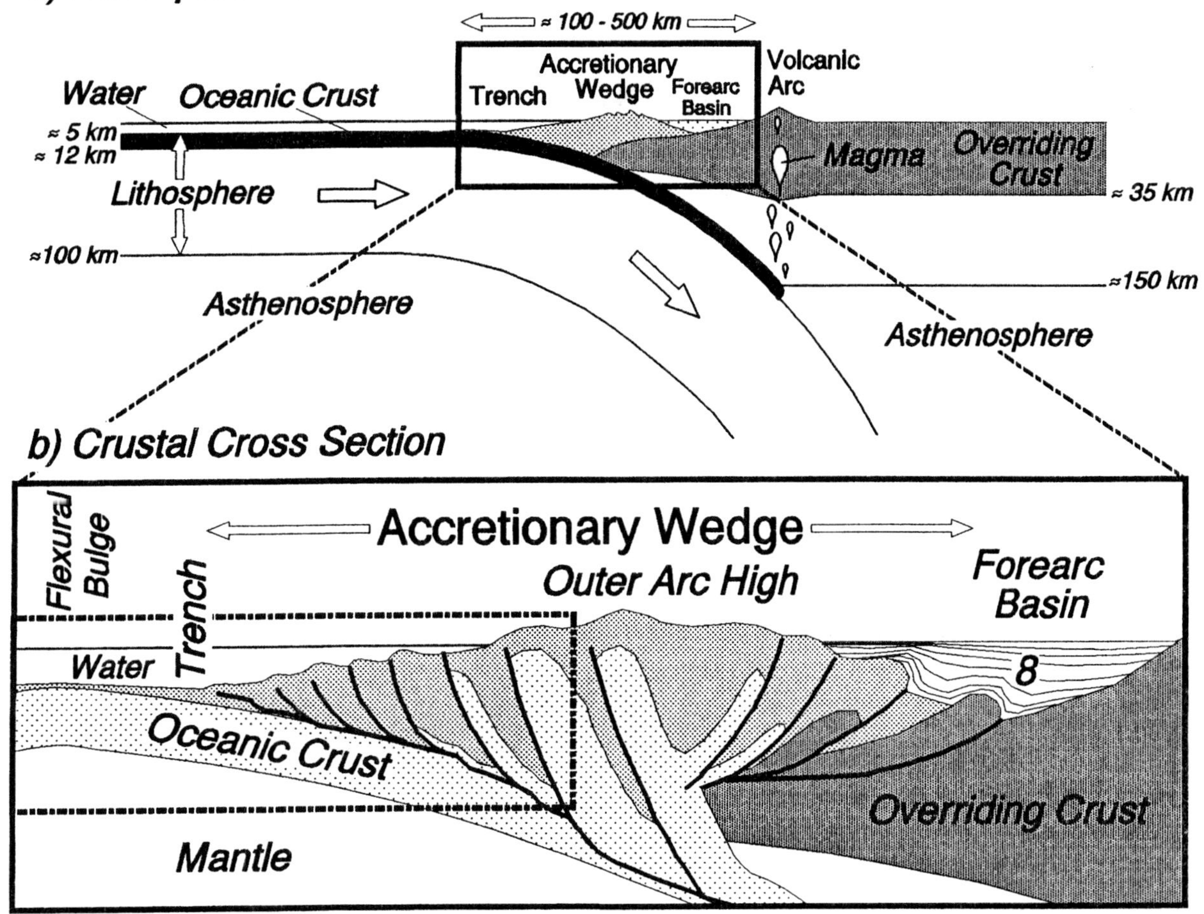

FIGURE 6.32 Hypothetical lithospheric (a) and crustal (b) cross sections of a subduction zone. Dashed rectangle in (b) is the region of Fig. 6.33.

3. Deep Ocean Basin Sediments Before reaching the vicinity of the trench, the basalt of oceanic layer 2 is normally covered with a thin layer of deep-marine (hemipelagic) sediments (Fig. 6.34a). The zone of thrust detachment (or décollement) commonly lies within the trench fill sediments, so that the hemipelagic sediments are imaged beneath the toe of the accretionary wedge (Moore, 1975; Moore and Shipley, 1988a, 1988b). Farther arcward, many styles of deformation are observed (Scholl et al., 1980); the décollement may cut downsection, so that the hemipelagic sediments and oceanic basement are incorporated into the accretionary wedge (Fig. 6.33).

4. Flexural normal faults offsetting basement and sediments A subducting lithospheric plate bends as it is weighted by topography developed on the overriding plate. A convex-upward shape results, producing a flexural bulge seaward of the trench (Fig. 6.32b). The upper part of the bulge undergoes extension, producing normal

a) Crustal Cross Section of Outer Forearc Region

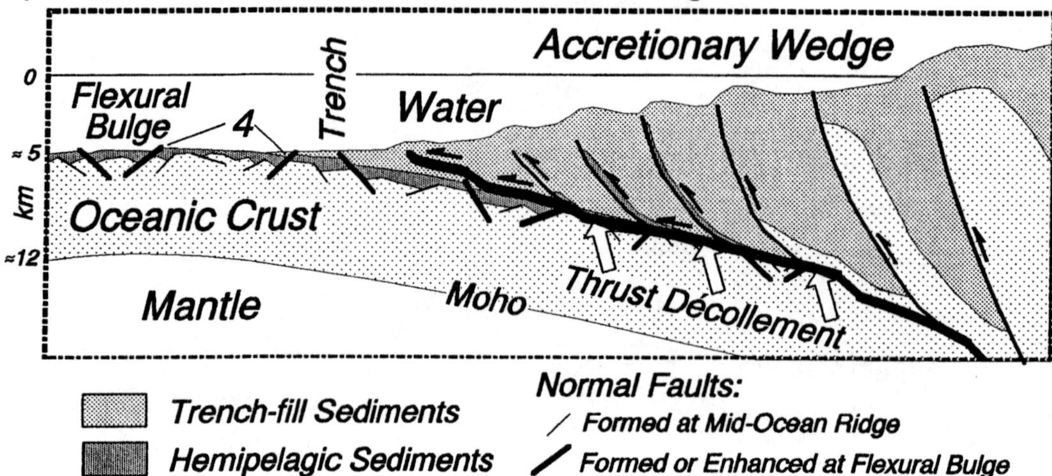

b) Reflection Signature

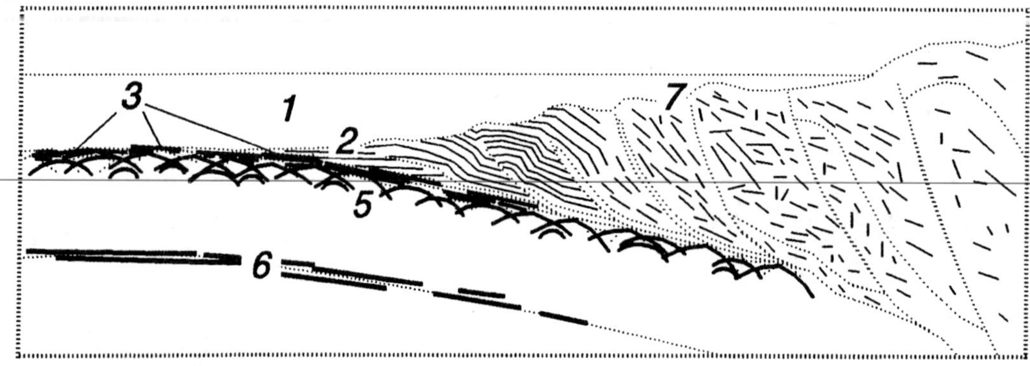

1. *Transparent water layer.*
2. *Trench-fill sediments (undeformed and deformed).*
3. *Hemipelagic sediments from deep-ocean basin.*
4. *Flexural normal faults offsetting basement and sediments (in "a" above).*
5. *Hyperbolic events (unmigrated) from top of oceanic layer 2.*
6. *Flat-lying Moho reflections.*
7. *Chaotic expression of highly deformed accretionary wedge.*
8. *Forearc basin sediments (previous figure).*

FIGURE 6.33 Cross section (a) and reflection sequences (b) commonly observed in the outer forearc region of a subduction zone (Fig. 6.32b).

faults (Fig. 6.33). This younger set of normal faults differs from the older normal faults that were produced soon after oceanic crust formed at a mid-ocean ridge: mid-ocean ridge faults cut the oceanic basement and are generally covered by hemipelagic sediments (Figs. 6.27, 6.28a); flexural faults offset the oceanic basement, hemipelagic sediments, and some of the trench-fill sediments (Figs. 6.33a, 6.34a).

5. Hyperbolic Events Defining Top of Oceanic Crust In addition to migrated sections, it is useful to look at unmigrated seismic profiles from accretionary wedges. The characteristic expression of hummocky, oceanic layer 2, developed at a mid-ocean ridge (Figs. 6.27, 6.28b), can often be traced beneath at least the toe areas of wedges. In some instances, like the Barbados Ridge Complex (Moore et al., 1982; Westbrook and Smith, 1983; Westbrook et al., 1988; Bangs and Westbrook, 1991; Fig. 6.34b) and the Nankai Trough (Nasu et al., 1982; Fig. 6.4a), the top of oceanic crust can be traced laterally for tens of km beneath the accretionary wedge.

6. Flat-lying Moho Reflections On reflection profiles across newly-formed oceanic crust, Moho events are approximately 2 s (two-way travel time) beneath the top of oceanic layer 2 (Fig. 6.28c). Reflection profiles from some deep-sea trench/accretionary wedge regions show flat-lying reflections, also about 2 s beneath the top of oceanic basement, indicating the position of the Moho (Fig. 6.34c).

7. Chaotic Expression of Deformed Accretionary Wedge Sediments and Basement At some distance within the accretionary wedge, commonly tens of km from the trench axis, materials dewater and are structurally deformed. Seismic velocity and density contrasts become less pronounced than they were at the toe of the wedge, leading to less contrast in acoustic impedance. Boundaries are also steeper and less continuous than they were at the toe, so that reflections are chaotic; it is therefore difficult to image the complex structure in the rearward portions of accretionary wedges (right side, Fig. 6.33b). Offsets of trench strata and thrust faults are often well imaged in the toe regions of accretionary wedges (right side, Fig. 6.4b). Farther arcward, however, more intense deformation leads to poor seismic resolution (left side, Fig. 6.4b).

8. Forearc Basin Sediments Fig. 6.32a shows that two parallel mountain ranges commonly form on top of a subduction zone, one structural (accretionary wedge) and one volcanic (arc). The low region that accumulates sediments between the two ranges is the forearc basin (Fig. 6.32b). Acoustic impedance contrasts within forearc basin strata (sometimes undeformed but often deformed) can lead to strong reflections.

Collisional Mountain Range

Reflections from collisional mountain ranges span a wide variety because structural styles change during collision (Figs. 6.35, 6.36). Key reflection signatures, nonetheless, are characteristic of a traverse from the craton of the downgoing plate ("foreland"), across the mountains ("hinterland").

Erosion of the mountains results in deposition of clastic ("synorogenic") strata over the approaching continental margin, as deep-marine flysch in the early stages of collision, then as molasse as the region uplifts above sea level (Fig. 6.35a, b). A basin forms in the foreland, initially because deposition is on relatively thin, oceanic or transitional crust (Stockmal et al., 1986). As the craton approaches, the

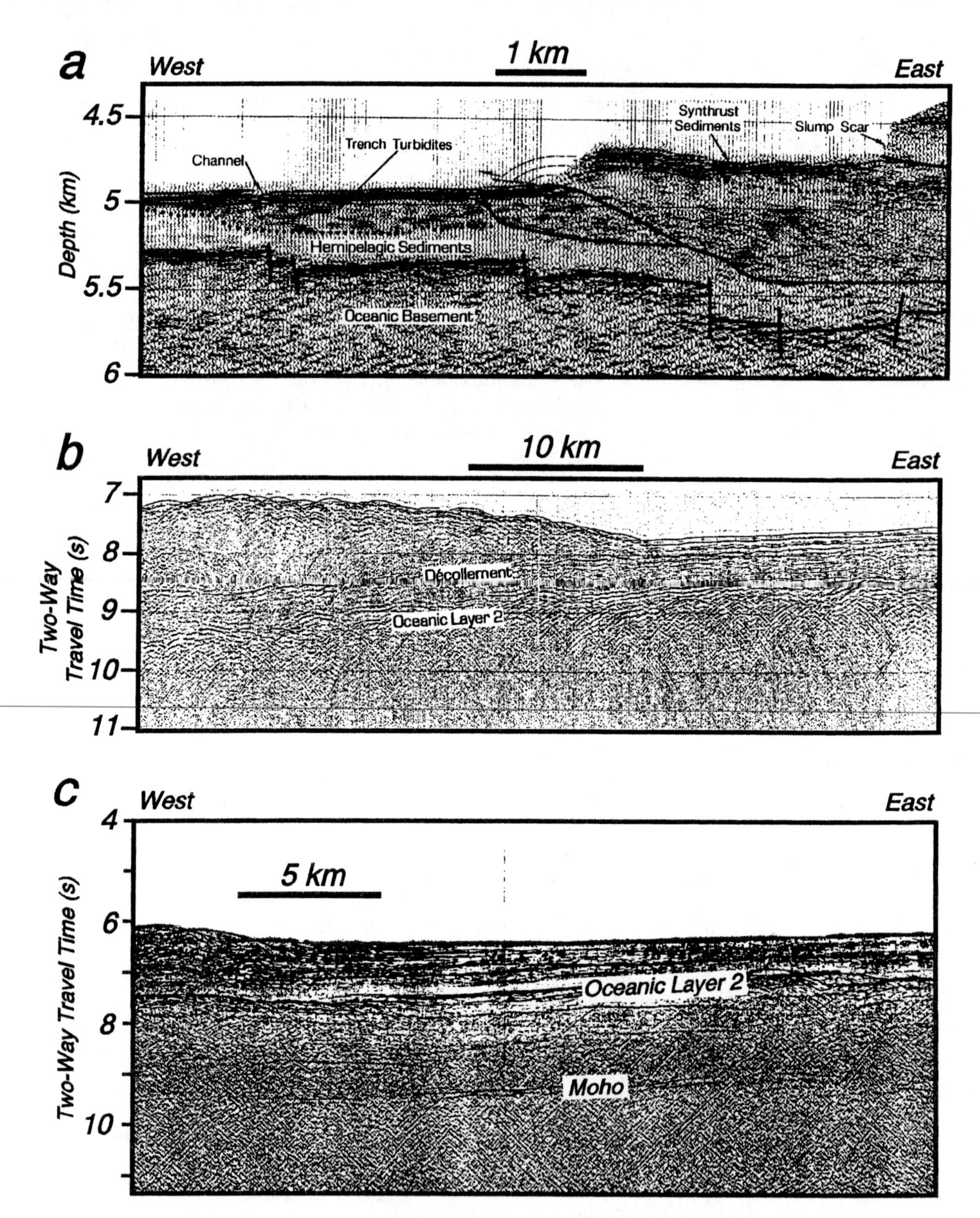

FIGURE 6.34 *Seismic reflection profiles across deep-sea trenches.* a) Migrated depth section from the Middle America Trench off Mexico (Line Mex-12). From G. Moore and T. Shipley, *Journal of Geophysical Research,* vol. 93, pp. 8911–8927, © 1988. Redrawn with permission of the American Geophysical Union, Washington, D.C. Water layer is transparent zone from surface to nearly 5 km depth. Trench turbidites are flat lying at trench axis, but deformed at toe of accretionary wedge, on the right. Hemipelagic sediments (transparent zone beneath trench turbidites) extend beneath the thrust décollement at the toe of the accretionary wedge. Flexural normal faults offset hemipelagic sediments and top of oceanic basement. Upthrown normal fault blocks act as rigid basement buttress, causing ramping of thrust décollement. b) Unmigrated time section from Barbados ridge complex in the Caribbean Sea. From Long décollements and mud volcanoes: Evidence from the Barbados Ridge Complex for

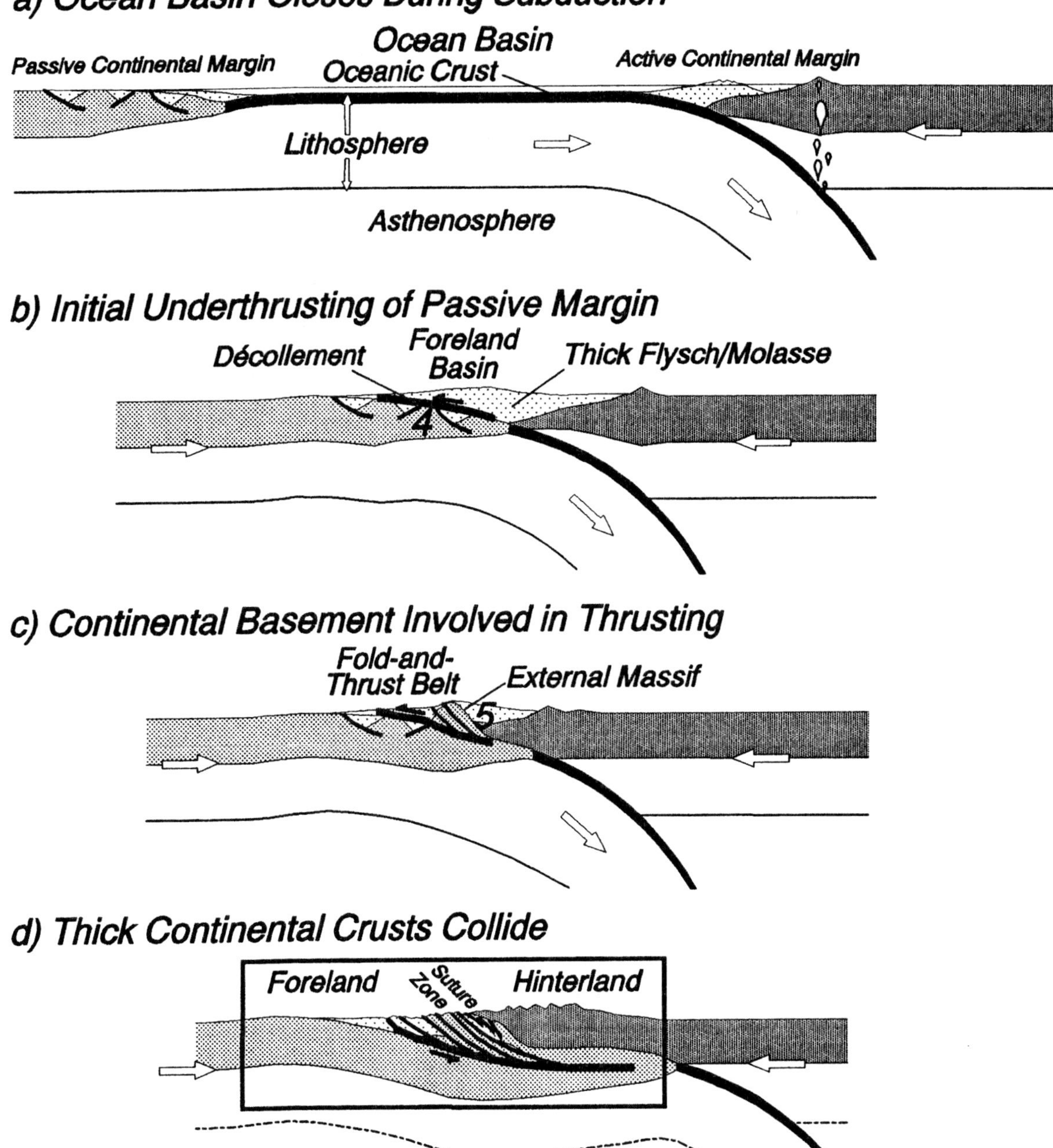

FIGURE 6.35 *Lithospheric cross sections illustrating stages of ocean basin closure and continental collision.* Rectangle in (d) is region portrayed in Fig. 6.36.

FIGURE 6.34 (continued) the role of high pore-fluid pressure in the development of an accretionary wedge, by G. Westbrook and M. Smith, *Geology,* vol. 11, pp. 279–283, © 1983. Redrawn with permission of the Geological Society of America, Boulder, Colorado, USA. Hyperbolic diffraction pattern representing the top of oceanic layer 2 extends from about 8.5 s on east, to 9.5 s on west. A décollement surface that separates deformed accretionary wedge sediments from hemipelagic sediments (oceanic layer 1) lies about 0.5 s above the diffractions. c) Unmigrated time section from Nanki Trough off Japan (Line N55-1). From N. Nasu and others, Multi-channel seismic reflection data across Nanki Trough, *IPOD, Japan Basic Data Series #4,* © 1982, Ocean Research Institute, University of Tokyo, Tokyo, Japan. Moho reflections lie about 2 s below the hyperbolic pattern representing the top of oceanic layer 2.

Collisional Mountain Range (Advanced)

a) Crustal Cross Section

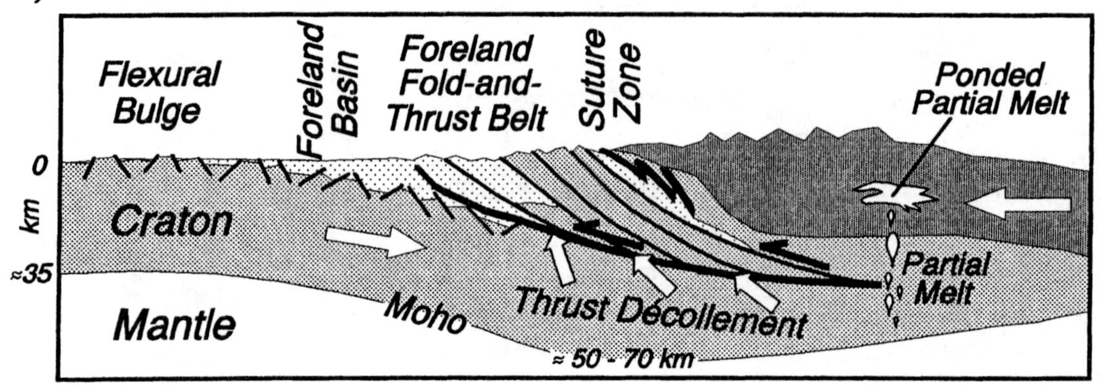

b) Reflection Signature

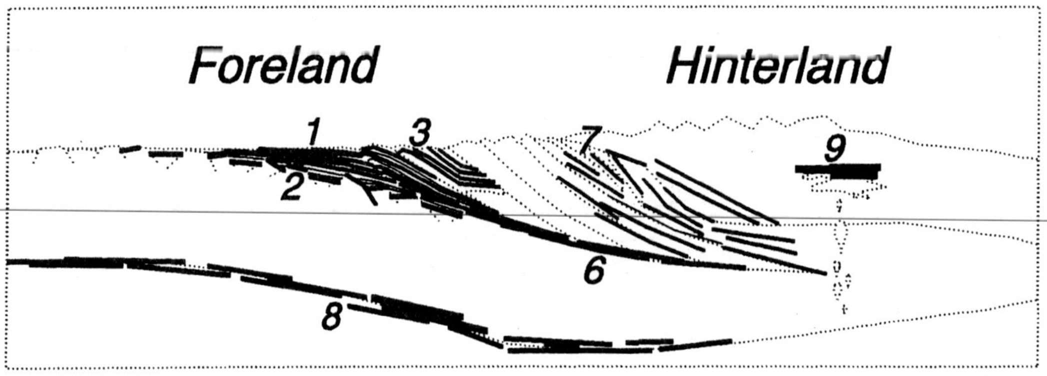

1. Foreland basin strata.
2. Flexural normal faults.
3. Foreland fold-and-thrust belt.
4. Underthrusted strata and structures (previous figure).
5. Basement involved structures (previous figure).
6. Thrust décollement.
7. Suture zone reflections.
8. Moho reflections.
9. Ponded partial melt.

FIGURE 6.36 Crustal scale cross section (a) and prominent reflection sequences (b) observed at an advanced stage of continental collision.

foreland basin is maintained because the rigid, downgoing plate is flexed by the weight of the mountains (Karner and Watts, 1983; see Chapter 8). The flexural bulge results in normal faults offsetting the basement and cover strata (Fig. 6.36a), similar to the normal faults at subduction zones (Figs. 6.33a, 6.34a).

Sedimentary strata are deformed and pushed toward the craton, as the accretionary wedge evolves into a foreland fold-and-thrust belt (Fig. 6.35a-c). Near the deformation front, the thrust décollement is normally within synorogenic strata, similar to the décollement at the toe of an accretionary wedge (Figs. 6.33, 6.34). Preorogenic strata and structures remain intact beneath the décollement, where they may be imaged seismically (Fig. 6.35b). Toward the hinterland, the décollement steps downward, cutting older strata. Deformation of crystalline rocks commonly occurs where the top of basement deepens to more than 10 to 15 km (Figs. 6.35c,d; 6.36a). Structures or strata in the zone of suturing between the two continents may also be imaged, as well as Moho reflections (Fig. 6.36b). Reflections in the mid-crust of active collision zones may be caused by the ponding of partial melts.

1. Foreland Basin Strata During initial stages of collision, sediments in the foreland basin can be thick, because deposition is still on the thin oceanic or transitional crust of the passive margin (Fig. 6.35b). Seismic reflection profiles suggest that deformed flysch strata are at least 12 km thick in the foreland basins of the Carpathian Mountains of Slovakia (Tomek et al., 1987), and the Ouachita Mountains of Arkansas (Nelson et al., 1982; Lillie et al., 1983); undeformed molasse and passive margin strata are at least 10 km thick in front of the Sulaiman Range in western Pakistan (Humayon et al., 1991; Jadoon et al., 1994; Fig. 6.37a).

As collision continues, the buoyancy of the thicker crust of the continental craton limits the thickness of foreland basin strata (Fig. 6.35c,d). Seismic profiles show less than 7 km thickness for the Molasse Basin in front of the Alps (Bachmann et al., 1982) and less than 5 km for the Himalayan foreland in Pakistan and India (Lillie et al., 1987; Raiverman, et al., 1994; Powers et al., 1998).

2. Flexural Normal Faults Beneath the foreland of the Ouachita Mountains in Arkansas, drilling and reflection profiles reveal normal faults offsetting continental basement, preorogenic strata, and synorogenic molasse (VanArsdale and Schweig, 1990; Fig. 6.37b). Only the synorogenic strata show growth across the faults, distinguishing them from older faults formed during passive margin development. The young normal faults are interpreted as due to lithospheric flexure, active during the advance of thrust sheets. Flexural normal faults are also interpreted from reflection profiles of the foreland in the incipient collision at the Banda Arc (Montecci, 1979), as well as advanced collisions in the Alps (Bachmann et al., 1982) and Himalayas (Lillie and Yousuf, 1986). If the thrust décollement is near the base of synorogenic strata, a normal fault may act as a buttress, causing thrust fault ramping (Lillie et al., 1987; Baker et al., 1988; Jaumé and Lillie, 1988; see similar ramping beneath accretionary wedge, Fig. 6.34a).

3. Foreland Fold-and-Thrust Belt Commonly (although with exceptions), the frontal portions of foreland fold-and-thrust belts involve deformation of only synorogenic strata. Basement and older sedimentary cover remain undeformed for some distance beneath a gently-dipping décollement zone (Fig. 6.37c). As with accretionary wedges (Fig. 6.32b), this decoupling surface commonly steps down-section, so that older and older rocks are involved in the deformation toward the

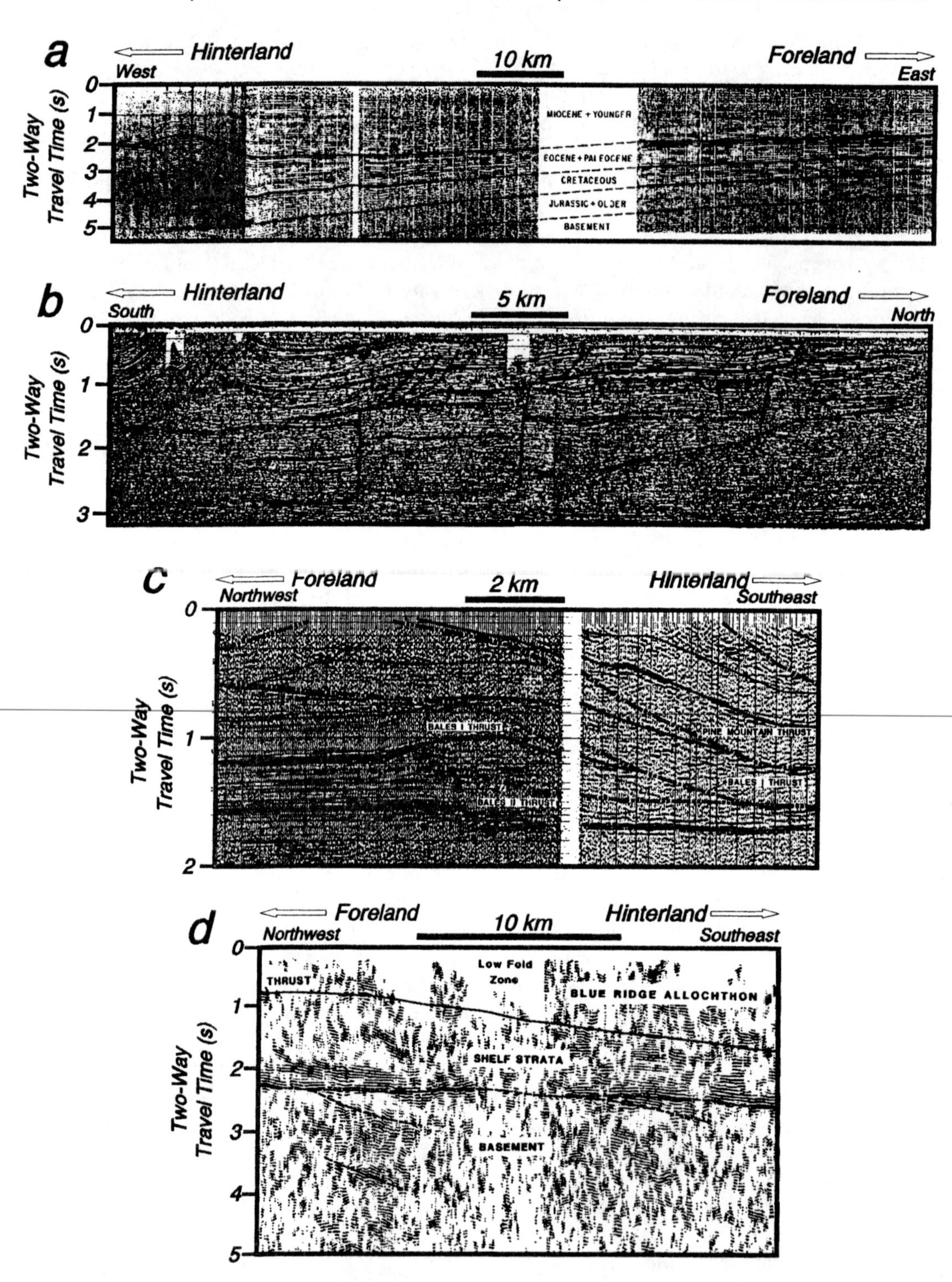

FIGURE 6.37 *Seismic reflection profiles from the foreland of mountain ranges.* a) Interpreted reflection profile from the Sulaiman foredeep, on the western edge of the Indian subcontinent in Pakistan. From M. Humayon, R. Lillie, and R. Lawrence, *Tectonics,* vol. 10, pp. 229–324, © 1991. Redrawn with permission of the American Geophysical Union, Washington, D.C. Sedimentary strata are more than 10 km thick (5 s two-way travel time) because they rest on thin, transitional or oceanic crust at an early stage of collision (Fig. 6.35b). Eocene and older rocks are a postrift sequence on a passive continental margin (Fig. 6.29);

hinterland. Reflection profiles reveal a great variety of structures within the overthrust wedge of strata (see Bally et al., 1966).

4. Underthrusted Strata and Structures Near the deformation front of a collisional mountain range, strata and structures from the continental margin may be intact beneath overthrust rocks (Fig. 6.35b,c). In the southern Appalachian foreland, reflection data show that shelf strata extend for more than 100 km beneath sedimentary rocks of the Valley and Ridge Province (foreland fold-and-thrust belt); the shelf strata extend even farther, beneath crystalline rocks of the Blue Ridge and Inner Piedmont (Cook et al., 1979, 1981). Dipping, wedge-shaped reflections (Fig. 6.37d), completely beneath the shelf strata, are interpreted as due to rift-graben fill from the Late Proterozoic/Early Cambrian passive margin (Harris et al., 1981; Lillie, 1984; Hubbard et al., 1991).

5. Basement Involved Structures Seismic reflection profiles suggest that continental basement begins to be deformed when the top of the underthrusting basement reaches 10 to 15 km depth. Along the Appalachian/Ouachita orogenic belt, lines reveal large, basement-cored structures, including the Notré Dame Anticlinorium in Quebec (St. Julien et al., 1983; Stewart et al., 1986), the Green Mountains in Vermont (Ando et al., 1983), the Benton Uplift in the Ouachita Mountains of Arkansas (Nelson et al., 1982; Lillie et al., 1983), and the Waco and Devils River uplifts in Texas (Nicholas and Rozendal, 1975); seismic profiles of the Alps reveal the antiformal nature of the basement-cored, Aar Massif (Fig. 6.38).

6. Thrust Décollement A décollement is a gently dipping zone separating overthrust from underthrust materials. Seismic reflection profiles suggest a décollement in the southern Appalachian Mountains extends for considerable distance (more than 100 km) beneath crystalline rocks (Cook et al., 1979; see Fig. 6.37d). Recent data from Tibet suggest that the décollement zone near the top of the Indian craton extends completely beneath crystalline rocks of the Lesser and High Himalaya (Fig. 6.39).

7. Suture Zone Reflections The suturing of continents involves a zone of deformation of sedimentary strata and crystalline rocks (Fig. 6.36a). Dipping reflections beneath the southern Appalachians in Georgia (Cook et al., 1979; 1981; Iverson and Smithson, 1982; Nelson et al., 1985) have been interpreted to mark the zone of suturing of the African continent to North America in the late Paleozoic (Fig. 6.36b). Similar reflections are interpreted to mark suture zones beneath the

FIGURE 6.37 (continued) Miocene and younger molasse is a superimposed foreland basin deposit. b) Flexural normal faults on migrated time section from the Arkoma Basin in the southern United States. From R. VanArsdale and E. Schweig, III, Subsurface structure of the eastern Arkoma Basin, *AAPG Bulletin,* vol. 74, pp. 1030–1037, © 1990. Redrawn with permission of the American Association of Petroleum Geologists, Tulsa, Oklahoma, USA. The flexural faults are overridden by thrust faults of the Ouachita orogen. c) Interpreted, migrated time section from the Valley and Ridge Province, the foreland fold-and-thrust belt of the southern Appalachian Mountains. From Three-dimensional geometry and kinematic evolution of the Pine Mountain thrust system, southern Appalachians, by S. Mitra, *GSA Bulletin,* vol. 100, pp. 72–99, © 1988. Redrawn with permission of the Geological Society of America, Boulder, Colorado, USA. The basal décollement, at 0.6 s on the northwest, cuts downsection to 1.5 s on the southeast. Note undeformed strata beneath the décollement on the northwest. The top of Precambrian crystalline basement, at 1.6 s on the northwest, is offset by a flexural normal fault. d) Portion of seismic profile from the Blue Ridge region of the southern Appalachian Mountains in Georgia. From S. Hubbard, C. Çoruh, and J. Costain, *Tectonics,* vol. 10, pp. 141–170, © 1991. Redrawn with permission of the American Geophysical Union, Washington, D.C. Note underthrusted shelf strata, overlying top of Precambrian basement at 2.0 s on northwest, 2.6 s on southeast. Southeastward dipping reflections beneath shelf strata are interpreted as due to rift graben fill (Figs. 6.25; 6.30a), associated with the late Precambrian passive continental margin.

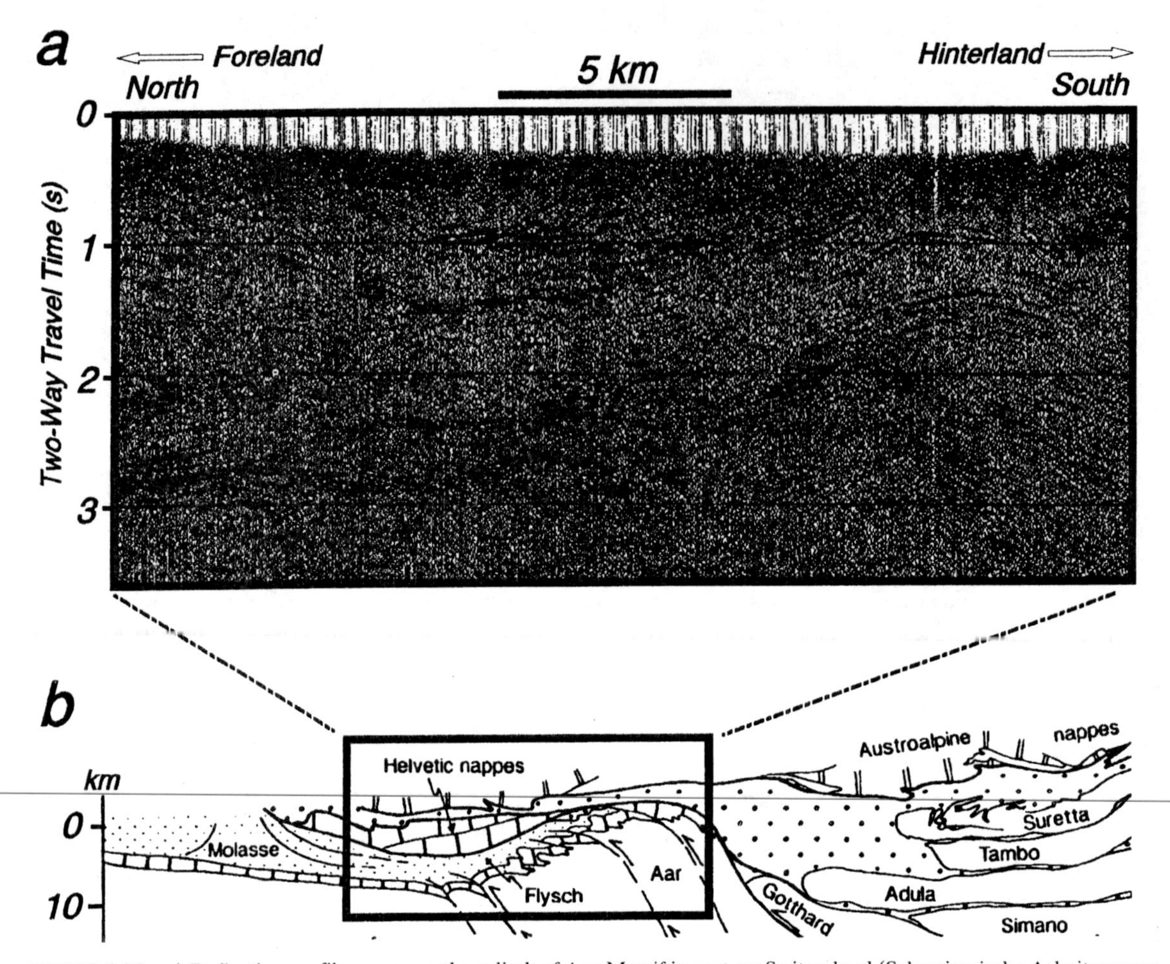

FIGURE 6.38 a) Reflection profile across northern limb of Aar Massif in eastern Switzerland (Schweizerische Arbeitsgruppe für Reflexionsseismik, 1988). The top of basement is at 2.8 s on the north side of the section. It rises southward toward the center of the massif. b) Interpretation of longer reflection traverse reveals that the crystalline basement becomes involved in thrusting when it extends southward from the Molasse Basin to about 10 km depth. From O. Piffner, W. Frei, P. Valased, M. Stauble, L. Levato, L. DuBois, S. Schmid, and S. Smithson, *Tectonics,* vol. 9, pp. 1327–1355, © 1990. Redrawn with permission of the American Geophysical Union, Washington, D.C. Box shows approximate position of seismic profile in (a).

northern Appalachians (Ando et al., 1983), the Alps (Pfiffner et al., 1990), and the British Caledonides (Beamish and Smythe, 1986).

8. Moho Reflections The crust/mantle boundary is difficult to image beneath collisional mountain ranges, because of the great thickness of the crust and the problems of resolving reflectors beneath complex structure (Figs. 6.18–6.20). Seismic profiles, nonetheless, reveal reflections from Moho depths beneath portions of collisional mountain ranges, including the Appalachians (Cook et al., 1979, 1981; Nelson et al., 1985), the Alps, and the Himalayas (Zhao, et al., 1993; Fig. 6.39).

9. Partial Melt Where the crust becomes nearly twice normal continental thickness, granitic rocks may become hot enough to melt partially (Fig. 6.36a). Reflection profiles from the Tibetan Plateau reveal strong reflections that may indicate the ponding of molten material at mid-crustal levels (Fig. 6.36b; Makovsky et al., 1996).

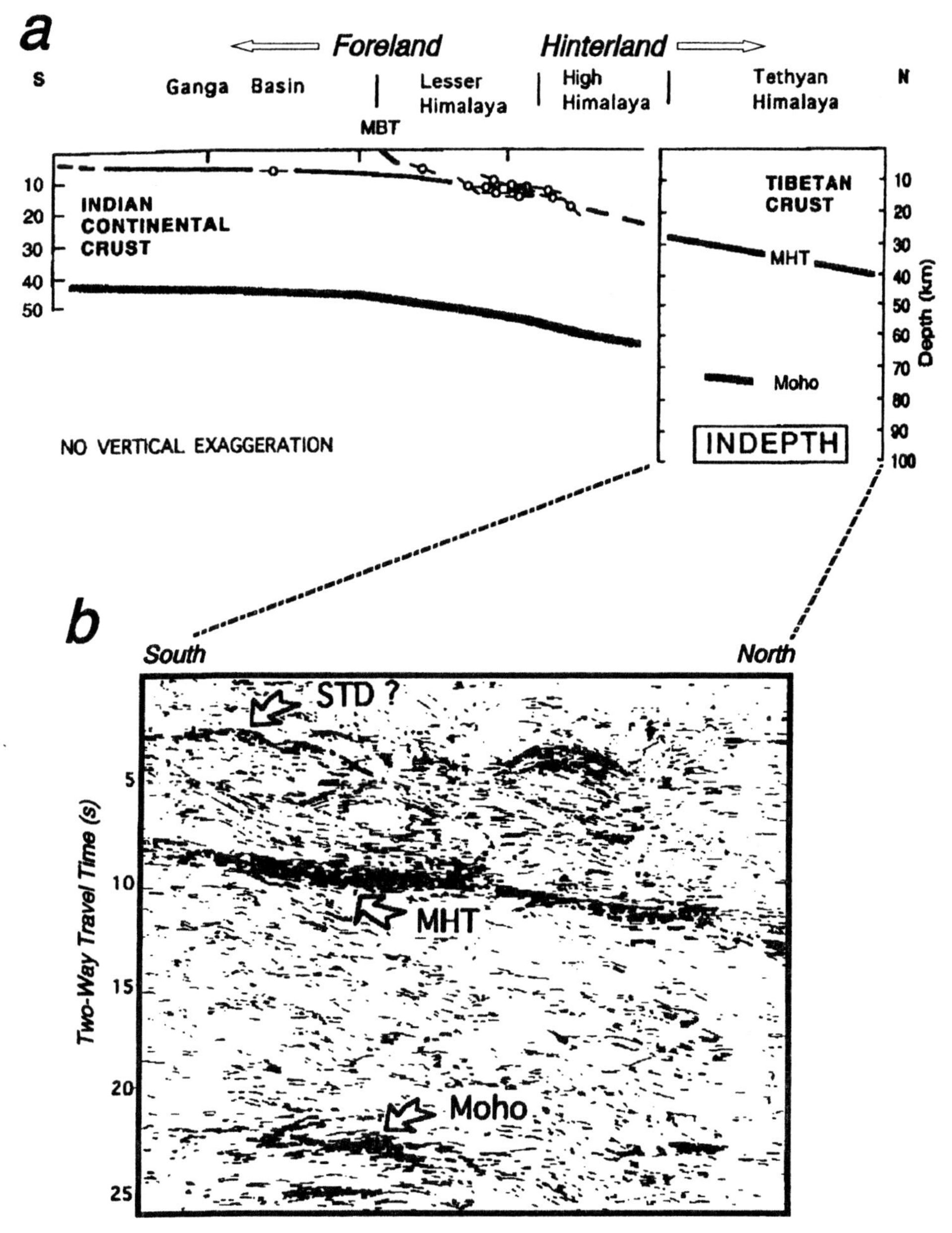

FIGURE 6.39 a) Cross section showing position of INDEPTH seismic reflection profile, relative to model of underthrusting Indian continental crust. b) INDEPTH profile in southern Tibet. The MHT is the Main Himalayan Thrust, believed to be the décollement separating the underthrusting Indian craton from the overriding Himalaya. Note deep reflections at about 22 s two-way travel time (TWT), interpreted to image the Moho. From Deep seismic reflection evidence for continental underthrusting beneath southern Tibet, by W. Zhao, K. Nelson, and INDEPTH team, *Nature,* vol. 366, pp. 557–559, © 1993. Redrawn with permission of Macmillan Magazines, Ltd., London.

EXERCISES

6-1 The model below shows a two-dimensional structure, with seismic velocities between interfaces.

a) Sketch *normal incidence raypaths* (to both reflectors) for several common source/receiver positions on the surface. *(Hint: A normal incidence raypath follows the same path upward as downward. For the lower reflector, it is easiest to trace raypaths upward. Start by drawing a vertical line from the lower interface to the dipping segment of the upper reflector. Then calculate how the ray refracts into the upper layer, according to Snell's Law).*

b) Change the vertical scale to two-way travel time, then compute and draw the general appearance of an *unmigrated time section.*

c) Compute and draw the general appearance of a *migrated time section.*

d) Draw the *migrated depth section.*

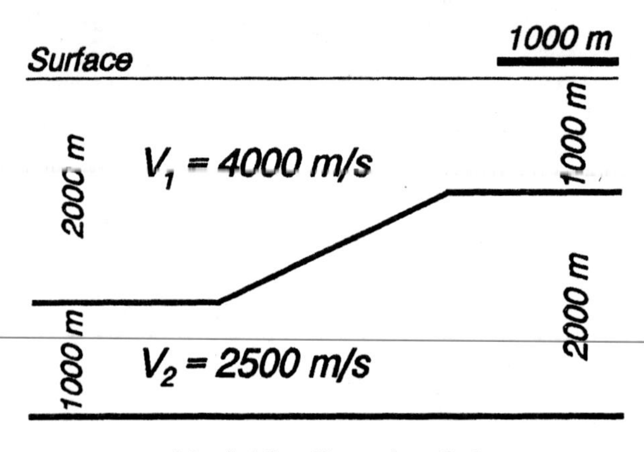

Model for Exercise 6-1

6-2 Given the following model of a synformal structure (Reflector 1) underlain by a horizontal interface (Reflector 2):

a) Draw *normal incidence raypaths* for both interfaces, to several source/receiver positions on the surface.

b) Draw the *unmigrated time section* which results from the model.

c) Draw the *migrated time section*, overlain on exactly the same plot as in (b).

d) Repeat (a), (b) and (c), but with each reflector 3000 m deeper.

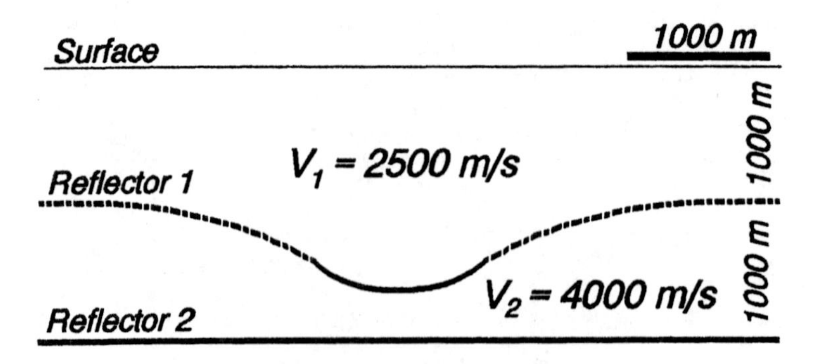

Model for Exercise 6-2

6-3 For the two-dimensional model below, calculate and draw:

a) *Normal incidence raypaths* (to both reflectors) for several source/receiver positions on the surface.

b) The general appearance of an *unmigrated time section.*

c) The general appearance of a *migrated time section.*

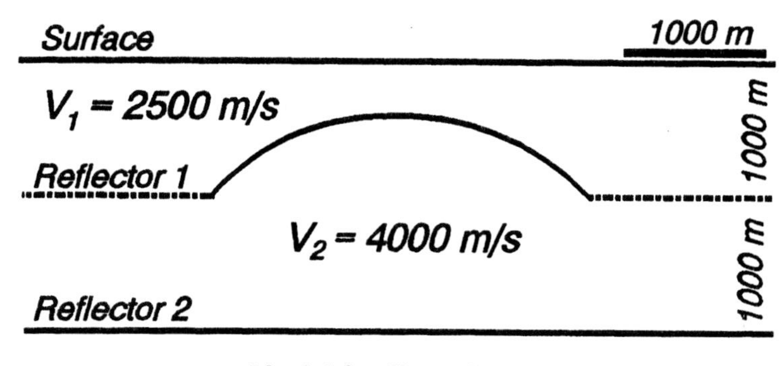

Model for Exercise 6-3

6-4 Interpret the unmigrated and migrated sections in Fig. 6.11. The profile is from an area where the crystalline basement is overlain by an incompetent unit. Above the incompetent unit is a thin (approximately one km thick) section of competent rocks, overlain by a thick section of clastic sediments.

a) Overlay the sections with tracing paper and draw your *interpretation* with colored pencils.

b) Make some assumptions about the velocities of the rocks and sketch a *migrated depth section.*

c) Draw a possible *geologic interpretation* (that is, a structural cross section), with all units identified.

d) Discuss a possible *geologic history* responsible for the formation of the structure.

6-5 **a)** Draw cross section B-B′ from Ex. 2-2 at a *crustal scale.*

b) Directly on your cross section, sketch different *seismic signatures* that might be observed at various places.

6-6 Discuss the *resolution* of the following types of geophysical data: seismic refraction; earthquake seismic; gravity; seismic reflection.

a) Be specific about: i) the size and geometry of bodies that can be resolved with the method; ii) zones of the Earth that can be studied with each method.

b) Discuss how the four methods might be used together to study the structure of the entire lithosphere/asthenosphere system at a *continental rift zone.*

SELECTED BIBLIOGRAPHY

Allmendinger, R. W., J. W. Sharp, D. Von Tish, L. Serpa, L. Brown, S. Kaufman, J. Oliver, and R. B. Smithson, 1983, Cenozoic and Mesozoic structure of the eastern Basin and Range province, Utah, from COCORP seismic reflection data, *Geology*, v. 11, pp. 532–536.

Allmendinger, R. W., T. A. Hague, E. C. Hauser, C. J. Potter, S. L. Klemperer, K. D. Nelson, P. Knuepfer, and J. Oliver, 1987, Overview of the COCORP 40° N transect, western United States: The fabric of an orogenic belt, *Geological Society of America Bulletin*, v. 98, pp. 308–319.

Anderson, R. E., M. L. Zoback, and G. A. Thompson, 1983, Implications of selected subsurface data on the structural form and evolution of some basins in the northern Basin and Range province, Nevada and Utah, *Geological Society of America Bulletin*, v. 94, pp. 1055–1072.

Ando, C. J., F. A. Cook, J. E. Oliver, L. D. Brown, and S. Kaufman, 1983, Crustal geometry of the

Appalachian orogen from seismic reflection studies, in: *Contributions to the Tectonics and Geophysics of Mountain Chains*, edited by R. D. Hatcher, H. Williams, and I. Zietz, Boulder, Colorado: Geological Society of America, Memoir 158, pp. 113–124.

Bachmann, G. H., G. Dohr, and M. Muller, 1982, Exploration in a classic thrust belt and its foreland: Bavarian Alps, Germany, *American Association of Petroleum Geologists Bulletin*, v. 66, pp. 2529–2542.

Baker, D. M., R. J. Lillie, R. S. Yeats, G. D. Johnson, and M. Yousuf, 1988, Development of the Himalayan frontal thrust zone: Salt Range, Pakistan, *Geology*, v. 16, pp. 57–70.

Bally, A. W. (editor), 1983, *Seismic Expression of Structural Styles*, Tulsa: American Association of Petroleum Geologists, Studies in Geology Series 15 (3 volume set).

Bally, A. W., P. L. Gordy and G. A. Stewart, 1966, Structure, seismic data and orogenic evolution of southern Canadian Rockies, *Bulletin of Canadian Petroleum Geology*, v. 14, pp. 337–381.

Bangs, N. L. B., and G. K. Westbrook, 1991, Seismic modelling of the décollement zone at the base of the Barbados Ridge accretionary complex, *Journal of Geophysical Research*, v. 96, pp. 3853–3866.

Barazangi, M., and L. Brown (editors), 1986, *Reflection Seismology: A Global Perspective*, Washington: American Geophysical Union, Geodynamics Series, v. 13, 311 pp.

Barazangi, M., and L. Brown (editors), 1986, *Reflection Seismology: The Continental Crust*, Washington: American Geophysical Union, Geodynamics Series, v. 14, 339 pp., 1986.

Beamish, D., and D. K. Smythe, 1986, Geophysical images of the deep crust: the Iapetus suture, *Journal Geological Society of London*, v. 143, pp. 489–497.

Brewer, J. A., and D. K. Smythe, 1986, Deep structure of the foreland to the Caledonian Orogen, NW Scotland: Results of the BIRPS winch profile, *Tectonics*, v. 5, pp. 171–194.

Brown, L. D., C. E. Chapin, A. R. Sanford, S. Kaufman, and J. Oliver, 1980, Deep structure of the Rio Grande Rift from seismic reflection profiling, *Journal of Geophysical Research*, v. 85, pp. 4773–4800.

Cape, C. D., S. McGeary, and G. Thompson, 1983, Cenozoic normal faulting and the shallow structure of the Rio Grande rift near Socorro, New Mexico, *Geological Society of America Bulletin*, v. 94, pp. 3–14.

Collins, B. P., and J. S. Watkins, 1986, The Middle America Trench, in: *Seismic Images of Modern Convergent Margin Tectonic Structure*, edited by R. von Huene, Tulsa: American Association of Petroleum Geologists, Studies in Geology 26, pp. 30–32.

Cook, F. A., D. S. Albaugh, L. D. Brown, S. Kaufman, J. E. Oliver, and R. D. Hatcher, Jr., 1979, Thin-skinned tectonics in the crystalline southern Appalachians: COCORP seismic-reflection profiling of the Blue Ridge and Piedmont, *Geology*, v. 7, pp. 563–567.

Cook, F. A., L. D. Brown, S. Kaufman, J. E. Oliver, and T. A. Peterson, 1981, COCORP seismic profiling of the Appalachian orogen beneath the Coastal Plain of Georgia, *Geological Society of America Bulletin*, v. 92, pp. 738–748.

Costain, J. K., and C. Çoruh, 1989, Tectonic setting of Triassic half-grabens in the Appalachians: Seismic data acquisition, processing, and results, in: *Extensional Tectonics and Stratigraphy of the North Atlantic Margins*, edited by A. J. Tankard and H. R. Balkwill, Tulsa: American Association of Petroleum Geologists, Memoir 46, pp. 155–174.

Falvey, D. A., 1974, The development of continental margins in plate tectonic theory, *APEA Journal*, v. 14, pp. 95–106.

Grow, J., R. Mattlick, and J. Schlee, 1979, Multichannel seismic depth sections and interval velocities over outer continental slope between Cape Hatteras and Cape Cod, in: *Geological and Geophysical Investigations of Continental Margins*, edited by J. S. Watkins, L. Montadert, and P. W. Dickinson, Tulsa: American Association of Petroleum Geologists, Memoir 29, pp. 65–83.

Hague, T. A., R. W. Allmendinger, C. Caruso, E. C. Hauser, S. L. Klemperer, S. Updyke, C. J. Potter, W. Sanford, L. Brown, S. Kaufman, and J. Oliver, 1987, Crustal structure of western Nevada from COCORP deep seismic-reflection data, *Geological Society of America Bulletin*, v. 98, pp. 320–329.

Harris, L. D., A. G. Harris, W. deWitt, Jr., and K. C. Bayer, 1981, Evaluation of southern eastern overthrust belt beneath Blue Ridge-Piedmont thrust, *American Association of Petroleum Geologists Bulletin*, v. 65, pp. 2497–2505.

Herman, A., R. Anania, J. Chun, C. Jacewitz, and R. Pepper, 1982, A fast three-dimensional modelling technique and fundamentals of 3-D frequency domain migration, *Geophysics*, v. 47, pp. 1627–1644.

Hinz, K., 1981, A hypothesis on terrestrial catastrophes. Wedges of very thick oceanward dipping layers beneath passive continental margins—their origin and paleoenvironmental significance, *Geol. Jahrb. Reihe E*, v. 22, pp. 3–28.

Hubbard, S. S., C. Çoruh, and J. K. Costain, 1991, Paleozoic and Grenvillian structures in the southern Appalachians: Extended interpretation of seismic reflection data, *Tectonics*, v. 10, pp. 141–170.

Humayon, M., R. J. Lillie, and R. D. Lawrence, 1991, Structural interpretation of the eastern Sulaiman foldbelt and foredeep, Pakistan, *Tectonics*, v. 10, pp. 229–324.

Hutchinson, D. R., K. D. Klitgord, and R. S. Detrick, 1986, Rift basins of the Long Island Platform, *Geological Society of America Bulletin*, v. 97, pp. 688–702.

Iverson, W. P., and S. B. Smithson, 1982, Master décollement root zone beneath the southern Appalachians and crustal balance, *Geology*, v. 10, pp. 241–245.

Jadoon, I. A. K., R. D. Lawrence, and R. J. Lillie, 1994, Seismic data, geometry, evolution, and shortening in the active Sulaiman fold-and-thrust belt of Pakistan, southwest of the Himalayas, *American Association of Petroleum Geologists Bulletin*, v. 78, pp. 758–774.

Karner, G. D., and A. B. Watts, 1983, Gravity anomalies and flexure of the lithosphere at mountain ranges, *Journal of Geophysical Research*, v. 88, pp. 10449–10477.

Klemperer, S. L., T. A. Hague, E. C. Hauser, J. E. Oliver, and C. J. Potter, 1986, The Moho in the northern Basin and Range, Nevada, along the COCORP 40° N seismic-reflection transect, *Geological Society of America Bulletin*, v. 97, pp. 603–618.

Klemperer, S. L., P. D. Ryan, and D. B. Snyder, 1991, A deep seismic reflection transect across the Irish Caledonides, *Journal of the Geological Society, London*, v. 148, pp. 149–164.

Larner, K., B. Gibson, and R. Chambers, 1983, Imaging beneath complex structure, in: *Seismic Expression of Structural Styles*, edited by A. W. Balley, Tulsa: American Association of Petroleum Geologists, Studies in Geology Series 15, v. 1, pp. 26–39.

Lillie, R. J., K. D. Nelson, B. deVoogd, J. A. Brewer, J. E. Oliver, L. D. Brown, S. Kaufman and G. W. Viele, 1983, Crustal structure of Ouachita Mountains, Arkansas: A model based on integration of COCORP reflection profiles and regional geophysical data, *American Association of Petroleum Geologists Bulletin*, v. 67, pp. 907–931.

Lillie, R. J., 1984, Tectonic implications of subthrust structures revealed by seismic profiling of Appalachian-Ouachita orogenic belt, *Tectonics*, v. 3, pp. 619–646.

Lillie, R. J., 1985, Tectonically buried continent/ocean boundary, Ouachita Mountains, Arkansas, *Geology*, v. 13, pp. 18–21.

Lillie, R. J., and M. Yousuf, 1986, Modern analogs for some midcrustal reflections observed beneath collisional mountain belts, in: *Reflection Seismology: The Continental Crust*, edited by M. Barazangi and L. Brown, Washington: American Geophysical Union, Geodynamics Series, v. 14, pp 55–65.

Lillie, R. J., G. D. Johnson, M. Yousaf, A. S. Hamid Zamin, and R. S. Yeats, 1987, Structural development within the Himalayan foreland fold-and-thrust belt of Pakistan, in: *Sedimentary Basins and Basin-Forming Mechanisms*, edited by C. Beaumont and A. J. Tankard, Calgary: Canadian Society of Petroleum Geologists, Memoir 12, pp. 379–392.

Makovsky, Y., S. L. Klemperer, and L. Rathschbacher, 1996, Mid-crustal reflector truncating the India-Asia suture and magma beneath the Tibetan rift system from wide-angle seismic data project INDEPTH (abstract), 11th Annual Himalayan-Karakorum-Tibet workshop, Flagstaff, Arizona, pp. 91–92.

Matthews, D., and C. Smith (editors), 1987, *Deep Seismic Reflection Profiling of the Continental Lithosphere*, Special Issue, Geophysical Journal of the Royal Astronomical Society, v. 89, No. 1, 495 pp.

McCarthy, J., and G. A. Thompson, 1988, Seismic imaging of extended crust with emphasis on the western United States, *Geological Society of America Bulletin*, v. 100, pp. 1361–1374.

May, B. T., and F. Hron, 1982, Synthetic seismic sections of typical petroleum traps, *Geophysics*, v. 43, pp. 1119–1147.

Mitra, S. 1988, Three-dimensional geometry and kinematic evolution of the Pine Mountain thrust system, southern Appalachians, *Geological Society of America Bulletin*, v. 100, pp. 72–95.

Mohriak, W. U., M. R. Mello, G. D. Karner, J. F. Dewey, and J. R. Maxwell, 1989, Structural and stratigraphic evolution of the Campos Basin, offshore Brazil, in: *Extensional Tectonics and Stratigraphy of the North Atlantic Margins*, edited by A. J. Tankard and H. R. Balkwill, Tulsa: American Association of Petroleum Geologists, Memoir 46, pp. 577–598.

Montecci, P. A., 1979, Some shallow consequences of subduction and their meaning to the hydrocarbon explorationist, in: *Circum-Pacific Energy and Mineral Resources*, edited by M. T. Halbouty, J. C. Maher, and H. M. Lian, Tulsa: American Association of Petroleum Geologists, Memoir 25, pp. 189–202.

Moore, G. F., and T. H. Shipley, 1988a, Mechanisms of sediment accretion in the Middle America Trench off Mexico, *Journal of Geophysical Research*, v. 93, pp. 8911–8927.

Moore, G. F., and T. H. Shipley, 1988b, Behavior of the décollement at the toe of the Middle America Trench, *Sonderbruck aus Geologische Rundschau*, v. 77, pp. 275–284.

Moore, G. F., T. H. Shipley, P. L. Stoffa, and D. E. Karig, 1990, Structure of the Nankai Trough Accretionary Zone from multichannel seismic reflection data, *Journal of Geophysical Research*, pp. 8753–8765.

Moore, J. C., 1975, Selective subduction, *Geology*, v. 3, pp. 530–532.

Moore, J. C., and 17 others, 1982, Offscraping and underthrusting of sediment at the deformation front of the Barbados Ridge: Deep Sea Drilling Project Leg 78A, *Geological Society of America Bulletin*, v. 93, pp. 1065–1077.

Mutter, J. C., M. Talwani, and P. L. Stoffa, 1982, Origin of seaward-dipping reflectors in oceanic crust off the Norwegian margin by "subaerial sea-floor spreading," *Geology*, v. 10, pp. 353–357.

Mutter, J. C., M. Talwani, and P. L. Stoffa, 1984, Evidence for a thick oceanic crust adjacent to the Norwegian margin, *Journal of Geophysical Research*, 89, pp. 483–502.

Nasu, N., et. al., 1982, *IPOD, Japan basic data series #4, Multichannel seismic reflection data across Nanki Trough*, Tokyo: Ocean Research Institute, University of Tokyo.

Nelson, K. D., R. J. Lillie, B. DeVoogd, J. A. Brewer, J. E. Oliver, S. Kaufman, and L. Brown, 1982, COCORP seismic reflection profiling in the Ouachita Mountains of western Arkansas: Geometry and geologic interpretation, *Tectonics*, v. 1, pp. 413–430.

Nelson, K. D., J. A. Arnow, J. H. McBride, J. H. Willemin, J. Huang, L. Zheng, J. E. Oliver, L. D. Brown, and S. Kaufman, 1985, New COCORP profiling in the southeastern United States. Part I: Late Paleozoic suture and Mesozoic rift basin, *Geology*, v. 13, pp. 714–718.

Nicholas, R. L., and R. A. Rozendal, 1975, Subsurface positive elements within Ouachita foldbelt in Texas and their relationship to Paleozoic cratonic margin, *American Association of Petroleum Geologists Bulletin*, v. 59, pp. 193–216.

Petty, C. P., and R. W. Hobbs, 1987, Lower crustal reflectivity of the continental margin southwest of Britain, *Annales Geophysicae*, v. 5B, pp. 331–338.

Piffner, O. A., W. Frei, P. Valased, M. Stauble, L. Levato, L. DuBois, S. M. Schmid, and S. B. Smithson, 1990, Crustal shortening in the Alpine orogen, Results from deep seismic reflection profiling in the eastern Swiss Alps, line NFP 20-east, *Tectonics*, v. 9, pp. 1327–1355.

Potter, C. J., C. Liu, J. Huang, L. Zheng, T. A. Hague, E. C. Hauser, R. W. Allmendinger, J. Oliver, S. Kaufman and L. Brown, 1987, Crustal structure of north-central Nevada: Results of COCORP deep seismic profiling, *Geological Society of America Bulletin*, v. 98, pp. 330–337.

Powers, P. M., R. J. Lillie, and R. S. Yeats, 1998, Structure and shortening of the Kangra and Dehra Dun Reentrants, Sub-Himalya, India, *Geological Society of American Bulletin,* in press.

Raiverman, V., A. K. Srivastava, and D. N. Prasad, 1994, Structural style in northwestern Himalayan foothills, *Himalayan Geology*, v. 15, pp. 263–280.

Reston, T. J., 1990, The lower crust and the extension of the continental lithosphere: Kinematic analysis of BIRPS deep seismic data, *Tectonics*, v. 9, pp. 1235–1248.

Riddihough, R. P., M. E. Beck, R. L. Chase, E. E. Davis, R. D. Hyndman, S. H. Johnson, and G. C. Rogers, 1983, Geodynamics of the Juan de Fuca Plate, in: *Geodynamics of the Eastern Pacific Region, Caribbean and Scotia Arcs*, edited by S. J. Ramón Cabré, Washington: American Geophysical Union, Geodynamics Series 9, pp. 5–21.

Robinson, E. A., and S. Treitel, 1980, *Geophysical Signal Analysis*, Englewood Cliffs, N.J.: Prentice Hall, Inc., 466 pp.

Rohr, K. M. M., 1988, Asymmetric deep crustal structure across the Juan de Fuca Ridge, *Geology*, v. 16, pp. 533–537.

Schlee, J. S., J. C. Behrendt, J. A. Grow, J. M. Robb, R. E. Mattick, P. T. Taylor, and B. J. Lawson, 1976, Regional geologic framework off northeastern United States, *American Association of Petroleum Geologists Bulletin*, v. 60, pp. 926–951.

Scholl, D. W., R. von Huene, T. L. Vallier, and D. G. Howell, 1980, Sedimentary masses and concepts about tectonic processes at underthrust ocean margins, *Geology*, v. 8, pp. 564–568.

Schweizerische Arbeitsgruppe für Reflexionsseismik, 1988, First results of the traverses across the Alps of NFF-20, *Vierteljahrsschift der Naturforschenden Gesellschaft in Zürich*, v. 133, pp. 61-98.

Serpa, L., T. Setzer, H. Farmer, L. Brown, J. Oliver, S. Kaufman, J. Sharp, and D. W. Steeples, 1984, Structure of the southern Keweenawan Rift from COCORP surveys across the midcontinent geophysical anomaly in northeastern Kansas, *Tectonics*, v. 3, pp. 367–384.

St. Julien, P., A. Slivitsky, and T. Feininger, 1983, A deep structural profile across the Appalachians of southern Quebec, in: *Contributions to the Tectonics and Geophysics of Mountain Chains*, edited by R. D. Hatcher, H. Williams, and I. Zietz, Boulder, Colorado: Geological Society of America, Memoir 158, pp. 103–111.

Stewart, D. B., J. D. Unger, J. D. Phillips, and R. Goldsmith, W. H. Poole, C. P. Spencer, A. G. Green, M. C. Loiselle, and P. St. Julien, 1986, The Quebec-western Maine seismic reflection profile: setting and first year results, in: *Reflection Seismology: The Continental Crust*, edited by M. Barazangi and L. Brown, Washington: American Geophysical Union, Geodynamics Series 14, pp. 189–199.

Stockmal, G. S., C. Beaumont, and R. Boutilier, 1986, Geodynamic models of convergent margin tectonics: Transition from rifted margin to overthrust belt and consequences for foreland-basin development, *American Association of Petroleum Geologists Bulletin*, v. 70, pp. 181–190.

Tomek, Č., L. Dvořáková, I. Ibrmajer, R. Jiřîček, and T. Koráb, 1987, Crustal profiles of active continental collisional belt: Czechoslovak deep seismic reflection profiling in the west Carpathians, *Geophysical Journal of the Royal Astronomical Society*, v. 89, pp. 383–388.

Trehu, A. M., K. D. Klitgord, D. S. Sawyer, and R. T. Buffler, 1989, Atlantic and Gulf of Mexico continental margins, in: *Geophysical Framework of the Continental United States*, edited by L. C. Pakiser and W. D. Mooney, Boulder, Colorado: Geological Society of America, Memoir 172, pp. 349–382.

VanArsdale, R. B., and E. S. Schweig, III, 1990, Subsurface structure of the eastern Arkoma Basin, *American Association of Petroleum Geologists Bulletin*, v. 74, pp. 1030–1037.

Vera, E. E., J. C. Mutter, P. Buhl, J. A. Orcutt, A. J. Harding, M. E. Kappus, R. S. Detrick, and T. M. Brocher, 1990, The structure of 0 to 2 m.y. old oceanic crust at 9° N on the East Pacific Rise from expanded spread profiles, *Journal of Geophysical Research*, v. 95, pp. 15529–15556.

Westbrook, G. K., and M. J. Smith, 1983, Long décollements and mud volcanoes: Evidence from the Barbados Ridge Complex for the role of high pore-fluid pressure in the development of an accretionary wedge, *Geology*, v. 11, pp. 279–283.

Westbrook, G. K., J. W. Ladd, P. Buhl, N. Bangs, and G. J. Tiley, 1988, Cross section of an accretionary wedge: Barbados Ridge complex, *Geology*, v. 16, pp. 631–635.

Welsink, H. J., J. D. Dwyer, and R. J. Knight, 1989, Tectono-stratigraphy of the passive margin off Nova Scotia, in: *Extensional Tectonics and Stratigraphy of the North Atlantic Margins*, edited by A. J. Tankard and H. R. Balkwill, Tulsa: American Association of Petroleum Geologists, Memoir 46, pp. 215–231.

Wilson, J. T., 1966, Did the Atlantic close and then reopen?, *Nature*, v. 211, pp. 676–681.

Zhao, W., K. D. Nelson, and INDEPTH team, 1993, Deep seismic reflection evidence for continental underthrusting beneath southern Tibet, *Nature*, v. 366, pp. 557–559.

CHAPTER 7

Earthquake Seismology

Earth *('ərth)* ***n.,*** [*OE.* eorthe], ***1****. the planet we live on* **2**. *land, as distinguished from sea or sky.*
quake *(kwāk)* ***vi,*** [*OE.* cwacian], *to tremble or shake.*
earthquake *('ərthkwāk)* ***n.,*** *vibrations within the Earth caused by the rupture and sudden movement of rocks that have been strained beyond their elastic limit.*
seismo- *(sīz mō),* [< *Gr.* seismos, *an earthquake* < seiein, *to shake*], *a combining form, pertaining to vibrations.*
-logy *(ləjē),* [*ult.* < *Gr.*, logos], *a combining form, meaning the scientific study of.*
seismology *(sīz mäl′ ə jē)* ***n.,*** [*SEISMO-+-LOGY*], *the study of vibrations that move as waves through the Earth.*
earthquake seismology *('ərthkwāk sīz mäl′ ə jē)* ***n.,*** *the study of vibrations within the Earth caused by the sudden movement along faults or other natural processes.*

The study of earthquakes is important for scientific, social, and economic reasons. Earthquakes attest to the fact that dynamic forces are operating within the Earth. Stress builds up through time, storing strain energy; earthquakes represent sudden release of the strain energy.

Plate tectonic theory relies heavily on observations from earthquakes. Most tectonic activity occurs due to interaction between plates; the distribution of earthquakes thus dramatically outlines lithospheric plate boundaries. Locations of earthquakes in three dimensions reveal the depths of stresses built up as a result of plate interaction. There are only shallow earthquakes at divergent and transform plate boundaries, but earthquakes occur over a broad range from shallow to deep where plates converge. The type of earthquake faulting shows relative motion between plates; rocks are generally subjected to normal faulting at divergent plate boundaries, strike-slip faulting at transform boundaries, and reverse faulting (with significant normal and strike-slip faulting) at convergent boundaries.

Earthquakes also provide crucial data on the deep interior of the Earth, because seismic waves travel through the entire Earth and are recorded by a worldwide network of seismometers. Interpretations of the thickness, structure and composition of the crust, mantle, and core can be made from the types and speeds of waves that travel through each zone.

Earthquakes are important from a human and economic point of view. In some years, earthquakes kill thousands of people and cause damage totalling billions of dollars. It is useful to understand *how* earthquakes occur, *where* they are likely to occur, and *when* they might occur. We can minimize earthquake effects by mapping hazardous regions, predicting the time and place of future earthquakes, designing buildings that will withstand earthquakes, and by not building in areas prone to intense shaking.

CHARACTERISTICS OF EARTHQUAKES

Elastic Rebound Theory

When Earth material is stressed beyond its elastic limit, failure could be through ductile flow or brittle fracture (Figs. 3.1, 3.2). The latter situation results in earthquakes. For earthquakes to occur, two factors are thus necessary: 1) there must be some sort of movement that will stress the material beyond its elastic limit; and 2) the material must fail by brittle fracture.

The region of the Earth that fits the above criteria is the lithosphere. Other regions, such as the asthenosphere and outer core, behave ductilly and fluidly, respectively, when large stresses are applied over long periods of time. The lower mantle (mesosphere) and the inner core are solid, but they are not subjected to large differential stresses. Earthquakes are, therefore, almost exclusively confined to the moving, rigid lithosphere, particularly where stresses are concentrated near the boundaries of plates.

Elastic rebound theory states that rock can be stressed, obeying Hooke's Law, until it reaches its elastic limit (Fig. 3.1c). If the rock fails in a brittle fashion, it rebounds (snaps) into a new position as the stored strain energy is released (Fig. 7.1). The sudden release of strain energy is an earthquake, which sends off vibrations as seismic waves.

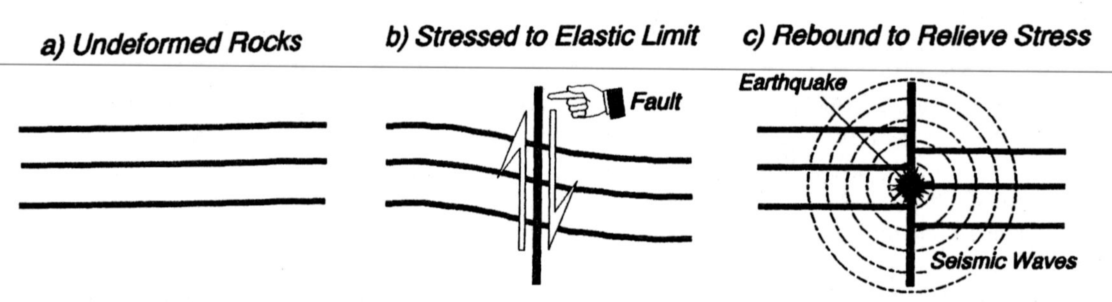

FIGURE 7.1 *Elastic rebound.* Figure can represent either map view of a strike-slip fault, or cross-sectional view of a dip-slip (normal or reverse) fault. a) Sequence of rocks in undeformed state. b) Rocks initially behave elastically as stress is applied (Fig. 3.1b). c) Elastic limit of the rocks is reached (Fig. 3.1c). If brittle failure occurs, stored energy is released as an earthquake. The rocks rebound to new positions across the fault, as seismic waves radiate from the rupture zone (earthquake focus).

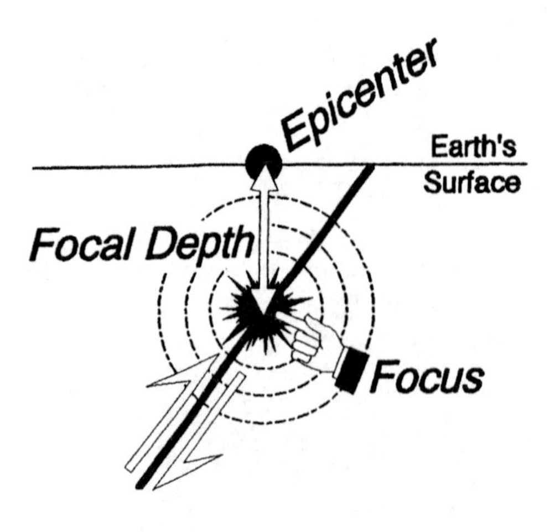

FIGURE 7.2 Cross section of a rupturing fault, illustrating terminology used to describe the location and depth of an earthquake.

Location of Earthquakes

Focus and Epicenter The location of an earthquake can be described by the latitude, longitude, and depth of the zone of rupture (Fig. 7.2). The *focus* (or *hypocenter*) is the actual "point" (relatively small volume) within the Earth where the earthquake energy is released. The *epicenter* is the point on Earth's surface directly above the focus. The *focal depth* is the distance from the epicenter to the focus.

Focal Depth Earthquakes occur in the *upper 700 km* of the Earth, because they are confined to the rigid *lithosphere*, which can undergo *brittle failure*. Focal depths are normally related to the type of plate boundary (Fig. 7.3). Most earthquakes are *shallow focus*, from the surface to 70 km depth; shallow focus earthquakes occur at all types of plate boundaries.

Most *intermediate focus* (70 to 300 km depth) and virtually all *deep focus* earthquakes occur in *convergent* (subduction) settings, where lithosphere extends deeply through the asthenosphere (Fig. 7.3c).

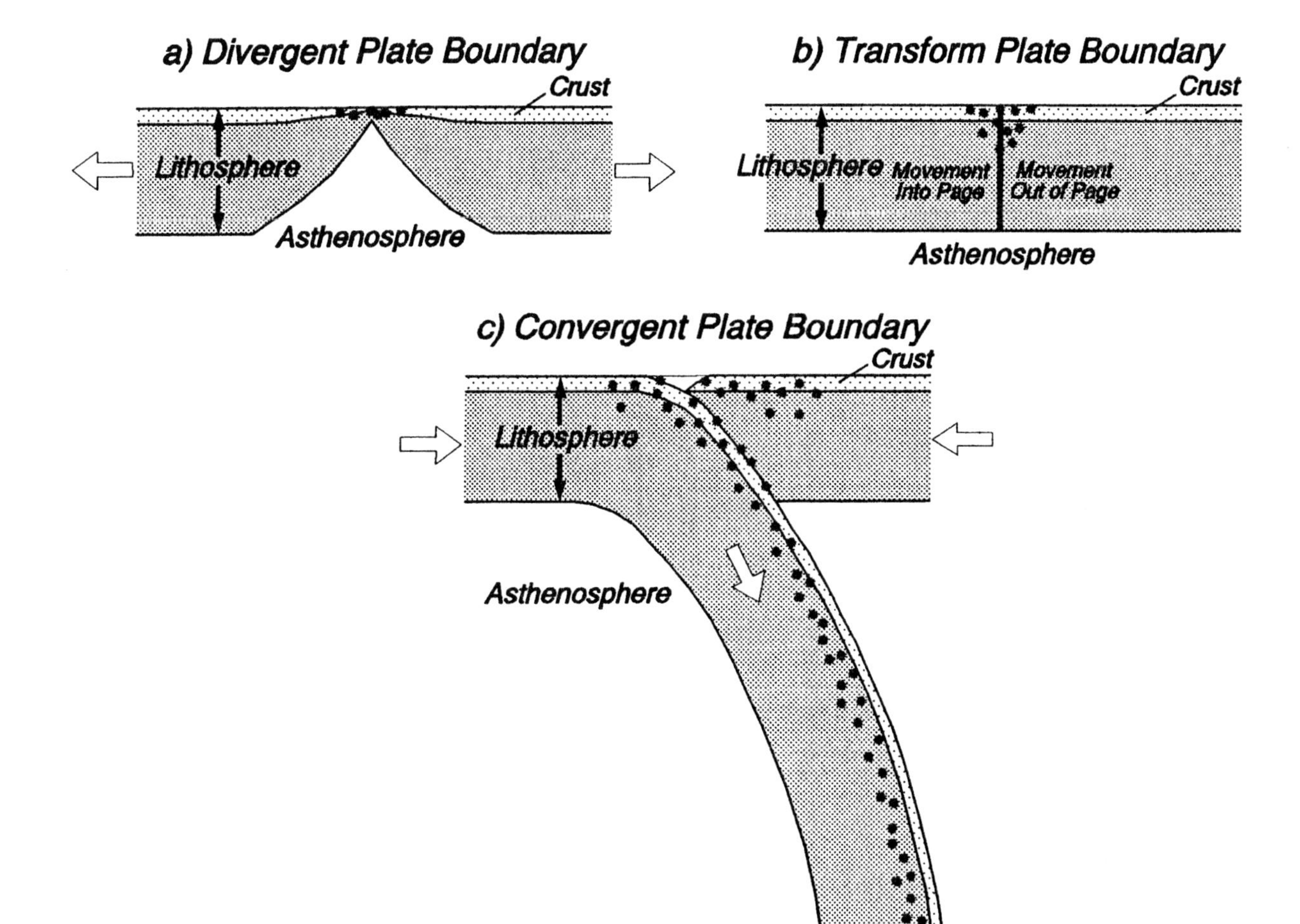

FIGURE 7.3 *Positions and depths to earthquakes (black dots) along different types of plate boundaries.* a) Only crust and uppermost mantle are cold and brittle enough to produce earthquakes along divergent boundaries. b) Earthquakes occur in the upper (cold) part of the lithosphere at transform boundaries. c) At convergent boundaries, shallow earthquakes occur in the upper plate due to compression. A zone of shallow to deep earthquakes (Wadati-Benioff Zone) occurs in the upper (cold) portion of the downgoing plate.

Location of Epicenter The initial compressional waves from an earthquake travel faster than the initial shear waves; P-waves therefore arrive at seismic stations some time before the S-waves (Fig. 7.4a). The amount of time that elapses between the P and S arrivals ($T_S - T_P$) is a function of the distance between the epicenter and station:

$$T_S - T_P = X/V_S - X/V_P$$

$$\boxed{\mathbf{T_S - T_P = X(1/V_S - 1/V_P)}}$$

where:

T_S = arrival time of initial S-wave
T_P = arrival time of initial P-wave
X = distance from earthquake to station (epicentral distance)
V_S = S-wave velocity
V_P = P-wave velocity.

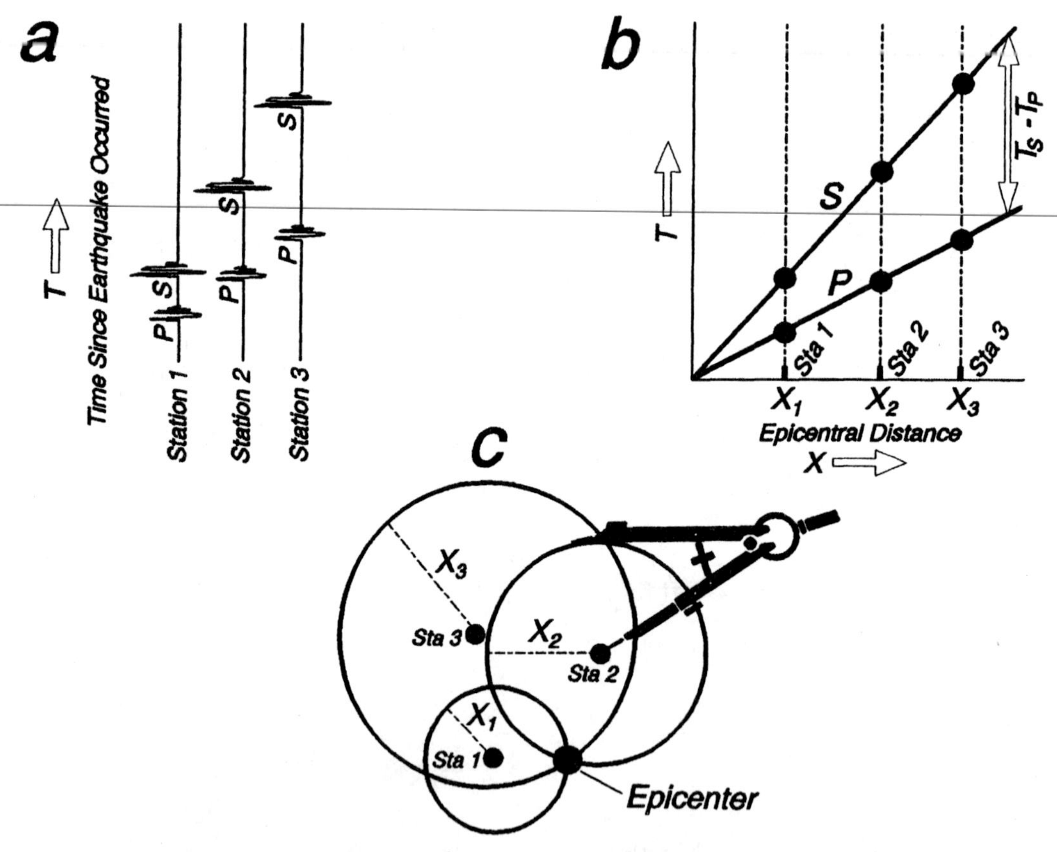

FIGURE 7.4 *Location of earthquake epicenter.* a) Seismograms from three stations, showing arrivals of compressional (P) and shear (S) waves. b) Graph of arrival times of P and S waves, plotted as a function of distance from the earthquake epicenter, for the region of the earthquake. The $T_S - T_P$ time gets longer with increasing distance from the epicenter. For each of the three seismograms in (a), the $T_S - T_P$ times are fit to the two curves; the distance from each station to the epicenter (epicentral distance) is thus approximated. c) The intersection of circles representing the epicentral distance for each station approximates the epicenter. (Note: This is an oversimplified example. In practice there are numerous stations, and three dimensional effects of focal depth and Earth's spherical shape are taken into account. Nonetheless, the concept is illustrated).

The epicenter for a local, shallow earthquake can be determined by using $T_S - T_P$ times to calculate the distances (X) from the earthquake to three or more stations (Fig. 7.4b). Circles are drawn on a map around each station; the radius of each circle represents the epicentral distance (X; Fig. 7.4c). The intersection of the circles approximates the location of the epicenter.

Strength of an Earthquake

There are two terms used to describe the strength of an earthquake: *magnitude* is quantitative, related to the *amount of energy released* by the earthquake; *intensity* is qualitative, describing the *severity of ground motion at a given location*.

Magnitude Magnitude is based on precise measurements of the *amplitude of seismic waves*, at known distances from the epicenter, using seismographs. It is expressed according to a *logarithmic* scale, whereby an increase in magnitude by one unit corresponds to a *10-fold increase in amplitude* of the seismic waves. For example, consider a magnitude 6 earthquake, compared to a magnitude 8 earthquake, occurring in exactly the same place (Fig. 7.5). At the same seismic station, the ground moves up and down 100 times as high for the magnitude 8 event as it did during the magnitude 6 ($10^8/10^6 = 10^2 = 10 \times 10 = 100$).

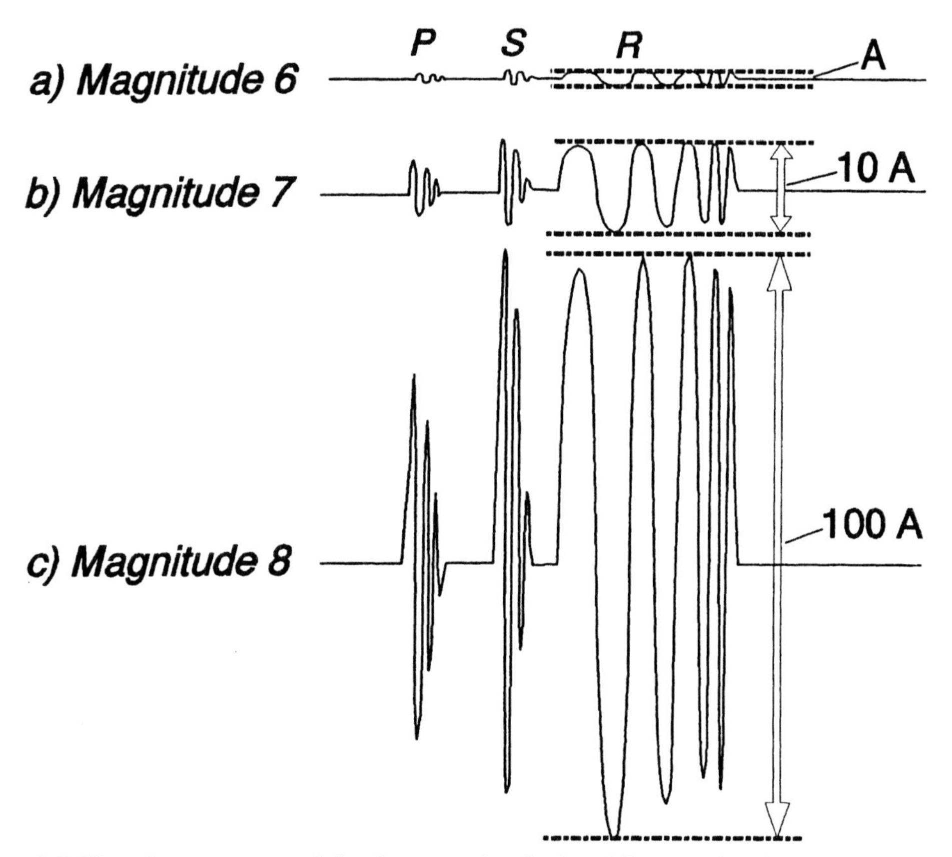

FIGURE 7.5 *Seismograms recorded at the same station, for three different earthquakes radiating from the same focus.* a) A magnitude 6 earthquake causes the ground surface to move with an amplitude (A). b) A magnitude 7 earthquake causes ten times the amplitude as the magnitude 6 ($A \times 10 = 10A$). c) Amplitude for a magnitude 8 earthquake is ten times again the magnitude 7, equivalent to 100 times the magnitude 6 amplitude ($A \times 10 \times 10 = 100A$).

There are different ways to determine the magnitude of an earthquake, depending on whether body or surface waves are used. *Body wave magnitude* depends on the amplitude of a particular compressional or shear wave arrival:

$$\boxed{\mathbf{m = \mathit{log}_{10}(A/T) + \alpha}}$$

where:

m = body wave magnitude
A = ground displacement in microns (10^{-6} m)
T = period of selected (P or S) wave in seconds
α = factor correcting for epicentral distance, focal depth, and type of wave. Correction factors can be read from published tables (for example, p. 35 of Gutenberg, 1945).

The *Richter scale*, popular in news media reports of earthquake magnitude, is based on P-wave amplitudes.

Surface wave magnitude is a function of the observed amplitude of Rayleigh waves:

$$\boxed{\mathbf{M_S = \mathit{log}_{10}A + 1.656\mathit{log}_{10}\Delta + 1.818}}$$

where:

M_S = surface wave magnitude
A = amplitude of horizontal component of 20 s period Rayleigh wave
Δ = epicentral distance in degrees.

Body and surface wave magnitudes correspond approximately as:

$$\boxed{\mathbf{m \approx 0.56M_S + 2.9}}$$

The *energy released by an earthquake* (E, in units of *Joules*) relates approximately to the surface wave magnitude as:

$$\log_{10}E = 5.24 + 1.44\,M_S$$

$$E = 10^{5.24 + 1.44Ms}$$

$$\boxed{\mathbf{E = 10^{5.24} \times 10^{1.44Ms}}}$$

The factor $10^{1.44} = 27.54$. This means that each unit increase in surface wave magnitude represents about a 27.5-fold increase in energy. In other words, to make the ground amplitude increase by a factor of 10, the energy released by the earthquake must increase approximately 27½ times. It would therefore take about 750 magnitude 6 earthquakes to produce the same amount of seismic energy as the magnitude 8 earthquake in Fig. 7.5 ($27.5^8/27.5^6 = 27.5^2 = 27.5 \times 27.5 \approx 750$). This reveals that, although there are thousands of times more small earthquakes each year than large ones, *more than half of all the seismic energy is released by a few great earthquakes* ($M_S \geq 8$); about 90% of the energy comes from earthquakes of magnitude 7 and above (Table 7.1).

For very large earthquakes, body and surface wave magnitudes saturate; that is, amplitudes cease to increase dramatically with increasing energy. It is therefore more useful to consider *moment magnitude*:

TABLE 7.1 *Worldwide number of earthquakes per year for different magnitude ranges, with compilations of energy released.* a) Magnitude range. b) Approximate number of earthquakes per year within range (after Gutenberg and Richter, 1954). c) Computation of approximate energy released for one earthquake in magnitude range (8.5 used for >8; 7.5 for 7–7.9; 6.5 for 6–6.9, etc.), using $E = 10^{5.24} \times 10^{1.44Ms}$ Joules. d) Approximate energy released per year for earthquakes in range (column b × column c). e) Percentage of total energy per year released by magnitude range.

(a) Earthquake Magnitude Range	(b) Number of Earthquakes per Year	(c) Energy (Joules) Released by 1 Earthquake	(d) Total Energy per Year for Magnitude Range	(e) Percent of Yearly Total
≥ 8	1.1	3.0×10^{17}	3300×10^{14}	55.9%
7 – 7.9	18	1.1×10^{16}	2000×10^{14}	33.2%
6 – 6.9	120	4.0×10^{14}	480×10^{14}	8.0%
5 – 5.9	800	1.5×10^{13}	120×10^{14}	2.0%
4 – 4.9	6,200	5.3×10^{11}	33×10^{14}	0.6%
3 – 3.9	49,000	1.9×10^{10}	9.3×10^{14}	0.2%
2 – 2.9	300,000	6.9×10^{8}	2.1×10^{14}	< 0.1%

$$\mathbf{M_W = [(\mathit{log}_{10}\, M_0)/1.5] - 10.73}$$

where:

M_W = moment magnitude
M_0 = seismic moment = $(A)(u)(\mu)$
A = area of fault that ruptures during earthquake
u = average displacement across the fault during earthquake
μ = shear modulus (rigidity) of the rock.

The equation illustrates that the larger the area of rupture and fault displacement, the larger the earthquake. Magnitudes for very large earthquakes (that is, those with large fault areas and displacements) can be expressed as moment magnitudes by examining amplitudes of low frequency (long period) seismic waves (Lay and Wallace, 1995).

Intensity Intensity of an earthquake is based on *effects at the surface*, as witnessed by people. For a given location, intensity is reported as Roman Numerals according to the *Mercalli Scale* (Table 7.2). Three factors that tend to *increase intensity* at a given location are: 1) magnitude of the earthquake; 2) proximity to the earthquake focus; and 3) loose soil as opposed to firm bedrock. A large magnitude earthquake would have higher intensities everywhere, compared to a small one. For a given magnitude, there is an approximate maximum intensity that might be expected (Table 7.3).

Immediately after an earthquake, survey forms are circulated to residents of the affected region to determine what people actually observed during and after the quake. Based on those surveys, intensities can be inferred for specific areas. Contour maps are drawn showing the observed intensity relative to the location of the earthquake epicenter (Fig. 7.6). Seismic wave amplitudes get smaller with increasing

TABLE 7.2 *Mercalli Scale of earthquake intensity.* From *Earth's Dynamic Systems*, 7th ed. by W. K. Hamblin and E. H. Christiansen, © 1995. Reprinted by permission of Prentice-Hall, Inc., Upper Saddle River, NJ.

Intensity	Observed Effects
I	Not felt except by very few people, under special conditions. Detected mostly by instruments.
II	Felt by a few people, especially those on upper floors of buildings. Suspended objects may swing.
III	Felt noticeably indoors. Standing automobiles may rock slightly.
IV	Felt by many people indoors, by a few outdoors. At night, some are awakened. Dishes, windows, and doors rattle.
V	Felt by nearly everyone. Many are awakened. Some dishes and windows are broken. Unstable objects are overturned.
VI	Felt by everyone. Many people become frightened and run outdoors. Some heavy furniture is moved. Some plaster falls.
VII	Most people are in alarm and run outside. Damage is negligible in buildings of good construction.
VIII	Damage is slight in specially designed structures, considerable in ordinary buildings, great in poorly built structures. Heavy furniture is overturned.
IX	Damage is considerable in specially designed structures. Buildings shift from their foundations and partly collapse. Underground pipes are broken.
X	Some well built wooden structures are destroyed. Most masonry structures are destroyed. The ground is badly cracked. Considerable landslides occur on steep slopes.
XI	Few, if any, masonry structures remain standing. Rails are bent. Broad fissures appear in the ground.
XII	Virtually total destruction. Waves are seen on the ground surface. Objects are thrown in the air.

TABLE 7.3 *Highest intensity expected for different earthquake magnitudes.* From *The Dynamic Earth: An Introduction To Physical Geology*, 3rd ed. by B. J. Skinner and S. C. Porter, © 1995. Used with permission of John Wiley and Sons, Inc., New York.

Earthquake Magnitude Range	Maximum Intensity
≥ 8	XII
7.4 – 7.9	XI
7.0 – 7.3	X
6.2 – 6.9	VII – IX
5.5 – 6.1	VI – VII
4.9 – 5.4	V
4.3 – 4.8	IV
3.5 – 4.2	II – III
≤ 3.4	I

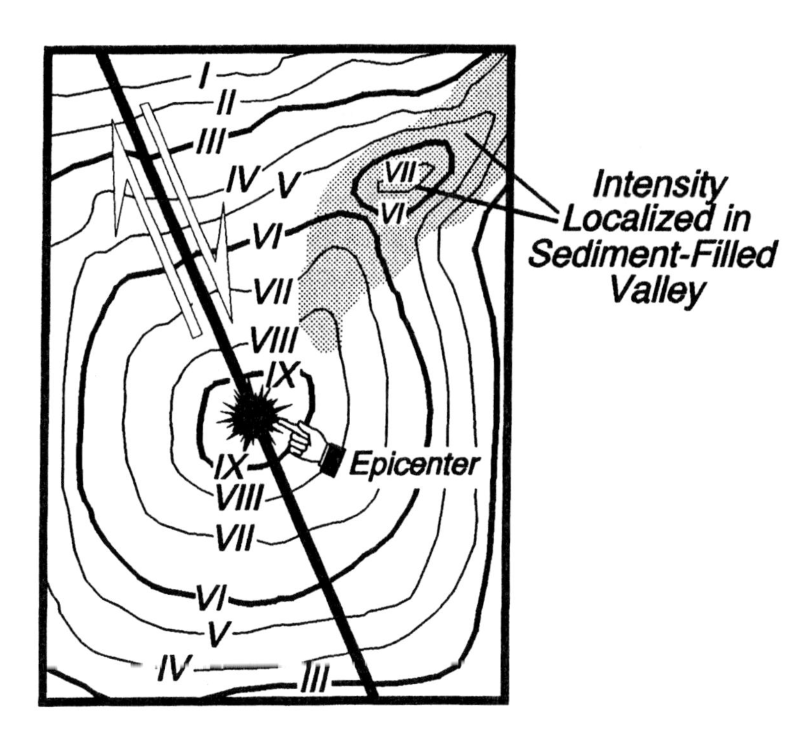

FIGURE 7.6 *Contour map of observed intensities in the region of an earthquake.* Intensities generally decrease with distance from the earthquake focus (epicenter in this map view). Anomalously high intensities on the northeast corner of the map may be associated with loose surficial material, such as sedimentary deposits or landfill.

distance from the earthquake source, so that intensities generally decrease with distance from the focus. Exceptions occur due to local ground conditions. Seismic waves are amplified in areas where there is unconsolidated sediments or landfill, locally increasing the intensity.

For a given earthquake, there is *only one magnitude*; a certain amount of energy was released by the earthquake, expressed by the magnitude. Measurements at different seismic stations around the world should yield approximately the same magnitude. For the same earthquake, however, there is a *range of intensities* observed (Fig. 7.6). The intensity observed at a given location depends on the three factors listed above; one must be specific about *where the effects were observed* when discussing intensities.

Effects of Earthquakes People often report hearing a rumble just prior to the violent shaking of the ground during an earthquake. They are initially hearing and feeling the effects of the compressional (P) waves, which commonly have smaller amplitudes and less severe ground motion than the later shear (S) and surface waves (Fig. 7.7). The S-waves can be large amplitude, shearing the ground back and forth (Fig. 3.8), causing more damage. Likewise, surface waves can be large with a shearing (Love) or rolling (Rayleigh) motion that can be destructive (Fig. 3.12).

Violent shaking in moderate earthquakes (magnitude 6 to 7) generally lasts for only a few seconds, but large (magnitude 8) earthquakes can last up to 3 or 4 minutes, causing buildings and bridges to shear and collapse, dams to break, etc. Areas of loose soil, landfill, or sediment are especially vulnerable, as those materials can amplify seismic waves, localizing high intensity (Fig. 7.6). Unless reinforced, the more rigid and massive the building material, the more likely a structure will collapse during an earthquake. For example, heavy concrete buildings and freeway overpasses break rather than bending, while light-weight wood houses are flexible and more likely to survive an earthquake.

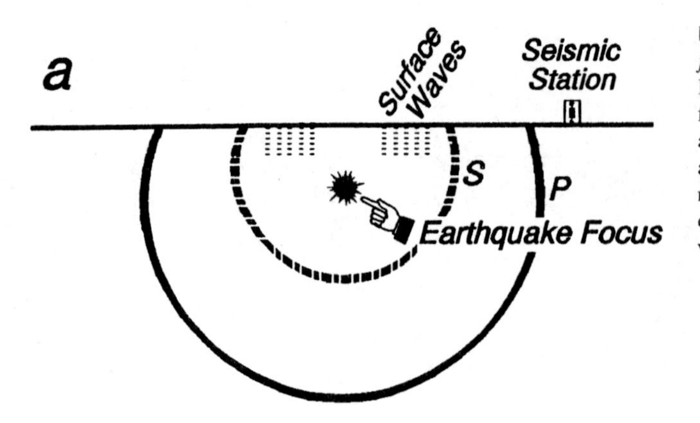

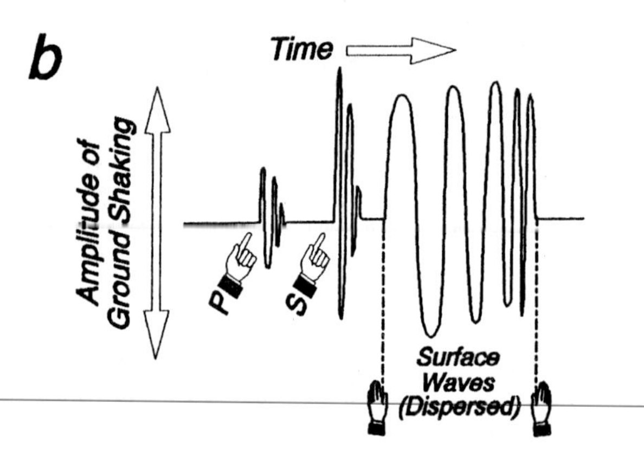

FIGURE 7.7 *Effects of ground motion from earthquake waves.* a) Wavefronts of P, S, and surface waves radiating from focus toward seismic station. b) Higher amplitude S and surface waves, arriving after the initial P-waves, are commonly more violent. Surface wave energy is dispersed (Fig. 3.17), prolonging the violent shaking during large earthquakes.

Secondary effects of earthquakes often cause more damage and loss of life than the shaking itself. Fires from the breaking of gas and electrical lines can cause large portions of cities to burn. A low-cost and effective solution is to install switches and cut-off valves that automatically shut down the electricity and gas when a building starts to shake. Sometimes earthquake faulting uplifts or downwarps large sections of the seafloor very abruptly (Fig. 7.8a), generating a water wave traveling at about 1000 km/hr, taking several hours to cross the ocean. These *tsunamis* (sometimes mistakenly called "tidal waves") have low amplitudes (a few cm) in deep ocean water, but amplitudes can grow to several m in shallow water approaching beaches (Fig. 7.8b).

Elastic rebound during an earthquake can cause tilting, uplifting, or downdropping of the land surface or sea bottom. Uplift can turn marshes and shallow bays into dry ground; downdropping causes sinking and flooding of coastal lowlands (Fig. 7.9 a,b). A special case involves a lithospheric plate that is flexed during subduction (Fig. 7.9c). Elastic strain causes bulging of the plate in the forearc region; sudden rebound of strain during an earthquake relaxes the flexure, raising or lowering coastal regions (Fig. 7.9d).

Earthquake Seismograms

Seismic stations typically have at least three seismometers, each sensitive to a different direction of ground motion (Fig. 7.10). The directions are perpendicular to one another (orthogonal), responding to vertical, north-south, and east-west motions.

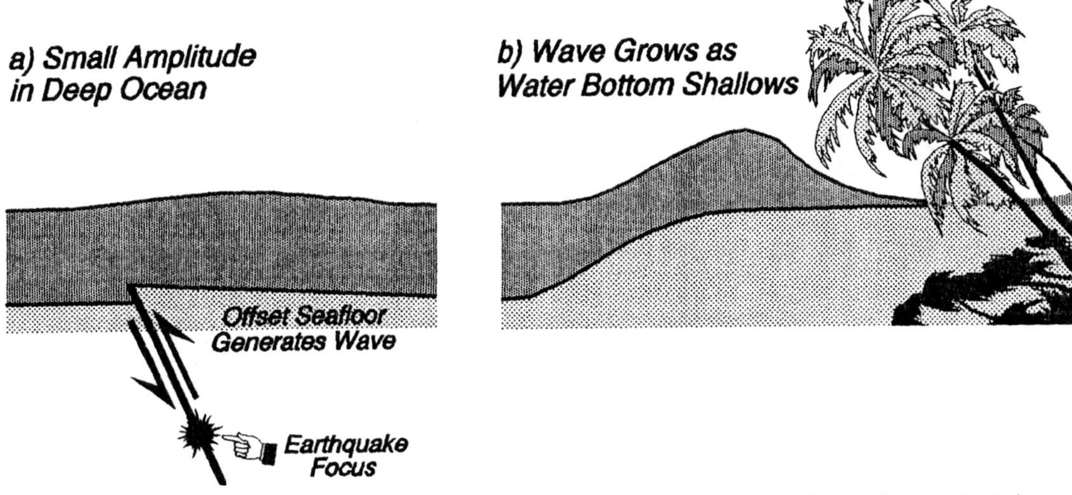

FIGURE 7.8 *Tsunami.* a) Abrupt offset of seafloor during earthquake causes water wave. Wave is low amplitude (a few cm) in deep ocean. b) Interaction with sea bottom in shallow area channels energy into thinner water layer. Wave amplitude grows considerably, rapidly (and often violently) inundating coastal areas.

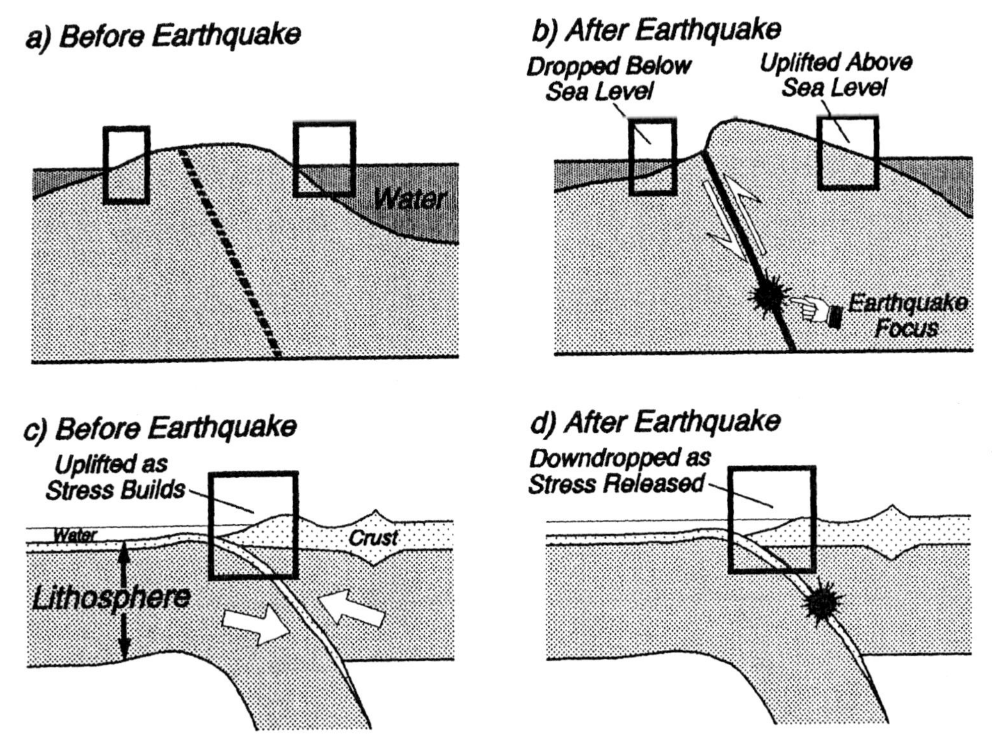

FIGURE 7.9 *Effects of tilting due to elastic rebound during earthquake.* a) Configuration of sea bottom and land surface before earthquake. b) Rebound during earthquake may lift some areas out of the sea and submerge others. c) Bulge in flexed plate uplifts coastal region above sea level. d) Plate suddenly unflexed during earthquake, downdropping and flooding region.

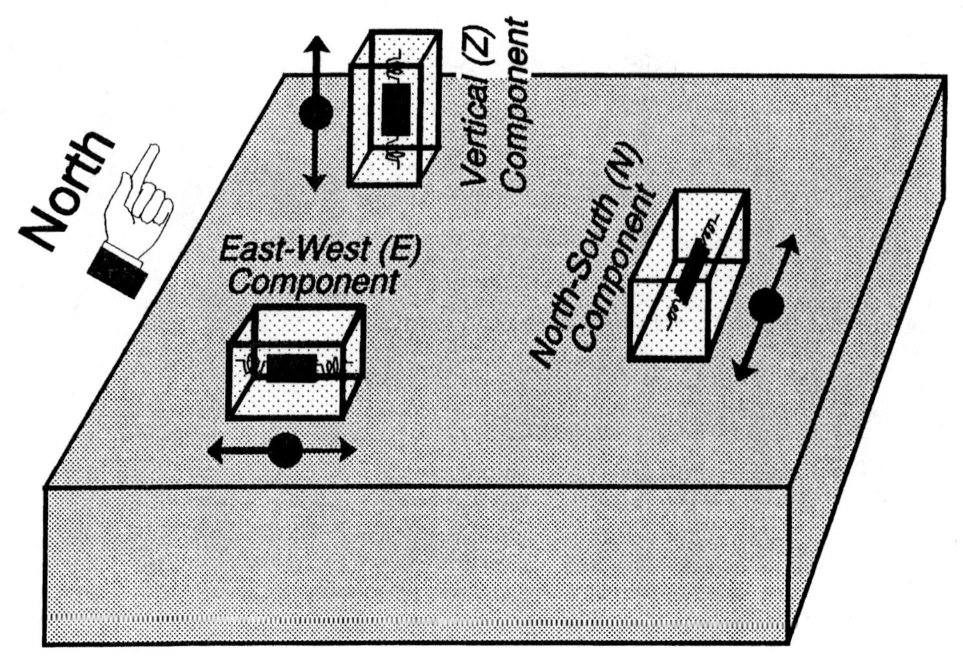

FIGURE 7.10 Seismographs measuring the three orthogonal components of ground motion at a seismic station.

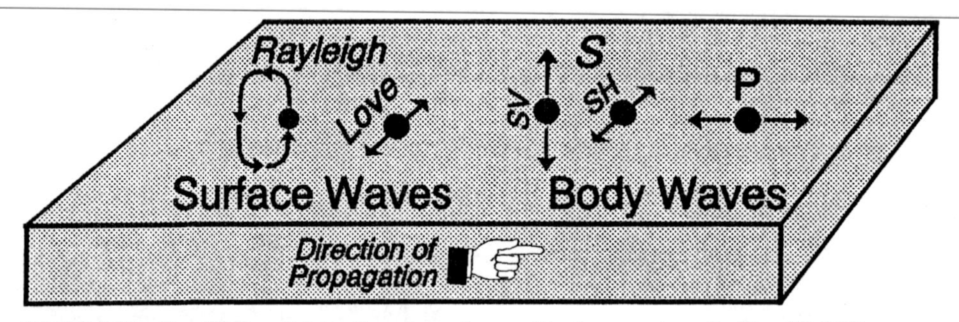

FIGURE 7.11 Simplified particle motions for surface and body waves (see also Figs. 3.8, 3.12). S = shear wave; SV = vertical shear wave; SH = horizontal shear wave; P = compressional wave.

Fig. 7.11 shows a simplified view of the particle motions of body and surface waves explained in Chapter 3. The three seismometers respond differently as seismic waves arrive at the station, due to the different paths taken by waves from the focus to the station, as well as the different particle motions. For an earthquake occurring south-southeast of the station, recordings (seismograms) for the three seismometers would show the patterns illustrated in Fig. 7.12. The initial response (first motion) on the Z-component is upward, revealing that the P-wave came out of the ground as a *compression*, pushing the ground upward. First motions of the P-wave on the N-S and E-W components show that the compression pushed the ground more northward than westward; the wave therefore must have come from a SSE direction.

Notice that there is no up-down motion (no Z-component) for Love waves (Fig. 3.12); the only type of surface wave appearing on the Z-component is the

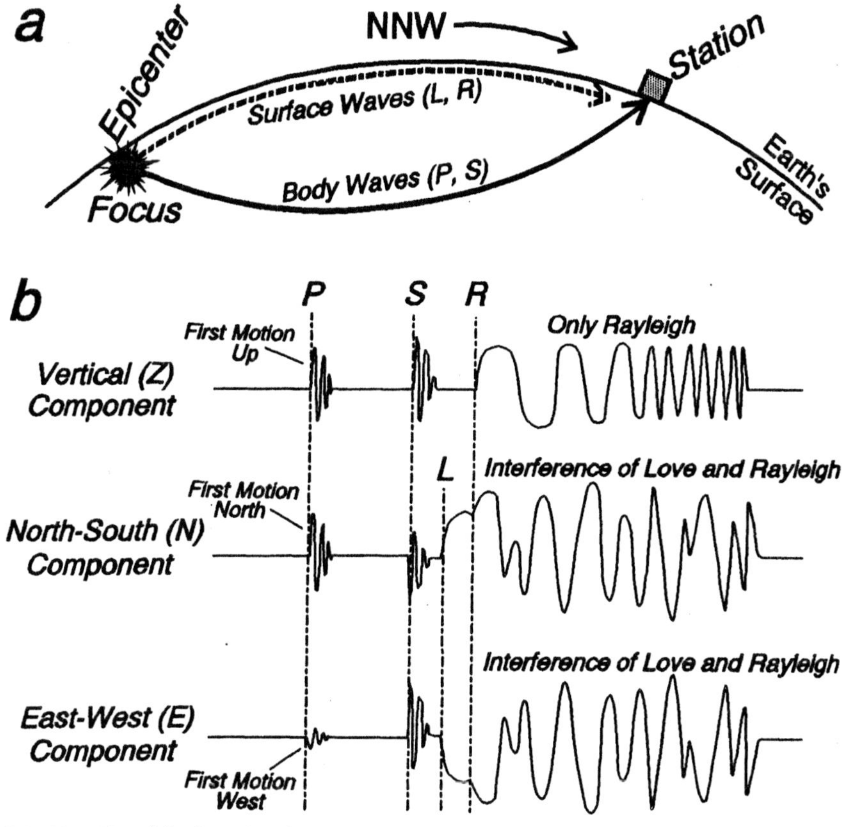

FIGURE 7.12 a) Body and surface wave paths from an earthquake located south-southeast of a station. b) Seismograms from each of the three seismometers, responding to arrivals of the body (P = compressional, S = shear) and surface (L = Love, R = Rayleigh) waves.

Rayleigh wave (Fig. 7.12b). Love waves travel slightly faster than Rayleigh waves, so that only Love waves appear initially on the north-south and east-west components. The later surface wave pattern is more complex, due to interference between Love and Rayleigh waves.

Earthquake Focal Mechanisms

When an earthquake occurs, P-waves radiate outward from the focus. In a given direction, the wave might first compress material in front (compression), followed by stretching apart of the material (dilatation). In another direction, the wave might stretch and then compress. The first P-waves coming out from the Earth at a station thus make the ground move either upward (compression) or downward (dilatation) (Fig. 7.13).

The pattern of initial compressions and dilatations (*first motions*) characterizes the nature of fault movement that produced the earthquake. Fig. 7.14a shows the pattern of compressions and dilatations radiating from a reverse fault. The seismograph at station 1 would initially move downward, at station 2 upward, and at station 3 downward. Seismograms recorded at the three stations would have first motions consistent with the radiation pattern shown in Fig 7.14b.

A shallow earthquake occurring along the San Andreas Fault in California (a *right-lateral*, strike-slip fault) would be recorded by seismic stations in the pattern

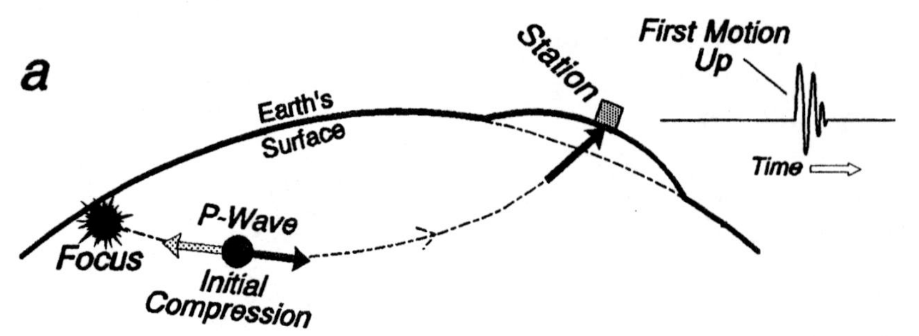

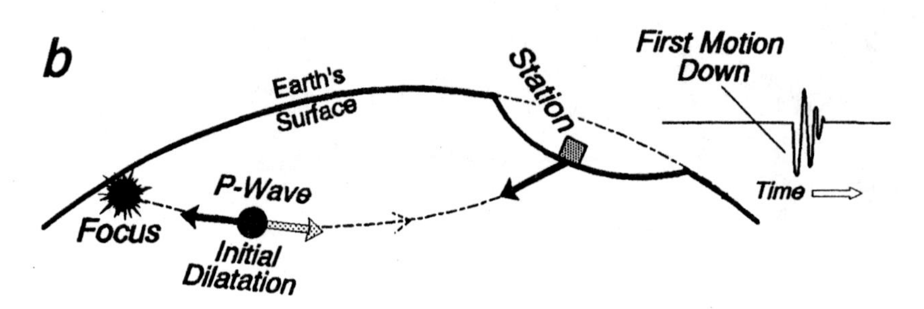

FIGURE 7.13 *Initial ground motions and Z-component seismograms for P-wave arriving at a seismic station.* (a) Initial arrival as a compression pushes the ground up; the Z-component shows an upward first motion. (b) Initial arrival as a dilatation pulls the ground down; the Z-component shows a downward first motion.

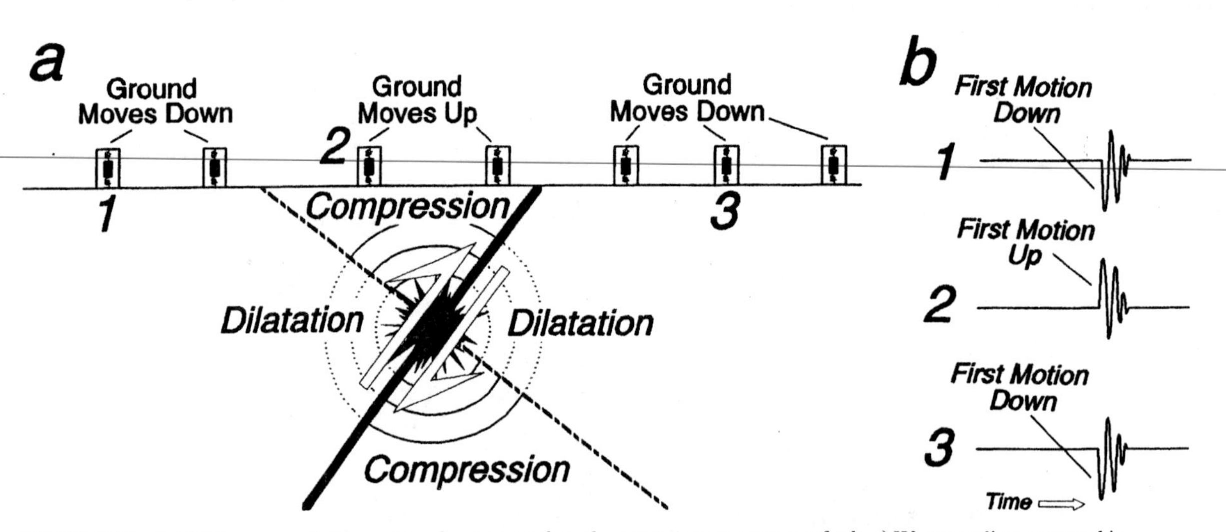

FIGURE 7.14 *Initial P-wave radiation pattern from an earthquake occurring on a reverse fault.* a) Waves radiate outward in quadrants of compression and dilatation. b) The Z-component seismograms for the three stations highlighted in (a).

shown on map view in Fig. 7.15a. The first-motion information can be synthesized by plotting arrivals from the stations on a circle, with quadrants of compression and dilatation represented by black and white, respectively (Fig. 7.15b).

Note that the observed pattern in Fig. 7.15a does not uniquely determine the nature of the fault that caused the earthquake. A fault perpendicular to the actual one (*auxiliary fault plane*; in this case a *left-lateral*, strike-slip fault) also could have been responsible (Fig. 7.15c).

Diagrams depicting quadrants of compression and dilatation are called *focal mechanism solutions*. For vertical, strike-slip faults, they are simply a map view of the initial arrivals at seismic stations (Fig. 7.15b,d). For faults that have *dip-slip* motion (normal and reverse faults), quadrants of compression and dilatation can be visualized by drawing a circle (*focal sphere*) around a cross section of the earthquake focus and fault (Fig. 7.16a).

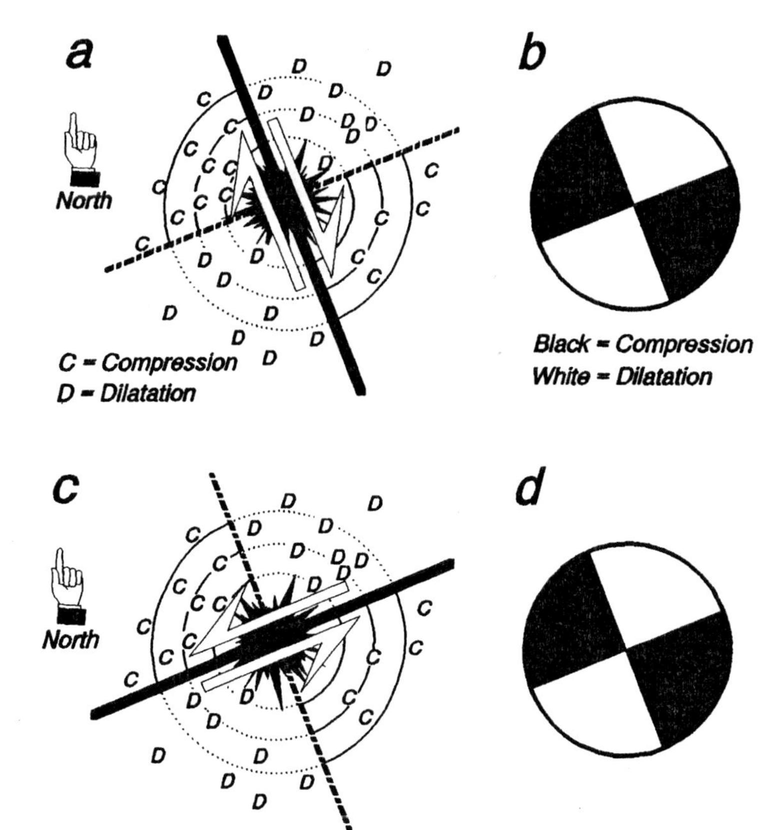

FIGURE 7.15 a) Map view of radiation pattern for *right-lateral*, strike-slip fault occurring along the San Andreas transform plate boundary (Fig. 2.20). b) First-motion information for arrival at stations indicated in (a), plotted as a focal mechanism solution. c) Auxiliary fault interpretation of the first-motion information in (a), showing that the same radiation pattern could have resulted from a *left-lateral* fault. d) Focal mechanism solution for (c) is exactly the same as that resulting from the right-lateral fault in (a).

First arrivals at stations on the surface make a compression/dilatation pattern that relates to the dip angle and sense of movement (normal or reverse) on the fault (Fig. 7.16b). Focal mechanism solutions for dip-slip faults depict a view of *the lower half of the focal sphere* that a bird would see (Fig. 7.16c). The focal mechanism solution for a reverse fault has compression surrounded by regions of dilatation (Fig. 7.17a); the opposite pattern represents a normal fault (Fig. 7.17b).

For such projections, Fig. 7.18 illustrates that: 1) the angle encompassed by the inside portion of the diagram equals 90°; 2) the distances (in degrees) from the edges to the inside portion of the diagram indicate the angles of dip of the fault and auxiliary planes; and 3) the trend of the inside portion of the diagram indicates the strike of the earthquake fault plane. Fig. 7.19 shows examples of focal mechanism solutions, along with cross sections of the earthquake and auxiliary faults, for normal, reverse, and strike-slip faults.

EARTHQUAKES AND PLATE TECTONICS

Fig. 7.20 shows that most seismic activity is associated with plate boundaries. Earthquakc focal depths reveal the configuration of relatively cold, brittle material comprising the upper part of the lithosphere (Figs. 2.7, 2.8, 7.3).

Divergent and Transform Plate Boundaries

Only shallow focus earthquakes occur at divergent (Fig. 7.3a) and transform (Fig. 7.3b) boundaries, because the brittle part of the lithosphere remains shallow (Figs. 2.11; 2.22). At divergent boundaries earthquakes are typically of small to moderate magnitude (up

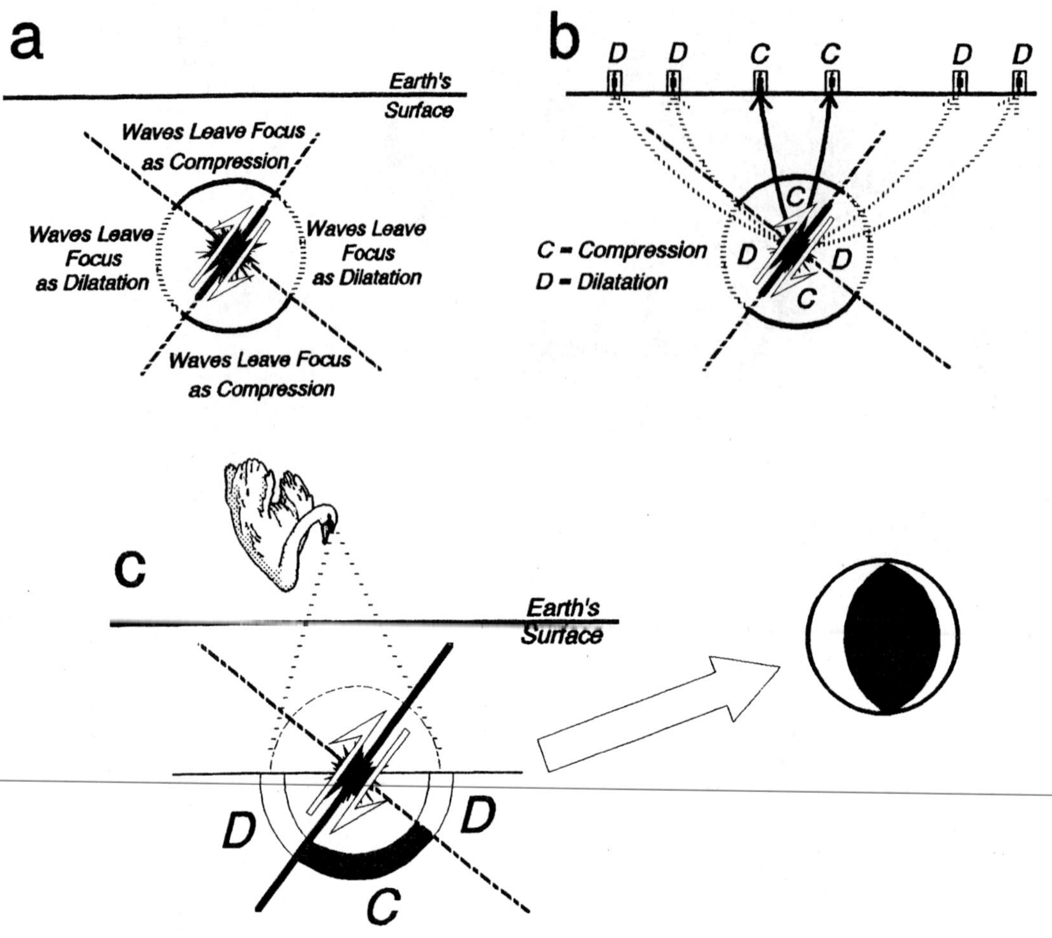

FIGURE 7.16 a) Cross section of reverse fault, earthquake focus, and quadrants of compression and dilatation. b) First motions observed at the surface reveal patterns of compression and dilatation. The patterns depend on whether rays left the region of the focus from quadrants of compression or of dilatation. c) The focal mechanism solution for a fault is commonly a *lower focal sphere* projection. For a dip-slip fault, this projection is equivalent to the compression/dilatation radiation pattern viewed by a bird flying over the earthquake focus. Black = compression; white = dilatation.

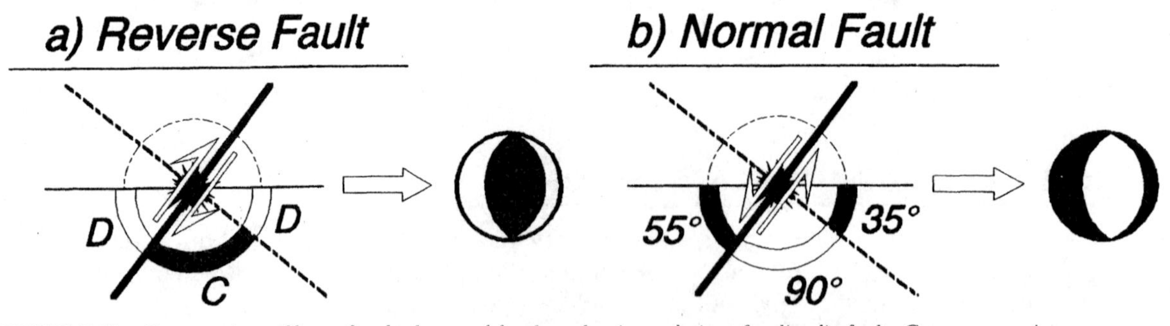

FIGURE 7.17 *Cross sections of lower focal sphere and focal mechanism solutions for dip-slip faults.* C = compression, D = dilatation. The focal mechanism solution for a reverse fault (a) has compression (black) in the inside portion of the circle, surrounded by regions of dilatation (white); the opposite pattern is observed for a normal fault (b).

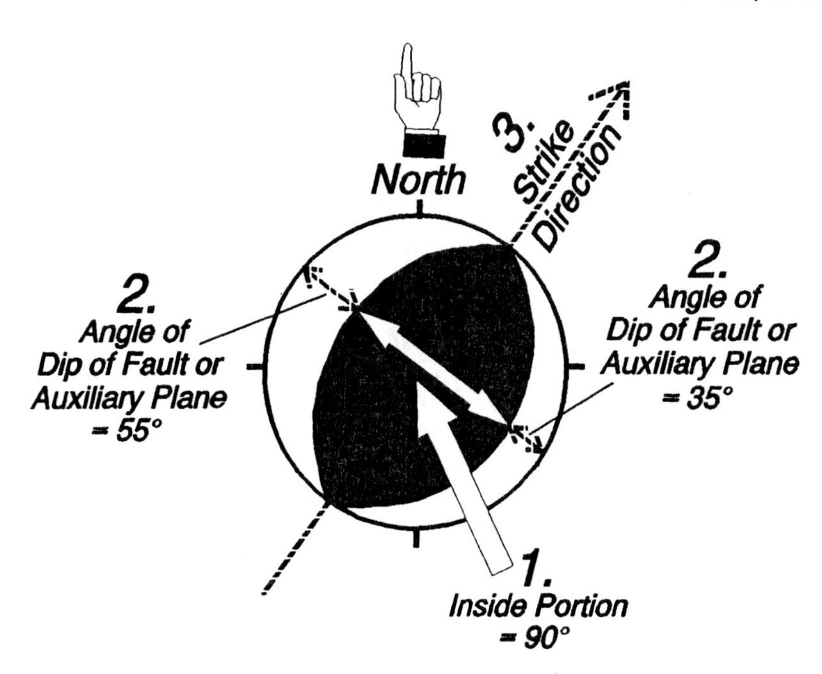

FIGURE 7.18 *Focal mechanism solution, illustrating strike and dip of earthquake and auxiliary fault planes.* Features 1, 2, and 3 discussed in text.

to about 7.5), because the shallow asthenosphere keeps the region hot (Figs. 2.13, 2.14). At mid-ocean ridges, earthquakes commonly coincide with the narrow zone following the *ridge axes* and adjoining *transform faults* (Fig. 7.21). The two types of boundaries can be distinguished on the basis of fault motions. At ridge axes (spreading centers), normal fault earthquakes occur as crustal blocks are downdropped during extension. Strike-slip fault earthquakes occur at transform offsets due to shearing as plates slide past one another. Most earthquakes occur along the active transform segments between the ridges. Beyond the ridge axes, on inactive fracture zones, plates are locked together; almost no earthquakes occur there.

The utility of focal mechanism solutions is illustrated in the map view of a fault connecting two mid-ocean ridge segments (Fig. 7.22). Prior to the 1960's ridge segments were interpreted to have moved apart from one another along *transcurrent faults* (Fig. 7.22a). According to this concept, seismicity would be as follows: 1) strike-slip earthquakes occur between the ridges and on the fracture zones extending beyond the ridges; 2) no earthquakes occur on the ridges axes; 3) for the specific example, earthquakes on the transcurrent fault have *right-lateral* motions.

The advent of plate tectonic theory led to the idea of new lithosphere created at ridge axes that are nearly stationary relative to nearby ridge axis segments (Fig. 7.22b). The ridge segments are connected by *transform faults*. This model exhibits a different pattern of seismicity: 1) strike-slip earthquakes occur only between the ridge segments, not on the fracture zones beyond the ridge segments; 2) normal fault earthquakes occur on the ridges axes; 3) for the specific example, the strike-slip earthquakes on the transform fault have *left-lateral* motions. Sykes (1967) showed that the observed distribution of earthquakes (Figs. 7.20, 7.21) and their focal mechanisms (Fig. 7.22b) are consistent with the transform fault hypothesis.

Convergent Plate Boundaries

Convergent plate boundaries are places where moving, brittle material descends to significant depths within the Earth. Shallow, intermediate and deep earthquakes occur at subduction zones (Fig 7.20), because the lithosphere extends from the surface to as

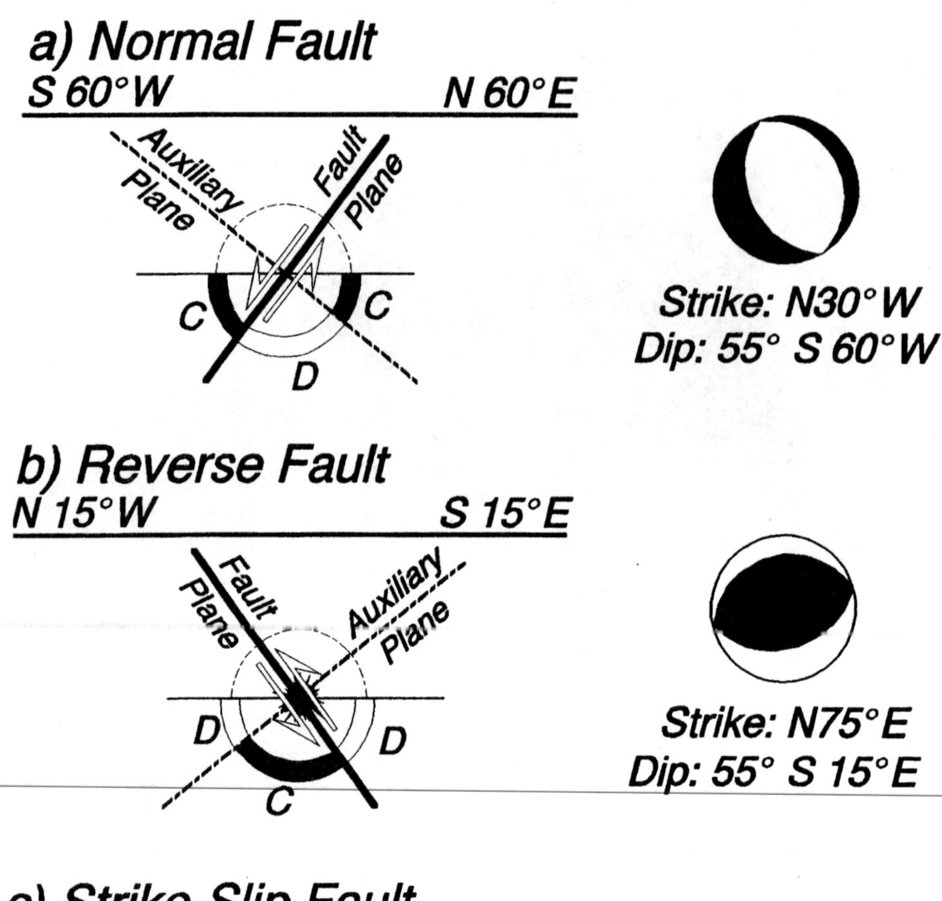

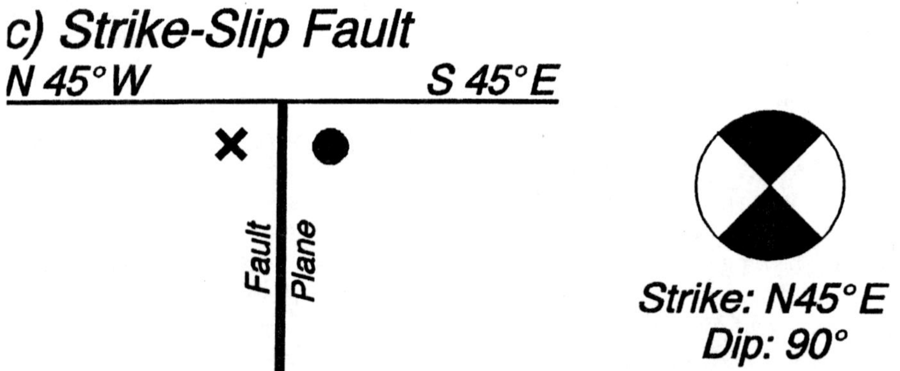

FIGURE 7.19 *Cross sections and focal mechanism solutions for three types of faults.* a) Normal fault striking north 30° west, dipping 55° toward the southwest. b) Reverse fault striking north 75° east, dipping 55° toward the south. c) Vertical, right-lateral strike-slip fault striking north 45° east. Block on the left side of the fault moving away from the reader is indicated by the "X." "Dot" shows other block moving toward the reader.

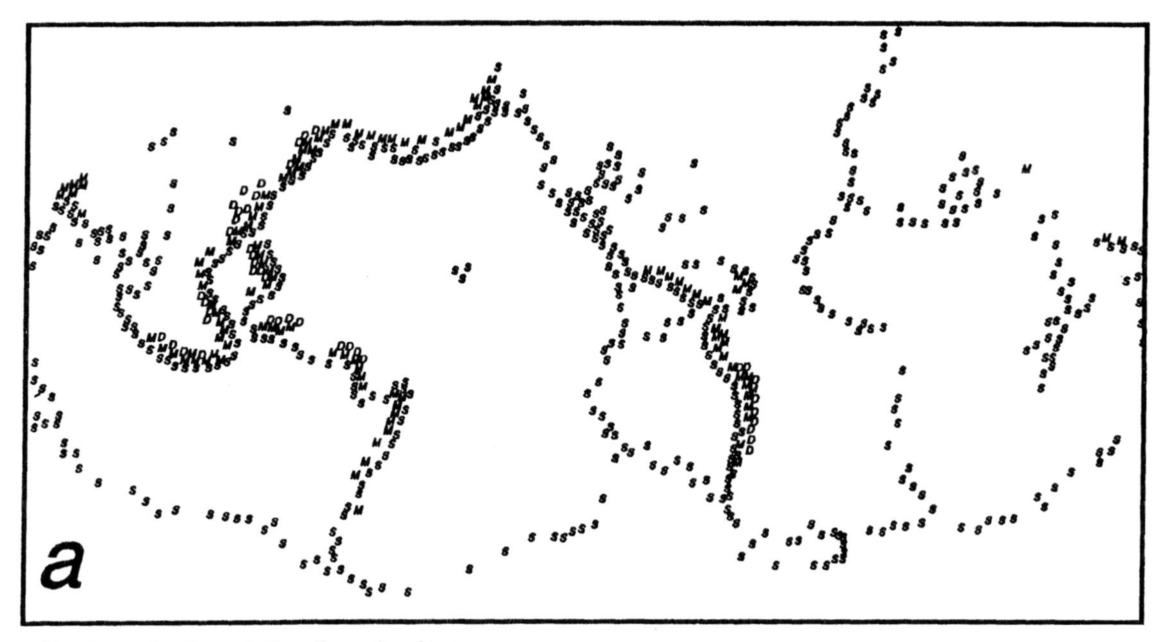

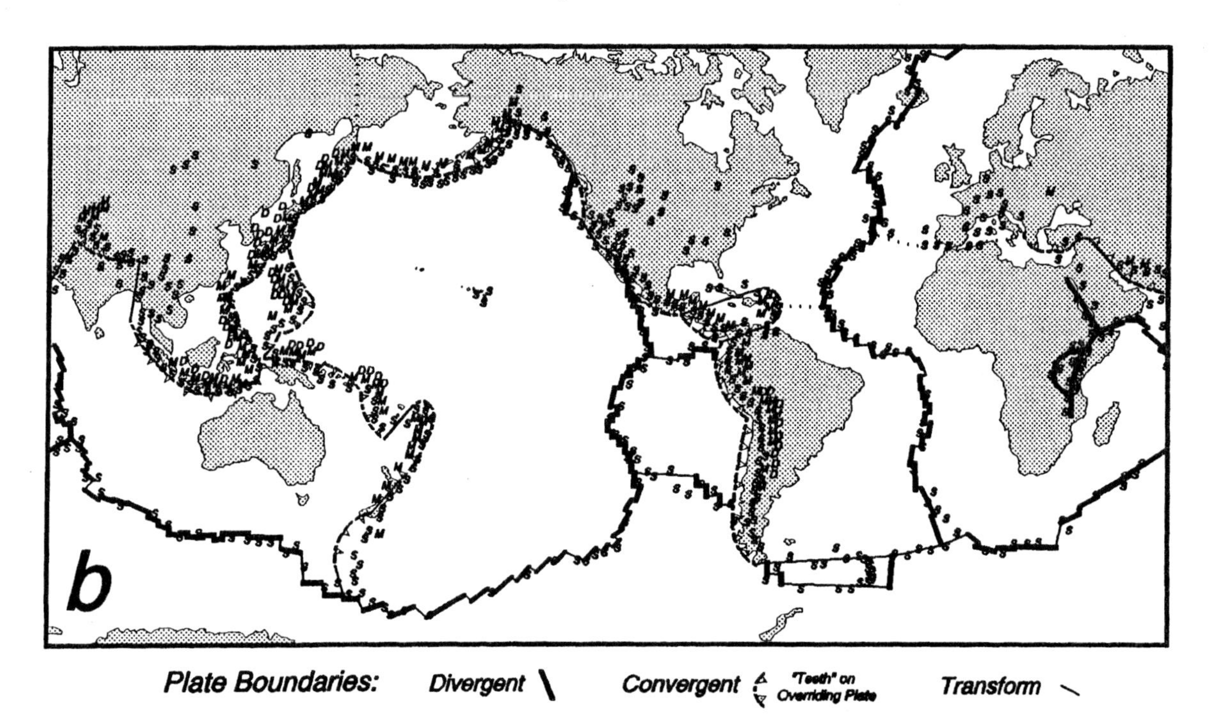

FIGURE 7.20 *Large earthquakes recorded over a five-year period.* From *Earth's Dynamic Systems*, 7/E by W. K. Hamblin and E. H. Christiansen, © 1995. Used with permission of Prentice-Hall, Inc., Upper Saddle River, NJ. a) Epicenters and focal depths, plotted without any geographic reference. The boundaries of major lithospheric plates are revealed (see Barazangi and Dorman, 1969). b) Same information as in (a), superimposed on outlines of continents and plate boundaries (Fig. 2.9). There are only shallow earthquakes at divergent and transform boundaries; shallow, intermediate, and deep earthquakes occur where plates converge.

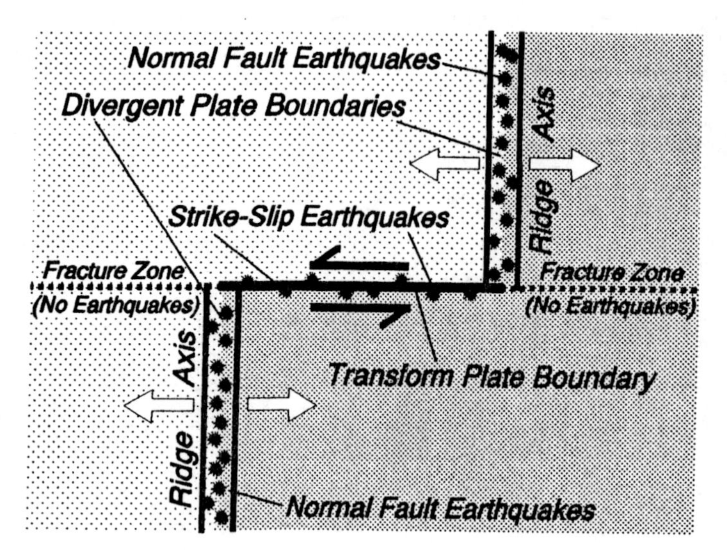

FIGURE 7.21 *Earthquakes at mid-ocean ridge system.* Compare with three-dimensional views in Fig. 2.22. Along ridge axes, very shallow (< 10 km depth) earthquakes are caused by normal faulting during extension of the brittle, upper part of the newly-formed lithosphere. At transform offsets connecting ridge-axis segments, earthquakes result from shearing stresses along strike-slip faults.

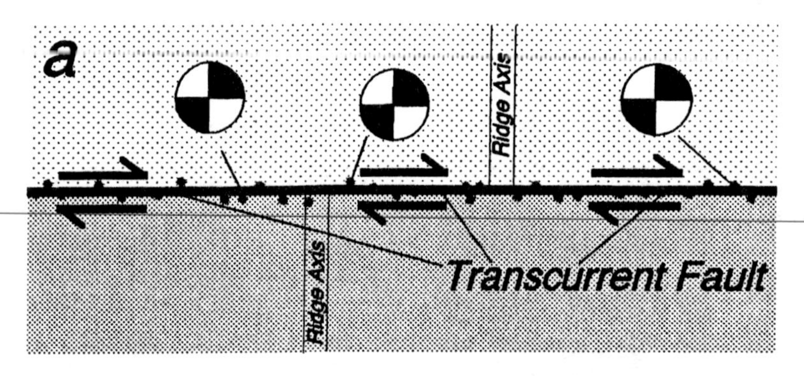

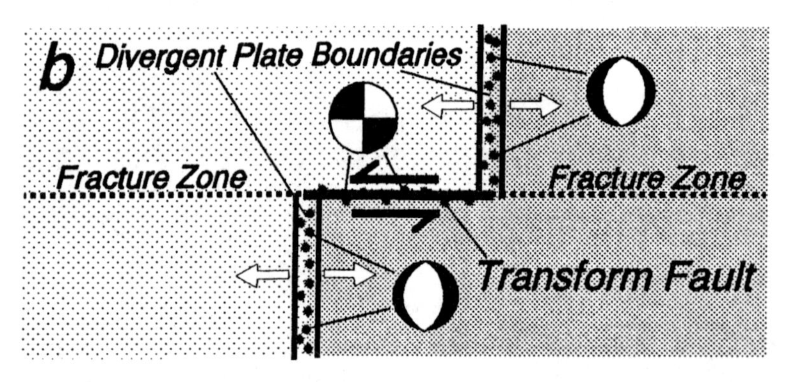

FIGURE 7.22 *Predicted earthquake distributions and focal mechanism solutions for two different interpretations of a mid-ocean ridge offset.* The two plates in each case are illustrated by light and dark shading. a) *Transcurrent fault.* The ridge segments are inactive, moving away from each other. Right-lateral, strike-slip focal mechanism solutions would be observed between and beyond the ridge segments. b) *Transform fault.* Normal faulting accompanies creation of lithosphere at the ridge axes. The two plates slide by one another on the transform fault between the ridge segments, producing left-lateral, strike-slip earthquakes. Few or no earthquakes occur on fracture zones beyond the ridge segments, where there is no plate boundary.

much as 700 km depth (Fig. 7.3c). All magnitudes of earthquakes can occur at subduction zones, up to the largest ever recorded ($M_W \approx 9.5$). In a continental collision, thick, buoyant crust on the downgoing plate may inhibit deep subduction of lithosphere (Fig. 2.18); earthquake foci are thus limited to shallow and intermediate depths, as observed in the Himalayan-Alpine collision zone in Fig 7.20.

At a subduction zone, one lithospheric plate dips beneath the other, resulting in a wide zone of epicenters from just seaward of the trench to beyond the volcanic arc (Fig. 7.20). A zone of earthquakes follows the upper portion of the downgoing, contorting plate (Wadati-Benioff zone; Fig. 7.23). The type of earthquake faulting

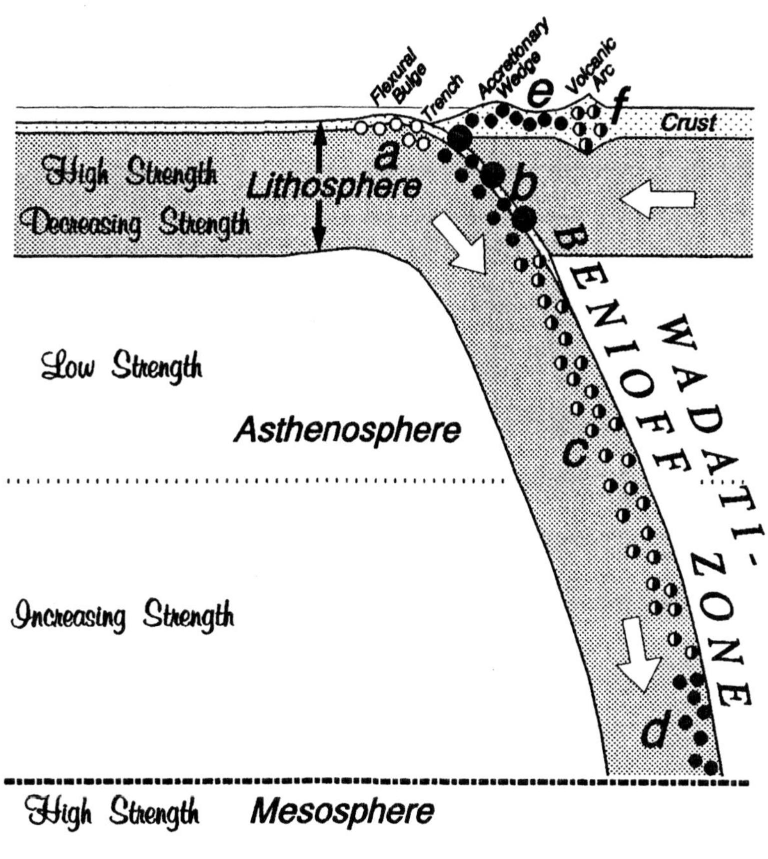

FIGURE 7.23 *Positions and types of earthquakes occurring at a subduction zone.* a) Extension at flexural bulge. b) Release of stress built up in zone where plates lock together. c) Extension or compression in intermediate to deep portion of subducting plate. d) Compression as descending plate encounters mesosphere. e) Compression or shearing in shallow portion of overriding plate. f) Earthquakes associated with magma migrating beneath volcanic arc.

varies with position along the subduction zone. Seaward of the trench, normal fault earthquakes sometimes occur due to tension on the upper part of the lithosphere as it flexes upward (Fig. 7.23a). Just landward of the trench, reverse faulting occurs due to friction along the plate boundary (Fig. 7.23b). Intermediate depth earthquakes occur beneath and beyond the volcanic arc due to extension if the plate sinks into the asthenosphere, or compression if the mantle resists downward motion of the plate (Fig. 7.23c). At great depth beyond the arc, there may be compression as the descending plate encounters the mesosphere (Fig. 7.23d).

Earthquakes can also occur within the overriding plate at convergent plate boundaries. In the crust of the forearc region (accretionary wedge, forearc basin), there may be reverse or strike-slip faulting as the plates converge (Fig. 7.23e). Stresses of all types occur during crustal readjustments as magma migrates beneath the volcanic arc (Fig. 7.23f).

SEISMIC WAVES AS PROBES OF EARTH'S INTERIOR

Most of the knowledge we have about Earth's deep interior comes from the fact that seismic waves penetrate the Earth and are recorded on the other side (Fig. 1.10). Years of study of travel times from earthquakes to stations at various distances suggests the speeds at which P- and S-waves traverse different regions of Earth's interior. These seismic velocities, along with density distributions from gravity data, can be used to interpret the physical state of regions of the Earth, according to the equations discussed in Chapter 3:

$$V_P = \sqrt{\frac{k + 4/3\,\mu}{\rho}}$$

and:

$$V_S = \sqrt{\frac{\mu}{\rho}}$$

In other words, if the compressional and shear-wave velocities (V_P and V_S) and the density (ρ) are known for a point within the Earth, the bulk modulus (k) and rigidity (μ) can be determined from the equations.

Travel-Time Curves for Earthquakes

Flat Layer Case The refraction theory developed in Chapter 4 can be used as a starting point to analyze the travel of earthquake body waves. A travel-time graph for horizontal layers shows refracting segments with slopes that indicate the increase in velocity with depth (Fig. 7.24a). When there are many, very thin layers, critically refracted rays emerge at steeper angles for deeper, higher velocity interfaces (Fig. 7.24b). Each angle of emergence relates to the velocity the ray traveled horizontally at the top of the critically refracting layer (Figs. 3.33, 3.34). The inverse of the slope for any of the line segments on the travel-time graph is thus a function of the velocity the wave encountered on the deepest part of its path. The first arrivals converge on a convex-upward curve, indicating increasing velocity with depth.

The layers may be so thin that velocities can be described as a function of depth, V(z). As in Figs. 7.24a,b, waves bottom out (travel horizontally) with the velocity at depth z (Fig. 7.24c). Instead of discrete line segments, the travel-time graph is a smooth curve, gradually decreasing in slope with distance from the source. The inverse of the slope of the curve at any point thus signifies the velocity that the wave encountered at its deepest point.

Curved Earth Surface The models and travel-time graphs in Fig. 7.24 represent a flat Earth. For longer travel paths of earthquake waves, Earth's curved surface must be taken into account. The models in Fig. 7.25 illustrate the effects of curvature as well as increasing velocity with depth. Travel times are plotted according to angular distance measured from the earthquake epicenter to recording stations on the surface.

If the Earth had *constant velocity*, body waves would follow linear raypaths (Fig. 7.25a). The resulting travel-time graph would be curved, because measurements are along Earth's curved surface. Notice how apparent velocity (inverse slope) changes with angular distance. Near the epicenter ($\Delta \approx 0°$) the inverse slope represents the true velocity. On the opposite side of the Earth ($\Delta = 180°$) rays emerge vertically, resulting in a slope of zero on the travel-time graph; infinite apparent velocity is thus recorded.

If the *velocity gradually increases* downward, all the way to the center of the Earth (Fig. 7.25b), the raypaths would bend outward, according to Snell's Law

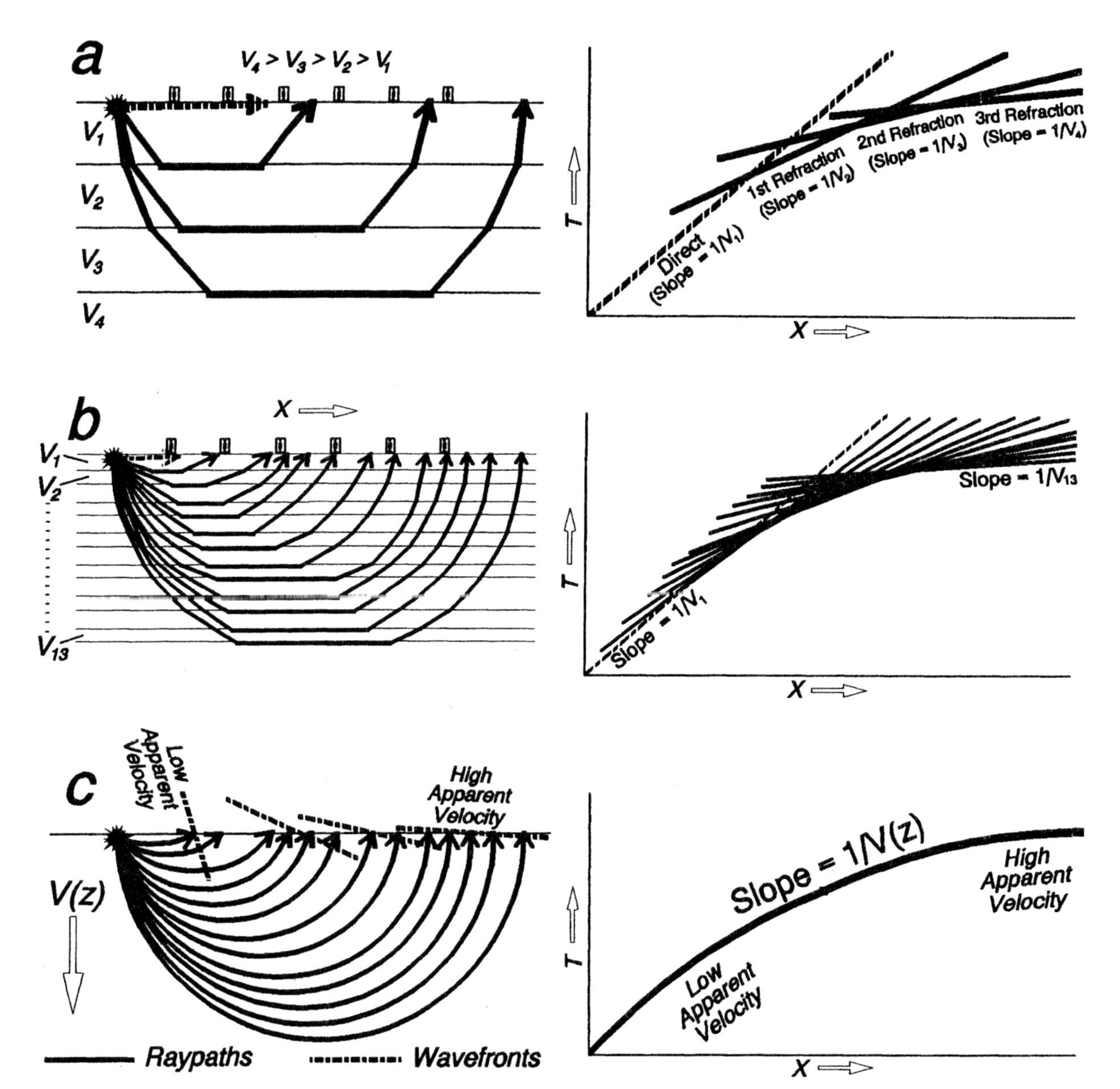

FIGURE 7.24 *Models and travel-time graphs of direct and critically refracted rays from horizontal interfaces separating layers that increase in velocity with depth.* a) Fairly thick layers result in distinct events on travel-time graph. b) Series of thin layers results in rays that emerge at steeper angles with increasing depth. Slopes of first arrivals change with distance, indicating the changing velocities with depth. c) Numerous layers so thin that the changing velocity with depth is modeled as a function V(z). Deep-penetrating rays, having encountered faster velocities, emerge at steeper angles than the shallower rays. The changing wavefront angles result in changing apparent velocity, evident as changing slope on the travel-time graph. (Note: The travel-time graphs in this chapter plot time (T) increasing *upward*, the convention commonly used in earthquake seismology).

(Fig. 3.23c). The travel-time graph would curve more abruptly than for the constant velocity case.

For an Earth with increasing V_p and V_s with depth, the travel-time graphs for P, S, and surface waves would have the forms shown in Fig. 7.25c. The travel-time graphs for the body waves are curved, convex-upward, for two reasons: 1) Earth's surface is curved (Fig. 7.25a); and 2) seismic velocities generally increase with depth (Fig. 7.25b). The surface wave travel-time graph is linear because the angular distance axis (Δ) follows Earth's curved surface.

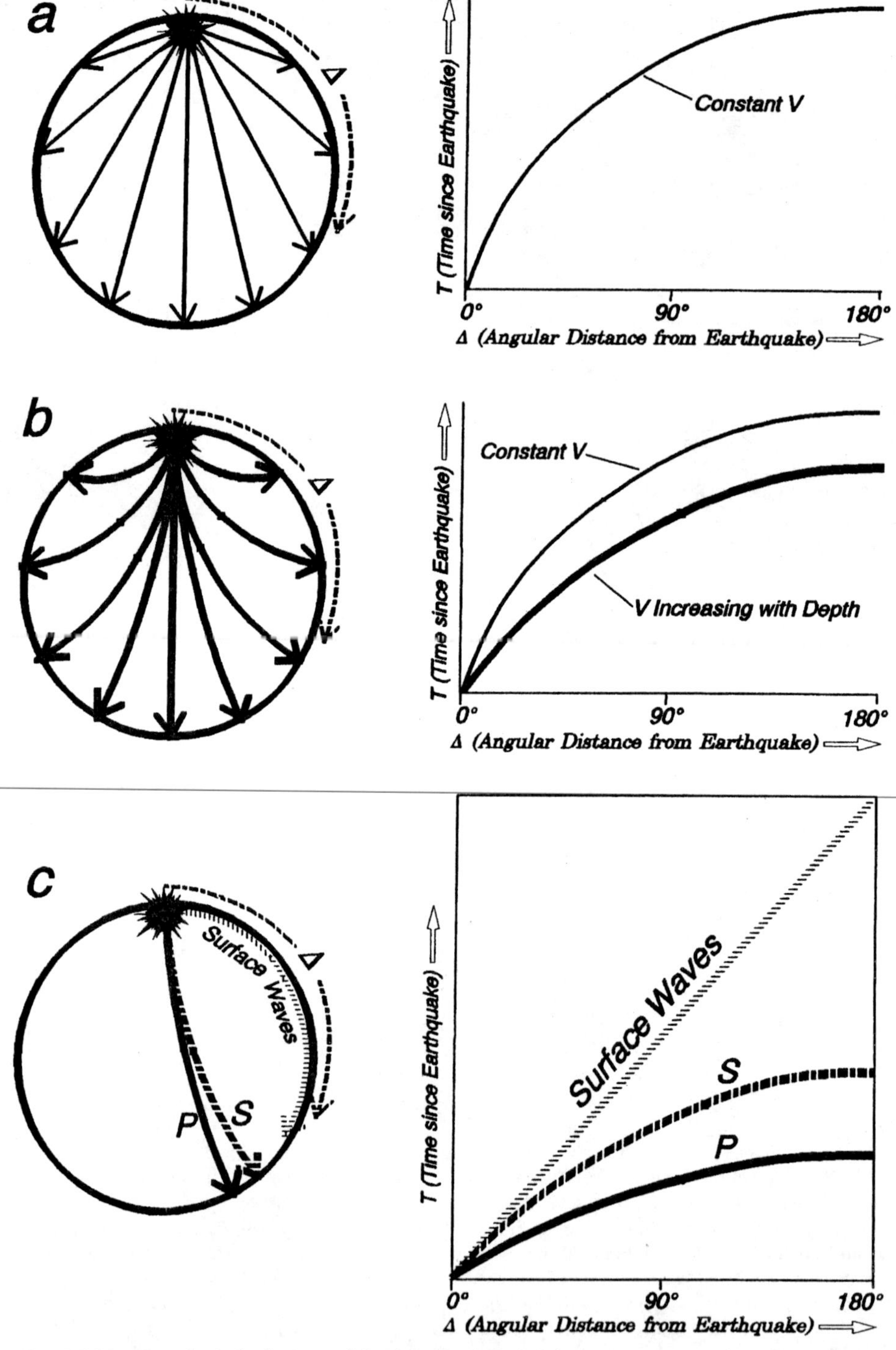

FIGURE 7.25 *Hypothetical velocity models of the Earth.* Travel times plotted according to the angular distance (Δ) from the earthquake epicenter to seismic stations on the surface. a) Earth with constant seismic velocity, V. Raypaths are linear because no refraction occurs. The convex-upward form of the graph is simply because measurements are along Earth's curved surface. b) Earth model with a gradual increase in velocity with depth, V(z). Rays bend away from the vertical, leading to steeper emergence angles (hence, higher apparent velocities) than for the constant velocity case in (a). Likewise, more curvature on the travel-time graph indicates higher apparent velocity than in (a). c) Selected raypaths and travel-time curves for body (P, S) and surfaces waves. As in (b), seismic velocities increase with depth.

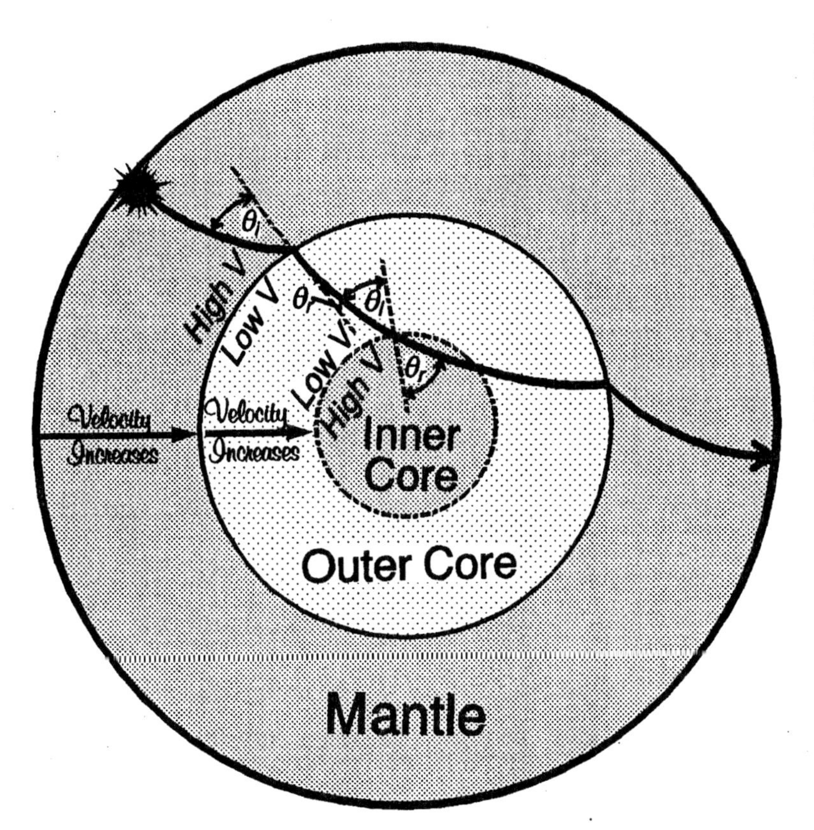

FIGURE 7.26 *Selected raypath for compressional wave penetrating the major zones of the Earth.* According to Snell's Law, $\sin\theta_i/V_i = \sin\theta_r/V_r$, where V_i = velocity of incident ray; V_r = velocity of refracted ray; θ_i = angle of incidence; θ_r = angle of refraction. Within each of the three layers, the seismic velocity increases with depth, so that the ray bends away from the vertical (outward). Velocity drops abruptly at the mantle/outer core boundary, refracting the ray inward. Abrupt outward bending results from velocity increase between the outer core and inner core.

Complex Earth Earthquake raypaths and arrival times are more complex than illustrated in Fig. 7.25, because velocity in the Earth does not simply increase with depth. Fig. 7.26 shows relative compressional wave velocities for Earth's major divisions (see Figs. 2.3 to 2.5). According to Snell's Law (Fig. 3.23), rays bend abruptly inward at the mantle/outer core boundary (sharp velocity decrease) and outward at the outer core/inner core boundary (sharp velocity increase). Within layers, velocities generally increase downward, bending rays away from the vertical on their downward journey; velocity generally decreases upward in layers, so that rays bend toward the vertical as they travel out of the Earth.

Particular arrivals of seismic energy, known as *phases*, are identified according to the following conventions:

P = compressional wave through the mantle;
K = compressional wave through the outer core;
I = compressional wave through the inner core;
S = shear wave through the mantle;
J = shear wave through the inner core;
L = Love wave (sometimes called LQ);
R = Rayleigh wave (sometimes called LR);
c = reflection of the wave off the top of the outer core.

Examples of raypaths for some common phases are illustrated in Fig. 7.27. Travel-time curves have been developed from observations of earthquakes occurring over a period of several years by plotting, according to epicentral distance, the average times of phases arriving at a worldwide network of seismic stations (Fig. 7.28).

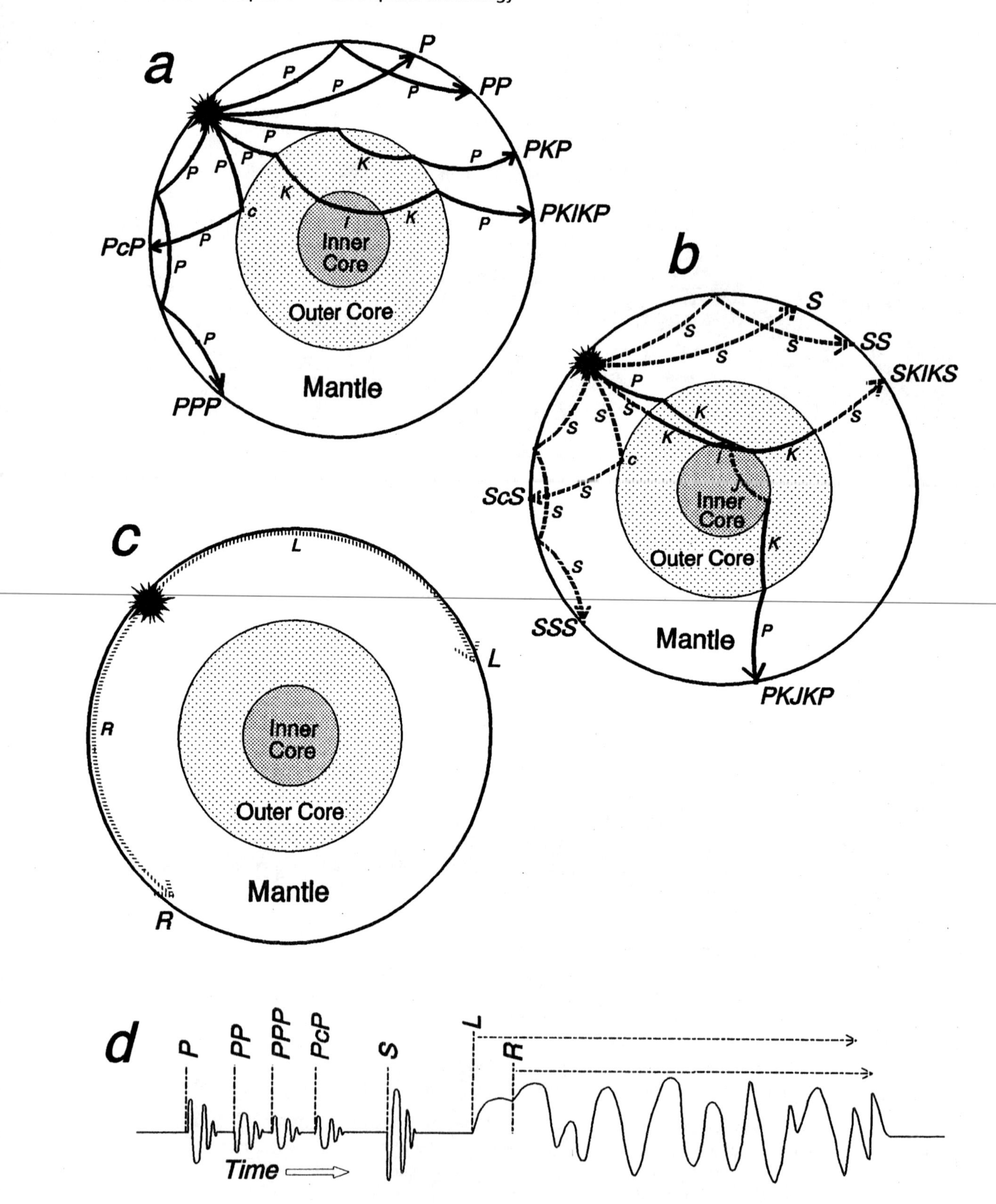
a
P
PP
PKP
PKIKP
PcP
PPP
P
K
I
c
Inner Core
Outer Core
Mantle
b
S
SS
SKIKS
ScS
SSS
PKJKP
S
K
I
J
c
Inner Core
Outer Core
Mantle
c
L
R
Inner Core
Outer Core
Mantle
d
P
PP
PPP
PcP
S
L
R
Time

Major Divisions of Crust, Mantle and Core

Based on travel-time curves (Fig. 7.28), compressional and shear wave velocities have been determined from the surface to the center of the Earth (Fig. 7.29a). When combined with approximations of density (Fig. 7.29b), the elastic properties (bulk modulus [k] and shear modulus [μ]) can be determined for the gross layers of the Earth by using the V_P and V_S equations presented above.

Three criteria help to interpret the gross nature of the Earth from seismic observations.

1) *Transmission of seismic waves.* The ability of materials to transmit seismic waves tells us about the physical state of the material. In order for an S-wave to travel through a material, for example, the material must be able to resist shearing (the shear modulus, μ, must not be zero). Fluids cannot resist shearing ($\mu = 0$), therefore they cannot transmit shear waves. The observation that S-waves don't travel through Earth's outer core ($V_s = 0$) thus tells us that the outer core is fluid (Fig. 7.30a).
2) *Velocity of seismic waves.* Seismic waves travel more slowly through soft (low μ) than through stiff (high μ) material. The observation that waves slow down as they encounter the asthenosphere thus suggests that the region is soft (Fig. 7.30b).
3) *Bending of seismic waves.* Upon encountering changes in velocity, seismic waves are bent according to Snell's Law, much as rays of light passing through a glass prism. The observation that waves are bent abruptly (refracted) suggests that there are drastic changes in seismic velocity across *discontinuities*. Three notable examples are the crust/mantle boundary (or "Moho"), the mantle/outer core boundary, and the outer core/inner core boundary (Fig. 7.31). Within layers, raypaths curve because of gradually changing velocity (Fig. 7.26).

FIGURE 7.27 (opposite) *Earthquake phases.*

a) Raypaths for selected compressional-wave phases.
- P = P-wave refracted through the mantle.
- PP = P-wave refracted through the mantle, reflected at Earth's surface, then refracted again.
- PPP = P-wave reflected twice at the surface.
- PKP = P-wave refracted from the mantle, through the outer core, then back through the mantle.
- PKIKP = P-wave refracted from the mantle, through the outer core, inner core and back out.
- PcP = P-wave reflected off the outer core.

b) Raypaths for selected shear-wave phases.
- S = S-wave refracted through the mantle.
- SS = S-wave refracted through the mantle, reflected at Earth's surface, then refracted again.
- SSS = S-wave reflected twice at the surface.
- PKJKP = Wave refracted through the mantle and outer core as P-energy, converted to S-energy through the inner core, then back to P-energy to the surface.
- SKIKS = Wave refracted through the mantle as S-energy, converted to P-energy through the outer core, inner core, and back into outer core, then as S-energy to the surface.
- ScS = S-wave reflected off the outer core.

c) Raypaths for selected surface wave phases.
- R = Rayleigh wave following Earth's surface.
- L = Love wave following Earth's surface.

d) Schematic illustration of seismogram showing some of the phases depicted in (a), (b), and (c).

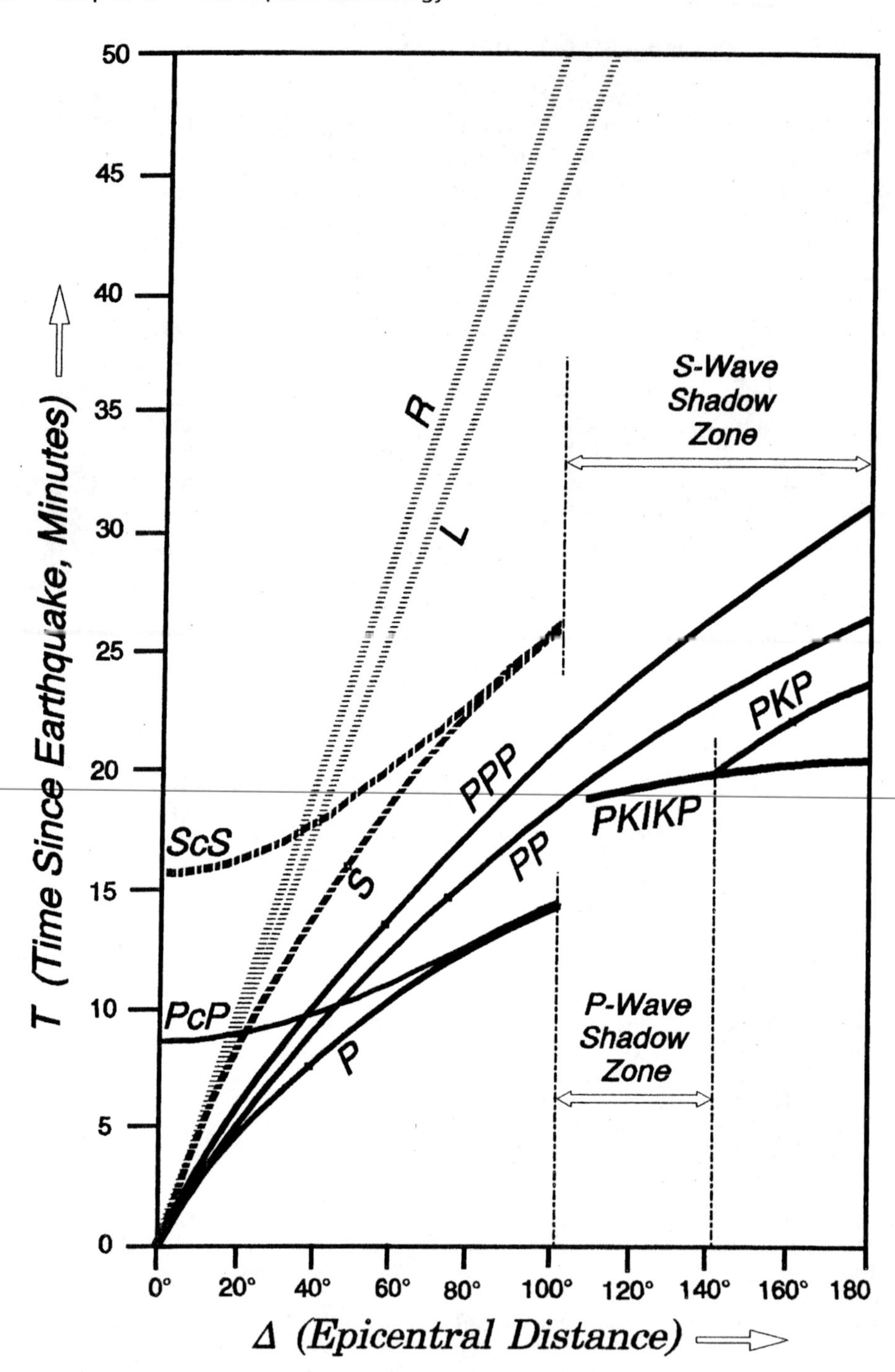

FIGURE 7.28 *Travel-time curves for selected earthquake phases.* See Fig. 7.27 for phase abbreviations and Fig. 7.35 for discussion of shadow zones.

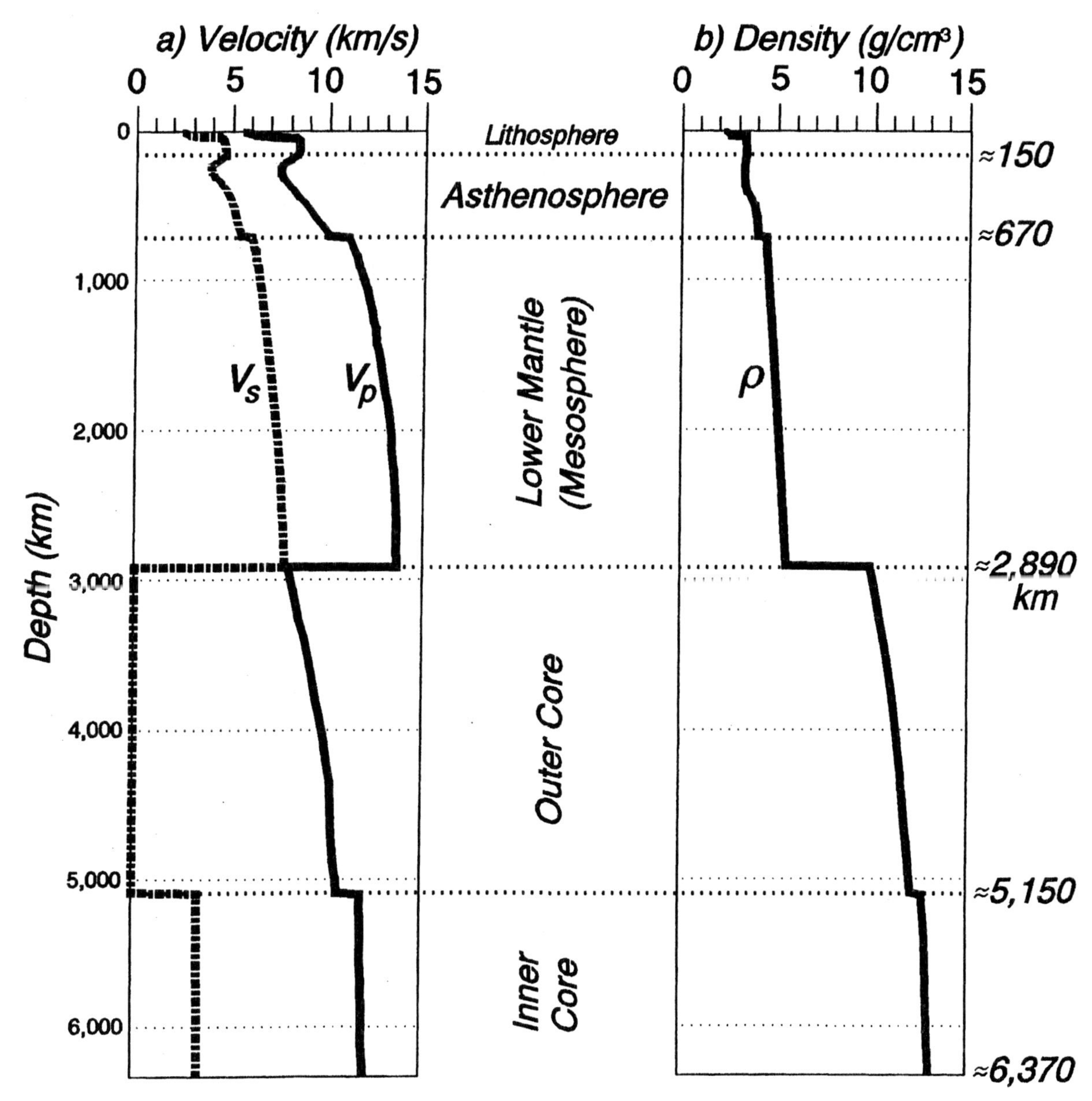

FIGURE 7.29 a) Velocities of compressional (V_p) and shear (V_s) waves, from the surface to the center of the Earth. Curves based on forward modeling of travel-time curves, like those in Fig. 7.28. b) Profile of gross density (ρ) of the Earth (from De Bremaecker, 1985; after Dziewonski and Anderson, 1981).

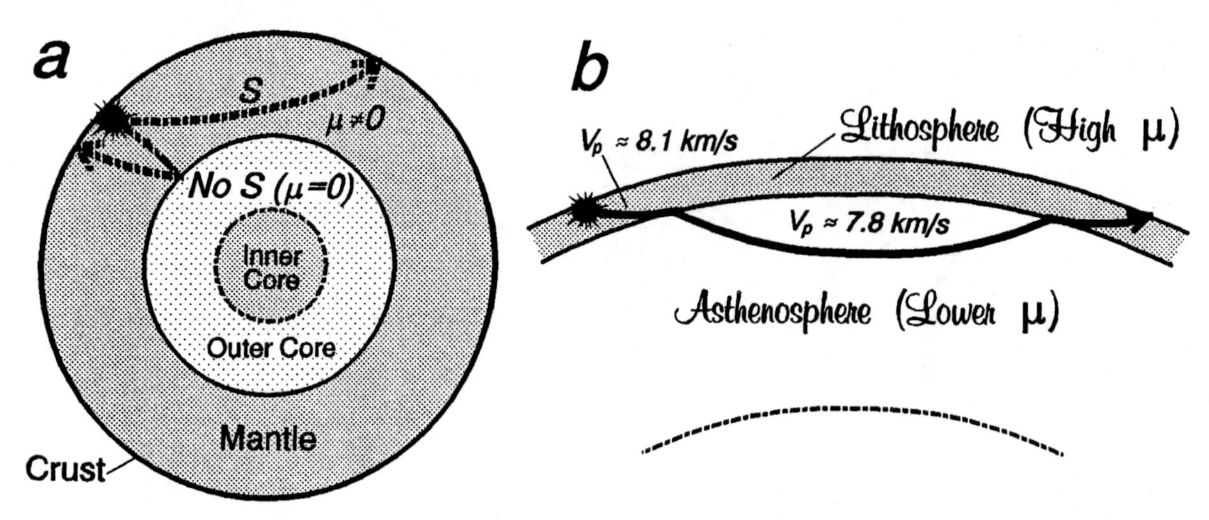

FIGURE 7.30 Effects of rigidity (μ). a) *Transmission*. S-waves travel through the crust and mantle because they have rigidity ($\mu \neq 0$). No S-waves travel through the outer core because it is a liquid, with no rigidity ($\mu = 0$). b) *Velocity*. Seismic waves slow down when traveling from the mantle part of the lithosphere to the asthenosphere (μ decreases).

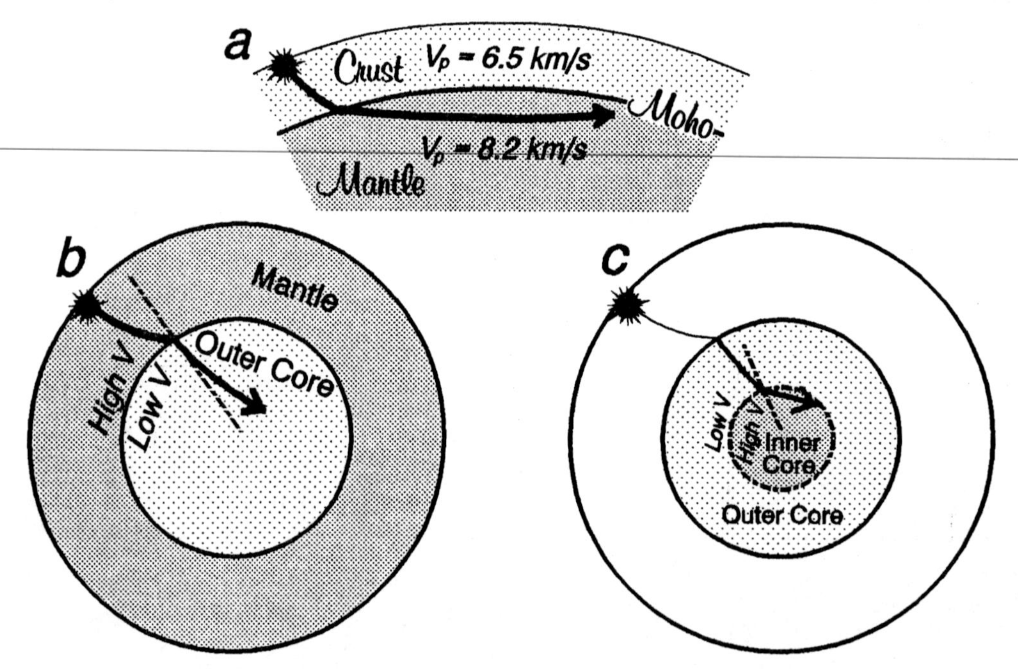

FIGURE 7.31 *Abrupt bending of raypaths at discontinuities.* a) *Crust/mantle boundary (Moho)*. Rays bend toward the horizontal because of the abrupt increase in velocity. b) *Lower mantle/outer core boundary*. Abrupt decrease in velocity causes rays to bend toward the vertical. c) *Outer core/inner core boundary*. Velocity increase causes bending away from the vertical.

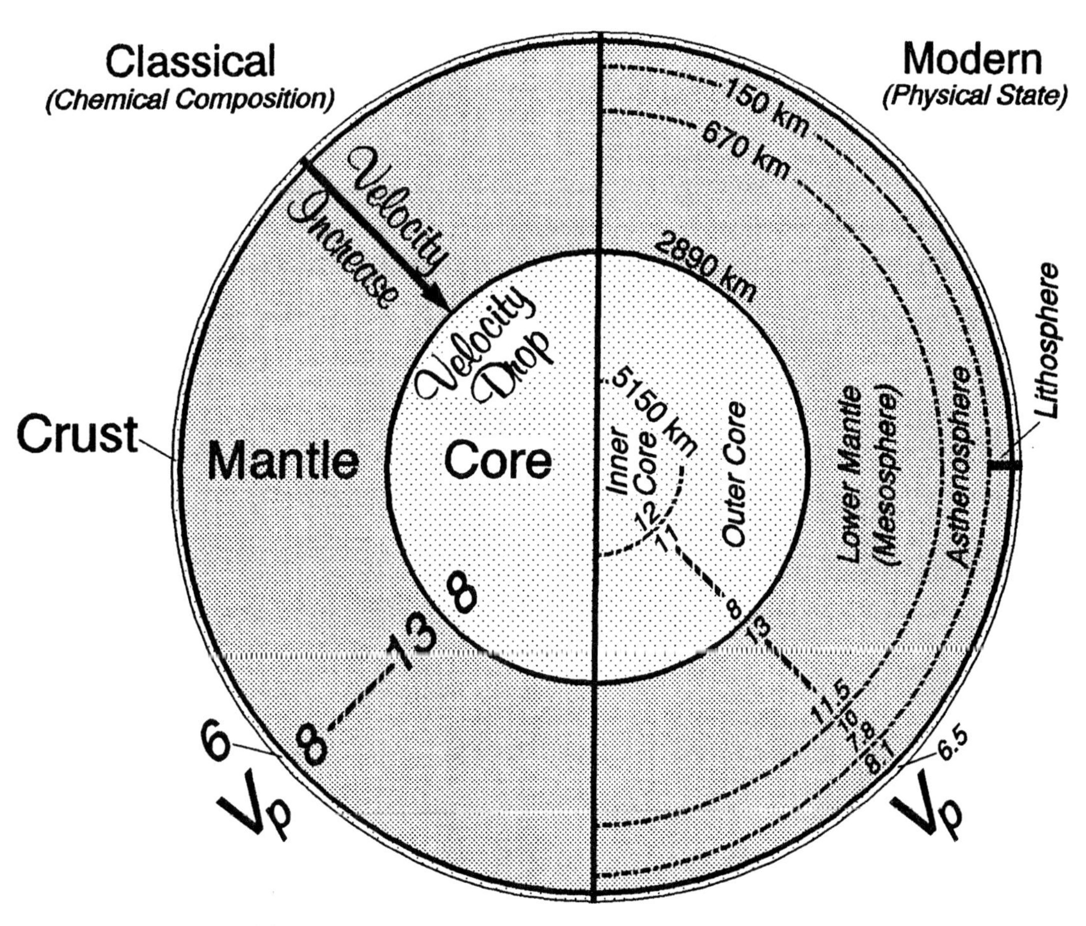

FIGURE 7.32 Classical divisions of Earth were based on gross P-wave velocity (V_p, in km/s), showing increase at crust/mantle boundary and decrease at mantle/core boundary. Refined seismic observations led to modern divisions, based on decrease in V_p across lithosphere/asthenosphere transition, abrupt increase at the top of the mesosphere, abrupt decrease at mesosphere/outer core boundary, and increase from the outer core to the inner core.

Based on velocities derived from travel-time curves (Fig. 7.29) and considerations of raypaths (especially the observations that seismic waves bend abruptly at about 35 and 2900 km depth, Fig. 7.31), three major zones of the Earth were recognized in the early 20th century (Fig. 2.3). These classical divisions are thought to correspond to changes in *chemical composition*, especially the amount of silica (SiO_4^{-2}): 1) *crust*, predominately silicates (50%–70% silica); 2) *mantle*, silicates high in iron and magnesium (peridotite, 30% silica); 3) *core*, heavy metals (iron and nickel). Refined seismic observations led to *modern divisions* based on *physical state* (Fig. 2.4): I) *Lithosphere*, hard solid (relatively *high* μ); II) *Asthenosphere*, softer solid (relatively *low* μ); III) *Mesosphere*, somewhat hard solid (relatively *high* μ); IV) *Outer Core*, liquid ($\mu = 0$); V) *Inner Core*, solid ($\mu \neq 0$). Fig. 7.32 shows the classical and modern divisions according to gross P-wave velocities.

Lithosphere The rigid lithospheric plates contain three components (Fig. 7.33). The *crust* has P-wave velocities generally below 7 km/s. In many areas the crust consists of sedimentary rock with velocities from about 2 to 5 km/s, underlain by igneous and metamorphic rocks with velocities slightly greater than 6 km/s. The *Mohorovičič Discontinuity* (or "Moho") separates the crust from the mantle. Across the Moho,

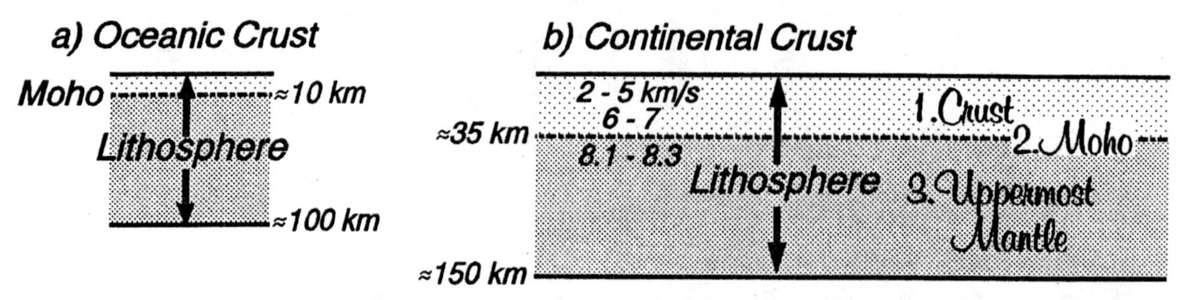

FIGURE 7.33 Rigid lithospheric plates in regions of (a) oceanic and (b) continental crust. Three components of the lithosphere are 1) crust; 2) Moho, separating crust from mantle; and 3) uppermost mantle.

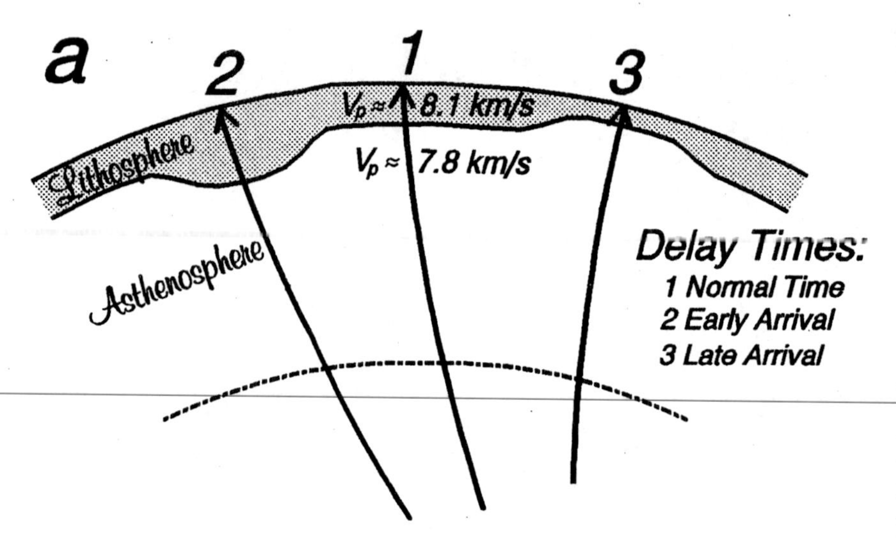

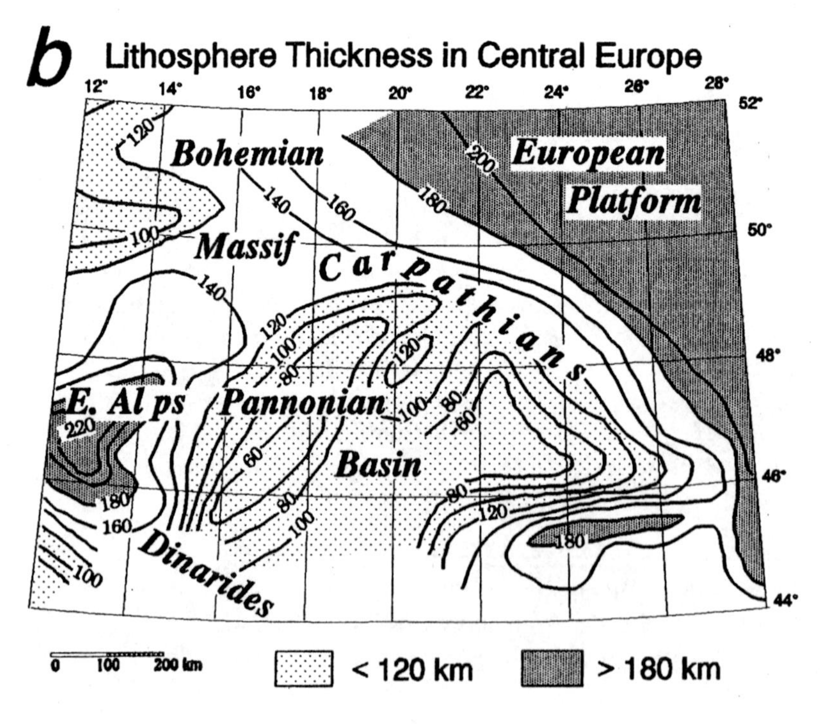

FIGURE 7.34 *Seismic delay-time method for mapping lithospheric thickness.* a) Relative to regions of normal lithosphere thickness (1), waves that arrive early (negative delay-time) encountered thick (high-velocity) lithosphere (2). Late arriving waves (positive delay-time) signify a thin lithosphere (3) that is underlain by an excess of low-velocity asthenosphere. b) Lithospheric thickness map of central Europe (Lillie et al., 1994 after Babuška et al., 1988). Typical lithospheric thickness in the region is about 140 km. Early arrivals in the Eastern Alps suggest a thick lithospheric root for the young collisional mountain range. Late arrivals in the Pannonian Basin indicate thin lithosphere (shallow asthenosphere) in a young continental rift zone.

P-wave velocities increase abruptly, from 6–7 km/s to about 8 km/s. The Moho is about 10 km deep beneath oceans, but much deeper (20 to 70 km) under continents. The *uppermost mantle* (or mantle lithosphere), with P-wave velocities just over 8 km/s, extends from the base of the crust down to 100 to 200 km depth.

Asthenosphere The asthenosphere is a soft, "plastic" solid. It can be broken into two parts. In the upper part (about 100 to 300 km depth), P and S wave velocities are about 6% lower than for the overlying lithosphere and underlying material. This *low-velocity zone* may be partially melted (1% to 10%; Fig. 2.7). The upper asthenosphere is thus a soft substratum (relatively low μ), over which the more rigid lithospheric plates ride. In the lower part of the asthenosphere (about 300 to 700 km depth), P and S wave velocities gradually increase, suggesting an *increase in shear strength* (Figs. 2.7, 2.8).

The thickness of lithospheric plates can be mapped by analyzing arrival times of seismic waves from the other side of the Earth (Fig. 7.34). If waves arrive late (positive delay time), they encountered an excess of low velocity material (asthenosphere?), suggesting that the lithosphere is thin. Early arriving waves (negative delay time) suggest a thick lithosphere.

Mesosphere Compared to the asthenosphere, the lower mantle (or mesosphere) is a more rigid solid. At the top (about 700 km depth) there is an abrupt increase in velocity; pressure is so great (Fig. 2.7) that the mantle reverts back to a harder solid (higher μ). Through the depth range of 700 to 2900 km, there is a gradual increase in P- and S-wave velocities (Fig. 7.29), as the shear and bulk moduli increase.

Outer Core The outer core extends from about 2900 to 5100 km depth. A region where no initial S-waves are recorded extends beyond 103° angular distance (Figs. 7.28, 7.35a); this "shadow zone" is evidence that S-waves are not transmitted through the outer core, implying it is fluid ($\mu = 0$). At the mantle/core boundary P-wave velocity drops from 13.5 to 8 km/s (Fig. 7.29). A P-wave shadow zone develops between 103° and 143° due to abrupt, inward bending of seismic rays (Fig. 7.35b).

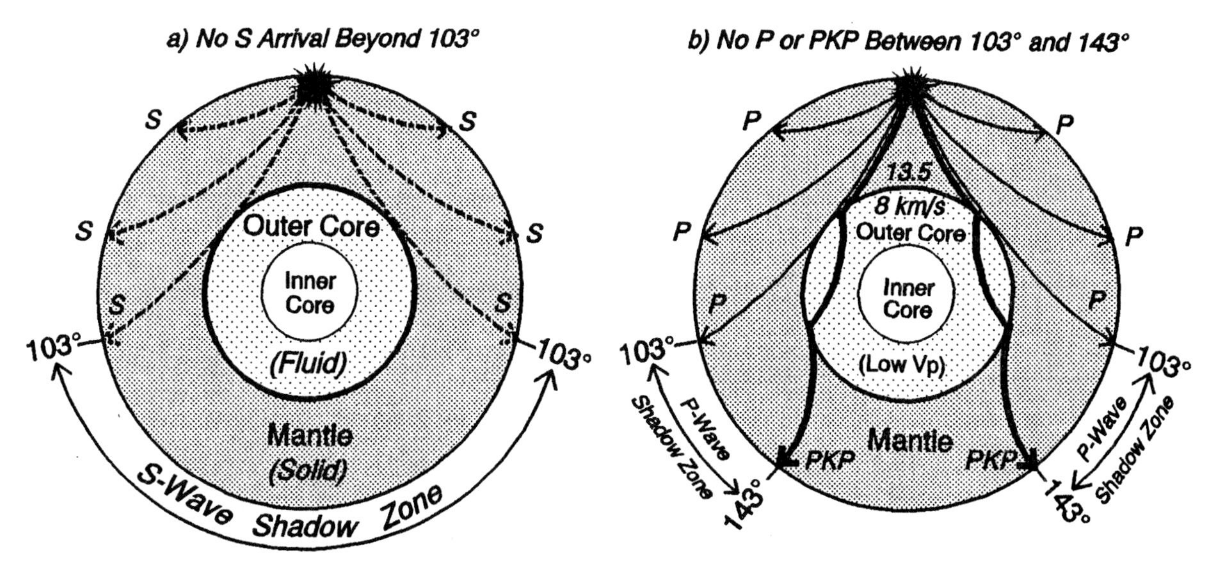

FIGURE 7.35 *Shadow zones provide evidence for a liquid outer core.* a) Initial S-waves extend continuously from the earthquake to 103° angular distance, where they no longer appear (Fig. 7.28). b) P-waves bend abruptly inward due to the velocity drop across the mantle/core interface, resulting in a shadow zone between 103° and 143°.

Inner Core Some weak P-waves arrive on the other side of the Earth earlier than expected ("PKIKP" phase in Figs. 7.27 and 7.28), suggesting an abrupt increase in velocity at about 5100 km depth (Fig. 7.36). This higher velocity suggests that the inner core is solid ($\mu \neq 0$). Some weak arrivals have been interpreted as waves that travel through the inner core as (converted) S-wave energy ("PKJKP" phase in Fig. 7.27). The inner core has the same chemical composition as the outer core, but is solid because it is under greater pressure. The phase change from the liquid outer core to the solid inner core is thus analogous to the change of phase from the relatively soft asthenosphere to the stiffer mesophere (Figs. 2.7, 2.8).

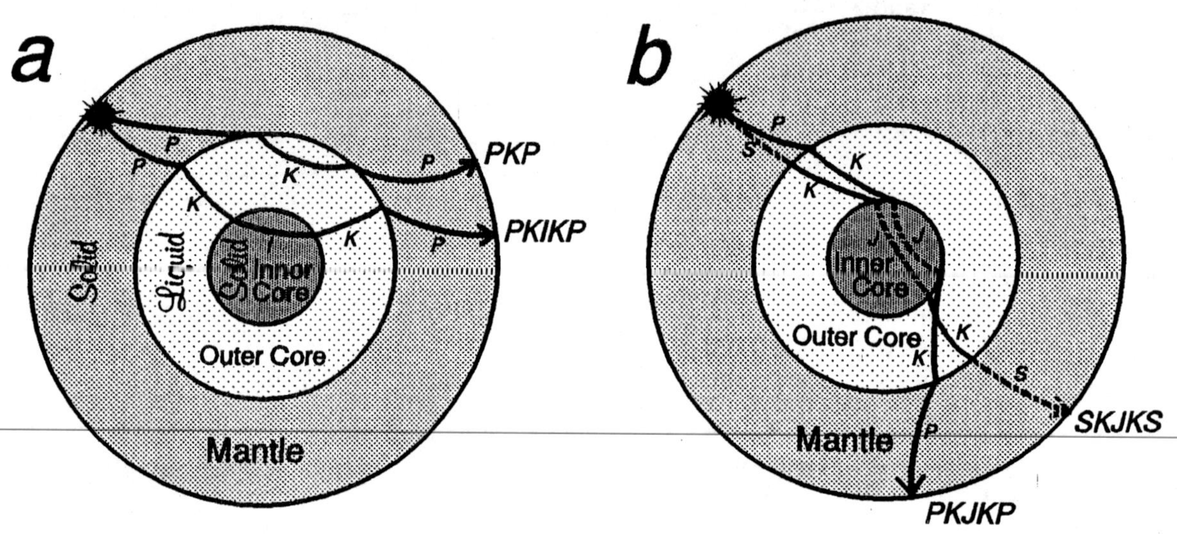

FIGURE 7.36 *Evidence for a solid inner core.* a) The PKIKP phase arrives sooner than expected (compared to the PKP phase, Fig. 7.28), because the "I" portion is through solid material. b) The PKJKP phase travels as compressional wave energy through the mantle and outer core ("P" and "K"), is partially converted to shear wave energy (Fig. 3.19) through the solid inner core ("J"), then returns to the surface as compressional wave energy ("K" and "P" again). The SKJKS phase travels as S-wave energy in both the mantle and inner core.

EXERCISES

7-1 An earthquake of magnitude 4 occurs in Turkey and is recorded in London. A month later, a magnitude 8 earthquake, with the same type of fault motion, occurs in exactly the same place.

a) How much larger will the waves recorded in London be for the second event compared to the first?

b) A magnitude 8 earthquake would be equivalent to how many magnitude 4 earthquakes occurring simultaneously?

7-2 An earthquake occurs along the San Andreas fault, directly south of a seismic station in Bend, Oregon (map, next page). The origin time is 6:00 GMT (Greenwich Mean Time), and the distance from Bend to the epicenter is 1000 km. On the map, illustrate: **a)** the sense of offset along the fault; **b)** the corresponding quadrants of compression and dilatation; **c)** the focal mechanism solution for the fault. Draw Z, N-S and E-W seismograms portraying (only) the following arrivals: **d)** first P-wave; **e)** first S-wave; **f)** Love wave; **g)** Rayleigh wave. Be sure your seismograms include the origin time, estimates of arrival time for each of the four phases, correct first motions for the P arrival on all three seismograms, and a portrayal of dispersion of the surface waves.

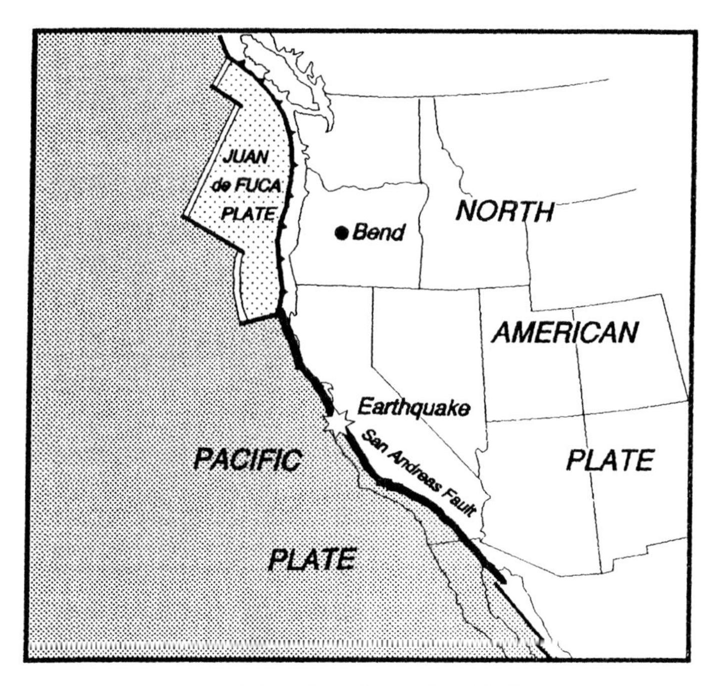

Map for Exercise 7-2

7-3 An earthquake occurs on one side of the Earth and is recorded by a seismograph on the other side (see diagram). The seismograph can detect compressional and shear wave arrivals. Compute travel times and sketch what seismograms would look like if: **a)** the Earth were entirely liquid; **b)** the Earth were the same material as in (a), but in a solid state; **c)** the Earth were solid, but with a liquid core starting halfway to the center. For parts (b) and (c), be sure your arrival times compare correctly, relative to arrival times for part (a).

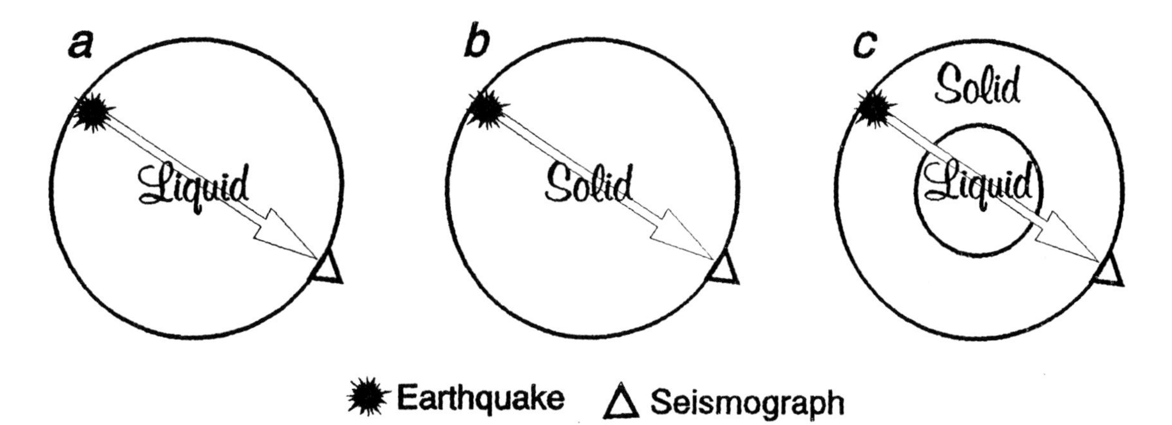

Diagram for Exercise 7-3

7-4 An earthquake is recorded in New York City, *7,500 km* from the epicenter. Seismic waves are traveling *due east* when they arrive at a station. The *P-phase* arrives as a *compression.*

a) Sketch the general appearance of waves recorded on the *Z, N-S, and E-W seismograms*. Include only *P, PcP, PP, S, Love and Rayleigh* phases. Use a time scale starting at the time of the earthquake (origin time), and be precise about the arrival time for each phase. Assume a shallow focal depth. Be sure to show the correct *first motions for the initial P-wave arrival*, and *dispersions for the surface waves*.

b) What was the *latitude and longitude* of the earthquake? (*Don't* use any mathematical equations to solve this one. Simply *use a globe and string*).

7-5 The graphs in Fig. 7.29 show body wave velocities and densities from the surface to the center of the Earth.

a) Using the equations relating seismic velocities, densities, and elastic constants, compute the bulk and shear moduli from the surface to the center of the Earth. Draw graphs of those two properties using the same vertical scale as in Fig. 7.29.

b) From the V_P and V_S equations and the graphs developed in (a), explain the following observations/interpretations. Be specific about chemical and/or phase changes across boundaries, and how they relate to changes in physical properties (ρ, k, and μ). **i)** The sharp increase in velocities across the crust/mantle boundary (Moho). **ii)** The lower seismic velocities in the upper part of the asthenosphere compared to the mantle part of the lithosphere. **iii)** The dramatic change in both P- and S-wave velocities across the mantle/core boundary. **iv)** The interpretation that the inner core is solid.

7-6 Sketch *focal mechanism solutions* for the following faults. Next to each of the sketches, draw *geologic cross sections* illustrating the *earthquake fault* and the *complimentary fault* that would produce the same focal mechanism solution.

a) *Reverse fault*. Dip 75° E, strike N30° W.

b) *Normal fault*. Dip 50° W, strike N-S.

c) *Overthrust*. Dip 15° W, strike N30° W.

d) *Right-lateral strike-slip fault*. Dip 90°, strike N20° W.

e) *Left-lateral strike-slip fault*. Dip 90°, strike N20° E.

7-7 **a)** Using Figs. 4.17 and 7.34, draw a cross section along 47° N latitude, illustrating the crust/mantle and lithosphere/asthenosphere boundaries.

b) Draw in expected P-wave and S-wave velocities along the cross section.

c) Based on tectonic evolution, explain the changes in crustal and lithospheric thickness on your cross section.

7-8 Note on the travel-time curves (Fig. 7.28) that: 1) the P phase arrives until 103° distance from the epicenter; 2) the PKP phase first appears at 143°; and 3) the PKIKP phase first appears at 110°. Draw an Earth model and sketch raypaths illustrating those three arrivals. Include seismic velocities and accurate portrayals of raypath bending within layers and across discontinuities.

7-9 Draw cross sections along the lines of section A-A′, B-B′, and C-C′ shown on the map.

a) Include a portrayal of how the depths to the following boundaries change along the length of the cross section: **i)** the topography and bathymetry; **ii)** the crust/mantle boundary; **iii)** the lithosphere/asthenosphere boundary.

b) Put a series of X's on the cross section to represent zones where earthquakes would be expected to occur.

c) In critical places, draw expected focal mechanism solutions for earthquakes.

d) Put Δ's on the cross section to represent where volcanoes would be expected to occur.

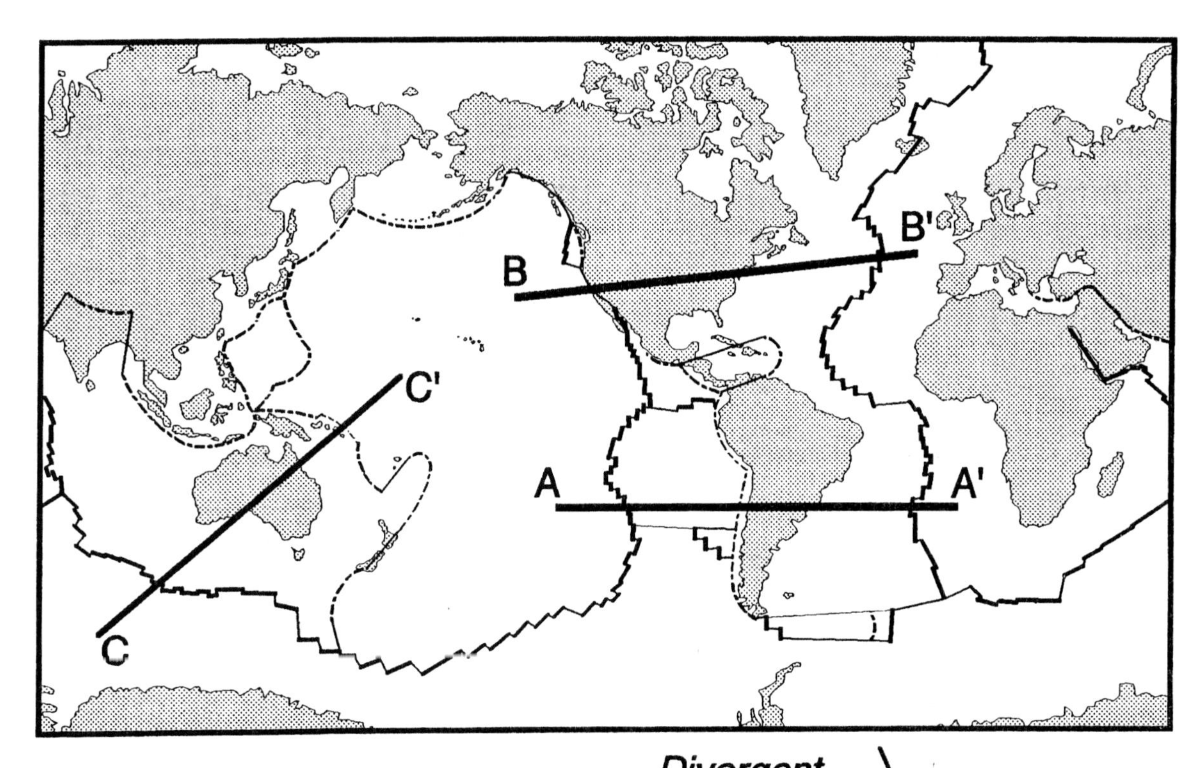

Plate Boundaries: *Divergent* *Convergent* *Transform*

Map for Exercise 7-9

SELECTED BIBLIOGRAPHY

Babuška, V., J. Plomerová, and P. Pajdušák, 1988, Lithosphere-asthenosphere in central Europe: Models derived from P residuals, In: *Proceedings of the 4th EGT Workshop: The Upper Mantle*, Commission of the European Communities, European Science Foundation, pp. 37–48.

Babuška, V., J. Plomerová, and M. Granet, 1990, The deep lithosphere in the Alps: A model inferred from P residuals, *Tectonophysics*, v. 176: pp. 137–165.

Barazangi, M., and J. Dorman, 1969, World seismicity map compiled from ESSA Coast and Geodetic Survey epicenter data, 1961–1967, *Seismological Society of America Bulletin*, v. 59, pp. 369–380.

Bolt, B. A., 1988, *Earthquakes*, New York: W. H. Freeman and Company, 282 pp.

Bryan, C. J., and C. E. Johnson, 1991, Block tectonics of the island of Hawaii from a focal mechanism analysis of basal slip, *Bulletin Seismological Society of America*, v. 81, pp. 491–507.

Bullen, K. E., 1965, *An Introduction to the Theory of Seismology* (3rd ed.), Cambridge: Cambridge University Press, 381 pp.

De Bremaecker, J., 1985, *Geophysics: The Earth's Interior*, New York: John Wiley and Sons, 342 pp.

Dziewonski, A. M., and D. L. Anderson, 1981, Preliminary reference Earth model, *Physics of the Earth and Planetary Interiors*, v. 25, pp. 297–356.

Francis, T. J. G., 1968, Seismicity of mid-oceanic ridges and its relation to properties of the upper mantle and crust, Nature, v. 220, pp. 899–901.

Garland, G. D., 1979, *Introduction to Geophysics* (2nd ed.), Toronto: W. B. Saunders Company, 494 pp.

Griffiths, D. H., and R. F. King, 1981, *Applied Geophysics for Geologists and Engineers* (2nd ed.), New York: Pergamon Press, 230 pp.

Gutenberg, B., 1945, Amplitudes of P, PP and S and magnitudes of shallow earthquakes, *Bulletin Seismological Society of America*, v. 35, pp. 57–69.

Gutenberg. B., and C. F. Richter, 1954, *Seismicity of the Earth and Associated Phenomena*, Princeton, New Jersey: Princeton University Press.

Hamblin, W. K., and E. H. Christiansen, 1995, *Earth's Dynamic Systems* (7th ed.), Englewood Cliffs, N. J.: Prentice Hall, Inc., 710 pp.

Herrin, E., and J. Taggart, 1962, Regional variations in Pn velocity and their effect on the location of epicenters, *Bulletin Seismological Society of America,* v. 52, p. 1037.

Humphreys, E., R. W. Clayton, and B. H. Hager, 1984, A tomographic image of mantle structure beneath southern California, *Geophysical Research Letters*, v. 11, pp. 625–627.

Isaacs, B., J. Oliver, and L. R. Sykes, 1968, Seismology and the new global tectonics, *Journal of Geophysical Research*, v. 73, pp. 5855–5899.

Jeffreys, H., 1976, *The Earth* (6th ed.), Cambridge: Cambridge University Press.

Kinsler, L. E., A. R. Frey, A. B. Coppens, and J. V. Sanders, 1982, *Fundamentals of Acoustics* (3rd. ed.), New York: John Wiley and Sons, New York, 480 pp.

Lay, T., and T. C. Wallace, 1995, *Modern Global Seismology*, International Geophysics Series Number 58, New York: Academic Press, 521 pp.

Lillie, R. J., M. Bielik, V. Babuška, and J. Plomerová, 1994, Gravity modeling of the lithosphere in the Eastern Alpine-Western Carpathian-Pannonian Basin region, *Tectonophysics*, v. 23 pp. 215–235.

Molnar, P., and L. R. Sykes, 1969, Tectonics of the Middle America regions from focal seismicity, *Geological Society of America Bulletin*, v. 80, pp. 1639–1684.

Pajdušák, P., J. Plomerová, and V. Babuška, 1989, A model of the lithosphere thickness in the region of the Carpathians, *Studies of Geophysics and Geodesy*, v. 33, pp. 11–21.

Plomerová, J., V. Babuška, M. Dobrath, and R. Lillie, 1993, Structure of the lithosphere beneath the Cameroon Rift from seismic delay time analysis, *Geophysical Journal*, v. 115, pp. 381–390.

Skinner, B. J., and S. C. Porter, 1995, *The Dynamic Earth: An Introduction to Physical Geology* (3rd. ed.), New York: John Wiley and Sons, Inc., 567 pp.

Smith, R. B., R. T. Shuey, Jr, R. Pelton, and J. P. Bailey, 1977, Yellowstone Hotspot: Contemporary tectonics and crustal properties from earthquake and aeromagnetic data, *Journal of Geophysical Research*, v. 82, pp. 3665–3676.

Sykes, L. R., 1967, Mechanism of earthquakes and nature of faulting on the mid-ocean ridges, *Journal of Geophysical Research*, v. 72, pp. 2131–2153.

Telford, W. M., L. P. Geldart, R. E. Sheriff, and D. A. Keys, 1976, *Applied Geophysics,* Cambridge: Cambridge University Press, 860 pp.

Yeats, R. S., K. Sieh, and C. R. Allen, 1997, *The Geology of Earthquakes*, New York: Oxford University Press, 568 pp.

CHAPTER 8

Gravity and Isostasy

gravity *(grav' ə tē)* ***n.,*** [< L. gravis, *heavy*], *1. the force of attraction between masses 2. the force that tends to draw bodies in Earth's sphere toward Earth's center.*
isostasy *(ī sä stə' sē)* ***n.,*** [< Gr. isos, *equal;* < Gr. stasis, *standing*], *a state of balance whereby columns of material exert equal pressure at and below a compensating depth.*
gravity and isostasy *(grav' ə tē ənd ī sä stə' sē)* ***n.,*** *the study of spatial variations in Earth's gravitational field and their relationship to the distribution of mass within the Earth.*

Earth's *gravity* and *magnetic* forces are potential fields that provide information on the nature of materials within the Earth. Potential fields are those in which the strength and direction of the field depend on the position of observation within the field; the strength of a potential field decreases with distance from the source. Compared to the magnetic field, Earth's gravity field is simple. Lines of force for the gravity field are directed toward the center of the Earth, while magnetic field strength and direction depend on Earth's positive and negative poles (Fig. 8.1).

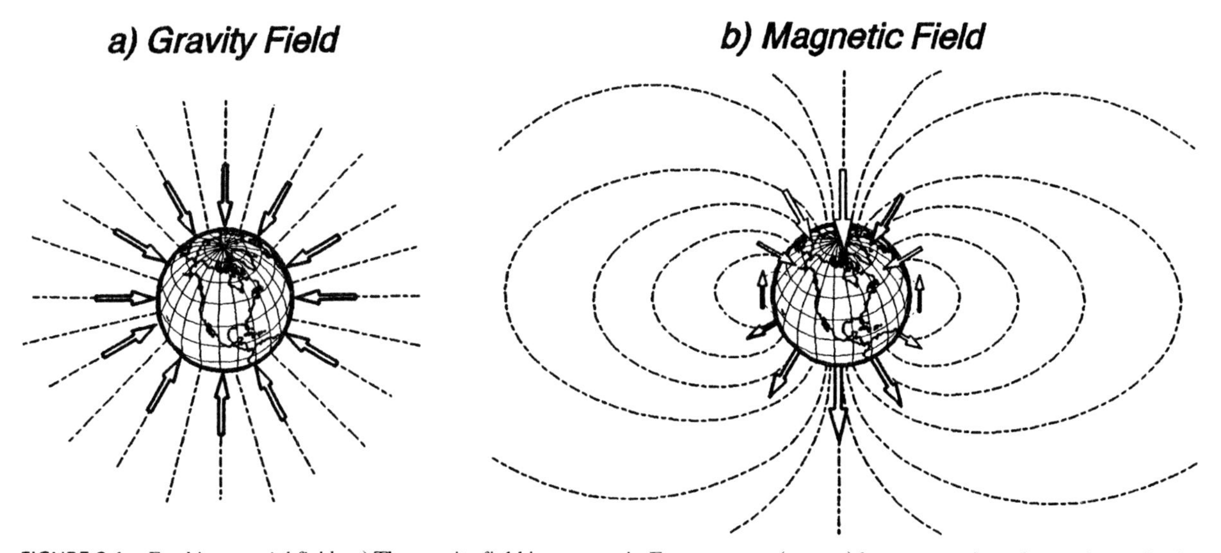

FIGURE 8.1 *Earth's potential fields.* a) The gravity field is symmetric. Force vectors (arrows) have approximately equal magnitude and point toward the center of the Earth. b) The magnitude and direction of the magnetic field is governed by positive (south) and negative (north) poles. Magnitude varies by a factor of two from equator to pole.

EARTH'S GRAVITY FIELD

Gravity is the attraction on one body due to the mass of another body. The force of one body acting on another is given by *Newton's Law of Gravitation* (Fig. 8.2a):

$$\mathbf{F = G\frac{m_1m_2}{r^2}}$$

where:

F = force of attraction between the two objects (N)
G = Universal Gravitational Constant (6.67×10^{-11} Nm^2/kg^2)
m_1, m_2 = mass of the two objects (kg)
r = distance between the centers of mass of the objects (m).

The force (F) exerted on the object with mass m_1 by the body with mass m_2, is given by *Newton's Second Law of Motion* (Fig. 8.2b):

$$F = m_1a$$

where:

a = acceleration of object of mass m_1 due to the gravitational attraction of the object with mass m_2 (m/s^2).

Solving for the acceleration, then combining the two equations (Fig. 8.2c):

$$a = \frac{F}{m_1} = \frac{1}{m_1}\frac{Gm_1m_2}{r^2}$$

$$a = \frac{Gm_2}{r^2}$$

For Earth's gravity field (Fig. 8.3a), let:

a = g = gravitational acceleration observed on or above Earth's surface;
m_2 = M = mass of the Earth;
r = R = distance from the observation point to Earth's center of mass;

so that:

$$\mathbf{g = \frac{GM}{R^2}}$$

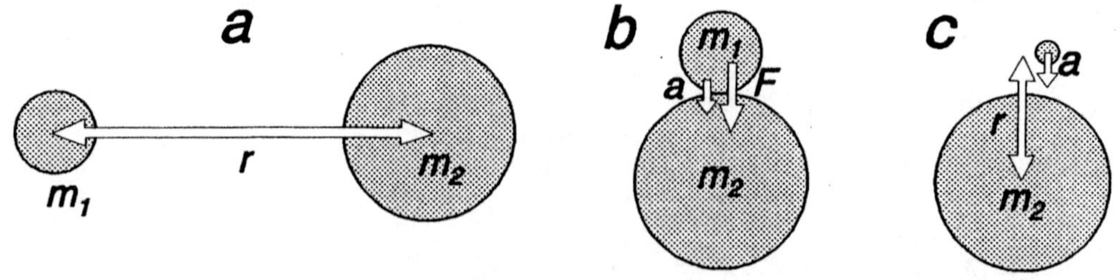

FIGURE 8.2 a) The gravitational force between two objects is directly proportional to their masses (m_1, m_2), and inversely proportional to the square of their distance (r). b) The mass (m_1), times the acceleration (a) due to mass (m_2), determines the gravitational force (F). c) The acceleration due to gravity (a) of a body depends only on the mass of the attracting body (m_2) and the distance to the center of that mass (r).

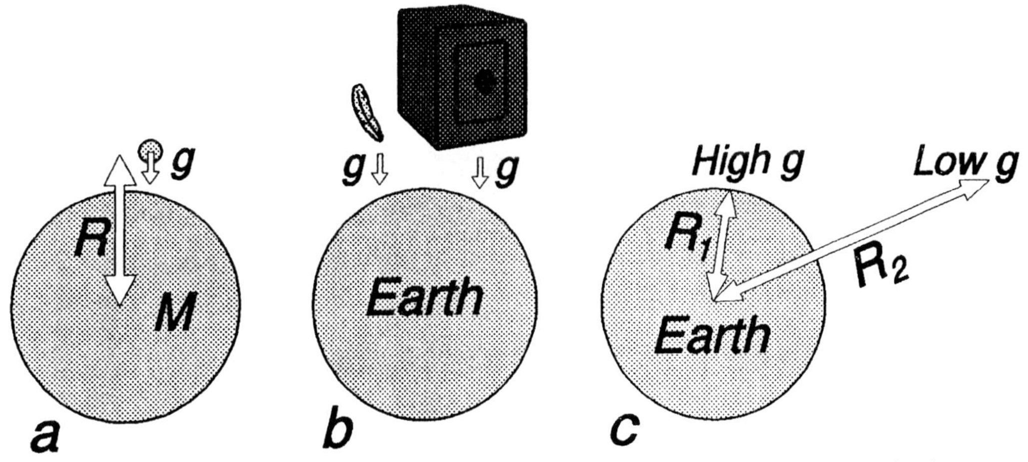

FIGURE 8.3 a) The mass (M) of the Earth and radius (R) to Earth's center determine the gravitational acceleration (g) of objects at and above Earth's surface. b) The acceleration is the same (g), regardless of the mass of the object. c) Objects at Earth's surface (radius R_1) have greater acceleration than objects some distance above the surface (radius R_2).

The above equation illustrates two fundamental properties of gravity. 1) Acceleration due to gravity (g) does *not* depend on the mass (m_1) attracted to the Earth (Fig. 8.3b); in the absence of air resistance, a small mass (feather) will accelerate toward Earth's surface at the same rate as a large mass (safe). 2) The farther from Earth's center of mass (that is, the greater the R), the smaller the gravitational acceleration (Fig. 8.3c); as a potential field, gravity thus obeys an *inverse square law*.

The value of the gravitational acceleration on Earth's surface varies from about *9.78 m/s² at the equator* to about *9.83 m/s² at the poles* (Fig. 8.4a). The smaller acceleration at the equator, compared to the poles, is because of the combination of three factors. 1) There is *less* inward acceleration because of *outward acceleration* caused by the spin of the Earth; the spin (rotation) is greatest at the equator but reduces to zero at the poles. 2) There is *less* acceleration at the equator because of the Earth's outward bulging, thereby increasing the radius (R) to the center of mass. 3) The added mass of the bulge creates *more* acceleration. Notice that the first two factors lessen the acceleration at the equator, while the third increases it. The net effect is the observed $-0.05\ m/s^2$ difference.

Gravitational acceleration (*gravity*) is commonly expressed in units of *milligals* (*mGal*), where:

$$1\ Gal = 1\ cm/s^2 = 0.01\ m/s^2$$

so that:

$$1\ mGal = 10^{-3}\ Gal = 10^{-3}\ cm/s^2 = 10^{-5}\ m/s^2.$$

Gravity, therefore, varies by about 5000 *mGal* from equator to pole (Fig. 8.4b).

GRAVITY ANOMALIES

Gravity observations can be used to interpret changes in mass below different regions of the Earth. To see the mass differences, the broad changes in gravity from equator to pole must be subtracted from station observations. This is accomplished

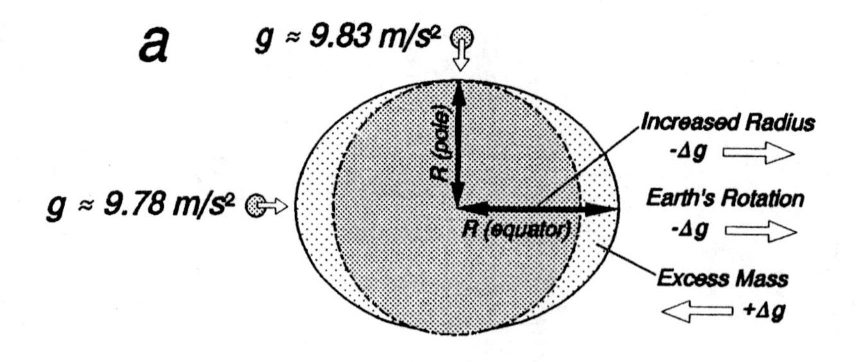

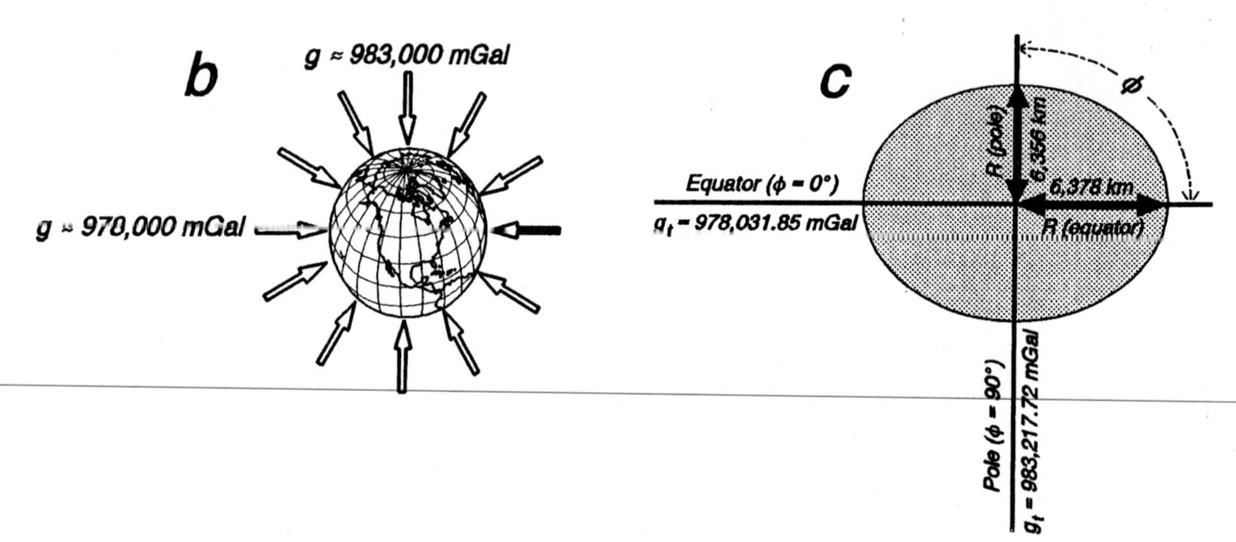

FIGURE 8.4 a) Three main factors responsible for the difference in gravitational acceleration at the equator compared to the poles. b) Gravity increases from about 978,000 *mGal* at the equator, to about 983,000 *mGal* at the poles. c) Variation in gravity from equator to pole, according to 1967 Reference Gravity Formula.

by predicting the gravity value for a station's latitude (*theoretical gravity*), then subtracting that value from the actual value at the station (*observed gravity*), yielding a *gravity anomaly*.

Theoretical Gravity

The average value of gravity for a given latitude is approximated by the *1967 Reference Gravity Formula,* adopted by the International Association of Geodesy:

$$g_t = g_e\,(1 + 0.005278895\sin^2\phi + 0.000023462\sin^4\phi)$$

where:

g_t = theoretical gravity for the latitude of the observation point (*mGal*)
g_e = theoretical gravity at the equator (978,031.85 *mGal*)
ϕ = latitude of the observation point (degrees).

The equation takes into account the fact that the Earth is an imperfect sphere, bulging out at the equator and rotating about an axis through the poles (Fig. 8.4a).

For such an *oblate spheroid* (Fig. 8.4c), it estimates that gravitational acceleration at the *equator* ($\phi = 0°$) would be *978,031.85 mGal*, gradually increasing with latitude to *983,217.72 mGal* at the *poles* ($\phi = 90°$).

Free Air Gravity Anomaly

Gravity observed at a specific location on Earth's surface can be viewed as a function of three main components (Fig. 8.5): 1) the *latitude* (ϕ) of the observation point, accounted for by the theoretical gravity formula; 2) the *elevation* (ΔR) of the station, which changes the radius (R) from the observation point to the center of the Earth; and 3) the *mass distribution* (M) in the subsurface, relative to the observation point.

The *free air correction* accounts for the second effect, the local change in gravity due to elevation. That deviation can be approximated by considering how gravity changes as a function of increasing distance of the observation point from the center of mass of the Earth (Fig. 8.6a). Consider the equation for the gravitational acceleration (g) as a function of radius (R):

$$g = \frac{GM}{R^2}$$

The first derivative of g, with respect to R, gives the change in gravity (Δg) with increasing distance from the center of the Earth (that is, increasing elevation, ΔR):

$$\lim_{\Delta R \to 0} \frac{\Delta g}{\Delta R} = \frac{dg}{dR} = -2\left(\frac{GM}{R^3}\right) = \frac{-2}{R}\left(\frac{GM}{R^2}\right) = \frac{-2}{R}(g)$$

$$\boxed{\mathbf{\frac{dg}{dR} = \frac{-2g}{R}}}$$

Assuming average values of $g \approx 980{,}625$ *mGal* and $R \approx 6{,}367$ km $= 6{,}367{,}000$ m (Fig. 8.4c):

$$\boxed{\mathbf{dg/dR \approx -0.308}\ \boldsymbol{mGal/m}}$$

where:

dg/dR = average value for the change in gravity with increasing elevation.

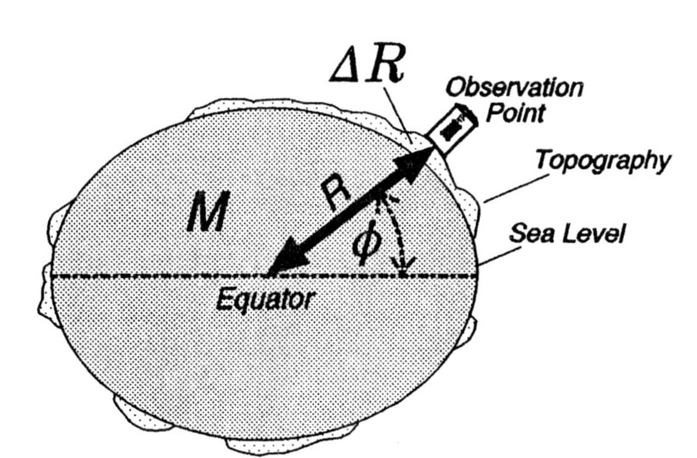

FIGURE 8.5 Three factors determining gravity at an observation point: a) latitude (ϕ); b) distance from sea-level datum to observation point (ΔR); c) Earth's mass distribution (M), relative to the station location (M includes material above as well as below sea level). ϕ is accounted for by subtracting the theoretical gravity from the observed gravity, and ΔR by the free air correction. The remaining value (free air anomaly) is thus a function of M.

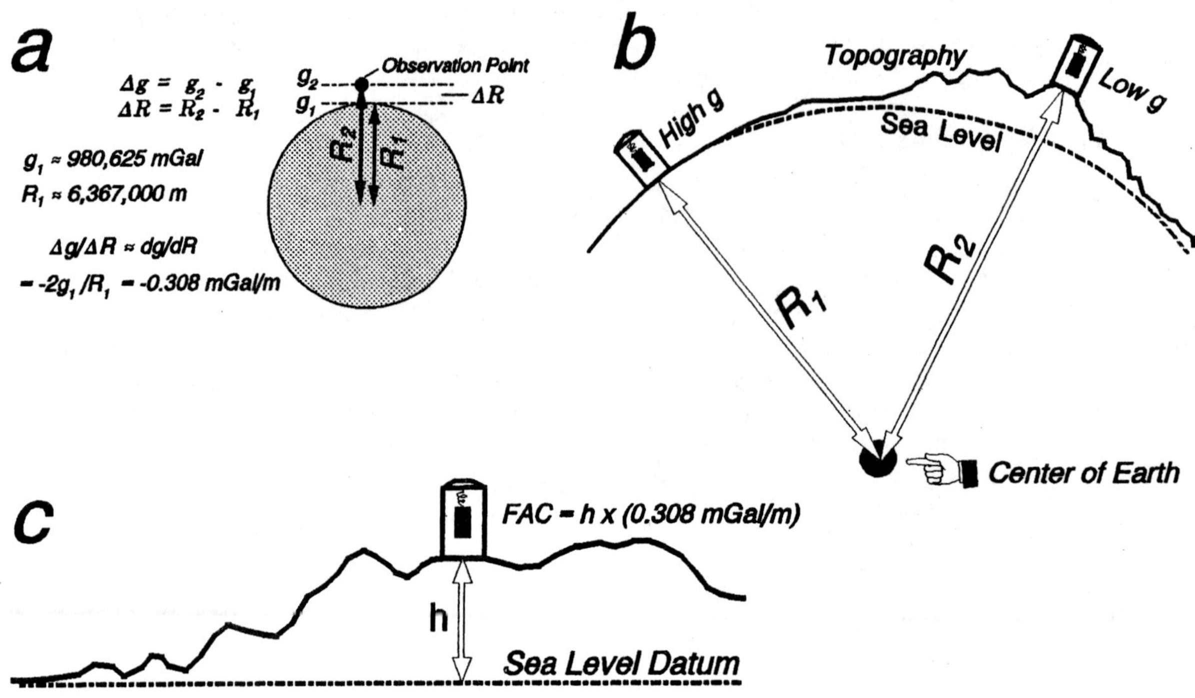

FIGURE 8.6 *Free air correction.* a) Rising upward from Earth's surface, gravitational acceleration decreases by about 0.308 *mGal* for every meter of height. b) A gravity station at high elevation tends to have a lower gravitational acceleration (g) than a station at lower elevation. c) The free air correction (FAC) accounts for the extended radius to an observation point, elevated h meters above a sea level datum.

The above equation illustrates that, for every 3 m (about 10 feet) upward from the surface of the Earth, the acceleration due to gravity decreases by about 1 *mGal*. Stations at elevations high above sea level therefore have lower gravity readings than those near sea level (Fig. 8.6b). To compare gravity observations for stations with different elevations, a *free air correction* must be added back to the observed values (Fig. 8.6c).

$$\text{FAC} = \text{h} \times (0.308\ mGal/\text{m})$$

where:

FAC = free air correction (*mGal*)
h = elevation of the station above a sea level datum (m).

The *free air gravity anomaly* is the observed gravity, corrected for the latitude and elevation of the station:

$$\boxed{\Delta \mathbf{g}_{\mathbf{fa}} = \mathbf{g} - \mathbf{g}_{\mathbf{t}} + \mathbf{FAC}}$$

where:

Δg_{fa} = free air gravity anomaly
g = gravitational acceleration observed at the station.

Notice in the above equation that: 1) subtracting the theoretical gravity (g_t) from the observed gravity (g) corrects for the *latitude*, thus accounting for the spin and bulge of the Earth; and 2) adding the free air correction (FAC) puts back the gravity lost to *elevation*, thereby correcting for the increased radius (R) to Earth's center.

The free air gravity anomaly is a function of *lateral mass variations* (M in Fig. 8.5), because the latitude and elevation effects (ϕ and ΔR in Fig. 8.5) have been corrected. Fig. 8.7 shows what a profile of changing free air anomalies might look like across bodies of excess and deficient mass. Notice that the anomaly shows relatively high readings near the mass excess, low readings near the mass deficiency; there are also abrupt changes that mimic sharp topographic features.

Bouguer Gravity Anomaly

Even after elevation corrections, gravity can vary from station to station because of differences in mass between the observation points and the sea-level datum. Relative to areas near sea level, mountainous areas would have extra mass, tending to increase the gravity (Fig. 8.8a).

The *Bouguer correction* accounts for the gravitational attraction of the mass above the sea-level datum. This is done by approximating the mass as an *infinite slab*, with thickness (h) equal to the elevation of the station (Fig. 8.8b). The attraction of such a slab is:

$$\mathbf{BC = 2\pi\rho Gh}$$

where:

BC = Bouguer correction
ρ = density of the slab
G = Universal Gravitational Constant
h = thickness of the slab (station elevation).

Substituting the values of G and 2π yields:

$$\mathbf{BC = 0.0419\rho h}$$

where BC is in *mGal* (10^{-5} m/s^2); ρ in g/cm^3 (10^3 kg/m^3); h in m.

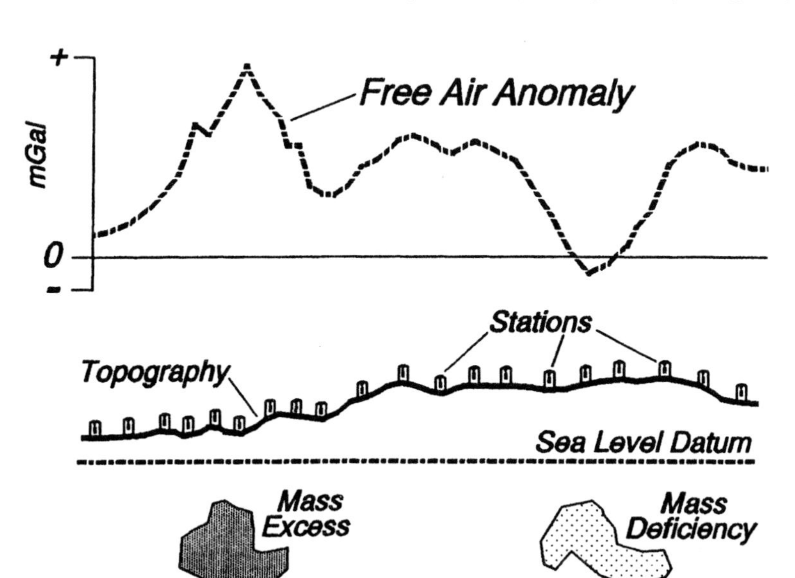

FIGURE 8.7 General form of free air gravity anomaly profile across areas of mass excess and mass deficiency.

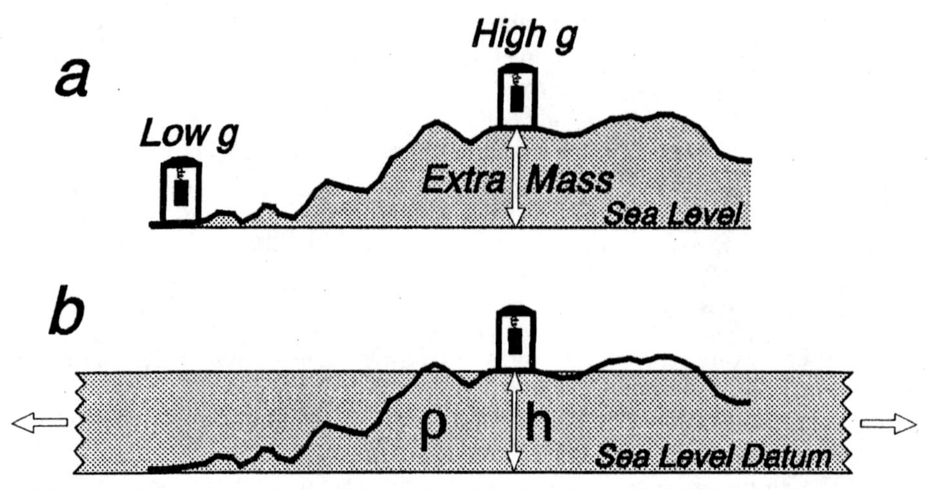

FIGURE 8.8 *Bouguer correction.* a) The extra mass of mountains results in higher gravity relative to areas near sea level. b) To account for the excess mass above a sea level datum, the Bouguer correction assumes an infinite slab of density (ρ), with thickness (h) equal to the station's elevation.

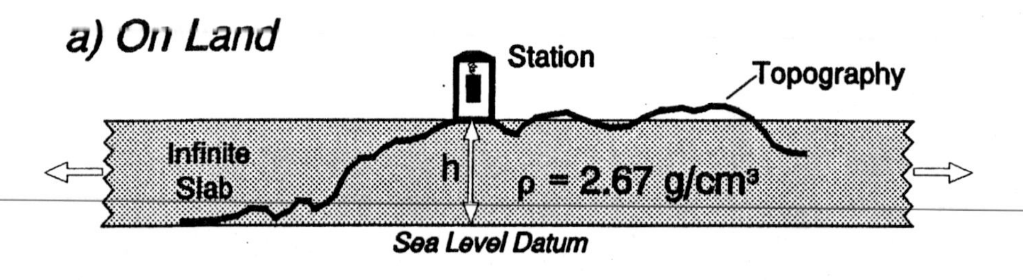

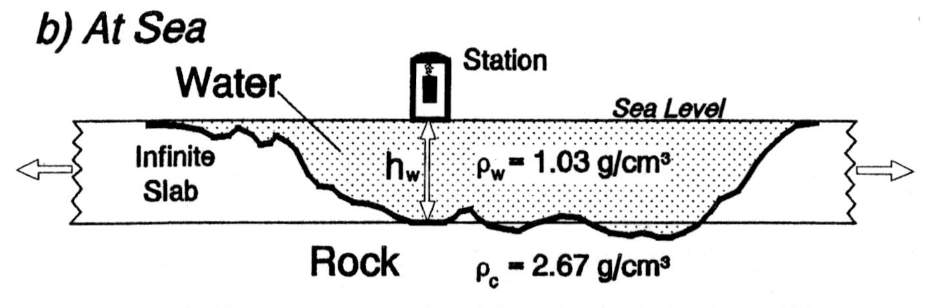

FIGURE 8.9 *Standard Bouguer correction values.* a) On land, the reduction density (ρ) is commonly taken as $+2.67\ g/cm^3$. The thickness of the infinite slab is equal to the station elevation (h). b) At sea, the reduction density ($-1.64\ g/cm^3$) is the difference between that of sea water (1.03 g/cm^3) and underlying rock (2.67 g/cm^3). The thickness of the slab is equal to the water depth (h_w).

Bouguer Gravity Anomaly on Land For regions above sea level (Fig. 8.9a), the *simple Bouguer gravity anomaly* (Δg_B) results from subtracting the effect of the infinite slab (BC) from the free air gravity anomaly:

$$\boxed{\Delta g_B = \Delta g_{fa} - \mathbf{BC}}$$

To determine the Bouguer correction, the density of the infinite slab (ρ) must be assumed (the *reduction density*). The reduction density is commonly taken as 2.67 g/cm^3, a typical density of granite (Figs. 3.9, 3.10).

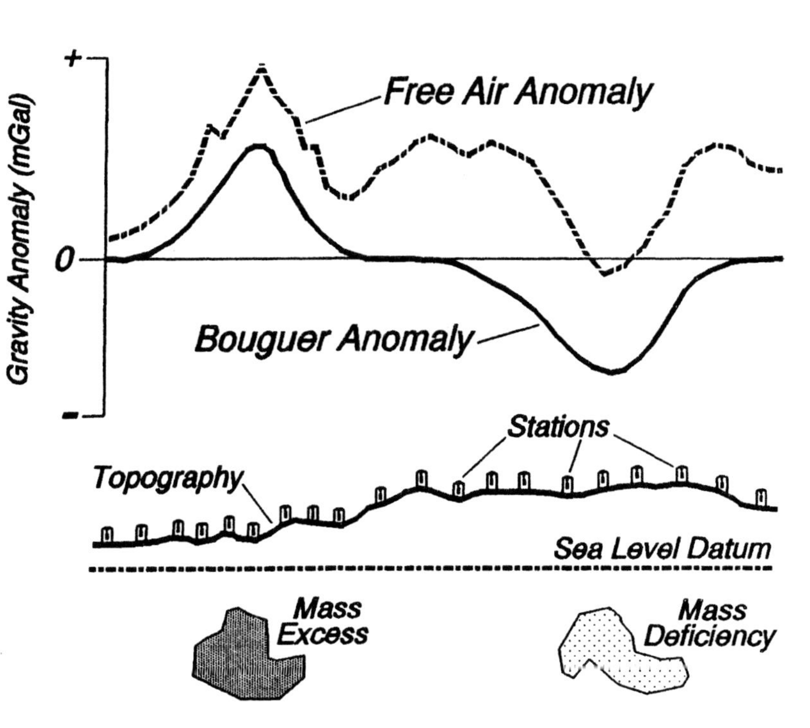

FIGURE 8.10 Bouguer correction applied to the free air gravity anomaly profile in Fig. 8.7.

The standard Bouguer correction for areas above sea level is thus:

$$BC = 0.0419\rho h = (0.0419)(2.67 g/cm^3)h$$
$$= (0.112\ mGal/m) \times h$$

where h is in m. The equation illustrates that, for about every 9 m of surface elevation, the increased mass below the observation point adds about 1 *mGal* to the observed gravity. Using the standard correction, the *simple Bouguer gravity anomaly on land* is computed from the free air gravity anomaly according to the formula:

$$\boldsymbol{\Delta g_B = \Delta g_{fa} - (0.112\ mGal/m)\ h} \qquad \text{(h in meters).}$$

Like the free air gravity anomaly, the Bouguer gravity anomaly reflects changes in mass distribution below the surface. The Bouguer anomaly, however, has had an additional correction, removing most of the effect of mass excess above a sea level datum (on land). Bouguer Corrections applied to the free air gravity profile (Fig. 8.7) would therefore yield a Bouguer gravity profile illustrated in Fig. 8.10. The two profiles illustrate three general properties of gravity anomalies. 1) For stations above sea level, the Bouguer anomaly is always less than the free air anomaly (the approximate attraction of the mass above sea level has been removed from the free air anomaly). 2) Short-wavelength changes in the free air anomaly, due to abrupt topographic changes, have been removed by the Bouguer correction; the Bouguer anomaly is therefore smoother than the free air anomaly. 3) Mass excesses result in positive changes in gravity anomalies; mass deficiencies cause negative changes.

Bouguer Gravity Anomaly at Sea In areas covered by the sea, gravity is generally measured on the surface of the water (Fig. 8.9b). In the strictest sense,

Bouguer anomalies at sea are exactly the same as free air anomalies, because station elevations (h) are zero:

$$\Delta g_B = \Delta g_{fa} - 0.0419\rho h; \quad h = 0, \text{so that: } \Delta g_B = \Delta g_{fa}$$

A type of Bouguer correction can be applied, however, because the density and depth of the water are well known. Instead of stripping the topographic mass away, as is done on land, the effect can be thought of as "pouring concrete" to fill the ocean. Thus, the Bouguer correction at sea can be envisioned as an infinite slab, equal to the depth of the water and with density equalling the difference between that of water and "concrete":

$$BC_s = 0.0419\rho h = 0.0419(\rho_w - \rho_c)h_w$$

where:

BC_s = Bouguer correction at sea
ρ_w = density of sea water
ρ_c = density of "concrete"
h_w = water depth below the observation point.

Assuming $\rho_w = 1.03 \text{ g/cm}^3$ and $\rho_c = 2.67 \text{ g/cm}^3$:

$$\mathbf{BC_s = 0.0419\,(-1.64\ g/cm^3)\ h_w = -0.0687\ (\mathit{mGal}/m) \times h_w}$$

where BC_s is in *mGal* and h_w in m.

Retaining the convention defined above, the Bouguer correction at sea is subtracted from the free air anomaly to yield the *Bouguer gravity anomaly at sea* (Δg_{Bs}):

$$\mathbf{\Delta g_{Bs} = \Delta g_{fa} - BC_s}$$

Notice that the water is a mass deficit when compared to adjacent landmasses of rock; the negative Bouguer correction at sea thus means that some value must be *added* to the free air anomaly to compute the Bouguer anomaly at sea:

$$\mathbf{\Delta g_{Bs} = \Delta g_{fa} + (0.0687\ \mathit{mGal}/m)\ h_w} \qquad (h_w \text{ in meters}).$$

Complete Bouguer Gravity Anomaly The infinite slab correction described above yields a simple Bouguer anomaly. That correction is normally sufficient to approximate mass above the datum in the vicinity of the station (Fig. 8.11a). In rugged areas, however, there may be significant effects due to nearby mountains pulling upward on the station, or valleys that do not contain mass that was subtracted (Fig. 8.11b). For such stations, additional *terrain corrections* (TC; see Telford et al., 1976) are applied to the simple Bouguer anomaly (Δg_B), yielding the *complete Bouguer gravity anomaly* (Δg_{Bc}):

$$\mathbf{\Delta g_{Bc} = \Delta g_B + TC}$$

Summary of Equations for Free Air and Bouguer Gravity Anomalies

Fig. 8.12 illustrates parameters used to determine free air and Bouguer gravity anomalies. The formulas below yield standard versions of the anomalies.

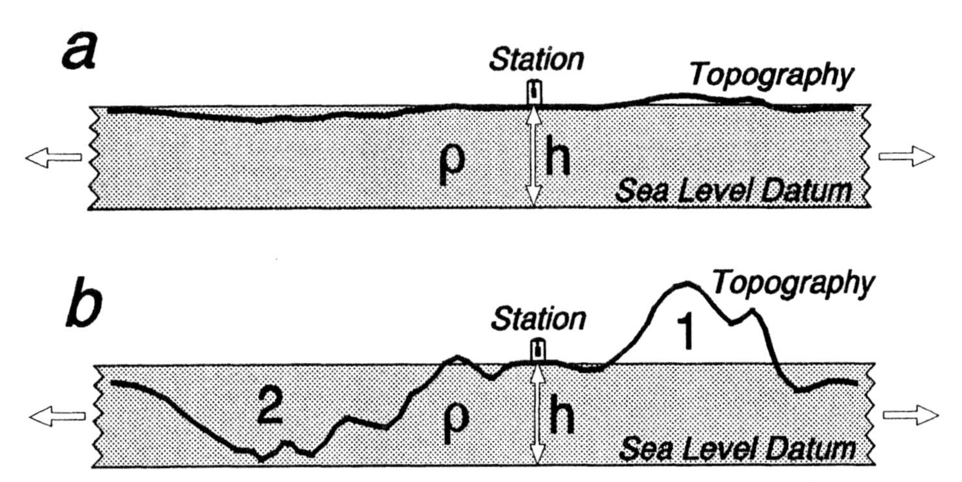

FIGURE 8.11 *Terrain correction.* a) In areas of low relief, the Bouguer slab approximation is adequate; terrain correction is unnecessary. b) High relief areas require terrain correction, to account for lessening of observed gravity due to mass of mountains above the slab (1), and overcorrection due to valleys (2). For both situations, the terrain correction is positive, making the complete Bouguer anomaly higher than the simple Bouguer anomaly.

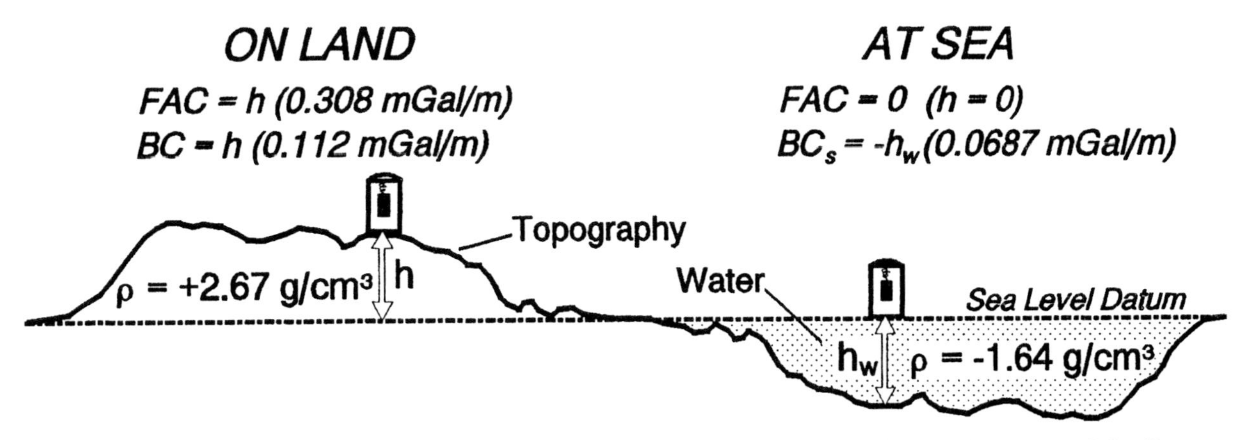

FIGURE 8.12 Standard parameters used to compute gravity anomalies on land and at sea. FAC = free air correction; BC = Bouguer correction; BC_S = Bouguer correction at sea; ρ = reduction density; h (elevation) and h_w (water depth) in meters.

Theoretical Gravity

$$\mathbf{g_t = g_e\,(1 + 0.005278895\mathit{sin}^2\phi + 0.000023462\mathit{sin}^4\phi)}$$

g_t = theoretical gravity for the latitude of the observation point (*mGal*)
g_e = theoretical gravity at the equator (978,031.85 *mGal*)
ϕ = latitude of the observation point (degrees).

Free Air Gravity Anomaly

$$\mathbf{\Delta g_{fa} = (g - g_t) + h(0.308\ \mathit{mGal}/m)}$$

Δg_{fa} = free air gravity anomaly (*mGal*)

g = observed gravity ($mGal$)
g_t = theoretical gravity ($mGal$)
h = elevation above sea level datum (m).

Bouguer Gravity Anomaly

$$\Delta g_B = \Delta g_{fa} - BC = \Delta g_{fa} - 0.0419\rho h$$

BC = Bouguer correction ($mGal$)
ρ = reduction density (g/cm^3)

a) On Land

$$\Delta g_B = \Delta g_{fa} - (0.112\ mGal/m)\ h \quad (\text{for } \rho = +2.67)$$

Δg_B = simple Bouguer gravity anomaly ($mGal$)
h = elevation above sea-level datum (m).

b) At Sea

$$\Delta g_{Bs} = \Delta g_{fa} + (0.0687\ mGal/m)\ h_w \quad (\text{for } \rho = -1.64)$$

Δg_{Bs} = Bouguer gravity anomaly at sea ($mGal$)
h_w = water depth below observation point (m).

c) In Rugged Terrain:

$$\Delta g_{Bc} = \Delta g_B + TC$$

Δg_{Bc} = complete Bouguer gravity anomaly ($mGal$)
TC = terrain correction ($mGal$).

MEASUREMENT OF GRAVITY

Gravitational acceleration on Earth's surface can be measured in absolute and relative senses (Fig. 8.13). *Absolute gravity* reflects the actual acceleration of an object as it falls toward Earth's surface, while *relative gravity* is the difference in gravitational acceleration at one station compared to another.

a) Absolute Gravity

b) Relative Gravity

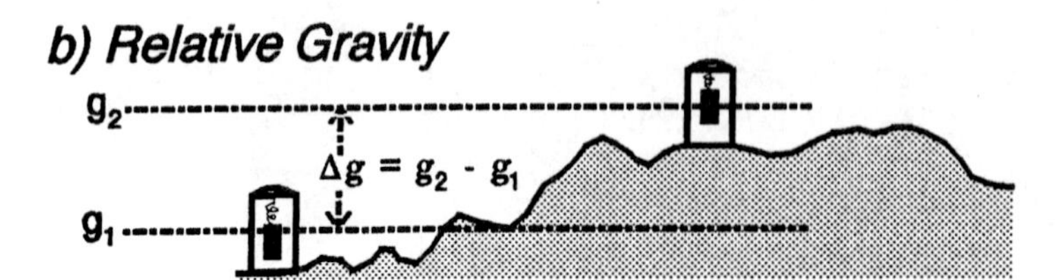

FIGURE 8.13 a) Absolute gravity is the true gravitational acceleration (g). b) Relative gravity reflects the difference in gravitational acceleration (Δg) at one station (g_1) compared to another (g_2).

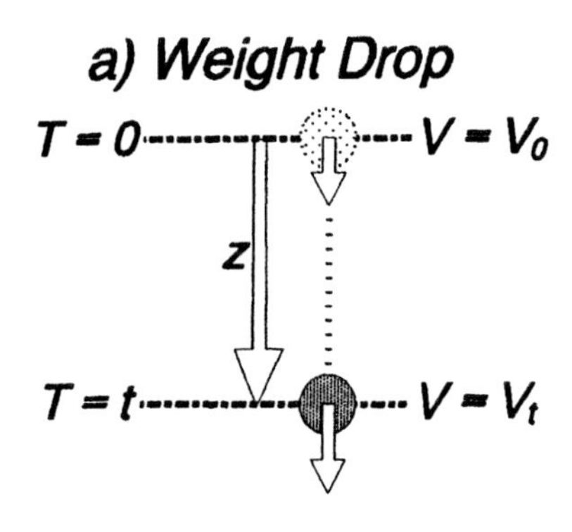

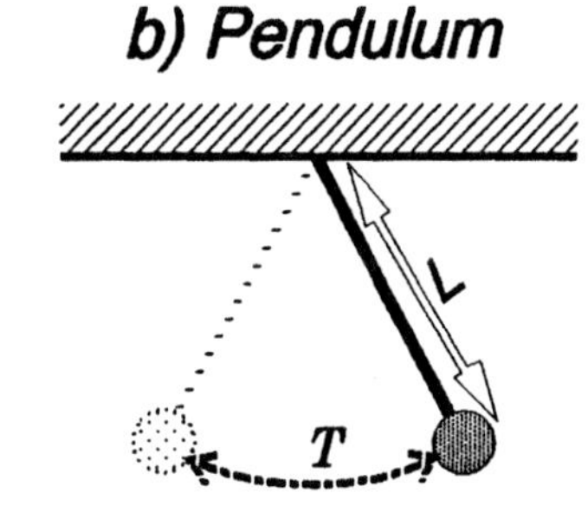

FIGURE 8.14 *Measurement of absolute gravity.* a) *Weight drop.* The object accelerates from an initial velocity of V_0 at time (T = 0), to a velocity of V_t at time (T = t), as it falls a distance (z). b) *Pendulum.* Gravitational acceleration is a function of the pendulum's length (L) and period of oscillation (T).

Absolute Gravity

There are two basic ways to measure absolute gravity. In the *weight drop* method (Fig. 8.14a), the velocity and displacement are measured for an object in free fall. The absolute gravity is computed according to:

$$z = v_0 t + \tfrac{1}{2} g t^2$$

where:

z = distance the object falls
t = time to fall the distance z
v_0 = initial velocity of the object
g = absolute gravity.

The absolute gravity is thus:

$$\boxed{\mathbf{g = 2\,(z - v_0 t)/t^2}}$$

Using the second method (Fig. 8.14b), a *pendulum* oscillates according to:

$$T = 2\pi\sqrt{L/g}$$

where:

T = period of swing of the pendulum
L = length of the pendulum.

The absolute gravity is computed according to:

$$\boxed{\mathbf{g = L(4\pi^2/T^2)}}$$

Relative Gravity

The precision necessary to obtain reliable, absolute gravity observations makes those measurements expensive and time consuming. Relative gravity measurements, however, can be done easily, with an instrument (*gravimeter*) that essentially measures the length of a spring (L; Fig. 8.15a). The *mass* of an object suspended from the spring remains constant. When the gravimeter is taken from one station location to another, however, the *force* (F) that the mass (m) exerts on the spring varies with the local gravitational acceleration (g):

$$F = mg$$

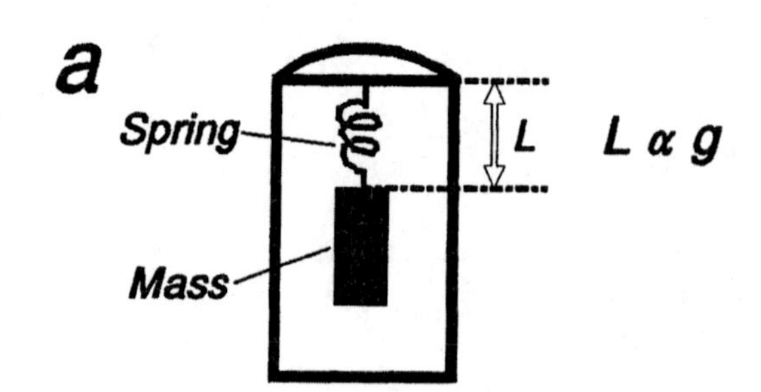

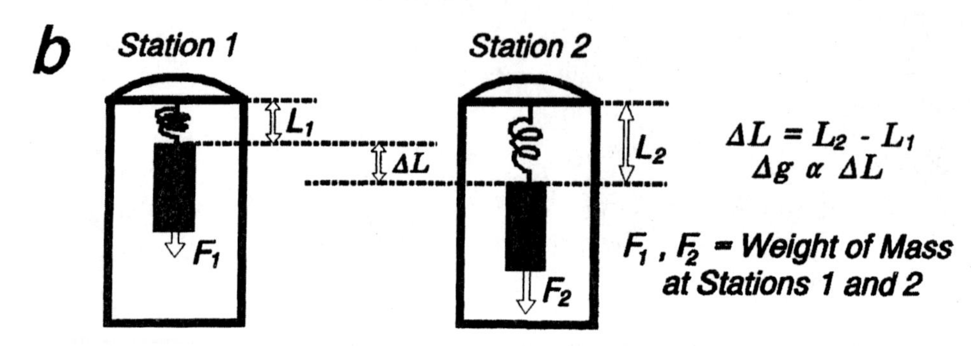

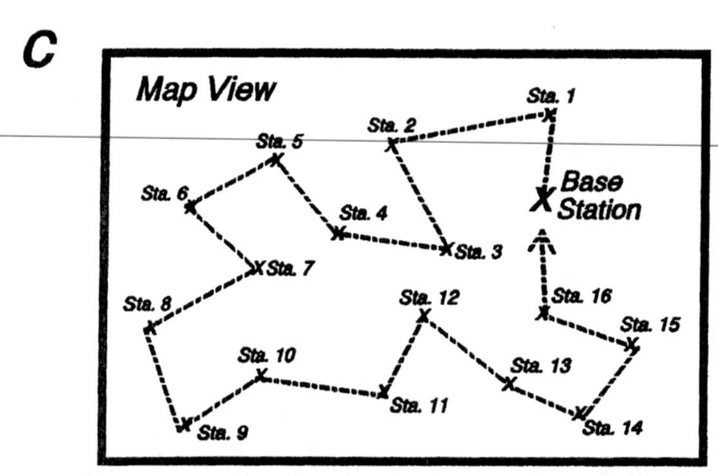

FIGURE 8.15 *Measurement of relative gravity*. a) A *gravimeter* measures the length of a spring (L), which is proportional to the gravitational acceleration (g). b) A force (F_1) at one station results in a spring length (L_1). The length may change to L_2 because of a different force (F_2) at another station. The force exerted by the mass is a function of g; the change in length of the spring (ΔL) is thus proportional to the change in gravitational acceleration (Δg). c) Map of relative gravity survey. The traverse starts with a measurement at the base station, then each of the 16 stations, followed by a re-measurement at the base station.

so that:

$$g = F/m$$

In other words, the mass will *weigh* more or less (exert more or less force), depending on the pull of gravity (g) at the station. A gravimeter is simply weighing the mass at different stations; the spring stretches ($+\Delta L$) where there is more gravity and contracts ($-\Delta L$) when gravity is less (Fig. 8.15b).

If we know the absolute gravity at a starting point (*base station*), we can use a gravimeter to measure points relative to that station (Fig. 8.15c). The initial reading (that is, the initial length of the spring) measured at the base station represents the absolute gravity at that point. Measurements are then taken at other stations, with the changes in length of the spring recorded. The gravimeter is calibrated so that a given change in spring length (ΔL) represents a change in gravity (Δg) by a certain amount (in *mGal*). The acceleration (g) can then be computed by adding the value of Δg to the absolute gravity of the base station.

At sea, gravity surveying is complicated by the fact that the measurement platform is unstable. Waves move the ship up and down, causing accelerations that add or subtract from the gravity. Also, like Earth's rotation, the speed of the ship over the water results in an outward acceleration; in other words, the ship's velocity adds to the velocity of Earth's rotation. An additional correction, known as the *Eötvös correction*, is therefore added to marine gravity measurements (Telford et al., 1976):

$$\mathbf{EC = 7.503\, V} \cos\phi \sin\alpha + \mathbf{0.004154\, V^2}$$

where:

EC = Eötvös correction (*mGal*)
V = speed of ship (knots; 1 knot = 1.852 km/hr = 0.5144 m/s)
ϕ = latitude of the observation point (degrees)
α = course direction of ship (azimuth, in degrees).

ISOSTASY

Until quite recently, surveyors leveled their instruments by suspending a lead weight (plumb bob) on a string. In the vicinity of large mountains, it was recognized that a correction must be made because the *mass excess* of the mountains standing high above the surveyor's location made the plumb bob deviate slightly from the vertical (Fig. 8.16a).

In the mid-1800's a large-scale survey of India was undertaken. Approaching the Himalaya Mountains from the plains to the south, the correction was calculated and applied. A systematic error was later recognized, however, as the plumb bob was not deviated toward the mountains as much as it should have been (Fig. 8.16b). This difference was attributed to *mass deficiency* within the Earth, beneath the excess mass of the mountains.

Pratt and Airy Models (Local Isostasy)

Scientists proposed two models to explain how the mass deficiency relates to the topography of the Himalayas. *Pratt* assumed that the crust of the Earth comprised blocks of different density; blocks of lower density need to extend farther into the air in order to exert the same pressure as thinner blocks of higher density (Fig. 8.17a). The situation is analogous to blocks of wood, each of different density, floating on water. By the Pratt model, the base of the crust is flat, so that the surface of equal pressure (depth of compensation) is essentially a flat crust/mantle boundary.

In the model of *Airy* (Fig. 8.17b), crustal blocks have equal density, but they float on higher-density material (Earth's mantle), similar to (low-density) icebergs floating on (higher-density) water. The base of the crust is thus an exaggerated, mirror image of the topography. Areas of high elevation have low-density "crustal roots" supporting their weight, much like a beach ball lifting part of a swimmer's body out of the water.

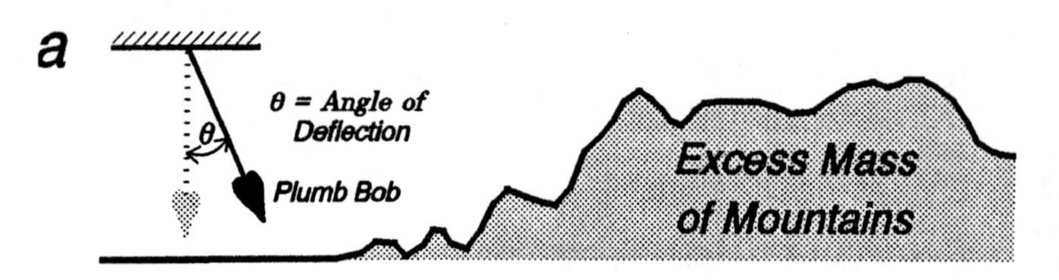

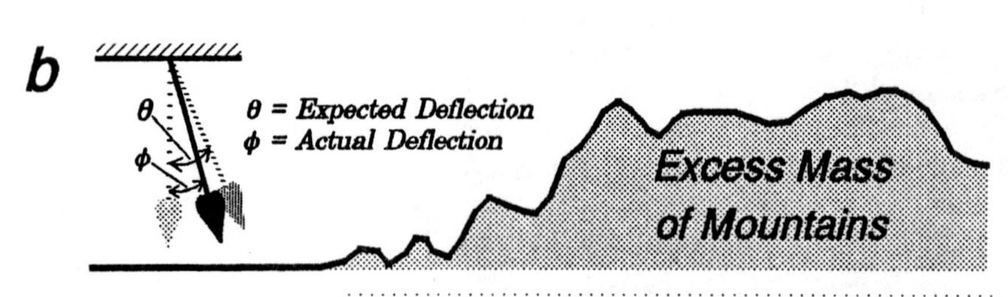

FIGURE 8.16 a) Expected deflection of a plumb bob (highly exaggerated), due to the attraction of the mass of a mountain range. b) The actual deflection for the Himalayas was less than expected, due to a deficiency of mass beneath the mountains.

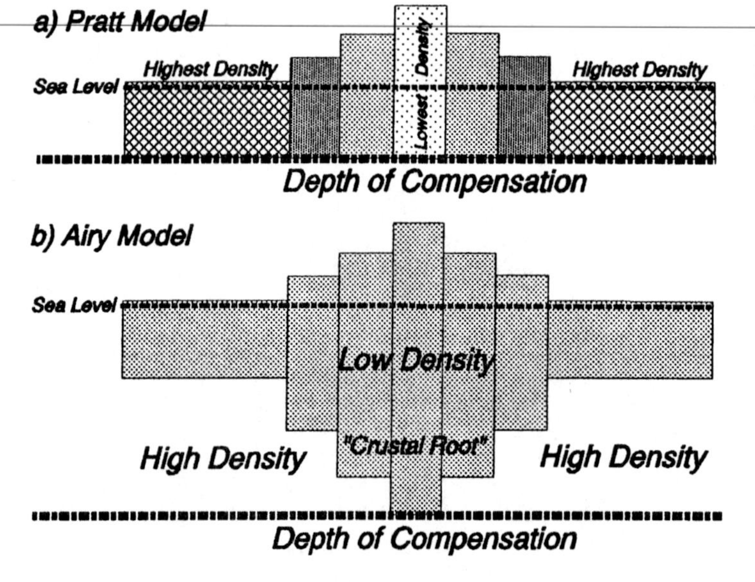

FIGURE 8.17 Pratt and Airy models of local isostatic compensation. In both models, pressure exerted by crustal columns is equal on horizontal planes at and below the depth of compensation.

Hydrostatic pressure is the pressure exerted on a point within a body of water. Similarly, pressure at a given depth within the Earth (Fig. 8.18a) can be viewed as lithostatic pressure, according to:

$$P = \rho g z$$

where:

P = pressure at the point within the Earth
ρ = average density of the material above the point

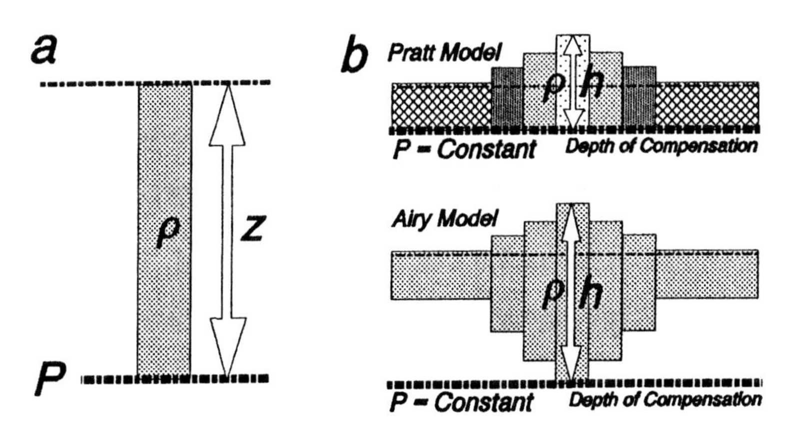

FIGURE 8.18 a) Pressure (P) at depth (z) is a function of the density (ρ) of the material above a point within the Earth. b) For the Pratt and Airy models, the pressure depends on the density and thickness (h) of crustal blocks. In both models, pressure equalizes at the depth of compensation.

g = acceleration due to gravity ($\approx 9.8\ \text{m/s}^2$)
z = depth to the point.

For the Pratt and Airy models (Fig. 8.18b), the pressure exerted by a crustal block can be expressed as:

$$P = \rho g h$$

where:

P = pressure exerted by the crustal block
ρ = density of the crustal block
h = thickness of the crustal block.

In both the Pratt and Airy models, the pressure must be the same everywhere at the *depth of compensation*. For the Pratt model, the base of each block is at the exact depth of compensation, so that:

$$P = \rho_2 g h_2 = \rho_3 g h_3 = \rho_4 g h_4 = \rho_5 g h_5$$

where:

$\rho_2, \rho_3, \rho_4, \rho_5$ = density of each block
h_2, h_3, h_4, h_5 = thickness of each block.

Dividing out a constant gravitational acceleration (g):

$$\mathbf{P/g = \rho_2 h_2 = \rho_3 h_3 = \rho_4 h_4 = \rho_5 h_5}$$

In the particular Pratt model shown in Fig. 8.19a, $\rho_5 < \rho_4 < \rho_3 < \rho_2 < \rho_1$, where ρ_1 is the density of the substratum (Earth's mantle).

In an Airy model the crustal density (ρ_2) is constant and less than the mantle density (ρ_1). Only the thickest crustal block extends to the depth of compensation. For the Airy isostatic model in Fig. 8.19b, the pressure exerted at the depth of compensation (divided by g) is:

$$\mathbf{P/g = \rho_2 h_5 = (\rho_2 h_4 + \rho_1 h_4') = (\rho_2 h_3 + \rho_1 h_3') = (\rho_2 h_2 + \rho_1 h_2')}$$

where:

h_2', h_3', h_4' = thickness of mantle column from the base of each crustal block to the depth of compensation.

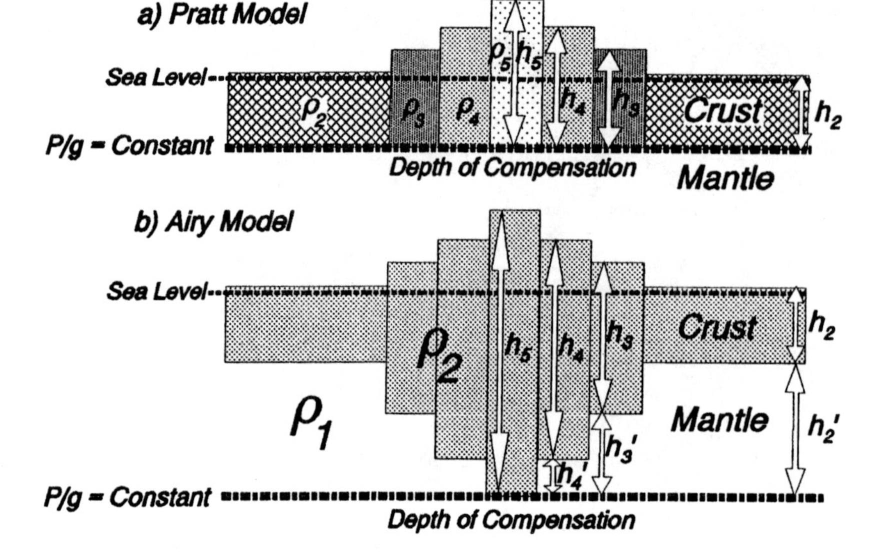

FIGURE 8.19 Density (ρ) and thickness (h, h′) relationships for Pratt and Airy isostatic models. P = pressure; g = gravitational acceleration.

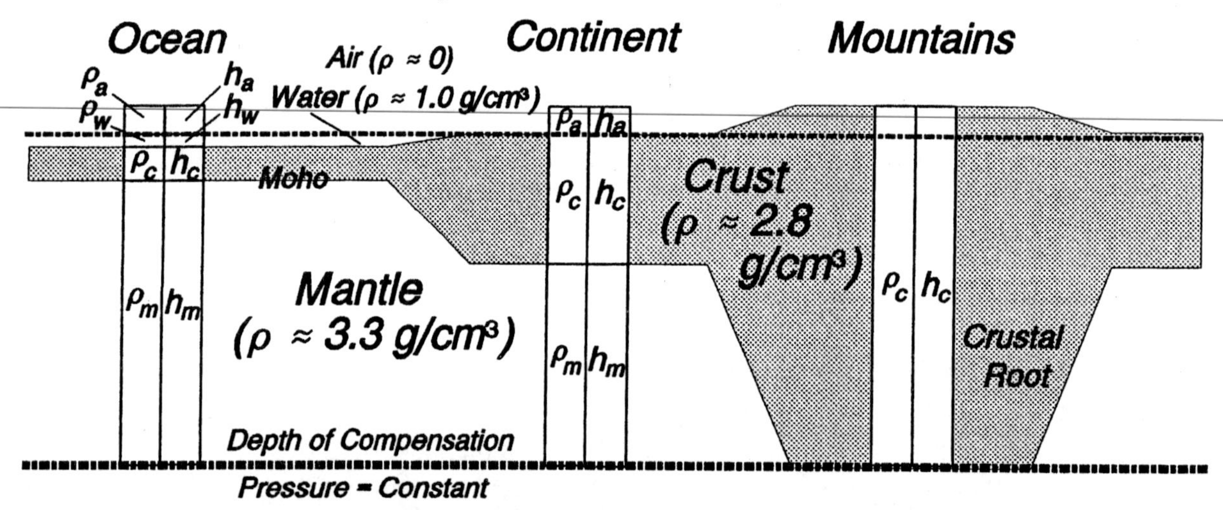

FIGURE 8.20 *Airy isostatic model.* Oceanic regions have thin crust, relative to continental regions. The weight of extra mantle material beneath the thin oceanic crust pulls downward until just enough depth of water fills the basin to achieve isostatic equilibrium. Mountainous regions have thick crust, relative to normal continental regions. The crustal root exerts an upward force until it is balanced by the appropriate weight of mountains.

While regions often exhibit components of both hypotheses, isostatic compensation is generally closer to the Airy than the Pratt model. Pure Airy isostatic compensation for regions with oceanic and continental crust, as well as thickened crust weighted down by mountains, might exhibit the form illustrated in Fig. 8.20. Notice that the crustal root beneath elevated regions is typically 5 to 8 times the height of the topographic relief. At the *depth of compensation* beneath each region, two equations hold true. 1) The total *pressure* (P) exerted by each vertical column, divided by the gravitational acceleration (g), is constant:

$$P/g = \rho_a h_a + \rho_w h_w + \rho_c h_c + \rho_m h_m = \text{Constant}$$

where:

ρ_a = density of the air ($\rho_a \approx 0$)
h_a = thickness of the air column, up to the level of the highest topography
ρ_w = density of the water
h_w = thickness of the water column
ρ_c = density of the crust
h_c = thickness of the crust
ρ_m = density of the mantle
h_m = thickness of the mantle column, down to the depth of compensation.

2) The total *thickness* (T) of each vertical column is constant:

$$T = h_a + h_w + h_c + h_m = \text{Constant}$$

If the isostatic column (P/g) can be determined or assumed for one area, then solving the two equations simultaneously can be used to estimate thicknesses (h) and/or densities (ρ) for vertical columns beneath other areas.

Lithospheric Flexure (Regional Isostasy)

Both the Pratt and Airy models assume *local isostasy*, whereby compensation occurs directly below a load (Fig. 8.21a); supporting materials behave like liquids, flowing to accommodate the load. In other words, the materials are assumed to have *no rigidity*. Most Earth materials, however, are somewhat rigid; the effect of a load is distributed over a broad area, depending on the *flexural rigidity* of the supporting material. Models of *regional isostasy* therefore take lithospheric strength into account (Fig. 8.21b).

A common model of regional isostatic compensation is that of an *elastic plate* that is bent by topographic and subsurface loads. The *flexural rigidity* (D) of the plate determines the degree to which the plate supports the load. The elastic plate model is analogous to a diving board, the *load* being the *diver* standing near the end of the board (Fig. 8.22). A thin, weak board (small D) bends greatly, especially near the diver. A thicker board of the same material behaves more rigidly; the diver causes a smaller deflection. The flexural rigidity (resistance to bending) thus depends on the *elastic thickness* of each board.

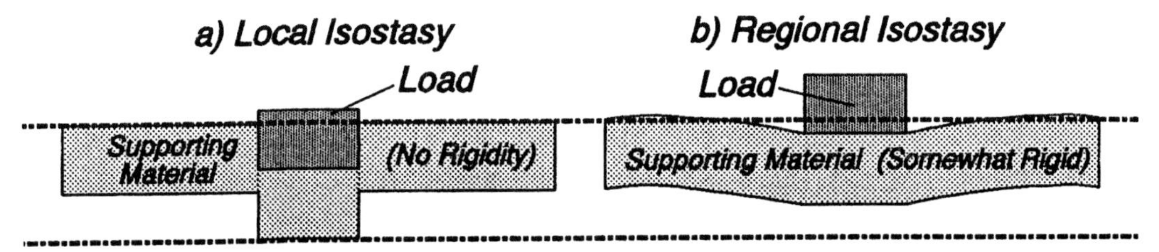

FIGURE 8.21 The type of isostatic compensation depends on the flexural rigidity of the supporting material. a) *Local isostasy*. Where there is no rigidity, compensation is directly below the load. b) *Regional isostasy*. Materials with rigidity are flexed, distributing the load over a broader region.

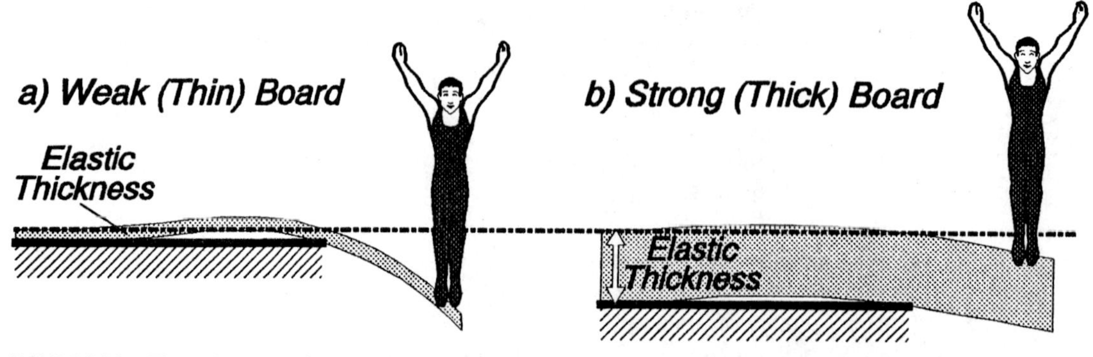

FIGURE 8.22 *Flexural rigidity.* a) A thin diving board (small elastic thickness) has low flexural rigidity. b) A thick board (large elastic thickness) has high flexural rigidity.

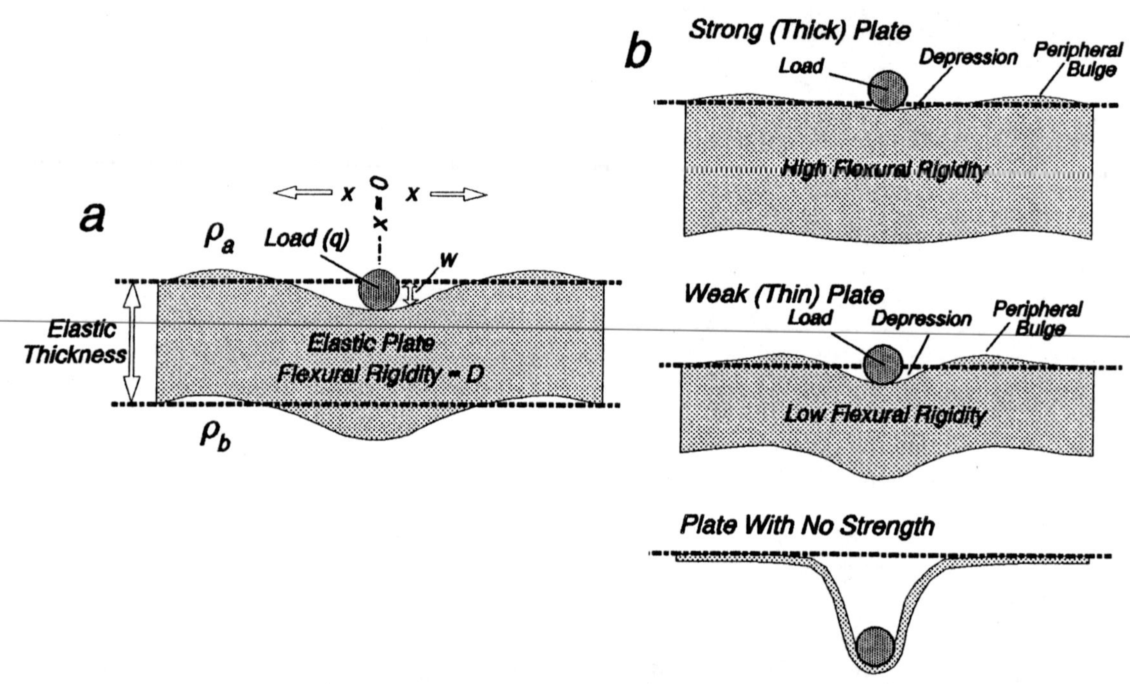

FIGURE 8.23 a) Parameters for two-dimensional model of a plate flexed by a linear load. Both the plate and load extend infinitely in and out of the page. See text for definition of variables. b) Positions of depressions and bulges formed on the surface of a flexed plate. A strong plate has shallow but wide depressions. The depressions and peripheral bulges have larger amplitudes on a weak plate, but are closer to the load. A very weak plate collapses into local isostatic equilibrium.

The deflection of a two-dimensional plate, due to a linear load depressing the plate's surface, is developed by Turcotte and Schubert (1982). The model (Fig. 8.23a) assumes that material below the plate is fluid. The vertical deflection of points along the surface of the plate can be computed according to:

$$\mathbf{D(d^4w/d^4x) + (\rho_b - \rho_a)gw = q(x)}$$

where:

D = flexural rigidity of the plate
w = vertical deflection of the plate at x

x = horizontal distance from the load to a point on the surface of the plate
ρ_a = density of the material above the plate
ρ_b = density of the material below the plate
g = gravitational acceleration
$q(x)$ = load applied to the top of the plate at x.

Four important concepts are illustrated by solutions to the above equation (Fig. 8.23b): 1) a strong lithospheric plate (large D) will have a small amplitude deflection (small w), spread over a long wavelength; 2) a weak lithospheric plate (small D) has large deflection (large w), but over a smaller wavelength; 3) where plates have significant strength, an upward deflection (*"peripheral" or "flexural" bulge*) develops some distance from the load, separated by a depression; 4) plates with no strength collapse into local isostatic equilibrium.

Two simplified examples of lithospheric flexure are shown in Fig. 8.24. At a subduction zone (Fig. 8.24a), flexure is analogous to the bending at the edge of a diving board (Fig. 8.22). The load is primarily the topography of the accretionary wedge and volcanic arc on the overriding plate. Flexure of the downgoing plate results in a depression (trench) and, farther out to sea, a bulge on the oceanic crust. The mass of high mountains puts a load on a plate that can be expressed in both directions (Fig. 8.24b). Depressions between the mountains and flexural bulges ("foreland basins") can fill with sediment to considerable thickness.

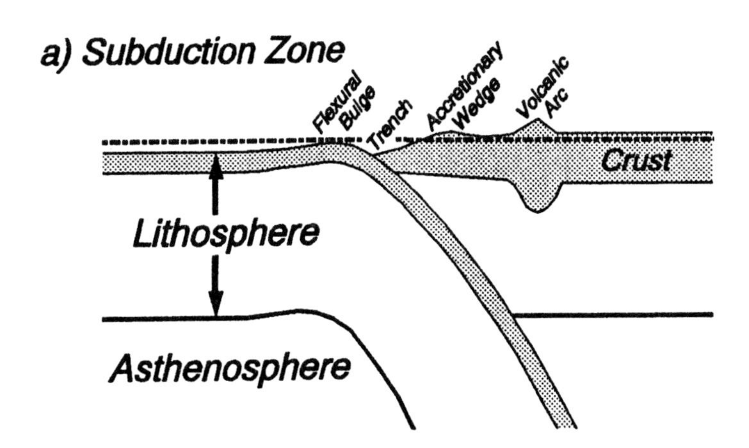

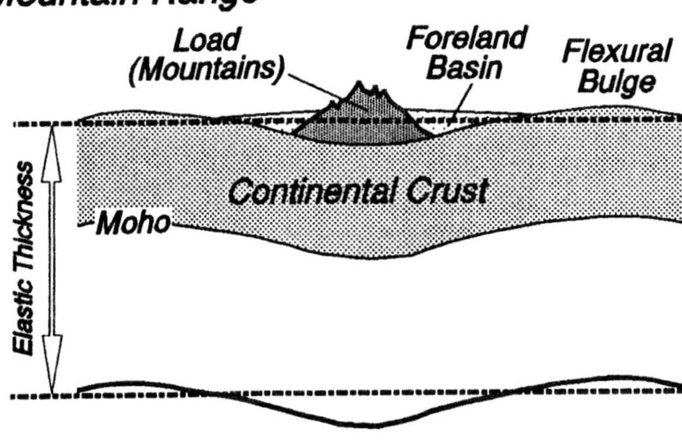

FIGURE 8.24 *Examples of lithospheric flexure.* a) A flexural bulge and depression (trench) develop as the downgoing plate is flexed at a subduction zone. b) The weight of a mountain range causes adjacent depressions that fill with sediment (foreland basins).

GRAVITY MODELING

Forward modelling of mass distributions is a powerful tool to visualize free air and Bouguer gravity anomalies that result from different geologic situations. For large tectonic features, gravity modeling can be even more insightful if considerations of the isostatic state of the region are incorporated.

A common method used to model gravity data is the two-dimensional approach developed by Talwani et al. (1959). The gravity anomaly resulting from a model is computed as the sum of the contributions of individual bodies, each with a given density (ρ) and volume (V) (that is, a mass, m, proportional to $\rho \times V$). The two-dimensional bodies are approximated, in cross section, as polygons (Fig. 8.25).

Gravity Anomalies from Bodies with Simple Geometries

To appreciate contributions from complex-shaped polygons, it is helpful to understand, first, the gravity expression of two simple geometric shapes: 1) a *sphere* and 2) a *semi-infinite slab*.

Sphere The attraction of a sphere buried below Earth's surface can be viewed in much the same way as the attraction of the entire Earth from some distance in space (Figs. 8.3; 8.26). The equation for both cases follows an inverse square law of the form:

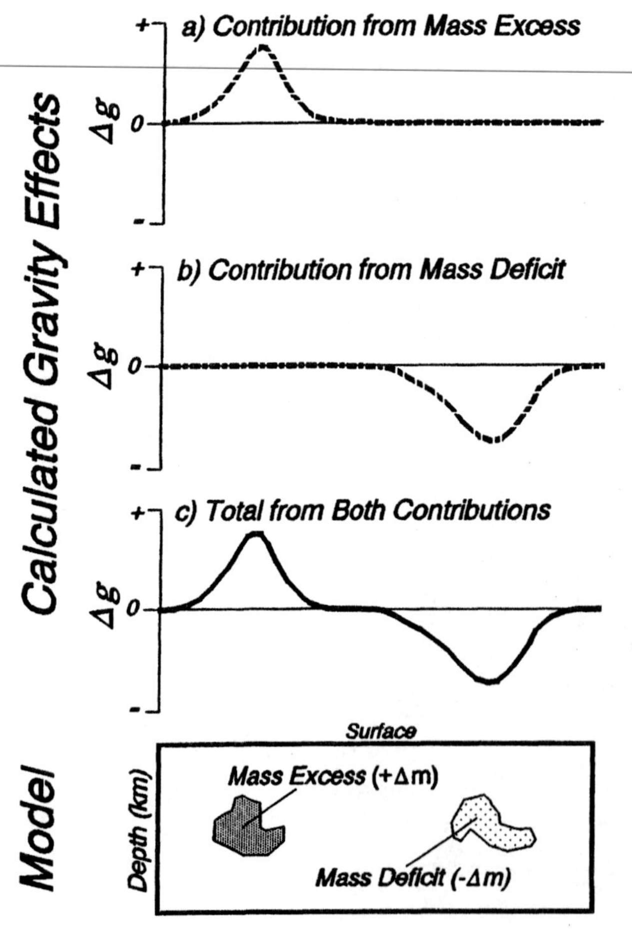

FIGURE 8.25 *Two-dimensional gravity modeling of subsurface mass distributions.* Bodies of anomalous mass are polygonal in cross section, maintaining their shapes to infinity in directions in and out of the page. a) Relative to surrounding material, a body with excess mass results in a positive contribution to the gravity anomaly profile (Δg). b) A negative contribution results from a body with a deficiency of mass. c) The gravity anomaly for the simple model is the sum of the contributions shown in (a) and (b).

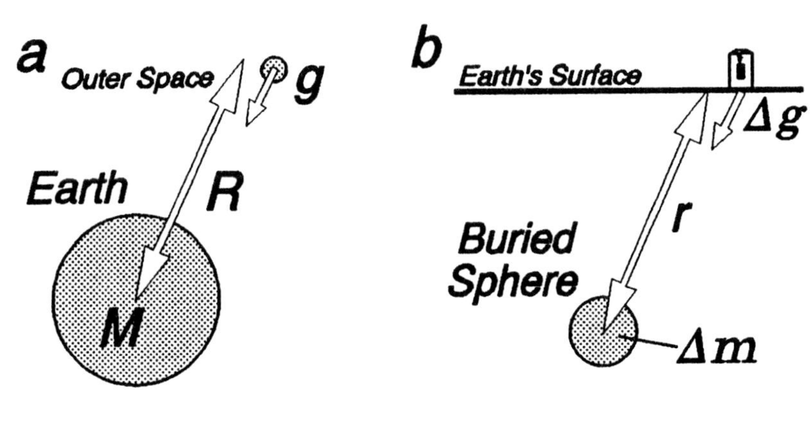

FIGURE 8.26 *Analogy between the gravitational attraction of the Earth from space and a sphere of anomalous mass buried beneath Earth's surface.* a) Earth's gravitational acceleration (g) at a distant observation point depends on the mass of the Earth (M) and the distance (R) from the center of mass to the observation point. b) The change in gravity (Δg) due to a buried sphere depends on the difference in mass (Δm, relative to the surrounding material), and the distance (r) from the sphere to an observation point on Earth's surface.

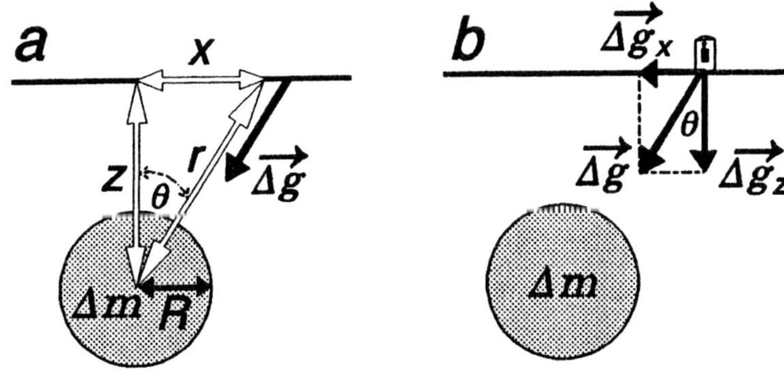

FIGURE 8.27 Gravitational effect ($\overrightarrow{\Delta g}$) of a buried sphere of radius (R) and anomalous mass (Δm). a) The distance (r) to the center of the sphere can be broken into horizontal (x) and vertical (z) components. b) The magnitude (Δg) of the gravitational attraction vector can be broken into horizontal (Δg_x) and vertical (Δg_z) components. For a perfect sphere with uniform Δm, the angle θ is the same as in (a).

$$g = \frac{GM}{R^2}$$

A buried sphere may have excess or deficient mass (Δm) relative to the surrounding material; its center lies a distance (r) from the observation point (Fig. 8.27a). The change in gravitational attraction (Δg) due to the sphere is:

$$\Delta g = \frac{G(\Delta m)}{r^2}$$

The density (ρ) of the material is defined as mass (m) per unit volume (V):

$$\rho = m/V$$

so that:

$$m = \rho V$$

The excess (or deficient) mass of the sphere, in terms of the density difference (Δρ) between the sphere and the surrounding material, is therefore:

$$\Delta m = (\Delta\rho)V$$

the change in gravity is thus:

$$\Delta g = \frac{G(\Delta\rho)(V)}{r^2}$$

The volume (V) of a sphere of radius R is:

$$V = 4/3\ \pi R^3$$

so that:

$$\Delta g = \frac{G(\Delta\rho)}{r^2}(4/3\ \pi R^3)$$

$$= \frac{4\pi R^3 G(\Delta\rho)}{3}\frac{1}{r^2}$$

Since $r^2 = x^2 + z^2$:

$$\boldsymbol{\Delta g = \frac{4\pi R^3 G(\Delta\rho)}{3}\frac{1}{(x^2 + z^2)}}$$

Δg is the magnitude of the *total attraction*, at the observation point, due to Δm (Fig. 8.27a). The total attraction is a vector sum of horizontal and vertical components (Fig. 8.27b):

$$\overrightarrow{\Delta g} = \overrightarrow{\Delta g_x} + \overrightarrow{\Delta g_z}$$

where:

$\overrightarrow{\Delta g}$ = vector expressing magnitude (Δg) and direction of total attraction due to the anomalous mass of the sphere
$\overrightarrow{\Delta g_x}$ = horizontal component of $\overrightarrow{\Delta g}$
$\overrightarrow{\Delta g_z}$ = vertical component of $\overrightarrow{\Delta g}$
$\Delta g_x = \Delta g(\sin\theta)$ = horizontal component of Δg
$\Delta g_z = \Delta g(\cos\theta)$ = vertical component of Δg
θ = angle between a vertical line and the $\overrightarrow{\Delta g}$ direction.

The magnitude can be expressed as the vector sum of horizontal and vertical components:

$$\Delta g = \sqrt{(\Delta g_x)^2 + (\Delta g_z)^2}$$

A gravimeter measures only the vertical component of the gravitational attraction (Fig. 8.27b):

$$\boldsymbol{\Delta g_z = \Delta g\ (\cos\theta)}$$

From Fig. 8.27a:

$$\cos\theta = z/r$$

so that:

$$\Delta g_z = \Delta g(z/r) = \frac{4\pi R^3 G(\Delta\rho)}{3}\frac{1}{(x^2 + z^2)}\frac{z}{r}$$

Again, using:

$$r^2 = x^2 + z^2,\ \text{meaning}\ r = (x^2 + z^2)^{1/2}$$

$$\Delta g_z = \frac{4\pi R^3 G(\Delta\rho)}{3}\frac{1}{(x^2 + z^2)}\frac{z}{(x^2 + z^2)^{1/2}}$$

Substituting the value for $4\pi/3$:

$$\Delta g_z = 4.1888\ R^3 G(\Delta\rho)\frac{z}{(x^2 + z^2)^{3/2}}$$

Using $G = 6.67 \times 10^{-11}\ Nm^2/kg^2$:

$$\boldsymbol{\Delta g_z = 0.02794\,(\Delta\rho)\,R^3 \frac{z}{(x^2 + z^2)^{3/2}}}$$

where the variables and units are:

Δg_z = vertical component of gravitational attraction measured by a gravimeter (*mGal*)
$\Delta\rho$ = difference in density between the sphere and the surrounding material (g/cm^3)
R = radius of the sphere (m)
x = horizontal distance from the observation point to a point directly above the center of the sphere (m)
z = vertical distance from the surface to the center of the sphere (m).

Fig. 8.28a shows the variables in the above equation. The buried sphere model illustrates some fundamental properties of gravity anomalies (Fig. 8.28b): 1) mass

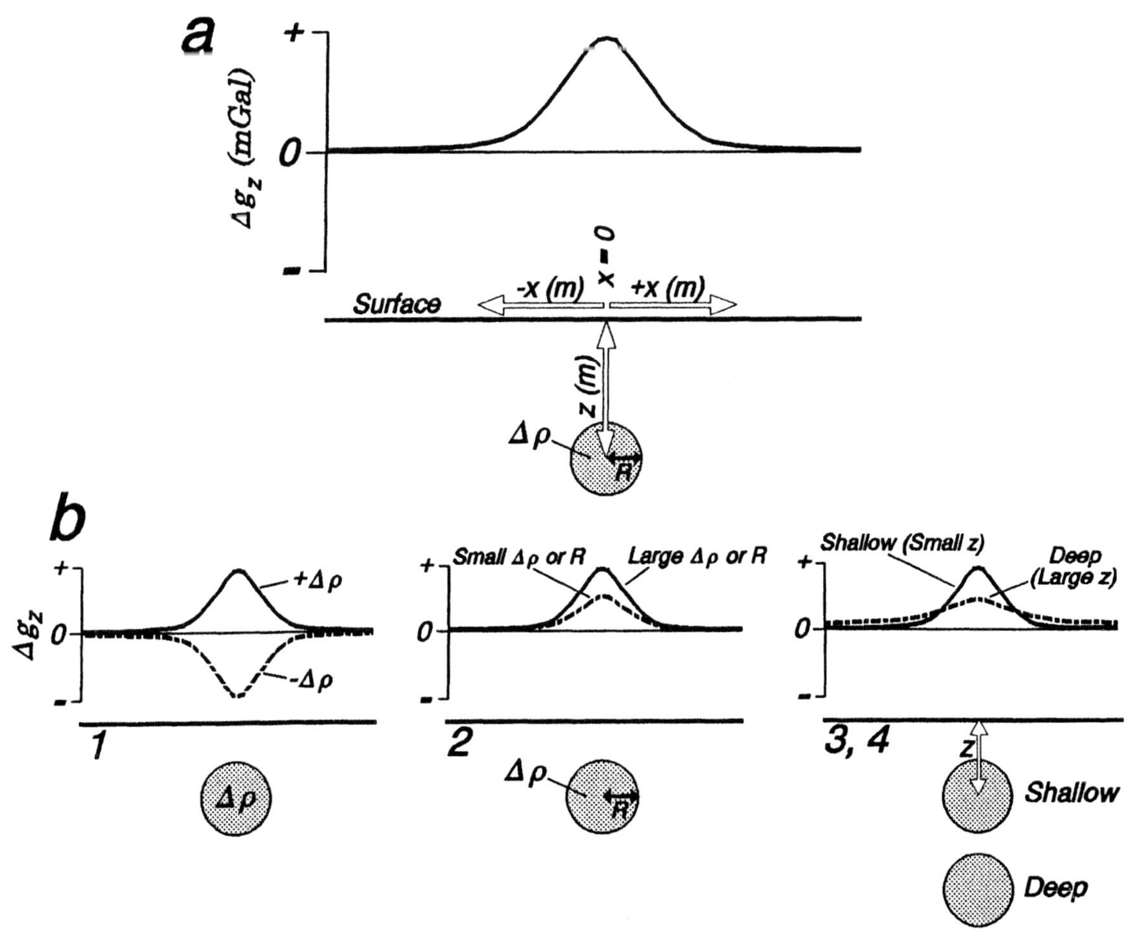

FIGURE 8.28 a) Gravity anomaly profile (Δg_z) attributable to a buried sphere of radius (R_1 in m), depth (z), and anomalous density ($\Delta\rho 1$ in g/cm^3). The horizontal distance (x) is measured in negative ($-x$) and positive ($+x$) directions from a point on the surface directly above the sphere. b) Form of gravity anomaly profiles due to (1) positive vs. negative density contrasts; (2) changing mass anomaly; and (3,4) changing depth.

excess ($+\Delta m$, implying $+\Delta\rho$) causes an increase in gravity ($+\Delta g_z$), while mass deficit ($-\Delta m$, implying $-\Delta\rho$) results in a gravity decrease ($-\Delta g_z$); 2) the more massive the sphere (larger $\Delta\rho$ and/or larger R), the greater the amplitude ($|\Delta g_z|$) of the gravity anomaly; 3) the anomaly is attenuated (smaller $|\Delta g_z|$) as the sphere is buried more deeply within the Earth; 4) the width of the gravity anomaly increases as the sphere is buried more deeply.

Semi-Infinite Slab Where there are density changes that can be approximated by horizontal layering, it is convenient to model lateral changes in gravity as the effects of abrupt truncations of infinite slabs. An infinite slab (Fig. 8.29a) that has excess mass ($+\Delta m$) will increase the gravity ($+\Delta g_z$), while mass deficit ($-\Delta m$) will cause the gravity to decline ($-\Delta g_z$). Truncating the slab (Fig. 8.29b) results in: 1) essentially no gravity effect in regions far from the slab; 2) an increase (or decrease) in gravity crossing the edge of the slab; and 3) the full (positive or negative) gravity effect in regions over the slab but far from the edge.

An infinite slab represents a mass anomaly (Δm) that is a function of the thickness of the slab (Δh) and its density ($\Delta\rho$) relative to surrounding materials (Fig. 8.29c). The amount the slab adds or subtracts to gravitational attraction (Δg_z) is exactly the same as that of the infinite slab used in the Bouguer correction (Fig. 8.9):

$$\Delta g_z = 0.0419(\Delta\rho)(\Delta h)$$

The gravity effect of a *semi-infinite slab*, however, changes according to position relative to the slab's edge (Fig. 8.29d): 1) far away from the slab, the contribution (Δg_z)

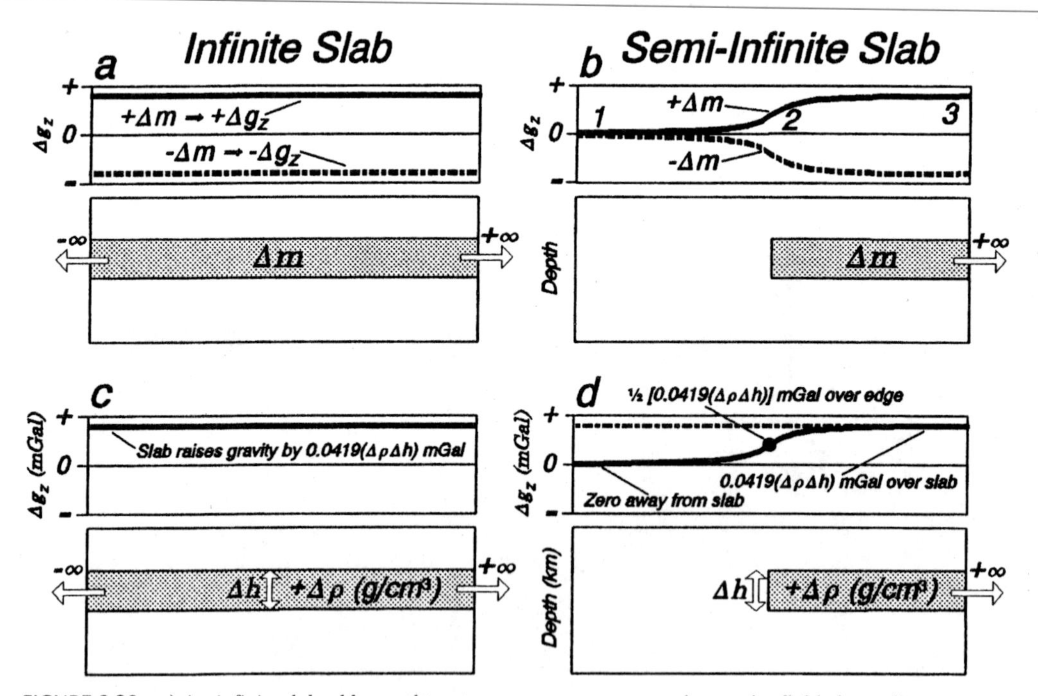

FIGURE 8.29 a) An *infinite slab* adds or subtracts a constant amount to the gravity field, depending on whether the slab represents a positive ($+\Delta m$) or negative ($-\Delta m$) mass anomaly. b) The gravity effect of a *semi-infinite slab* changes gradually as the edge of the slab is crossed. c) An infinite slab produces exactly the same gravity effect as the slab used for the Bouguer correction. d) The gravity effect of a semi-infinite slab is equal to the Bouguer slab approximation far out over the slab (right), ½ of that value directly over the slab's edge, and zero far away from the edge (left).

is zero; 2) above the edge of the slab, the contribution is exactly 1/2 the maximum value ($\Delta g_z = ½[0.0419\Delta\rho\Delta h]$); 3) over the slab, but far from the slab's edge, Δg_z is the same as for an infinite slab ($\Delta g_z = 0.0419\Delta\rho\Delta h$); 4) the rate of change in gravity (the *gradient* of Δg_z) depends on the depth of the slab.

Griffiths and King (1981) develop an equation for the anomaly caused by a semi-infinite slab (Fig. 8.30a,b):

$$\Delta g_z = G(\Delta\rho)(\Delta h)(2\phi)$$

where:

ϕ = angle (in radians) from the observation point, between the horizontal surface and a line drawn to the central plane at the slab's edge

G = Universal Gravitational Constant ($6.67 \times 10^{-11}\ Nm^2/kg^2$).

The angle ϕ can be expressed as:

$$\phi = \pi/2 + tan^{-1}(x/z)$$

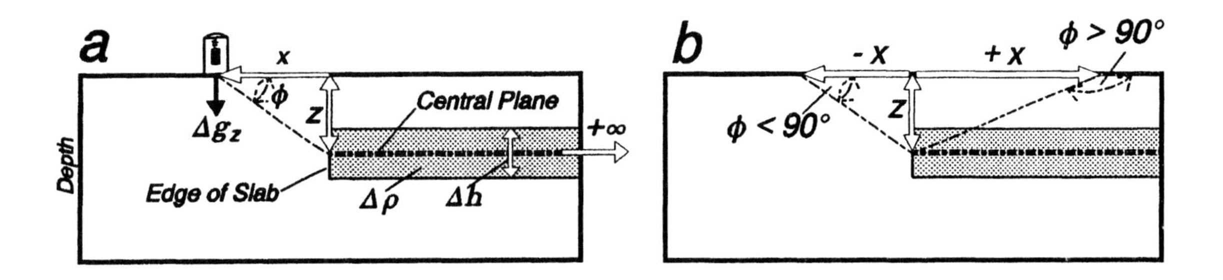

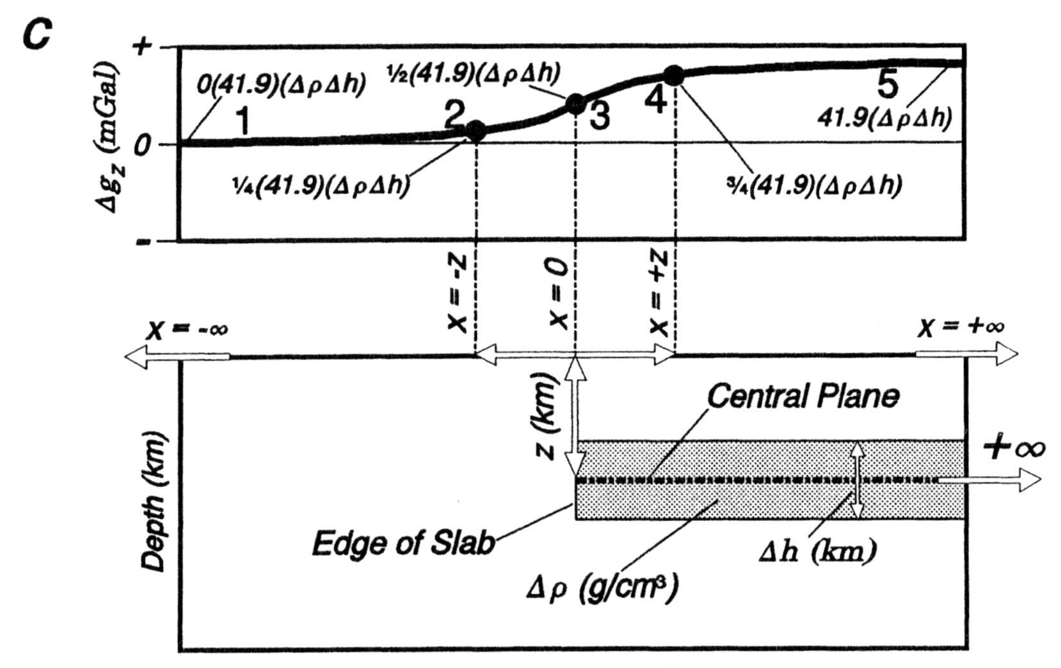

FIGURE 8.30 a) For a semi-infinite slab, the gravity anomaly measured at the surface is $\Delta g_z = G(\Delta\rho)(\Delta h)(2\phi)$, where ϕ is measured in radians. b) Away from the slab, $\phi < \pi/2$ (that is, $\phi < 90°$). Over the slab, $\phi > \pi/2$. c) Method to estimate the change in gravity anomaly (Δg_z) at five horizontal distances (x, in km) from the edge of a semi-infinite slab.

where:

x = horizontal distance from a point on the surface above the slab's edge
z = depth of a horizontal surface bisecting the slab (central plane).

The equation is thus:

$$\Delta g_z = 2G(\Delta\rho)(\Delta h)(\pi/2 + tan^{-1}[x/z])$$

or:

$$\boldsymbol{\Delta g_z = 13.34\,(\Delta\rho)\,(\Delta h)\,(\pi/2 + tan^{-1}[x/z])}$$

when the units are: Δg_z in *mGal*; $\Delta\rho$ in g/cm^3; Δh, x, z in km. Note five important points from the above equation, illustrated in Fig. 8.30c:

1. $x = -\infty \Rightarrow \Delta g_z =$ zero $\Rightarrow \Delta g_z = 0(41.9\Delta\rho\Delta h)$.
2. $x = -z \Rightarrow \Delta g_z =$ ¼ its full value $\Rightarrow \Delta g_z = ¼(41.9\Delta\rho\Delta h)$.
3. $x = 0 \Rightarrow \Delta g_z =$ ½ its full value $\Rightarrow \Delta g_z = ½(41.9\Delta\rho\Delta h)$.
4. $x = +z \Rightarrow \Delta g_z =$ ¾ its full value $\Rightarrow \Delta g_z = ¾(41.9\Delta\rho\Delta h)$.
5. $x = +\infty \Rightarrow \Delta g_z =$ its full value $\Rightarrow \Delta g_z = 1(41.9\Delta\rho\Delta h)$.

For layered cases, a quick estimate of the gravity change across the edge of an anomalous mass can be made by calculating and plotting those five points.

The semi-infinite slab approximation illustrates two fundamental properties of gravity anomalies (Fig. 8.31).

1. The *amplitude* (full value) of the anomaly reflects the *mass excess or deficit* (Δm). The mass excess or deficit depends on the product of density contrast ($\Delta\rho$) and thickness (Δh) of the anomalous body.
2. The *gradient* (rate of change) of the anomaly reflects the *depth of the excess or deficient mass below the surface* (z). The depth thus determines how abruptly the gravity anomaly changes from near zero to near its full value, according to the term ($\pi/2 + tan^{-1}[x/z]$). A body near the surface results in a gravity

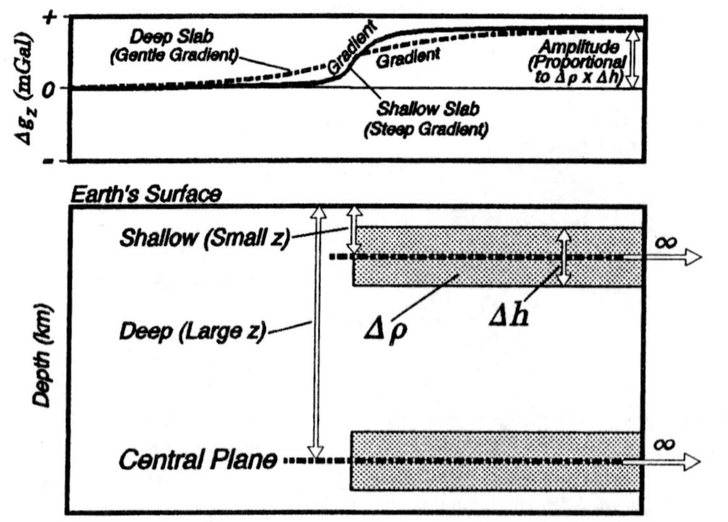

FIGURE 8.31 Lateral change in gravity due to a semi-infinite slab of density contrast ($\Delta\rho$) and thickness (Δh). The amount of change (*amplitude*) depends on the mass anomaly ($\Delta\rho \times \Delta h$), while the rate of change (*gradient*) depends on the depth (z) to the central plane of the slab. The greater the mass anomaly, the greater the amplitude; the more deeply buried the slab, the more gentle the gradient.

change with a steep gradient, while the same body deep within the Earth would produce a more gentle gradient.

Models Using Semi-Infinite Slab Approximations

Semi-infinite slab models can be used to approximate contributions to the free-air gravity anomaly at regions in isostatic equilibrium. Two insightful examples are the transition from continental to oceanic crust along a passive continental margin and the thickening of crust at a mountain range.

Passive Continental Margin Thin oceanic crust at passive margins is underlain by mantle at the same depth as the mid-to-lower crust of the adjacent continent. The *mass excess* ($+\Delta m$) of the mantle exerts a force that pulls the oceanic crust downward. By the Airy model, the resulting ocean basin subsides until it has exactly enough water ($-\Delta m$) so that the region is in isostatic equilibrium.

The model in Fig. 8.32 is in Airy isostatic equilibrium, according to parameters modified from Fig. 8.20:

Densities:

ρ_w = density of the water = $\mathbf{1.03\ g/cm^3}$
ρ_c = density of the crust = $\mathbf{2.67\ g/cm^3}$
ρ_m = density of the mantle = $\mathbf{3.1\ g/cm^3}$.

Thicknesses for the ocean side:

$\mathbf{h_w}$ = thickness of the water column = **5 km**
$\mathbf{(h_c)_O}$ = thickness of the oceanic crust = **8 km**
$\mathbf{h_m}$ = thickness of the extra mantle column = **?**

Thickness for the continent side:

$\mathbf{(h_c)_c}$ = thickness of the continental crust = **?**

The two unknowns [(h_m) and $(h_c)_c$] can be determined from equations expressing the two conditions for local isostatic equilibrium:

	Continent		Ocean
Equal Pressure:	$\rho_c\mathbf{(h_c)_c}$	=	$\rho_w(h_w) + \rho_c(h_c)_O + \rho_m\mathbf{(h_m)}$
Equal Thickness:	$\mathbf{(h_c)_c}$	=	$h_w + (h_c)_O + \mathbf{h_m}$

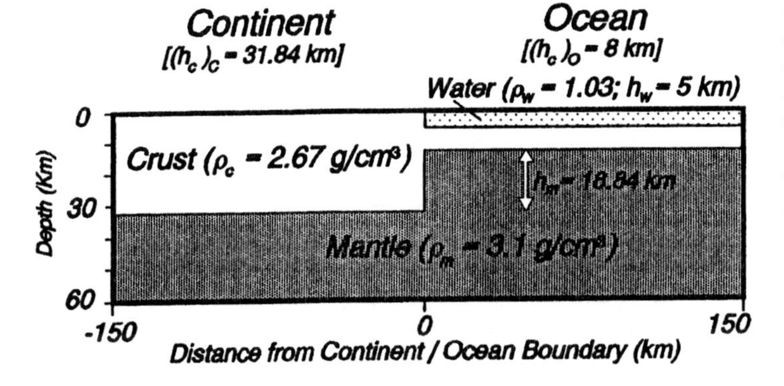

FIGURE 8.32 *Airy isostatic model of the transition from thick continental to thin oceanic crust at a passive continental margin.* Densities of crust and mantle are simplified so that reasonable contrasts result for the water vs. upper continental crust ($-1.64\ g/cm^3$) and the mantle vs. lower continental crust ($+0.43\ g/cm^3$). See text for definition of variables.

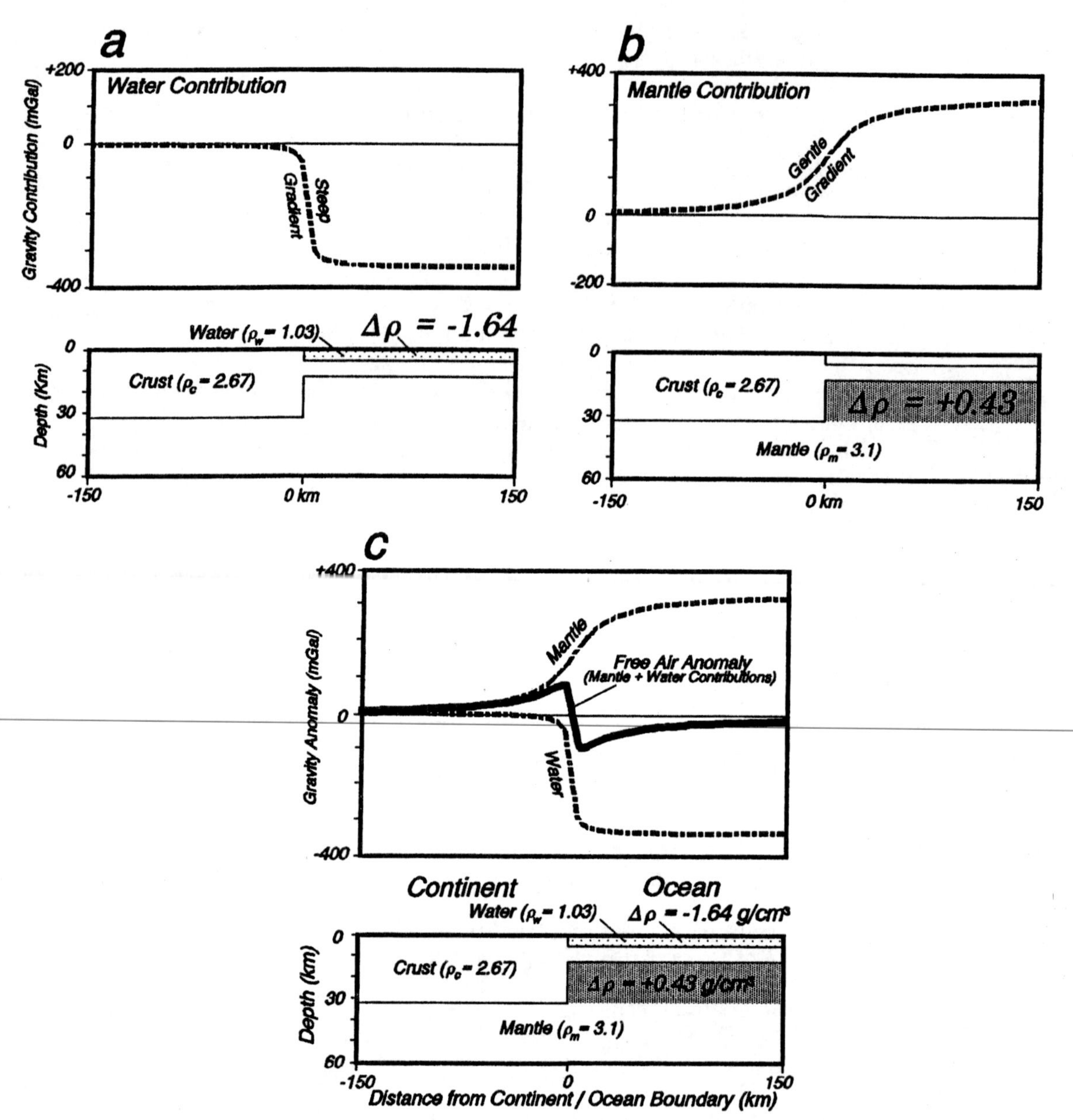

FIGURE 8.33 *The main gravity contributions at a passive continental margin have equal amplitude but different gradient.* a) The water effect is shallow, causing an abrupt change (steep gradient). b) The extra mantle beneath the oceanic crust is a deeper effect, giving a less abrupt change in gravity (gentle gradient). c) The free air gravity anomaly at a passive continental margin is a positive/negative "edge effect," due to the summing of contributions that have equal amplitudes but different gradients.

Solving the two equations for the two unknowns yields:

$$\boxed{(\mathbf{h_c})_\mathbf{c} = \mathbf{31.84\ km}} \qquad \text{and} \qquad \boxed{\mathbf{h_m} = \mathbf{18.84\ km}}$$

The water deepening seaward represents a mass deficit ($-\Delta m$), a function of the product of the water depth (h_w) times the density difference at upper crustal levels ($\Delta\rho = \rho_w - \rho_c = -1.64\ g/cm^3$). Fig. 8.33a shows that this negative contribution to the gravity anomaly is an abrupt change, along a steep gradient where the water deepens.

The mass excess ($+\Delta m$) that compensates the shallow water relates to the amount of shallowing of the mantle (h_m) times the difference between mantle and crustal densities ($\Delta\rho = \rho_m - \rho_c = +0.43$ g/cm^3; Fig. 8.33b). At great distance from the continental margin, the positive contribution to gravity (due to the mantle shallowing) has the same *amplitude* as the negative contribution (due to the water deepening), because the two effects represent compensatory mass excess and deficit, respectively. The *gradient* for the mantle contribution is more gentle, however, because the anomalous mass causing it is deeper.

The *free air gravity anomaly* (Δg_{fa}) for the simple, passive margin model is the sum of the contributions from the shallow (water) and deep (mantle) effects (Fig. 8.33c). Note that the anomaly is near zero over the interiors of the continent and ocean, but shows a *maximum* over the continental edge and a *minimum* over the edge of the ocean. This positive/negative couple, known as an *edge effect*, results because the contributions due to the shallow and deep sources have different gradients.

The passive margin model shows two important attributes of the *free air gravity anomaly* for a region in *isostatic equilibrium* (Fig. 8.34a): 1) values are *near zero* (except for edge effects), because the mass excess ($+\Delta m$) equals the mass deficit ($-\Delta m$); 2) at edge effects, the *area under the curve* of the gravity anomaly *equals zero*, because the integral of the anomaly, with respect to x, is equal to zero.

The second point is worth further discussion, because it provides a quick test for local isostatic equilibrium. The free air anomaly curve for the passive margin model is the sum of two contributions (Fig. 8.33):

$$\begin{aligned}\Delta g_z &= \Delta g_z(\text{bath}) + \Delta g_z(\text{moho})\\ &= 2G(\Delta\rho)_b(\Delta h)_b(\pi/2 + \tan^{-1}[x/z_b])\\ &\quad + 2G(\Delta\rho)_m(\Delta h)_m(\pi/2 + \tan^{-1}[x/z_m])\end{aligned}$$

where:

Δg_z = free air anomaly

Δg_z(bath) = contribution to the free air anomaly due to the mass deficiency of the water deepening seaward (bathymetry)

Δg_z(moho) = contribution to the free air anomaly due to the mass excess of the mantle shallowing seaward

$(\Delta\rho)_b$ = density contrast of the water compared to upper continental crust

$(\Delta\rho)_m$ = density contrast of the shallow mantle compared to lower continental crust

$(\Delta h)_b$ = thickness of semi-infinite slab of water

$(\Delta h)_m$ = thickness of semi-infinite slab of elevated mantle

x = horizontal distance from the continent/ocean boundary

z_b = vertical distance from sea level to the central plane of the semi-infinite slab of water

z_m = vertical distance from sea level to the central plane of the semi-infinite slab of elevated mantle.

The equation can be simplified:

$$\Delta g_z = 2G\left\{\begin{aligned}&(\Delta\rho)_b(\Delta h)_b(\pi/2 + \tan^{-1}[x/z_b])\\ &+ (\Delta\rho)_m(\Delta h)_m(\pi/2 + \tan^{-1}[x/z_m])\end{aligned}\right\}$$

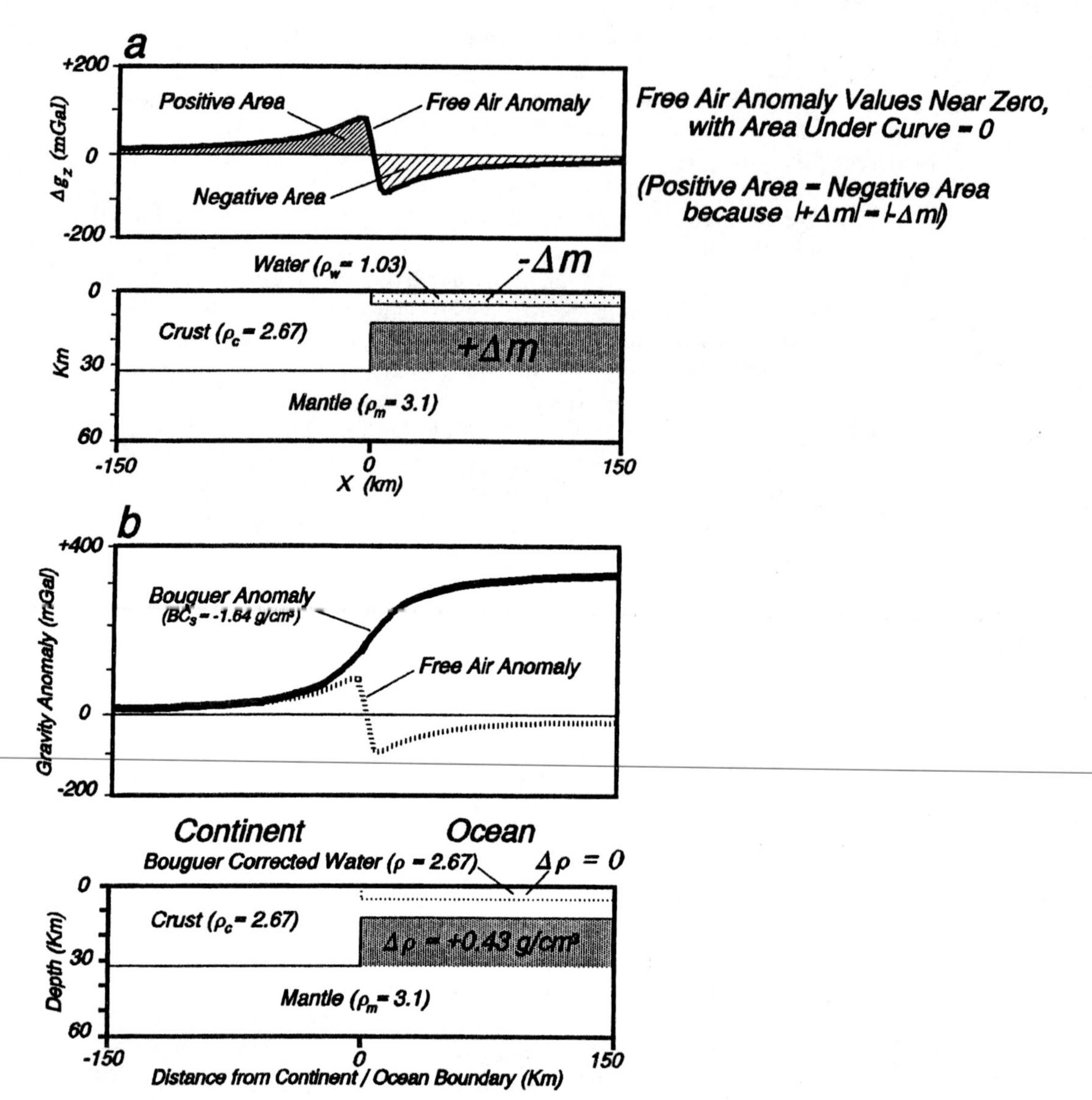

FIGURE 8.34 *Free air and Bouguer gravity anomalies for passive continental margin in local isostatic equilibrium.* a) Isostatic equilibrium means the absolute value of excess mass ($|+\Delta m|$) equals the absolute value of deficient mass ($|-\Delta m|$). With this equality, the integral of the change in gravity with respect to x ($\int \Delta g_z \, dx$) = 0. The zero integral means that the positive and negative areas under the free air anomaly curve sum to zero. b) The Bouguer correction at sea (Fig. 8.9b), applied to the free air anomaly in (a), yields the general form of the Bouguer anomaly at a passive continental margin.

Let:

A = horizontal surface area of a slab (same for each slab).

The density of a slab is:

$$\rho = m/V$$

where:

m = mass of a slab
V = volume of a slab = A(Δh).

therefore:

$$m = \rho V = \rho A(\Delta h)$$

For each slab:

$$\Delta m = (\Delta\rho)A(\Delta h)$$

$$(\Delta\rho)(\Delta h) = \Delta m/A$$

so that:

$$\Delta g_z = 2G \left\{ \begin{array}{l} ([\Delta m]_b/A)(\pi/2 + tan^{-1}[x/z_b]) \\ + ([\Delta m]_m/A)(\pi/2 + tan^{-1}[x/z_m]) \end{array} \right\}$$

where:

$[\Delta m]_b$ = mass deficit of the water slab
$[\Delta m]_m$ = mass excess of the mantle slab.

Airy isostatic equilibrium implies that:

$$[\Delta m]_b = -[\Delta m]_m$$

so that:

$$\Delta g_z = 2G([\Delta m]_b/A)\{(\pi/2 + tan^{-1}[x/z_b]) - (\pi/2 + tan^{-1}[x/z_m])\}$$

$$\Delta g_z = 2G([\Delta m]_b/A)(tan^{-1}[x/z_b] - tan^{-1}[x/z_m])$$

The area under the free air anomaly curve is the integral of Δg_z, with respect to x:

$$\int_{-\infty}^{+\infty} \Delta g_z \, dx = 2G([\Delta m]_b/A) \int_{-\infty}^{+\infty} (tan^{-1}[x/z_b] - tan^{-1}[x/z_m])\, dx$$

Standard integral tables show that, regardless of the depths of the slabs (z_b and z_m), the integral from $-\infty$ to $+\infty$ is zero:

$$\int_{-\infty}^{+\infty} (tan^{-1}[x/z_b] - tan^{-1}[x/z_m])\, dx = 0$$

therefore:

$$\boxed{\int_{-\infty}^{+\infty} \mathbf{\Delta g_z \, dx = 0}}$$

The last expression demonstrates that the area under the curve for the free air anomaly equals zero (Fig. 8.34a). This relationship is seen in each of the models below that are in a state of local isostatic equilibrium.

Fig. 8.35 shows an observed free air gravity anomaly profile, and a density model, for the passive continental margin on the east coast of the United States. The free air anomaly shows clearly the edge effect due to the water deepening as the mantle shallows (Fig. 8.33). Some isostatic imbalance is also evident, because the negative area (under the curve) is greater than the positive area.

The *Bouguer gravity anomaly* (Δg_B) for the simple, passive margin model results from correcting the mass deficit of the water to approximate that of the upper part of the crust (Fig. 8.34b). The passive margin model thus illustrates the general form of the Bouguer anomaly for a region in local isostatic equilibrium: 1) values are near zero over normal continental crust; 2) the Bouguer anomaly mimics

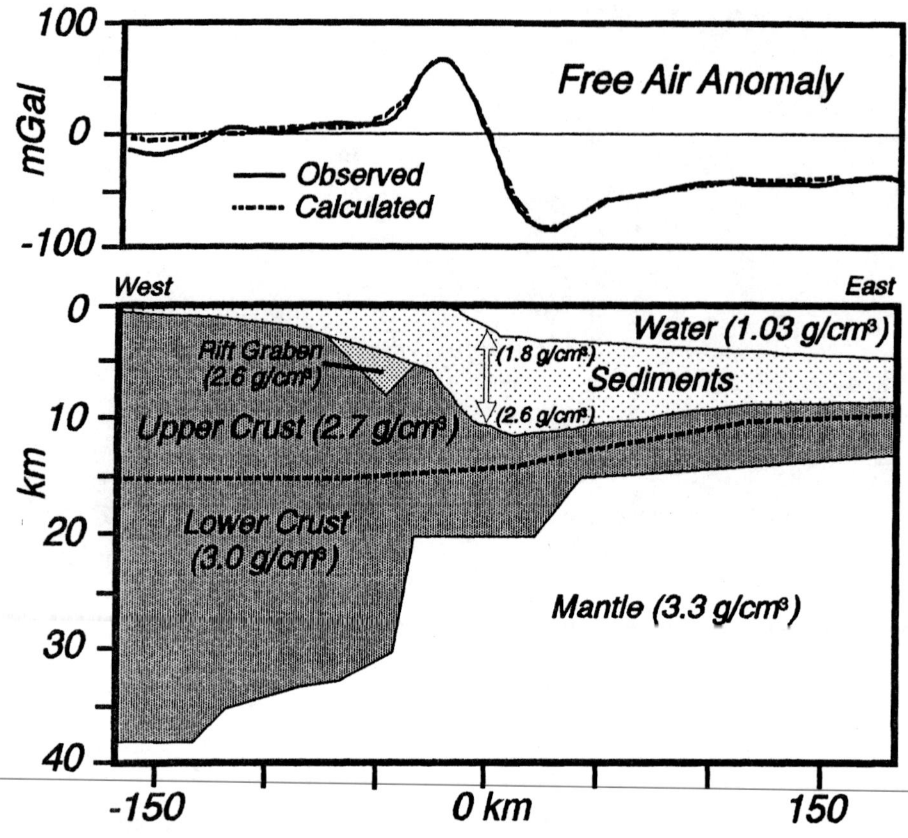

FIGURE 8.35 *Observed free air gravity anomaly from the passive continental margin off the Atlantic coast of the United States.* The dashed line is the anomaly calculated from the two-dimensional density model. Note the edge effect, with the high toward the continent, the low over the ocean. The zero crossing is near the edge of the continental shelf, where the water column deepens abruptly. Line IPOD off Cape Hatteras, North Carolina. From "Deep structure and evolution of the Carolina Trough," by D. Hutchinson, J. Grow, K. Klitgord, and B. Swift, AAPG Memoir, no. 34, pp. 129–152, © 1983. Redrawn with permission of the American Association of Petroleum Geologists, Tulsa, Oklahoma, USA.

the Moho, increasing to large positive values as the mantle shallows beneath the ocean; 3) the form of the Bouguer anomaly is somewhat a mirror image of the topography (or bathymetry); the increase in the anomaly thus correlates with deepening of the water.

Mountain Range As continental crust thickens during orogenesis (Figs. 2.18, 6.35), the crustal root exerts upward force, due to its buoyancy relative to surrounding mantle. By the Airy model, the topography ($+\Delta m$) grows until its weight exactly balances the effect of the low-density root ($-\Delta m$). The mountain range model (Fig. 8.36) is in Airy isostatic equilibrium, according to parameters modified from Fig. 8.20:

Densities:

ρ_a = density of the air = **0**
ρ_c = density of the crust = **2.67 g/cm³**

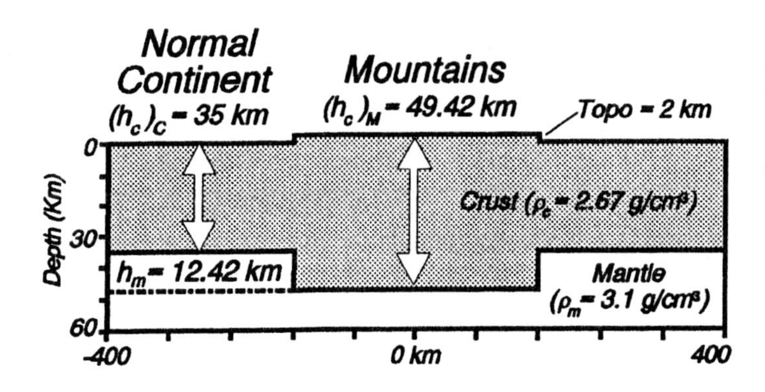

FIGURE 8.36 *Airy isostatic model of 2 km high mountain range.* Densities of crust and mantle are simplified so that reasonable contrasts result for the topography vs. air (+2.67 g/cm³) and the crustal root vs. mantle (−0.43 g/cm³).

ρ_m = density of the mantle = **3.1 g/cm³.**

Thicknesses at the Normal Continental Crust:

$\mathbf{h_a}$ = thickness of the air column = **2 km**
$\mathbf{(h_c)_C}$ = thickness of crust outside mountains = **35 km**
$\mathbf{h_m}$ = thickness of the extra mantle column = **?**

Thickness at the Mountains:

$\mathbf{(h_c)_M}$ = thickness of total crust at mountains = **?**

As with the passive margin model, the two unknowns can be determined from the two conditions for local isostatic equilibrium:

	Normal Continent		Mountains
Equal Pressure:	$\rho_a(h_a) + \rho_c(h_c)_C + \rho_m(\mathbf{h_m})$	=	$\rho_c(\mathbf{h_c})_\mathbf{M}$
Equal Thickness:	$h_a + (h_c)_C + \mathbf{h_m}$	=	$(\mathbf{h_c})_\mathbf{M}$

Solving the two equations for the two unknowns yields:

$\mathbf{h_m = 12.42\ km}$ and $\mathbf{(h_c)_M = 49.42\ km}$ (Moho depth = 47.42 km)

The contribution to the free-air anomaly due to *topography* of the mountains (Fig. 8.37a) results from the *mass excess* of the material above sea level (+Δm). This excess is a function of the product of the mountain height (equal to h_a) times the density difference at upper crustal levels ($\Delta\rho = \rho_c - \rho_a = +2.67$ g/cm³). Note that, as with the water effect for the passive margin, the contribution is abrupt, resulting in a steep gradient.

The *crustal root* provides a *mass deficit* (−Δm) that compensates the extra weight of the topography (Fig. 8.37b). The deficit relates to the product of the amount of deepening of the crust (equal to h_m) times the difference between crustal and mantle densities ($\Delta\rho = \rho_c - \rho_m = -0.43$ g/cm³). If the mountain range is wide (several hundred km), the negative contribution due to the crustal root has the same *amplitude* as the positive contribution due to topography. The *gradient* for the crustal root contribution is more gentle, however, because that anomalous mass is deeper.

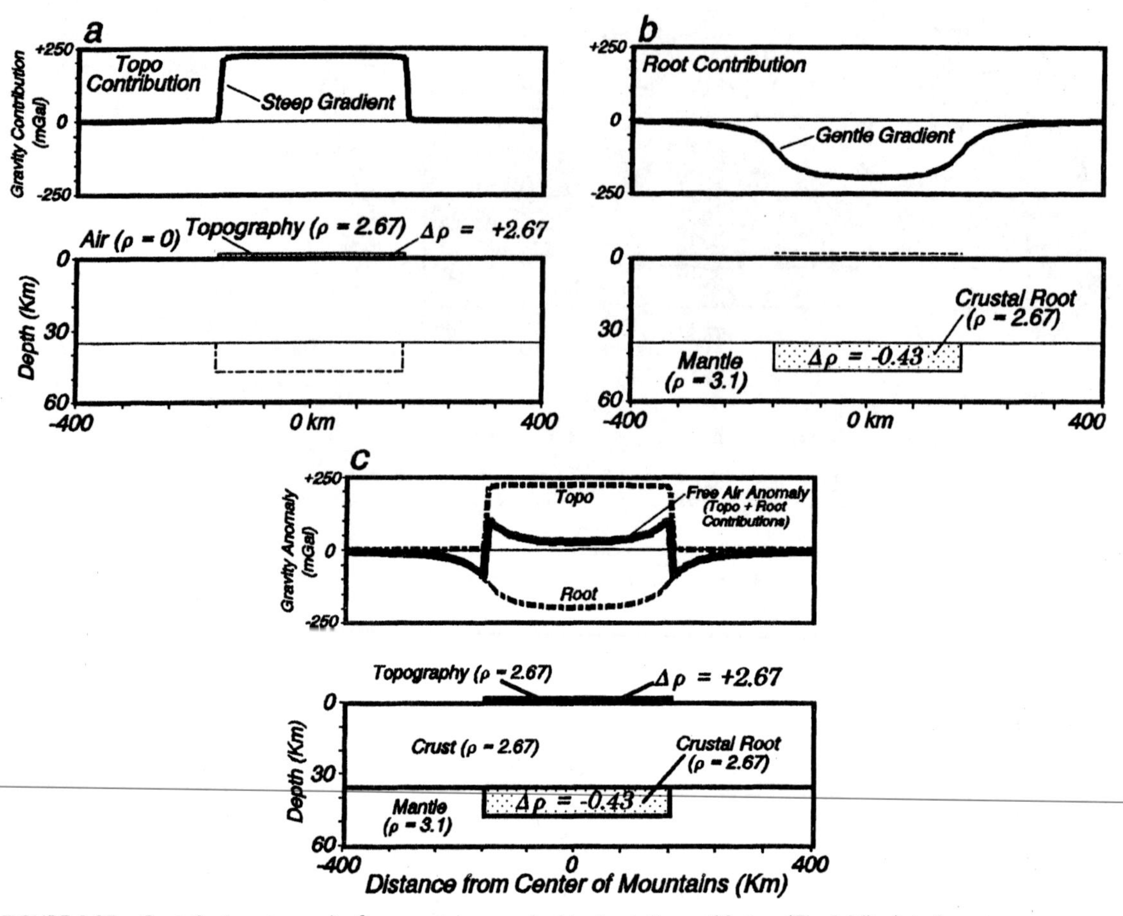

FIGURE 8.37 *Contributions to gravity for mountain range in Airy isostatic equilibrium* (Fig. 8.36). a) A sharp increase, with nearly full amplitude, results from mass excess of topography. b) Mass deficit of crustal root gives a more gradual decrease. c) The free air gravity anomaly profile for a mountain range in local isostatic equilibrium has edge effects due to the differing gradients of the shallow (a) and deep (b) contributions. Unless the range is very broad, so that the deep effect of the root approaches full amplitude, the free air anomaly has significant positive values over the range.

As in the passive margin model, the free air gravity anomaly (Δg_{fa}) for the mountain range is the sum of the contributions from the shallow and deep sources (Fig. 8.37c). The anomaly is zero over the normal thickness continent and approaches zero over the central part of the mountains. It shows edge effects, however, along the flanks of the range.

The free air gravity anomaly for a mountain range often illustrates some of the fundamental properties of a region in *local isostatic equilibrium* (Fig. 8.38a): 1) values are near zero because the mass excess ($+\Delta m$) of the topography equals the mass deficit ($-\Delta m$) of the crustal root; 2) significant edge effects occur because shallow and deep contributions have different gradients; 3) the area under the curve of the anomaly sums to zero.

Fig. 8.39 shows observed and modeled free gravity anomalies across western South America. Note that the observed free air gravity anomaly profile shows classic edge effects, suggesting that the region is close to local isostatic equilibrium (Fig. 8.38a). The model shows that the crust is very thick ($\approx$ 60 km), beneath the high topography of the Andes Mountains. Thinner crust flanks the mountains, as normal thickness

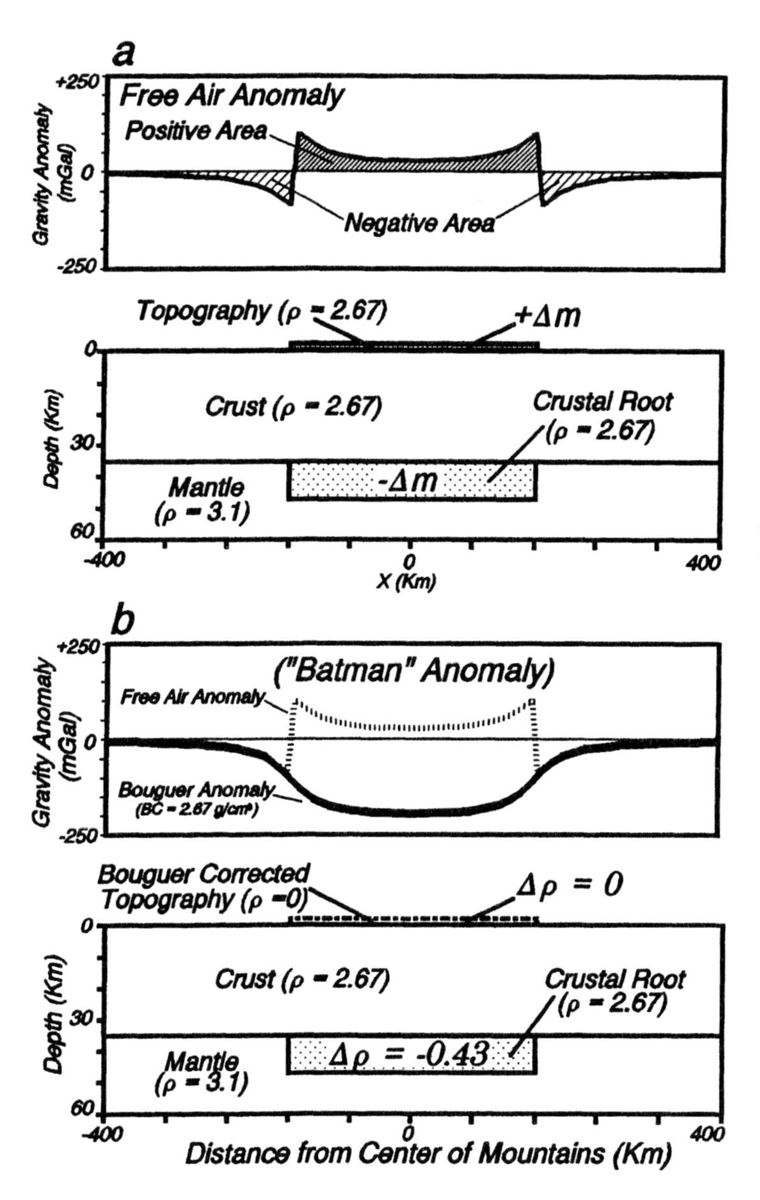

FIGURE 8.38 *Free air and Bouguer gravity anomalies for mountain range in local isostatic equilibrium.* a) Compensatory positive ($+\Delta m$) and negative ($-\Delta m$) mass anomalies mean that the integral, with respect to x, of the free air gravity anomaly is zero ($\int \Delta g_z\, dx = 0$). In other words, the "area under the curve" sums to zero. b) The Bouguer correction (BC) removes most of the contribution of the mass above sea level (Fig. 8.37a). Bouguer anomaly profiles of mountain ranges thus commonly show low values reflecting the contribution of the crustal root (Fig. 8.37b). Taken together, the free air and Bouguer gravity anomalies form the "Batman anomaly" characteristic of a mountain range in local isostatic equilibrium.

continental craton to the east and oceanic crust to the west. The broad region is thus a mountain range close to a state of Airy isostasy (Fig. 8.36). Deviations from local isostasy (discussed below) occur at the subduction zone on the west side, where the edge effect low is exaggerated at the trench and a flexural bulge high occurs over the adjacent oceanic crust.

The Bouguer gravity anomaly (Δg_B) for the mountain range results from subtracting the effect of the mass excess of the topography from the free air anomaly (Fig. 8.9a). Attributes of the Bouguer gravity anomaly that result from isostatic equilibrium are illustrated in Fig. 8.38b: 1) values are near zero over continental crust of normal thickness; 2) the form of the Bouguer anomaly mimics the root contribution; the anomaly decreases as the Moho deepens beneath the mountains; 3) the form of the Bouguer anomaly is almost a mirror image of the topography; the anomaly decreases where the topography of the mountains rises.

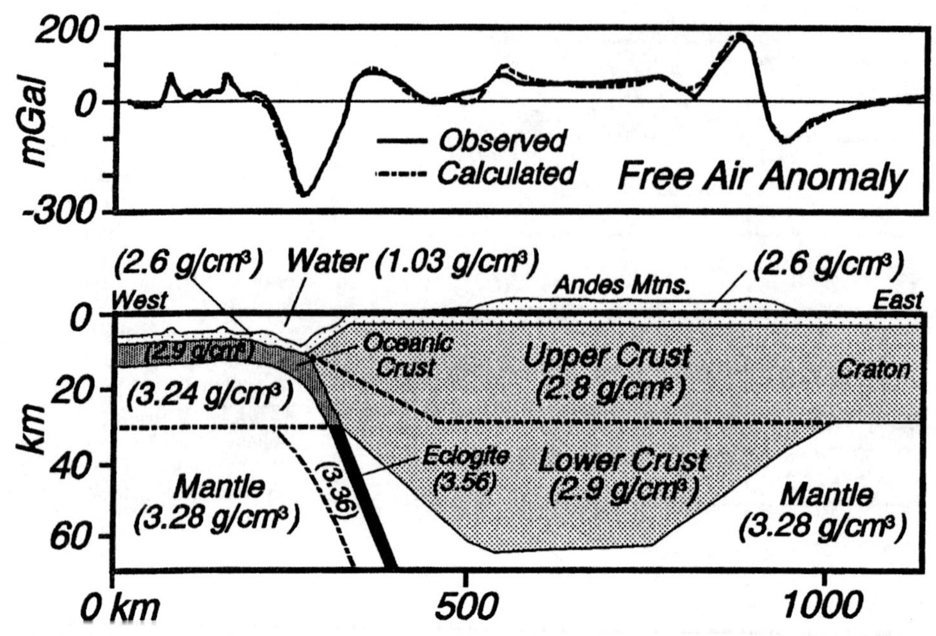

FIGURE 8.39 Observed and modeled free air gravity anomaly of the Andes Mountains and adjacent regions, showing the ears portion of the classic "Batman anomaly." The model shown results in a calculated anomaly in close agreement with the observed. Densities in g/cm^3 (= 10^3 kg/m^3). From J. Grow and C. Bowin, *Journal of Geophysical Research*, vol. 80, pp. 1449–1458, © 1975. Redrawn with permission of the American Geophysical Union, Washington, D.C.

TECTONIC SETTINGS AND THEIR GRAVITY EXPRESSIONS

The gross forms of free air and Bouguer gravity anomalies reflect: 1) the *density distribution* of Earth materials in a region; and 2) the *flexural strength* of the materials. The Airy isostatic model is an end-member case where supporting materials have no flexural strength. While an oversimplification, the Airy model is nonetheless useful in understanding the general form of gravity anomalies. Once those simple forms are appreciated, modeling of other parameters can be attempted (flexural strength, complex density distributions).

The Airy model suggests isostatic balance involving two boundaries: 1) the *topography* and/or *bathymetry*; and 2) the crust/mantle boundary (*Moho*). Those boundaries are important because they are universal and represent large density contrasts. In some regions, a third boundary is a significant element in isostatic balance: 3) the *lithosphere/asthenosphere boundary*.

The Airy isostatic model in Fig. 8.20 can be modified to incorporate all three boundaries, as summarized below and illustrated in Fig. 8.40. At the *depth of compensation* beneath each region, two equations hold true. 1) The total *pressure* (P) exerted by any vertical column (divided by g) is equal to that of any other vertical column:

$$P/g = \rho_a h_a + \rho_w h_w + \rho_c h_c + \rho_m h_m + \boldsymbol{\rho_A h_A} = \text{Constant}$$

where:

ρ_a = density of the air ($\rho_a \approx 0$)

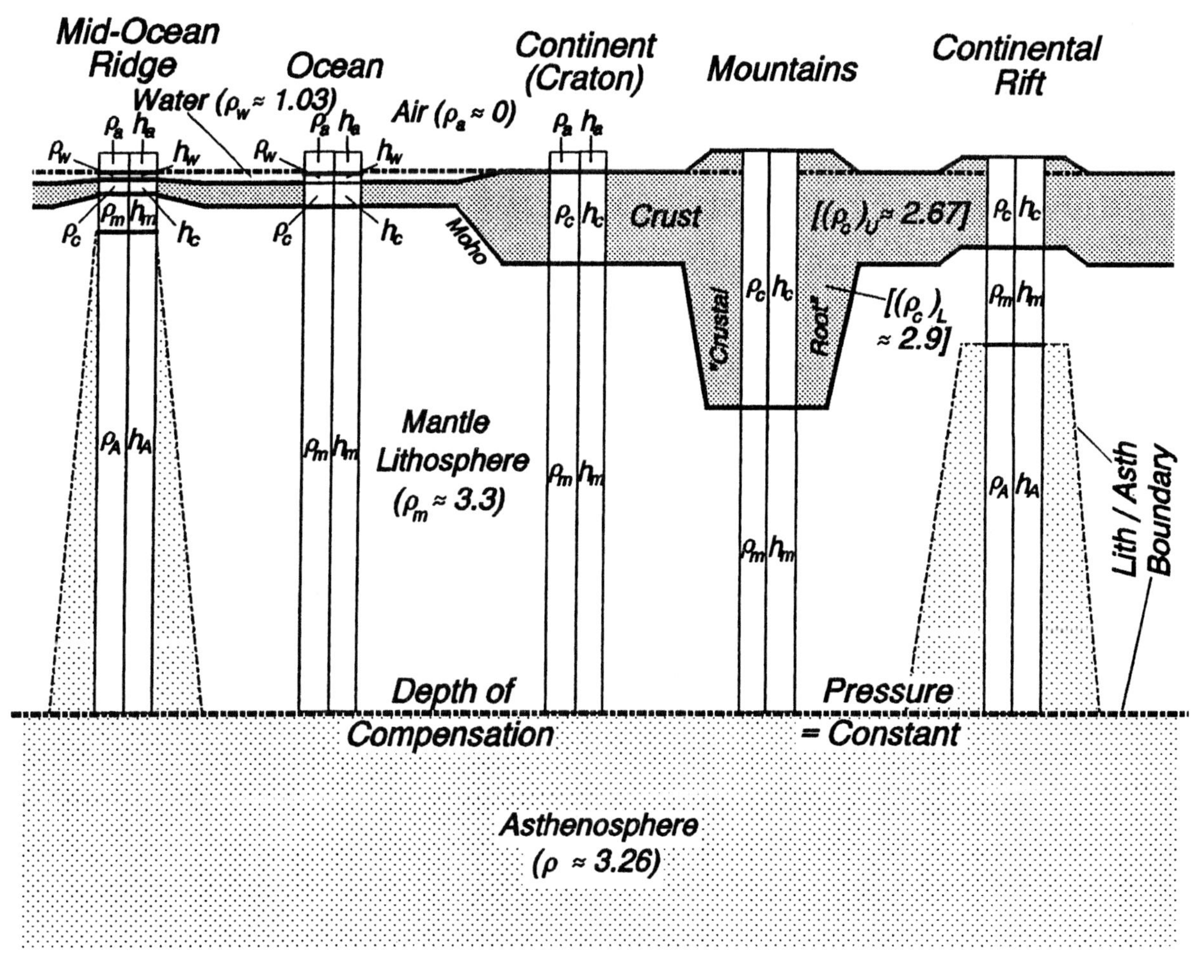

FIGURE 8.40 *Airy isostatic model, modified to include lithosphere/asthenosphere boundary.* See text and subsequent figure captions for definition of parameters. Note that h_c in continental areas is the depth to the Moho below the sea level datum, plus the height of the topography (h_t in subsequent figures). There is a slight drop in density across the lithosphere/asthenosphere boundary; elevated asthenosphere thus provides buoyancy, supporting the weight of uplifted topography and mantle at continental rifts and mid-ocean ridges.

h_a = thickness of the air column
ρ_w = density of the water
h_w = thickness of the water column
ρ_c = density of the crust
h_c = thickness of the crust
ρ_m = density of the mantle part of the lithosphere
h_m = thickness of the mantle part of the lithosphere
$\boldsymbol{\rho_A}$ = **density of the asthenosphere**
$\mathbf{h_A}$ = **thickness of the asthenosphere column.**

2) The total *thickness* (T) of any vertical column is equal to that of any other vertical column:

$$T = h_a + h_w + h_c + h_m + \mathbf{h_A} = \text{Constant}$$

Gravity Anomalies for Regions in Local Isostatic Equilibrium

The models below comprise five tectonic settings. Each setting has characteristic heights of topography (bathymetry) and thicknesses of the crust and entire lithosphere.

1. *Continental Craton*: Topography near sea level; crust and lithosphere of normal thickness.
2. *Continental Rift*: Uplifted topography; thin crust and thin lithosphere.
3. *Continental Margin*: Drop in topography; transition from thicker continental to thinner oceanic crust.
4. *Mid-Ocean Ridge*: Shallower water than normal ocean; thin oceanic crust and thin lithosphere.
5. *Mountain Range*: High topography; thick crust.

The settings are related, in that: a) as a *craton* rips apart the crust and the entire lithosphere thin, forming a *continental rift* zone (Fig. 2.13); b) rifting can continue to the point where new oceanic lithosphere is created, forming an ocean basin with a *mid-ocean ridge* in the center and passive *continental margins* on the sides (Fig. 2.14); c) the basin can close through subduction of the oceanic lithosphere, resulting in collision of the continental margins and thickening of the crust as a *mountain range* forms (Fig. 2.18).

The continental margin and mountain range on the model (Fig. 8.40) are similar to those in previous models (Figs. 8.32, 8.36). Their respective gravity expressions are also similar, because when those features are old, there may be no significant relief on the lithosphere/asthenosphere boundary. The same can be said for a continental craton. Gravity modeling assuming normal Airy isostasy, where compensation is achieved at the depth of the deepest Moho (Fig. 8.20), is often sufficient to illustrate the forms of free-air and Bouguer anomalies in those three settings.

In areas of active plate divergence (continental rifts; mid-ocean ridges), gravity anomalies cannot be explained adequately without considering relief on the lithosphere/asthenosphere boundary. Note that the asthenosphere is less dense than the overlying, mantle part of the lithosphere, so that: 1) the asthenosphere column represents mass deficit ($-\Delta m$), relative to the mantle part of the lithosphere ($+\Delta m$); 2) the density contrast between the asthenosphere and the mantle part of the lithosphere is small, so that large relief on the lithosphere/asthenosphere boundary is required to compensate mass excesses due to topography and crustal thinning.

In active convergent plate settings, the lithosphere/asthenosphere boundary also needs to be considered in modeling gravity anomalies. In the Alps, for example, a lithosphere root is an important component of isostatic balance, contributing 30 to 50 *mGal* to the observed gravity anomalies (Kissling et al., 1983; Lillie et al., 1994). At subduction zones, the mantle of the downgoing plate comprises a mass excess that needs to be accounted for in gravity models as well as models of isostatic equilibrium and lithosphere flexure (Grow and Bowin, 1975).

The models below show contributions to gravity anomalies due to compensatory changes in topography (bathymetry), Moho depth, and lithosphere/asthenosphere boundary depth, for simplified models of the five tectonic settings. Density assumptions are:

Air: $\boldsymbol{\rho_a = 0}$

Water: $\boldsymbol{\rho_w = 1.03\ g/cm^3}$

Upper crust: $(\rho_c)_U = \mathbf{2.67\ g/cm^3}$
Lower crust: $(\rho_c)_L = \mathbf{2.9\ g/cm^3}$
Mantle part of Lithosphere: $\rho_m = \mathbf{3.3\ g/cm^3}$
Asthenosphere: $\rho_A = \mathbf{3.26\ g/cm^3}$

In gravity modeling, important effects are due to *lateral* changes in mass (Δm), reflected by *density contrasts* across boundaries ($\Delta\rho$). Zones of significant density contrast, bounded by the three fundamental boundaries, are therefore:

1. *Shallow effects:*

a) *Mass above sea level vs. air* (***topography***):

$$\Delta\rho = (\rho_c)_U - \rho_a = +\mathbf{2.67\ g/cm^3}.$$

b) *Ocean water vs. upper crust* (***bathymetry***):

$$\Delta\rho = \rho_w - (\rho_c)_U = -\mathbf{1.64\ g/cm^3}.$$

2. *Mantle vs. lower crust* (***Moho***):

$$\Delta\rho = \rho_m - (\rho_c)_L = +\mathbf{0.4\ g/cm^3}.$$

3. *Asthenosphere vs. mantle part of lithosphere* (***lithosphere/asthenosphere boundary***):

$$\Delta\rho = \rho_A - \rho_m = -\mathbf{0.04\ g/cm^3}.$$

For each of the five models, contributions to gravity due to the change in depth to each of the boundaries are calculated. The *free air anomaly* is then computed as the sum of the three contributions. Density contrasts chosen for the topographic and bathymetric effects are exactly the same as those commonly used for Bouguer corrections (Fig. 8.12); the *Bouguer anomaly* is therefore the sum of the contributions without the topographic and bathymetric effects.

The models start with a continental craton, constructed with these simplifying assumptions: 1) the surface of the craton is at sea level; 2) the Moho is at a depth of 33 km; 3) the lithosphere/asthenosphere boundary is at 180 km depth. The series of models can be viewed as a progression of ripping the craton apart (continental rift zone), opening an ocean basin (continental margins and mid-ocean ridge), then closing the ocean and colliding the continental fragments (mountain range). This progression is the "Wilson Cycle," portrayed in Figs. 2.14 to 2.18.

The three fundamental boundaries change depth in each model so that, at the depth of compensation (180 km), the pressure is the same as it was for the starting craton. Contributions to gravity result from changes in depth of the three boundaries, relative to each contribution equal to zero for the craton. A composite model then shows a comparison of the amplitudes and forms of free air and Bouguer gravity anomalies in the different tectonic settings.

1. Continental Craton Fig. 8.41 is a simplified version of the lithosphere of a stable continental craton. The model incorporates a 180 km thick lithosphere; 180 km is thus chosen as the standard depth of compensation for the other models. The model surface is at sea level, so there is no topographic contribution to gravity. Likewise, the Moho and lithosphere/asthenosphere boundaries are flat, resulting in no form to their gravity contributions. Changes in depths to these three boundaries result in positive or negative contributions to gravity anomalies in the other models.

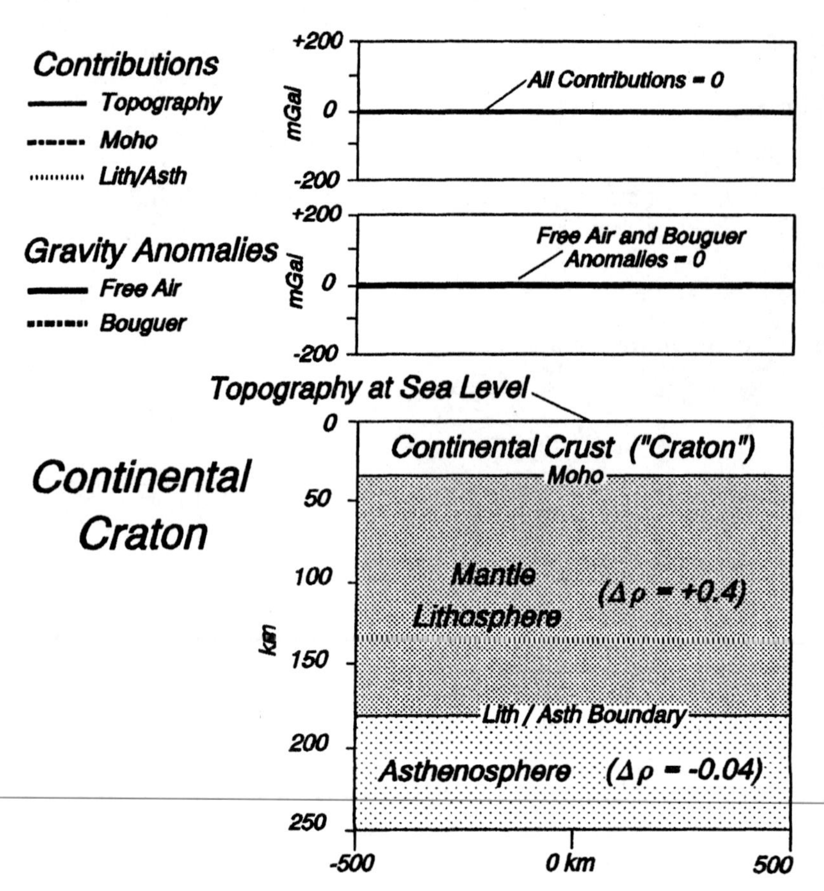

FIGURE 8.41 *Gravity expression of normal continental lithosphere.* Parameters for the model are (see Fig. 8.40):

h_t = height of topography = 0
h_c = thickness of crust = 33 km
h_m = thickness of mantle part of lithosphere = 147 km.

The model arbitrarily assumes that this configuration results in zero amplitude free air and Bouguer gravity anomalies. (In reality, the topography of a craton is tens to a few hundred meters above sea level, resulting in slightly negative Bouguer anomalies). The 180 km depth of compensation is maintained throughout the series of models in the next few figures.

2. Continental Rift At a continental rift (Fig. 8.42), both the crust and entire lithospheric plate thin. Elevated asthenosphere can be viewed as an "inflated balloon," its buoyancy supporting the weight of elevated mantle and, in some cases, topography. Note that the negative density contrast across the lithosphere/asthenosphere boundary ($\Delta\rho = -0.04$ g/cm^3) is far less than the positive contrasts for the topography ($\Delta\rho = +2.67$ g/cm^3) and the Moho ($\Delta\rho = +0.4$ g/cm^3). The relief on the lithosphere/asthenosphere boundary (130 km) is thus far greater than the combined relief of the other boundaries (4.5 km).

The topography and the elevated mantle beneath the thin crust result in positive gravity contributions of relatively steep gradient. The compensatory mass deficit of the elevated asthenosphere is a deep effect, giving a much broader gravity low that does not reach full amplitude. The free air gravity anomaly (sum of the three effects) is a high over the rift, with flanking lows. When the effect of mass above sea level (topographic contribution) is removed, the Bouguer anomaly is a broad low reflecting the elevated asthenosphere.

The Basin and Range Province in the western United States illustrates the pattern of free air and Bouguer gravity anomalies shown in Fig. 8.42. The asthenosphere is so shallow that there is very little mantle lithosphere, as indicated by low seismic velocities (Fig. 4.15, 4.19a) and low Bouguer anomalies (Fig. 8.43a). Hot asthenosphere at shallow depth is under such low pressure that partial melting occurs (Figs. 2.7, 2.24b). The resulting gabbroic magma can underplate the continental crust, establishing a new, flat Moho. This interpretation is consistent with obser-

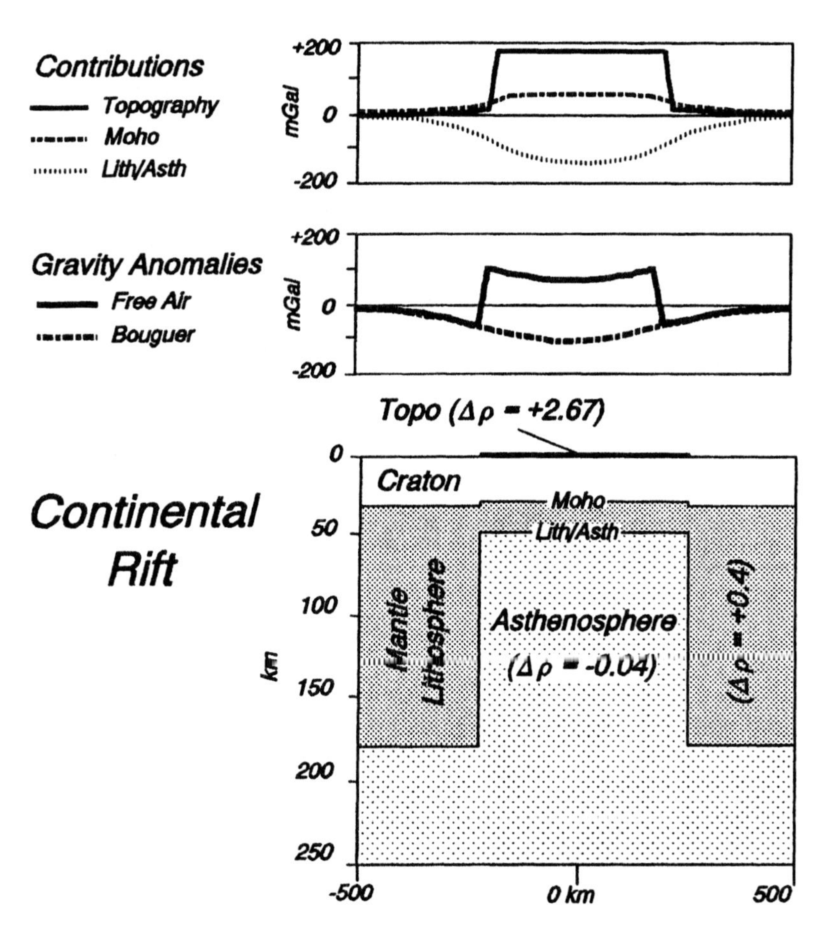

FIGURE 8.42 *Local isostatic model, showing free air and Bouguer gravity anomalies at a continental rift zone.* Density contrasts, in g/cm^3, are: +2.67 for the mass above sea level relative to air; +0.4 for the mantle lithosphere relative to the lower crust; and −0.04 for the asthenosphere relative to the mantle lithosphere. The depth of compensation (180 km) is the same as in the starting craton model (Fig. 8.41). Parameters for the model are:

At the rift:
h_t = height of topography = 1.5 km
h_c = thickness of crust = 31.5 km (Moho depth = 30 km)
h_m = thickness of mantle part of lithosphere = 19.875 km
h_A = thickness of asthenosphere column = 130.125 km.

Outside the rift (normal continental lithosphere):
h_t = 0
h_c = 33 km
h_m = 147 km
h_A = 0.

vations of prominent reflections from a horizontal Moho at about 30 km depth beneath much of the Basin and Range Province (Figs. 6.23, 6.24).

Continental rifts have a wide variety of crustal and lithospheric structure and, hence, a variety of free air and Bouguer gravity anomalies. The Pannonian Basin in central Europe is broad, like the Basin and Range Province, but it has thinner crust (≈ 25–30 km) and less relief on the lithosphere/asthenosphere boundary (Babuška et al., 1988; Šefara, 1986). The region is thus a topographic depression. Free air anomalies have a form similar to that shown in Fig. 8.42, but with much smaller amplitude; the Bouguer anomaly is a broad, low-amplitude high reflecting the Moho relief (Bielik, 1988; Lillie et al., 1994).

The East African rift zone is narrower than the Basin and Range Province or Pannonian Basin rifts. The Bouguer gravity anomaly low is prominent, resulting from very shallow asthenosphere (Fig. 8.43b). Superimposed on the low is a high of about 50 *mGal*, interpreted to be caused by high density (gabbroic?) material intruded throughout the rift valley crust.

At very old continental rift zones, cooling transforms the density of the elevated asthenosphere back to that of normal mantle lithosphere; the lithosphere/asthenosphere boundary thus flattens back to normal continental depths. Intruded gabbro cools and densifies, weighting down the crust, causing it to sag. An example is the 1.1 billion-year-old Keweenawan Rift in the central United States (Fig. 8.43c). The Bouguer gravity anomaly is a broad low reflecting the sagging crust and

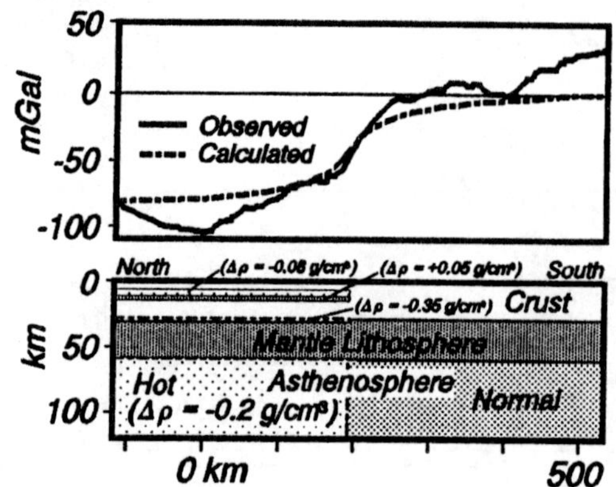

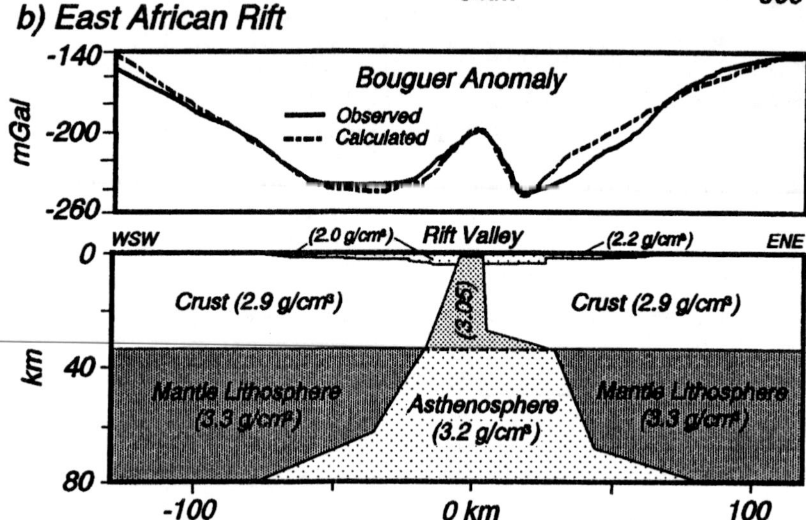

c) Keweenawan Rift

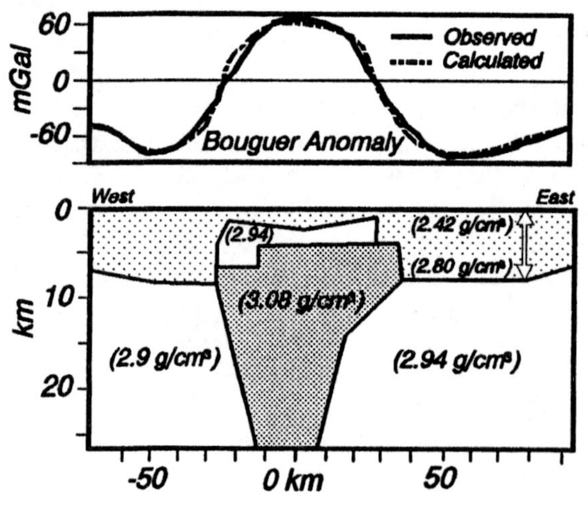

FIGURE 8.43 Bouguer gravity anomalies at continental rifts, along with interpreted density models. a) *Basin and Range Province* from southern California to central Nevada. From R. Saltus and G. Thompson, *Tectonics,* vol. 14, pp. 1235–1244, © 1995. Redrawn with permission of the American Geophysical Union, Washington, D.C. Observed curve is a residual anomaly, whereby effects of Cenozoic basins were removed from the complete Bouguer anomaly. Low values around −100 *mGal* reflect shallow, hot asthenosphere that supports the weight of the higher topography on the north side of the section. b) *East African Rift.* From "The geology of the Eastern Rift System of Africa," by B. Baker, P. Mohr, and J. Williams, GSA Special Paper, No. 136, 67 pp., © 1972. Redrawn with permission of the Geological Society of America, Boulder, Colorado, USA. A Bouguer gravity maximum, due to high density material intruding the crust, is superimposed on the asthenosphere minimum. c) *Keweenawan Rift* in Iowa and Nebraska, central United States. From L. Ocola and R. Meyer, *Journal of Geophysical Research*, vol. 78, pp. 5173–5194, © 1973. Redrawn with permission of the American Geophysical Union, Washington, D.C. The prominent high results from relatively high-density (3.08 g/cm^3) material extending through the crust, perhaps gabbro solidified within this 1.1 billion-year-old rift (compare with modern rift in b).

sedimentary infill of rift basins; a superimposed gravity high results from gabbro that intruded the crust along the rift axis (Serpa et al., 1984).

3. Passive Continental Margin In a fully evolved ocean basin, passive continental margins are far from a mid-ocean ridge (Fig. 2.14c). The region cools through time, so that there may be no significant relief left on the thermally controlled, lithosphere/asthenosphere boundary (Fig. 8.44). Substantial change in depth remains on the Moho, which is a chemical boundary (Figs. 2.3, 4.9). The free air and Bouguer gravity anomalies at a fully evolved passive margin are therefore of the form and amplitude shown in Figs. 8.33 to 8.35; isostatic compensation and gravity anomalies result from balance between mass excess of the extra mantle beneath the thin oceanic crust and the overlying, low-density water column.

4. Mid-Ocean Ridge The isostatic situation for a mid-ocean ridge (Fig. 8.45) is similar to that depicted for a continental rift zone (Fig. 8.42). The density deficiency

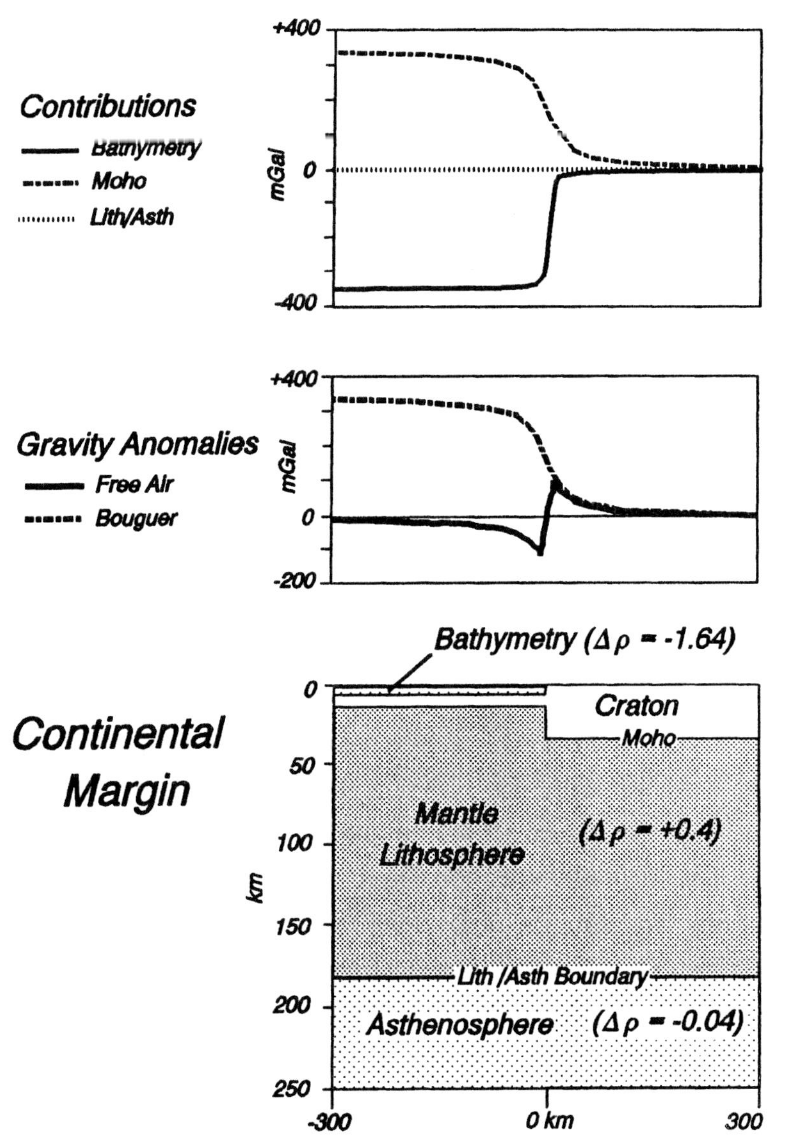

FIGURE 8.44 *Contributions of three fundamental boundaries and resulting gravity anomalies for a passive continental margin in local isostatic equilibrium.* Parameters used in the model are:

Normal oceanic lithosphere:
h_w = thickness of water column = 5 km
h_c = thickness of crust = 7.5 km
(Moho depth = 12.5 km)
h_m = thickness of mantle part of lithosphere = 167.5 km.

Normal continental lithosphere:
$h_w = 0$
h_c = 33 km
h_m = 147 km.

Density contrasts are nearly identical to those used in the earlier passive margin model (Figs. 8.32 to 8.34), and there is no relief on the lithosphere/asthenosphere boundary; free air and Bouguer gravity anomalies are essentially the same as in the earlier model.

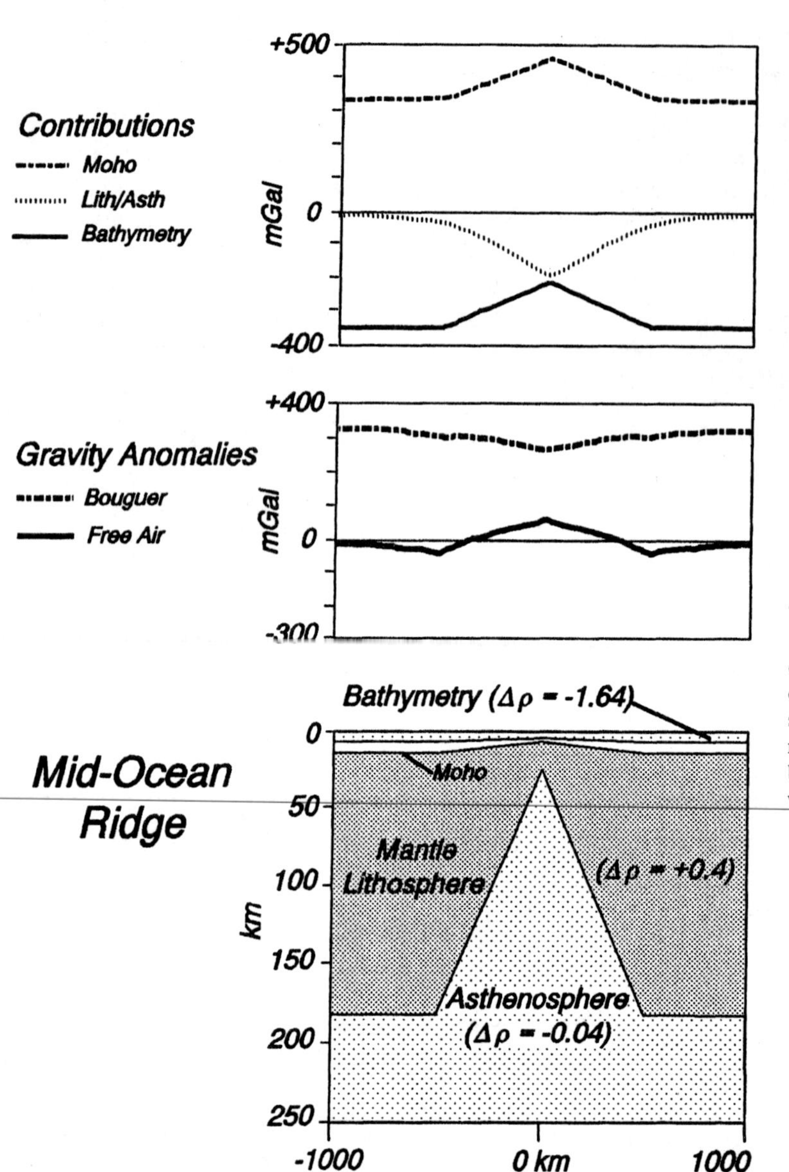

FIGURE 8.45 *Contributions of three fundamental boundaries and resulting gravity anomalies for a mid-ocean ridge in local isostatic equilibrium.* The depth of compensation is 180 km. Model parameters are:

On the ridge:
h_w = thickness of water column = 3 km
h_c = thickness of crust = 2 km (Moho depth = 5 km)
h_m = thickness of mantle part of lithosphere = 18.0 km
h_A = thickness of asthenosphere column = 157.0 km.

Away from the ridge (normal oceanic lithosphere):
h_w = 5 km
h_c = 7.5 km (Moho depth = 12.5 km)
h_m = 167.5 km
h_A = 0.

The Moho contribution is more than +300 *mGal*, because the region has thin crust, relative to the standard, continental craton (Figs. 8.40, 8.41; compare to the seaward portion of the passive margin model, Fig. 8.44). Likewise, the bathymetric contribution has background values less than −300 *mGal*.

of the elevated asthenosphere supports the weight of the ridge and elevated mantle. The positive changes in gravity across the ridge from the Moho and bathymetry are superimposed on a broad low resulting from the deep effect of the asthenosphere. The resulting free air gravity anomaly thus displays a high with flanking lows.

After effectively removing the bathymetric contribution, the Bouguer anomaly shows the more than 300 *mGal* background level associated with oceanic crust (see Bouguer anomaly for the passive continental margin in Figs. 8.34 and 8.44). There is a drop in Bouguer anomaly values over the ridge, due to the elevated asthenosphere.

Fig. 8.46 shows free air and Bouguer gravity anomalies observed across the Mid-Atlantic Ridge. Short-wavelength highs and lows mimic the topographic relief on the sea floor, which is uncompensated. The longer wavelength anomalies show widths and amplitudes that are in general agreement with those computed in

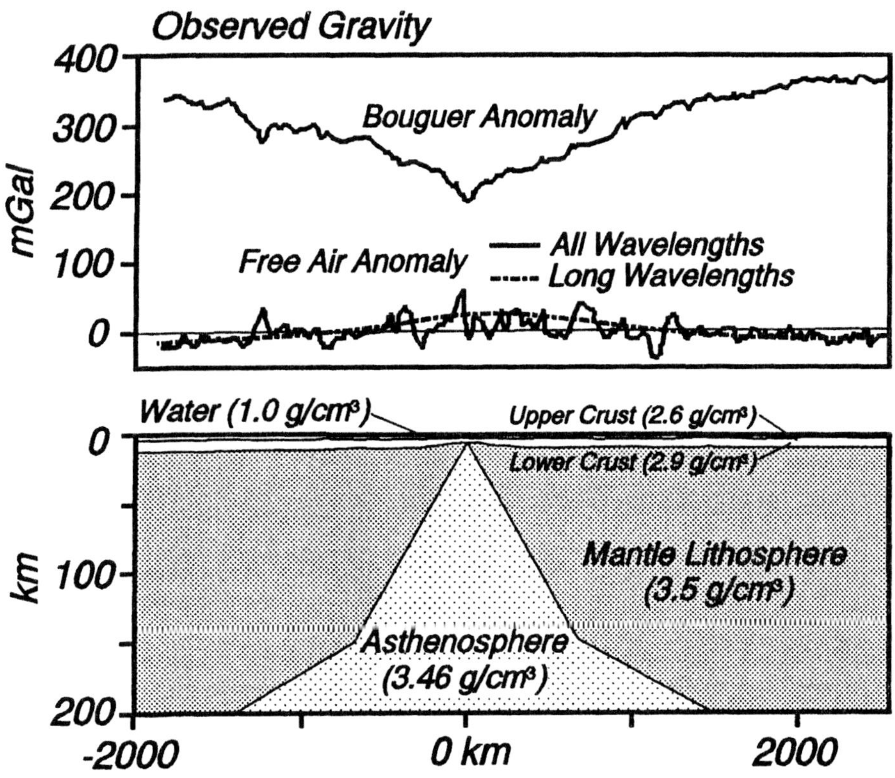

FIGURE 8.46 Free air and Bouguer gravity anomalies observed across the Mid-Atlantic Ridge (Talwani et al., 1965). Densities in model are in g/cm^3 (= 10^3 kg/m^3). Interpreted density distribution (Keen and Tramontini, 1970) shows shallow asthenosphere with density contrast of -0.04 g/cm^3, relative to adjacent mantle lithosphere. Compare observed gravity anomalies and lithosphere structure with Fig. 8.45. From M. Talwani, X. Le Pichon, and M. Ewing, *Journal of Geophysical Research*, vol. 70, pp. 341–352, © 1965. Redrawn with permission of the American Geophysical Union, Washington, D.C.

Fig. 8.45; the free air anomaly shows a broad central high with flanking lows, and the Bouguer anomaly reveals a decrease associated with the shallow asthenosphere. The modeled lithosphere structure for the Mid-Atlantic Ridge thus suggests local isostatic equilibrium, the hot asthenosphere supporting the weight of the topography and elevated mantle at the ridge.

5. Mountain Range The lithosphere/asthenosphere boundary generally has little relief left as a mountain range ages, so that the topography and supporting crustal root are the major contributions to gravity anomalies (Fig. 8.47). The density contrast of the topography compared to air ($\approx +2.67$ g/cm^3) is generally 5 to 8 times that of the crustal root compared to surrounding mantle (≈ -0.4 g/cm^3); crustal roots must therefore be 5 to 8 times as thick as the topographic relief of a mountain range. In the Himalaya Mountains and adjacent Tibetan Plateau, for example, the 4.5 km elevation of the region is supported by crust that is about 30 km thicker than normal.

For some mountain ranges it is essential to consider relief on the lithosphere/asthenosphere boundary (for example, the Alps in Europe and Sierra Nevada in the United States). Those ranges are so young that there is still a root of lithosphere left from the plate convergence that formed the mountains (Fig. 2.16b

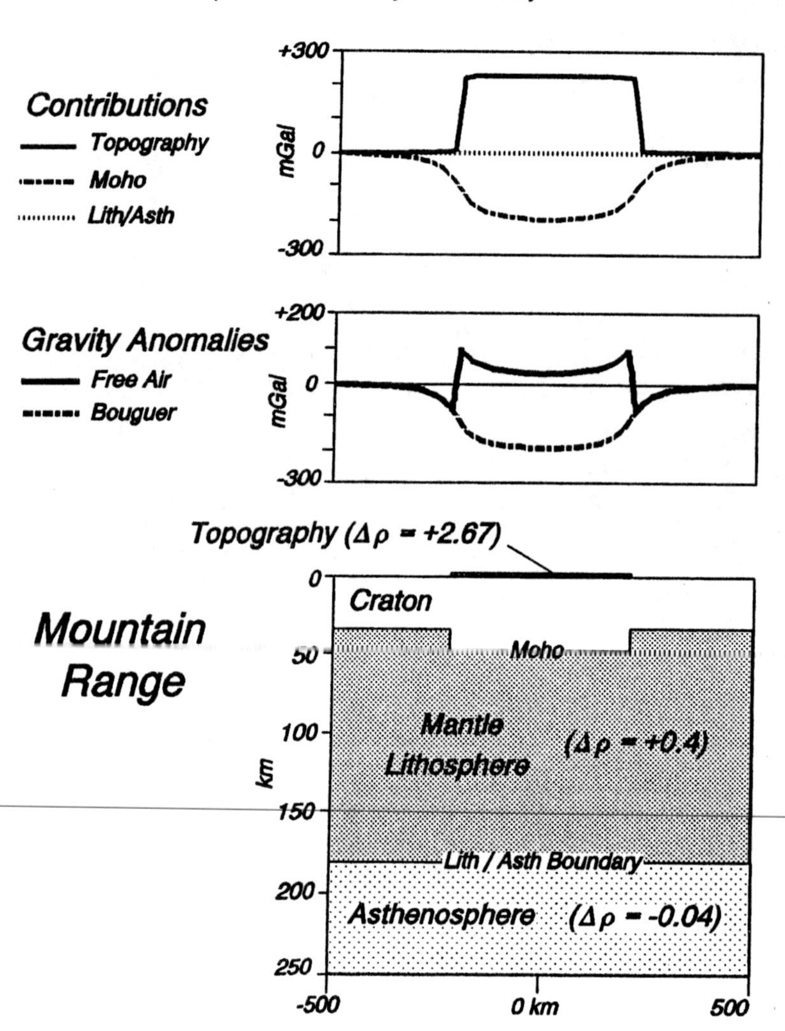

FIGURE 8.47 *General form of free air and Bouguer gravity anomalies at a mountain range.* Parameters are:

At the mountains:
h_t = height of topography = 2 km
h_c = thickness of crust = 48.35 km (Moho = 46.35 km deep)
h_m = thickness of mantle part of lithosphere = 133.65 km.

Outside the mountains (normal continental lithosphere):
$h_t = 0$
h_c = 33 km
h_m = 147 km.

In an old mountain range the thermally controlled, lithosphere/asthenosphere boundary equilibrates to a horizontal surface. Free air and Bouguer anomalies thus reveal the classic "Batman anomaly," because the major contributions are from the topography and its supporting crustal root (Fig 8.38)

for the Sierra Nevada; Fig. 2.18 for the Alps). In such cases, the lithosphere roots represent mass excesses that lessen the topographic relief.

Composite Model The composite model (Fig. 8.48) shows general forms of contributions, and free air and Bouguer gravity anomalies, at each of the five tectonic settings. The figure is important because it represents the "background" (or "regional") anomalies that one should appreciate before modeling and interpreting gravity anomalies in a particular setting. By understanding the major contributions for a local isostatic situation, one can then analyze problems of mass distribution and lithospheric strength in terms of deviation from local isostasy.

Gravity Anomaly Maps

Gravity anomaly maps present information useful to understanding density distribution and isostatic state of a region. An ideal situation would be to analyze, together, maps of: 1) *topography* (see front inside cover of this book); 2) *surface geology*; 3) *sediment thickness and intra-crustal structure* based on drilling and seismic reflection profiles (Figs. 6.23 to 6.39); 4) *Moho depths* from refraction surveys

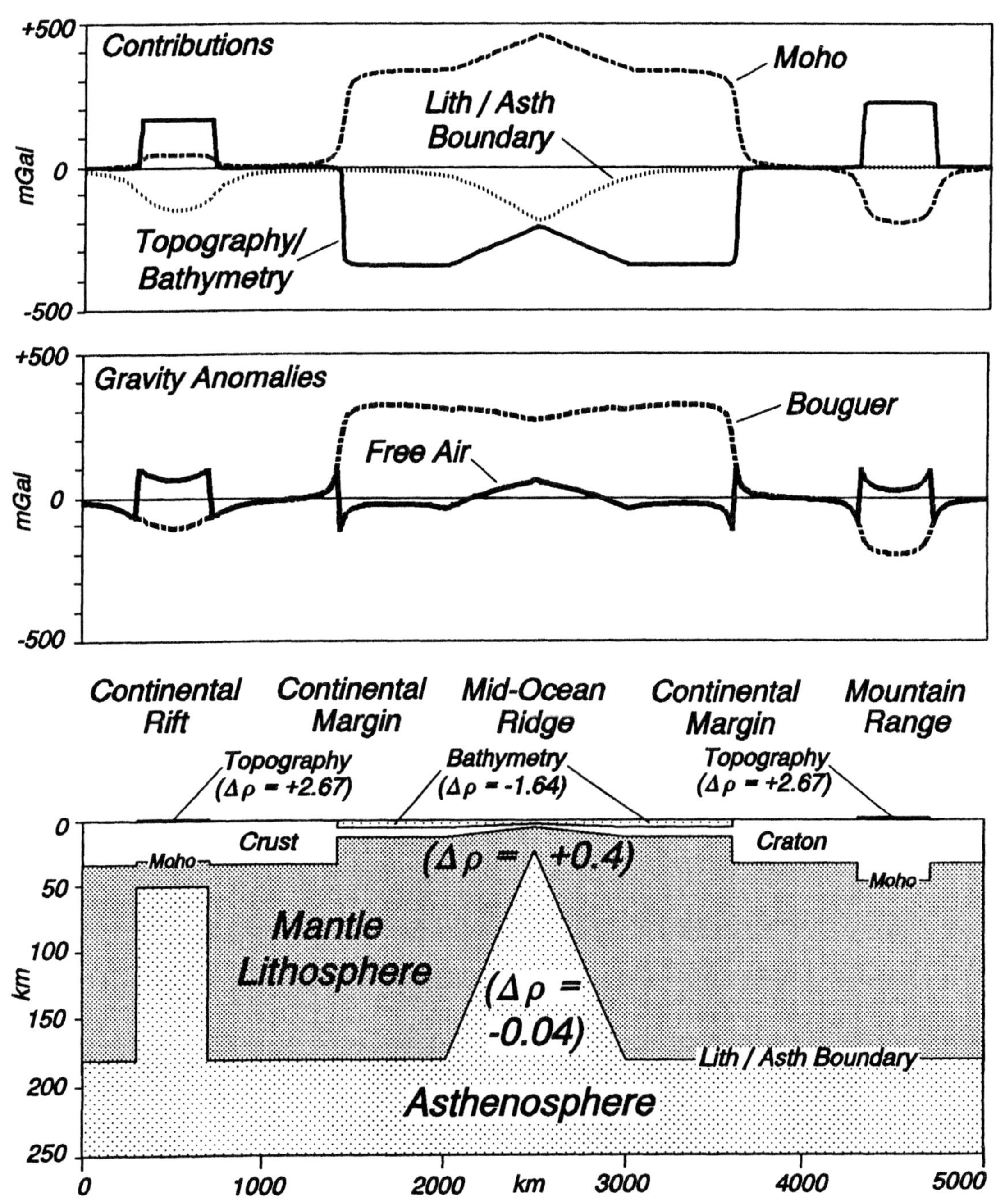

FIGURE 8.48 *Composite model of local isostatic equilibrium, showing all five tectonic settings from the previous series of illustrations.* The depth of compensation is the depth of the deepest lithosphere (180 km). Density contrasts ($\Delta\rho$ in g/cm^3) are for topography relative to air; bathymetry relative to upper crust; mantle lithosphere relative to lower crust; asthenosphere relative to mantle lithosphere. Note that the contributions from the three critical boundaries lead to free air anomalies near zero, except at edge effects. The fact that the positive and negative areas under the curve sum to zero illustrates local isostasy. The Bouguer gravity anomaly generally follows the form of the Moho, staying near zero over the cratons, rising over the thin oceanic crust, and falling where the crust is thick beneath the mountain range. Bouguer lows also reflect asthenosphere elevated beneath continental rifts and mid-ocean ridges. The inverse relationship between Bouguer anomaly and topography/bathymetry is apparent: high values occur where bathymetry is deep in the ocean; relative lows are associated with surface relief at the continental rift, mid-ocean ridge, and mountain range.

(Figs. 4.16 to 4.19); 5) *depths to the lithosphere/asthenosphere boundary* from seismic delay time/tomography methods (Fig. 7.34); 6) *free air gravity anomaly*; 7) *Bouguer gravity anomaly*. Constraints on densities come from surface sample and borehole measurements, as well as approximations based on seismic velocities (Fig. 3.10). Incorporation of all of these data would more likely result in cross-sectional interpretations that honor geological reality (first order constraints, Fig. 1.8), and that consider the state of isostasy.

North America The gravity anomaly map of North America (back inside cover of this book) presents free air anomalies at sea and Bouguer anomalies on land. The anomalies correspond at the shoreline, because the Bouguer correction at sea level is zero. Several prominent features are evident; most can be interpreted in terms of the isostatic model in Fig. 8.48.

1. Near-zero values occur over the continental craton, comprising most of the eastern parts of Canada and the United States (Fig. 8.41).
2. Low Bouguer values are coincident with the high topography and crustal root of the Rocky Mountains, extending from western Mexico to western Canada (Figs. 4.16, 8.38, 8.47).
3. A broad Bouguer low occurs over the Basin and Range Province, caused by shallow asthenosphere (Figs. 8.42, 8.43a).
4. A prominent gravity high runs through the central United States, caused by gabbro intruded into the crust along the failed, 1.1 billion-year-old Keweenawan rift (Fig. 8.43c).
5. Free air edge effects are prominent near the shelf edge off the east coast of Canada and the United States, representing the transition from thick continental to thinner oceanic crust of the passive margin (Figs. 8.34, 8.35).
6. Edge effects also occur along transitions from thicker (continental or island arc) to thinner (oceanic) crust at subduction zones, although modified by deviations from local isostasy (see left side, Fig. 8.39). At the Alaska Peninsula and Aleutian island arc, effects of a rigid lithosphere accentuate the highs and lows of the free air anomaly, and a flexural bulge high occurs seaward of the trench. Similar effects are observed in the Lesser Antilles arc in the Caribbean region, and along the Pacific coast of Mexico.
7. The Mid-Atlantic Ridge segment south of Iceland displays a broad low split by a central high, characteristic of free air anomalies across a mid-ocean ridge (Figs. 8.45, 8.46).

Central Europe Prominent features evident from the Bouguer gravity anomaly map of central Europe (Fig. 8.49) include: 1) a narrow, low-amplitude minimum associated with a downwarped "keel" on the Moho, formed during an early stage of continental collision in the Carpathians (Fig. 8.50b); 2) a broad low representing a crustal root beneath the eastern Alps, a consequence of a more advanced stage of continental collision (Fig. 8.50c); 3) a circular high representing shallow mantle beneath the Pannonian Basin (Fig. 8.51).

Deviations from Local Isostasy

Lithospheric plate strength partially supports the weight of topographic and subsurface loads. Gravity anomalies resulting from the two examples presented earlier in this chapter (Fig. 8.24) can be studied in the context of how they deviate from anomalies caused by simple Airy isostasy.

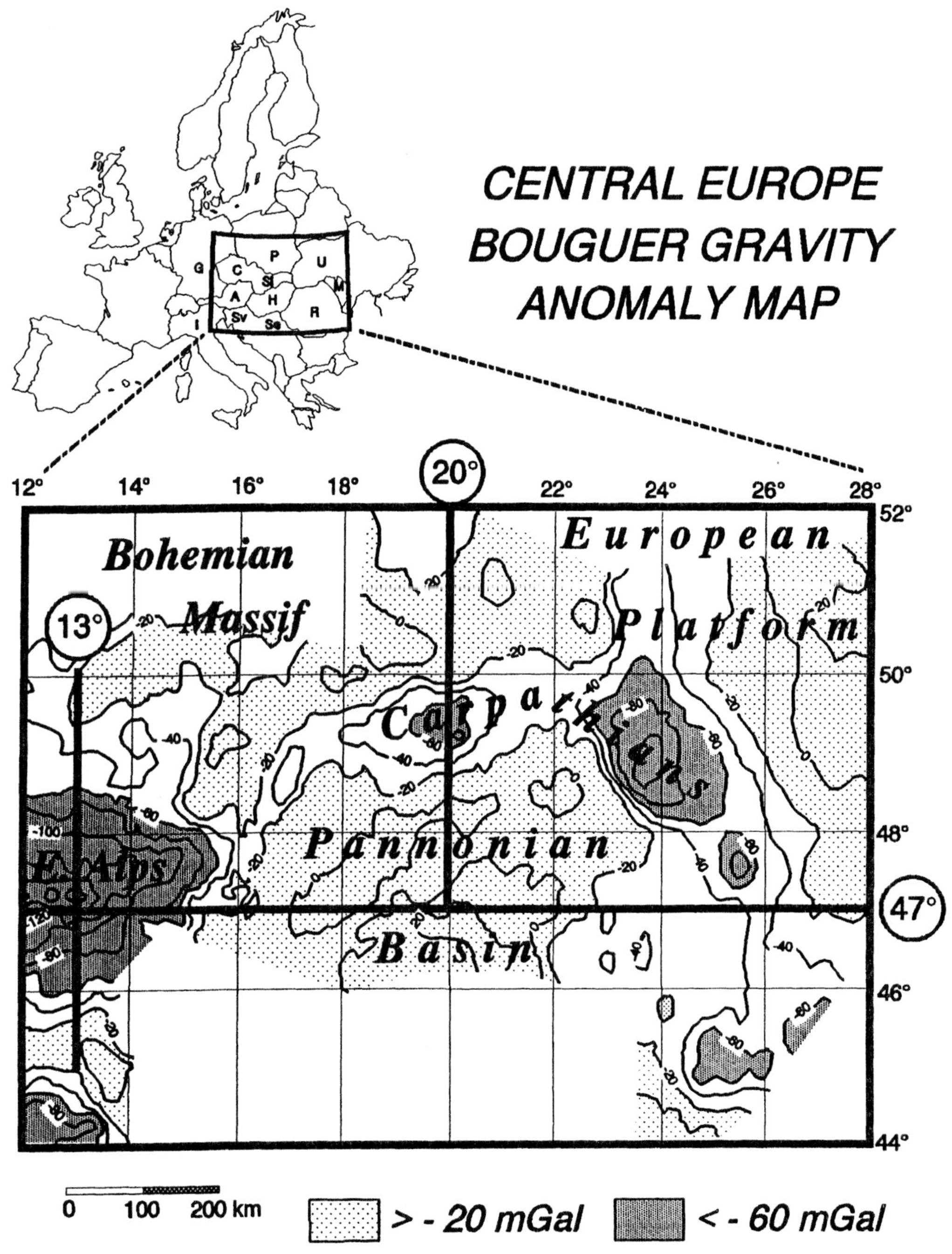

FIGURE 8.49 *Bouguer gravity anomaly map of central Europe* (Lillie et al., 1994). Gravity profiles along 13° E, 20° E, and 47° N shown in Figs. 8.50 and 8.51. Blank regions in southern and extreme northwestern and southeastern portions of map lacked sufficient station density to draw contours. A = Austria; C = Czech Republic; G = Germany; H = Hungary; I = Italy; M = Moldava; P = Poland; R = Romania; Se = Serbia; Sl = Slovakia; Sv = Slovenia; U = Ukraine.

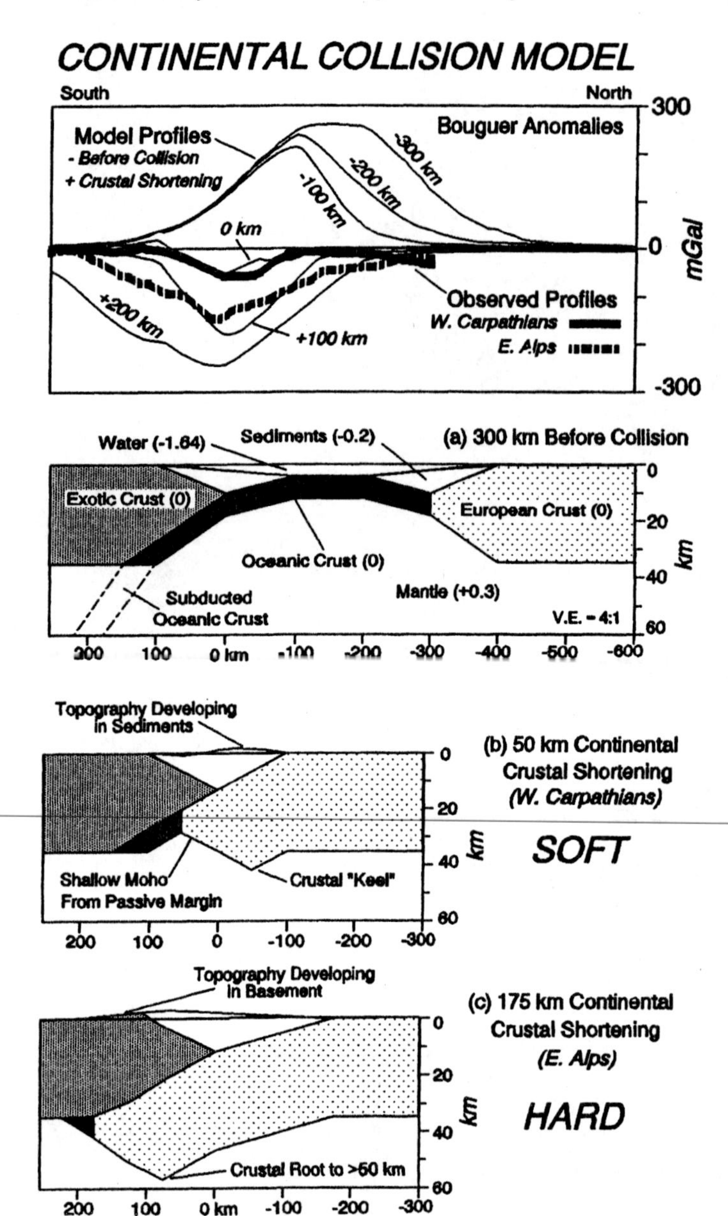

FIGURE 8.50 *Balanced crustal model of ocean basin closure, maintaining local isostatic equilibrium* (Lillie et al., 1994). The upper diagram shows the Bouguer gravity anomaly at different stages of continental collision. The −300 km profile represents the Bouguer anomaly when the ocean basin is 300 km wide; the 0 km profile shows the Bouguer anomaly when the continental crusts initially collide; the +200 km profile represents 200 of continued crustal convergence (shortening) after collision. Observed Bouguer anomaly profiles (Fig. 8.49) for the eastern Alps (13° E) and western Carpathians (20° E) are superimposed. a) Ocean basin and bordering passive and convergent margins, at a stage 300 km before the margins collide. Numbers in parenthases are density contrasts, in g/cm^3. A Bouguer anomaly high occurs over the ocean basin (Fig. 8.34b). b) *Soft Collision*. Crustal configuration after one continent has overridden the edge of the other by 50 km, as in the western Carpathians. A narrow Bouguer anomaly low results from a depressed Moho ("keel") at the edge of the continental crust. c) *Hard Collision*. Situation after 175 km of continental crustal shortening (eastern Alps). A broad low results from a crustal root supporting the high topography of the mountains (Fig. 8.38b).

Mountain Range If a lithospheric plate supports the topographic load of a mountain range, effects of the load may be distributed over a broad region (Fig. 8.24b). Three important perturbations to gravity anomalies result (Fig. 8.52). 1) In the region of the load (mountains), the Moho is not depressed as much for the flexural case as it would be for the Airy case, resulting in higher gravity anomalies. Relative to the Airy model, flexural strength thus means that the region of the mountains is *undercompensated*. 2) In front of the mountains, there is a depression (foreland basin) where none existed in the Airy case. The top of the continental basement and Moho are depressed, leading to lower Bouguer gravity anomalies. Flexural strength thus means that the region of the foreland basin is *overcompensated*. 3) At the flexural bulges both the top of basement and the Moho are upwarped, resulting in free air and Bouguer gravity anomaly highs.

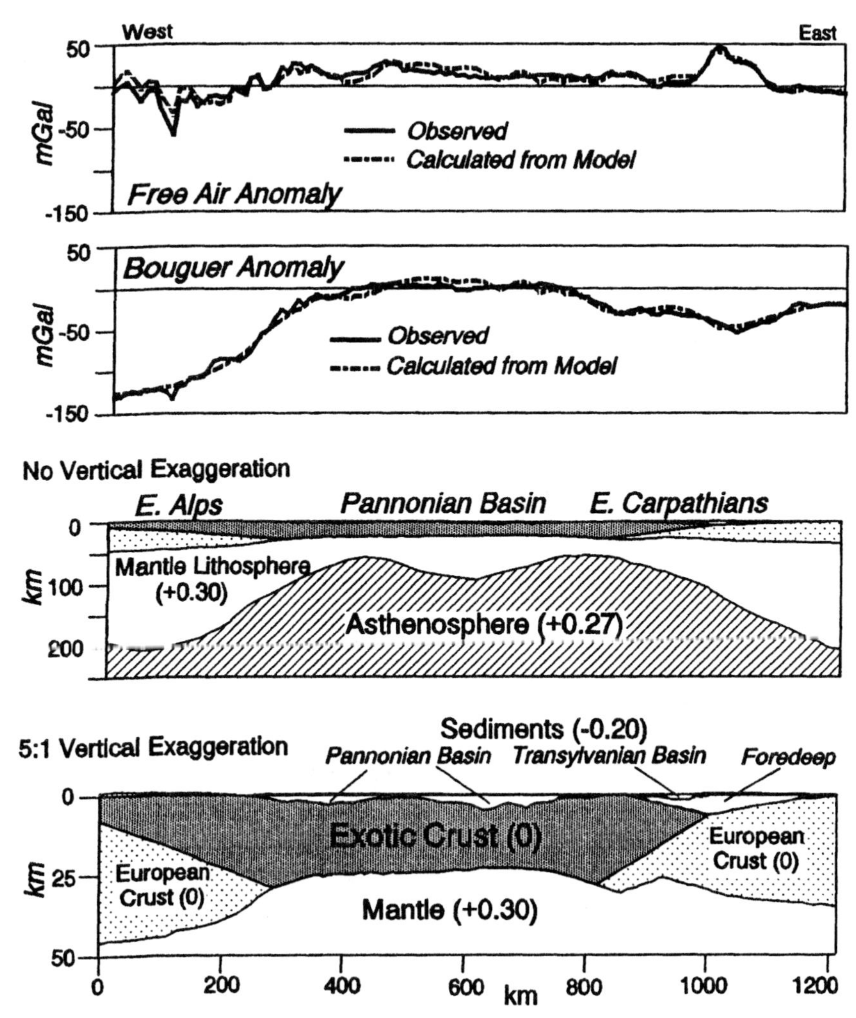

FIGURE 8.51 *Observed free air and Bouguer gravity anomalies, along with anomalies computed from density model, from the Pannonian Basin in central Europe* (47° N profile, Fig. 8.49). Assumed density contrasts (in g/cm^3) are relative to typical crustal materials. The model shows shallow mantle beneath thin crust, supported by elevated asthenosphere. Upper cross section shows entire model, down to 250 km depth, with no vertical exaggeration. Lower section is the upper 50 km of the model, at 5:1 vertical exaggeration, illustrating the crustal thinning. The Bouguer anomaly high, in this case, reflects the shallow mantle. From Lillie et al. (1994).

Subduction Zone (Convergent Continental Margin) At a convergent continental margin, the main form of the free air anomaly is an edge effect similar to that of a passive continental margin (Fig. 8.33). In other words, the main contributions are a shallow, negative effect due to the water deepening seaward, compensated by the mantle shallowing in the same direction. Where the downgoing plate has flexural strength, loads from the overriding plate are distributed over a broad region (Fig. 8.24a). Analogous to the mountain range (Fig. 8.52), three things result. 1) Distributing the load means less depression of the Moho in the area of the accretionary wedge, enhancing the edge effect high. 2) The water deepens and the Moho is depressed in the region of the trench, enhancing the low portion of the edge effect on the free air anomaly. 3) The water and Moho are shallow at the flexural bulge, producing both free air and Bouguer gravity highs.

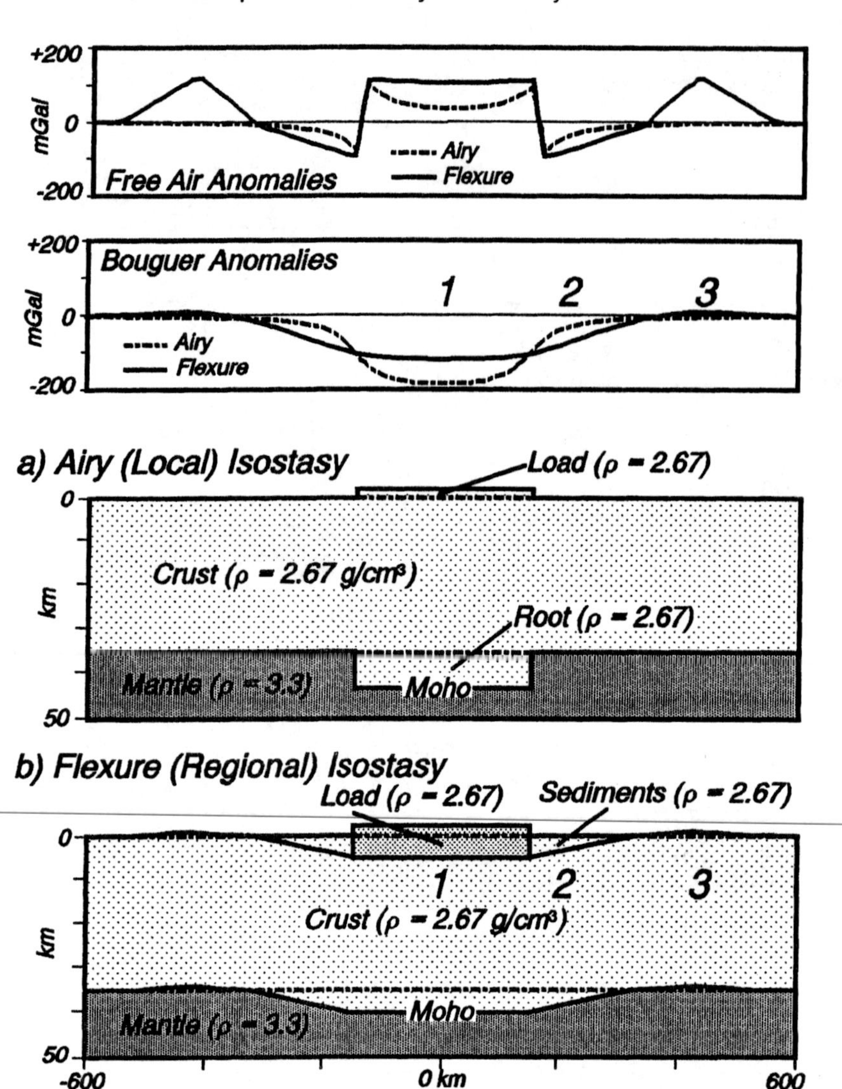

FIGURE 8.52 *Hypothetical model showing free air and Bouguer gravity anomalies for a mountain range in local (Airy) and regional (flexure) isostatic equilibrium.* a) Airy isostasy (Fig. 8.38) results in a free air anomaly profile with edge effects on both sides of the range, and values falling to near zero over the center. The Bouguer anomaly is the low due to the mountain root. b) Strength to the lithospheric plate results in higher gravity anomaly values over the range (1), with additional highs due to flexural bulging (3). Adjacent to the mountains (2), the downwarped Moho results in lower values compared to the Airy case; those values are even lower if there is significant density contrast between the sediments and crystalline crust.

The left portion of Fig. 8.39 shows this characteristic free air anomaly at the convergent continental margin of South America. The flexural bulge on the oceanic lithosphere results in a broad free air gravity high, adjacent to the edge-effect anomaly consisting of an enhanced low over the trench and a high over the continental edge. Similar high/low/high anomalies are evident at the Aleutian, Central American, and Caribbean subduction zones in the map on the back inside cover of this book.

EXERCISES

8-1 a) Using the equation for theoretical gravity, compute the difference in gravity expected at the equator compared to the poles.

b) Draw *sketches* and compute the *relative contributions* (in *mGal*) for each of the three factors responsible for the difference (Fig. 8.4a). **i)** Use an equation for *centripetal acceleration* to estimate the negative contribution due to Earth's rotation.

ii) Use the equation for the *free air correction* to estimate the negative contribution due to the increased equatorial radius. **iii)** Use a *Bouguer slab approximation* (assuming the average density of the Earth as the reduction density) to estimate the positive contribution due to the excess mass of the equatorial bulge.

c) How does your answer for part (a) compare with the sum of the three contributions in (b)? If there is a significant difference, explain the error in terms of the methods and assumptions used in (b).

8-2 Given the following data for a gravity station:

Latitude: 48.1195° N
Longitude: 12.1878° E
Elevation: 487.9 m
Observed gravity: 980,717.39 *mGal.*

a) Calculate the: **i)** theoretical gravity; **ii)** free air correction; **iii)** Bouguer correction; **iv)** free air gravity anomaly; and **v)** Bouguer gravity anomaly.

b) Suppose that the gravity station in (a) had exactly the same latitude, longitude and observed gravity, but it was located on the surface of the ocean, over a water depth of 487.9 m. Recompute the free-air and Bouguer gravity anomalies.

8-3 Draw a block diagram of a passive continental margin in Airy isostatic equilibrium (Figs. 8.20 and 8.32), assuming: topography at sea level over the continent; ocean water depth of 5.0 km; water density of 1.03 g/cm^3; crustal density of 2.67 g/cm^3 for both the oceanic and continental regions; mantle density of 3.1 g/cm^3; oceanic crustal thickness of 8.0 km; 60 km depth of compensation.

a) Calculate and show on the model the corresponding *continental crustal thickness* needed to achieve isostatic equilibrium.

b) Assuming *semi-infinite slab approximations* (Fig. 8.30), calculate then plot on a graph above the model: **i)** contribution to the free air anomaly due to the water deepening seaward; **ii)** contribution to the free air anomaly due to the mantle shallowing seaward; **iii)** free air anomaly; **iv)** Bouguer anomaly.

c) For this specific model, explain why the *free air anomaly* is equivalent to the *contribution due to the water deepening seaward*, added to the *contribution due to the mantle shallowing seaward.* (Discuss in terms of the *amplitude* and the *gradient* of each contribution). Based on your response, discuss the meaning of the term "edge effect."

d) For this specific model, explain why the *contribution due to the mantle shallowing seaward* is equivalent to the *Bouguer anomaly*. Is this always the case? Explain why or why not.

8-4 Introductory geology textbooks often state that the top of oceanic crust is lower than the top of continental crust because oceanic crust is more dense than continental crust.

a) Design, *compute* and *plot* an *isostatic model* to test this idea. The model should have the following assumptions: i) the top of oceanic crust is about 5 km lower than the top of continental crust; ii) the Moho is exactly the same depth beneath the continental and oceanic crusts. (In other words, compute the differences in *crustal densities* needed to achieve a *Pratt model*).

b) Design, compute and plot a modification of the isostatic model in (a), so that the only difference is in *crustal thicknesses*. (That is, assume that the continental and oceanic crustal densities are the same, as in an *Airy model*). Use the same assumption that the top of oceanic crust is 5 km lower than the top of continental crust.

c) Develop a more realistic model which combines elements of the models in (a) and (b). The model should involve changes in both *crustal density* and *crustal thickness*, as constrained by observations from data other than gravity. Does this refined model suggest that the actual situation is closer to the Airy or Pratt hypothesis? Explain.

d) Write a concise statement for an introductory geology textbook that explains why the top of oceanic crust is lower than the top of continental crust.

8-5 True or False: "*Oceanic crust subducts more readily than continental crust because oceanic crust is more dense than continental crust.*" Discuss your answer, including sketches and computations as support.

8-6 A model of a *basalt filled basin* is shown below. The basin is 5 km deep and the crust outside the basin is 40 km thick. The topography both at and outside of the basin is flat, exactly at sea level. Note that the basalt density (2.9 g/cm^3) is higher than that of the crust (2.7 g/cm^3). The mantle density is 3.3 g/cm^3.

a) Compute and illustrate on the model the change in Moho depth necessary to achieve isostatic equilibrium.

b) Compute and plot the profile for the contribution to gravity made by the basalt-filled basin.

c) Compute and plot the profile for the contribution to gravity caused by the change in Moho depth.

d) Compute and plot the free air gravity anomaly profile.

e) Compute and plot the Bouguer gravity anomaly profile.

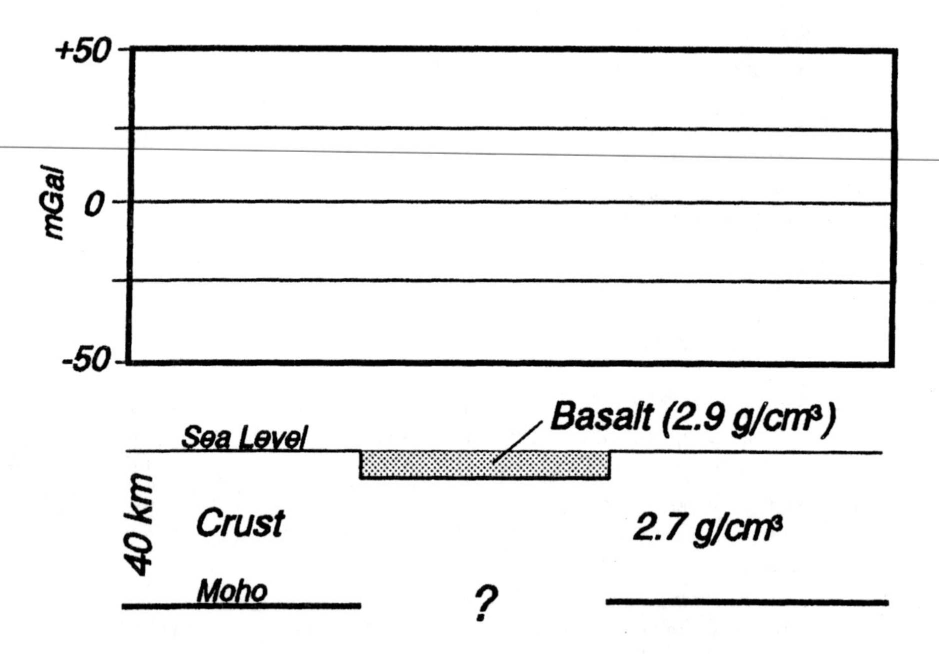

8-7 Densities and thicknesses for a highly simplified model of a *continental rift* are shown on the following page.

a) Calculate and sketch above the model: **i)** contribution to the free air anomaly due to the *mass above sea level*; **ii)** contribution to the free air anomaly due to the *shallow mantle beneath the thin crust.*

b) Calculate and sketch on the model the *change in depth of the lithosphere/asthenosphere boundary* needed to achieve isostatic equilibrium.

c) Calculate and sketch above the model: **iii)** contribution to the free air anomaly due to the change in depth of the *lithosphere/asthenosphere boundary*; **iv)** *free air anomaly* for entire model; **v)** *Bouguer anomaly* for entire model.

d) Discuss a geological interpretation of the model. How might the region have evolved to the lithosphere structure depicted in the model?

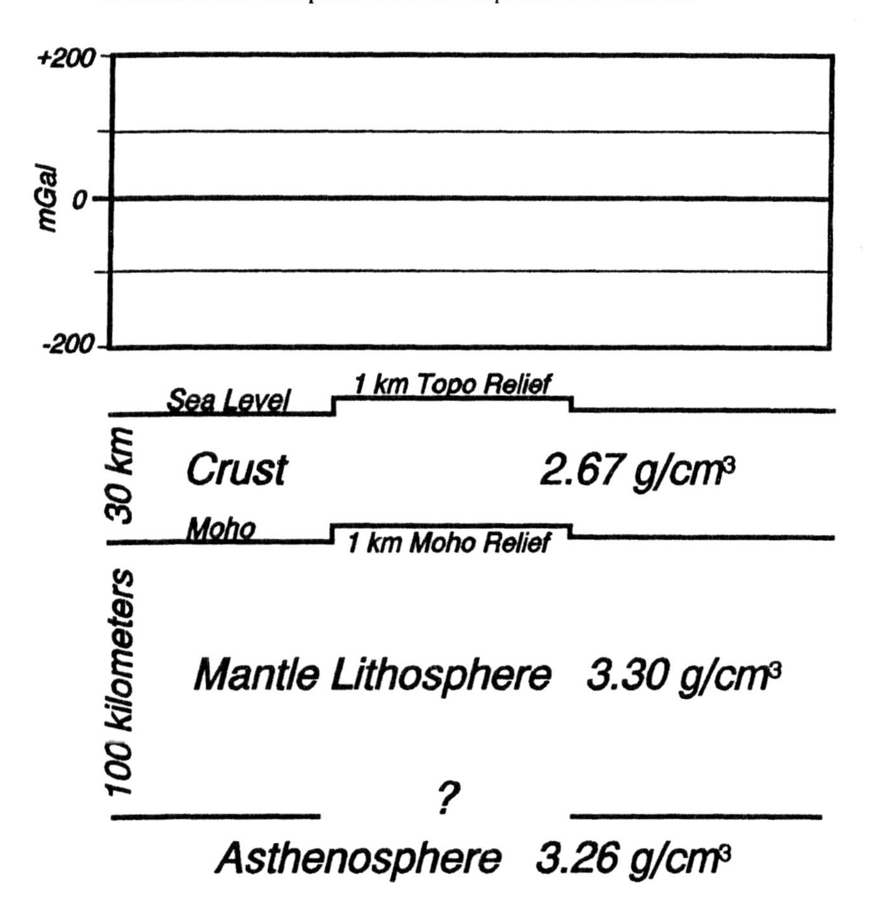

8-8 Draw a block diagram of a mountain range in Airy isostatic equilibrium (Figs. 8.20 and 8.36), assuming: the range is over 500 km wide; topographic relief of 2 km; crustal density of 2.8 gm/cm^3; mantle density of 3.3 gm/cm^3; crustal thickness outside the range of 35 km; 60 km depth of compensation.

a) Calculate and show on the model the *change in Moho depth* needed to compensate the topography.

b) Assuming *semi-infinite slab approximations* (Fig. 8.30), calculate and plot on a graph above the model: **i)** contribution to free air anomaly due to crustal root; **ii)** contribution to free air anomaly due to mass above sea level; **iii)** free air anomaly; **iv)** Bouguer anomaly.

8-9 Densities and thicknesses for a highly simplified model of a *mountain range* are shown on page 280.

a) Calculate and sketch above the model: **i)** contribution to free air anomaly due to *mass above sea level*; **ii)** contribution to free air anomaly due to 8 km thick *crustal root*.

b) Calculate and sketch on the model the *change in depth of lithosphere/asthenosphere boundary* needed to achieve isostatic equilibrium.

c) Calculate and sketch above the model: **iii)** contribution to free air anomaly due to change in depth of *lithosphere/asthenosphere boundary*; **iv)** *free air anomaly* for entire model; **v)** *Bouguer anomaly* for entire model.

d) Discuss a geological interpretation of the model. How might the region have evolved to the lithosphere structure depicted?

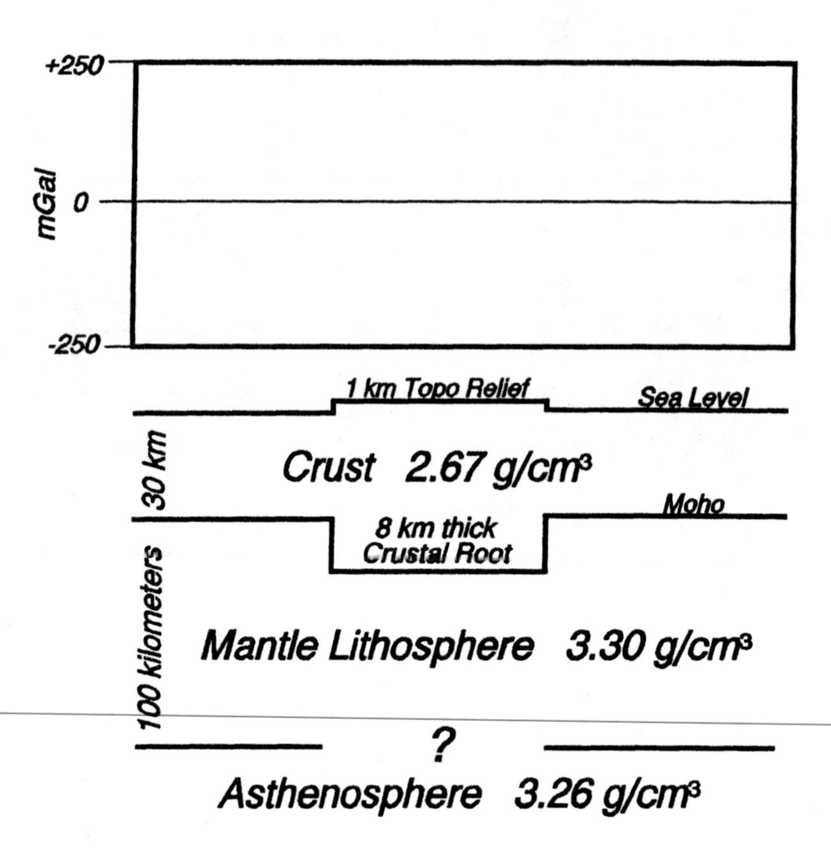

8-10 Draw a *cross section* along B-B′ shown on the map in Exercise 2-2. Notice that the line extends from the Pacific Plate, across the East Pacific Rise, Peru-Chile Trench, South America, the Mid-Atlantic Ridge, and all the way across the East African Rift.

a) Include a portrayal of how the depths to the following boundaries might be expected to change along the length of the cross section: **i)** *topography* and *bathymetry*; **ii)** *Moho*; **iii)** *lithosphere/asthenosphere boundary*.

b) Illustrate, on the cross section, the seismicity: **iv)** put X's where *earthquake activity* might be expected to occur; **v)** draw *focal mechanism diagrams*, illustrating types of faulting expected in various parts of the section;

c) Label in various parts of the cross section: **vi)** approximate *P-wave velocities* (km/s); **vii)** approximate *densities* (g/cm^3).

d) Draw *graphs* above the cross section of: **viii)** expected *free air* anomaly profile; **ix)** expected *Bouguer* anomaly profile. (Be sure to label each graph and include a scale).

8-11 a) From the maps on the inside covers of this book, draw the topography and gravity anomaly profiles across the United States along 40° N latitude, from 1000 km out in the Pacific Ocean to 1000 km out in the Atlantic Ocean.

b) Start a cross section by drawing the Moho along the same line of section from the map in Fig. 4.16.

c) Based on (a) and (b), draw an interpretation of the change in depth to the lithosphere/asthenosphere boundary. Write an explanation for your sketch of the boundary, in the context of isostasy for the overall model.

8-12 a) Using the maps of central Europe in Figs. 4.17, 7.34, and 8.49, draw profiles along 47° N latitude of the: i) Moho; ii) lithosphere/asthenosphere boundary; iii) Bouguer anomaly.

b) Explain relationships between the profiles in (a), in terms of mass distributions, isostasy, and tectonic evolution.

SELECTED BIBLIOGRAPHY

Abers, G. A., and H. Lyon-Caen, 1990, Regional gravity anomalies, depth of the foreland basin and isostatic compensation of the New Guinea highlands, *Tectonics*, v. 9, pp. 1479–1493.

Allis, R. G., 1981, Continental underthrusting beneath the Southern Alps in New Zealand, *Geology*, v. 9, pp. 303–307.

Anderson, R. E., M. L. Zoback, and G. A. Thompson, 1983, Implications of selected subsurface data on the structural form and evolution of some basins in the northern Basin and Range province, Nevada and Utah, *Geological Society of America Bulletin*, v. 94, pp. 1055–1072.

Babuška, V., J. Plomerová, and P. Pajdušák, 1988, Lithosphere-Asthenosphere in central Europe: Models derived from P residuals, 1988, In: *Proceedings of the 4th EGT Workshop: The Upper Mantle*, Commission of the European Communities, European Science Foundation, pp. 37–48.

Baker, B. H., P. A. Mohr, and J. A. Williams, 1972, *The geology of the Eastern Rift System of Africa*, Boulder, Colorado: Geological Society of America, Special Paper 136, 67 pp.

Bielik, M., 1988, A preliminary stripped gravity map of the Pannonian Basin, *Physics of the Earth and Planetary Interiors*, v. 51, pp. 185–189.

Bott, M. H. P., 1982, *The Interior of the Earth: Its Structure, Constitution and Evolution* (2nd ed.), New York: Elsevier Science Pub.Comp., 403 pp.

Braile, L. W., 1989, Crustal structure of the continental interior, in: *Geophysical Framework of the Continental United States*, edited by L. C. Pakiser and W. D. Mooney, Boulder, Colorado: Geological Society of America, Memoir 172, pp. 285–315.

Chamalaun, F. H., K. Lockwood, and A. White, 1976, The Bouguer gravity field and crustal structure of eastern Timor, *Tectonophysics*, v. 30, pp. 241–259.

Cook, F. A., and J. E. Oliver, 1981, The late Precambrian-early Paleozoic continental edge in the Appalachian orogen, *American Journal of Science*, v. 281, pp. 993–1008.

Couch, R. W., and S. Woodcock, 1981, Gravity and structure of the continental margins of southwestern Mexico and northwestern Guatemala, *Journal of Geophysical Research*, v. 86, pp. 1829–1840.

Couch, R. W., and R. P. Riddihough, 1989, The crustal structure of the western continental margin of North America, in: *Geophysical Framework of the Continental United States*, edited by L. C. Pakiser and W. D. Mooney, Boulder, Colorado: Geological Society of America, Memoir 172, pp. 103–128.

De Bremaecher, J. C., 1985, *Geophysics:The Earth's Interior*, New York: John Wiley and Sons, Inc., 342 pp.

Duroy, Y., A. Farah, and R. J. Lillie, 1989, Subsurface densities and lithospheric flexure of the Himalayan foreland in Pakistan, in: *Tectonics of the Western Himalayas*, edited by L. L. Malinconico, Jr., and R. J. Lillie, Boulder, Colorado: Geological Society of America, Special Paper 132, pp. 217–236.

Farah, A., M. A. Mirza, M. A. Ahmad, and M. H. Butt, 1977, Gravity field of the buried shield in the Punjab plain, Pakistan, *Geological Society of America Bulletin*, v. 88, pp. 1147–1155.

Fowler, C. M. R., 1990, *The Solid Earth: An Introduction to Global Geophysics*, Cambridge: Cambridge University Press, 472 pp.

Garland, G. D., 1979, *Introduction to Geophysics* (2nd ed.), Toronto: W. B. Saunders Comp., 494 pp.

Griffiths, D. H., and R. F. King, 1981, *Applied Geophysics for Geologists and Engineers:The Elements of Geophysical Prospecting* (2nd ed.), New York: Pergamon Press, 230 pp.

Grow, J. A., 1973, Crustal and upper mantle structure of the central Aleutian Arc, *Geological Society of America Bulletin*, v. 84, pp. 2169–2192.

Grow, J. A., and C. O. Bowin, 1975, Evidence for high-density crust and mantle beneath the Chile Trench due to the descending lithosphere, *Journal of Geophysical Research*, v. 80, pp. 1449–1458.

Grow, J., R. Mattlick, and J. Schlee, 1979, Multichannel seismic depth sections and interval velocities over outer continental slope between Cape Hatteras and Cape Cod, in: *Geological and Geophysical Investigations of Continental Margins*, edited by J. S. Watkins, L. Montadert, and P. W. Dickinson, Tulsa: American Association of Petroleum Geologists, Memoir 29, pp. 65–83.

Hall, S. A., J. F. Casey, and D. L. Elthon, 1986, A possible explanation of gravity anomalies over mid-ocean ridges, *Journal of Geophysical Research*, v. 91, pp. 3724–3738.

Hanna, W. F., R. E. Sweeney, T. G. Hildenbrand, J. G. Tanner, R. K. McConnell, and R. H. Godson, 1989, The gravity anomaly map of North America, in: *The Geology of North America—An Overview*, edited by A. W. Bally and A. R. Palmer, Boulder, Colorado: Geological Society of America, Decade of North American Geology, v. A, pp. 17–27.

Horváth, F., 1994, Towards a mechanical model for the formation of the Pannonian Basin, *Tectonophysics*, v. 226, pp. 333–357.

Hutchinson, D. R., J. A. Grow, K. D. Klitgord, and B. A. Swift, 1983, Deep structure and evolution of the Carolina Trough, in: *Studies in Continental Margin Geology*, edited by J. S. Watkins and C. L. Drake, Tulsa: American Association of Petroleum Geologists, Memoir 34, pp. 129–152.

Karner, G. D., and A. B. Watts, 1983, Gravity anomalies and flexure of the lithosphere at mountain ranges, *Journal of Geophysical Research*, v. 88, pp. 10449–10477.

Keen, C., and C. Tramontini, 1970, A seismic refraction survey on the mid-Atlantic ridge, *Journal of the Royal Astronomical Society of Canada*, v. 20, pp. 473–491.

Kissling, E., S. Mueller, and D. Werner, 1983, Gravity anomalies, seismic structure and geothermal history of Central Alps, *Annales Geophysics*, v. 1, pp. 37–46.

Kruger, J. M., and G. R. Keller, 1986, Interpretation of crustal structure from regional gravity anomalies, Ouachita Mountains area and adjacent Gulf Coastal Plain, *American Association of Petroleum Geologists Bulletin*, v. 70, pp. 667–689.

Lillie, R. J., K. D. Nelson, B. deVoogd, J. A. Brewer, J. E. Oliver, L. D. Brown, S. Kaufman, and G. W. Viele, 1983, Crustal structure of Ouachita Mountains, Arkansas: A model based on integration of COCORP reflection profiles and regional geophysical data, *American Association of Petroleum Geologists Bulletin*, v. 67, pp. 907–931.

Lillie, R. J., G. D. Johnson, M. Yousuf, A. S. H. Zamin, and R. S. Yeats, 1987, Structural development within the Himalayan foreland fold-and-thrust belt of Pakistan, in: *Sedimentary Basins and Basin-Forming Mechanisms*, edited by C. Beaumont and A. J. Tankard, Calgary: Canadian Society of Petroleum Geologists, Memoir 12, pp. 379–392.

Lillie, R. J., 1991, Evolution of gravity anomalies across collisional mountain belts: Clues to the amount of continental convergence and underthrusting, *Tectonics*, v. 10, pp. 672–687.

Lillie, R. J., M. Bielik, V. Babuška, and J. Plomerová, 1994, Gravity modelling of the lithosphere in the Eastern Alpine-Western Carpathian-Pannonian Basin region, *Tectonophysics*, v. 23, pp. 215–235.

Lowry, A. R., and R. B. Smith, 1994, Flexural rigidity of the Basin and Range-Colorado Plateau-Rocky Mountain transition from coherence analysis of gravity and topography, *Journal of Geophysical Research*, v. 99, pp. 20123–20140.

Lyon-Caen, H., and P. Molnar, 1983, Constraints on the structure of the Himalaya from an analysis of gravity anomalies and a flexural model of the lithosphere, *Journal of Geophysical Research*, v. 88, pp. 8171–8191.

Lyon-Caen, H., and P. Molnar, 1985, Gravity anomalies, flexure of the Indian plate, and the structure, support and evolution of the Himalaya and Ganga Basin, *Tectonics*, v. 4, pp. 513–538.

Madsen, J. A., and R. S. Detrick, 1990, A two- and three-dimensional analysis of gravity anomalies associated with the East Pacific Rise at 9° N and 13° N, *Journal of Geophysical Research*, v. 95, pp. 4967–4987.

Malinconico, L. L., 1986, The structure of the Kohistan Island-Arc terrain in northern Pakistan inferred from gravity data, *Tectonophysics*, v. 124, pp. 297–307.

Malinconico, L. L., 1989, Crustal thickness estimates for the western Himalaya, in: *Tectonics of the Western Himalayas*, edited by L. L. Malinconico, Jr., and R. J. Lillie, Boulder, Colorado: Geological Society of America, Special Paper 132, pp. 237–242.

McKenzie, D., and C. Bowin, 1976, The relationship between bathymetry and gravity in the Atlantic Ocean, *Journal of Geophysical Research*, v. 81, pp. 1903–1915.

Molnar, P., 1984, Structure and tectonics of the Himalaya: Constraints and implications of geophysical data, *Annual Reviews of Earth and Planetary Science*, v. 12, pp. 489–512.

Nettleton, L. L., 1971, *Elementary Gravity and Magnetics for Geologists and Geophysicists*, Tulsa: Society of Exploration Geophysicists, Monograph Series 1, 121 pp.

Nyblade, A., and H. N. Pollack, 1992, A Gravity Model for the Lithosphere in Western Kenya and northeastern Tanzania, *Tectonophysics*, v. 212, pp. 257–267.

Ocola, L. C., and R. P. Meyer, 1973, Central North American rift system, 1. Structure of the axial zone from seismic and gravimetric data, *Journal of Geophysical Research*, v. 78, pp. 5173–5194.

Pakiser, L. C., and Mooney, W. D., 1989, Introduction, in: *Geophysical Framework of the Continental United States*, edited by L. C. Pakiser and W. D. Mooney, Boulder, Colorado: Geological Society of America, Memoir 172, pp. 1–9.

Saltus, R. W., and G. A. Thompson, 1995, Why is it downhill from Tonopah to Las Vegas?: A case for mantle plume support of the high northern Basin and Range, *Tectonics*, v. 14, pp. 1235–1244.

Šefara, J., 1986, Various aspects of lithospheric interfaces modelling, *Sbor. Geol. Ved, Uzitá Geofyz.*, v. 21, pp. 9–28.

Serpa, L., T. Setzer, H. Farmer, L. Brown, J. Oliver, S. Kaufman, J. Sharp, and D. W. Steeples, 1984, Structure of the southern Keweenawan Rift from COCORP surveys across the midcontinent geophysical anomaly in northeastern Kansas, *Tectonics*, v. 3, pp. 367–384.

Simpson, R., R. Jachens, R. Blakely, and R. Saltus, 1986, A new isostatic residual gravity map of the conterminous United States with a discussion on the significance of isostatic residual anomalies, *Journal of Geophysical Research*, v. 91, pp. 8348–8372.

Stacey, F. D., 1992, *Physics of the Earth*, Brisbane, Australia: Brookfield Press, 513 pp.

Stockmal, G. S., C. Beaumont, and R. Boutilier, 1986, Geodynamic models of convergent margin tectonics: Transition from rifted margin to overthrust belt and consequences for foreland-basin development, *American Association of Petroleum Geologists Bulletin*, v. 70, pp. 181–190.

Swain, C. J., 1992, The Kenya Rift Axial Gravity High: A Re-Interpretation, *Tectonophysics*, v. 204, pp. 59–70.

Talwani, M., 1970, Gravity, in *The Sea* (volume 4), edited by A. E. Maxwell, New York: John Wiley and Sons, Inc. pp. 251–296.

Talwani, M., J. L. Worzel, and M. Landisman, 1959, Rapid gravity computation for two-dimensional bodies with application to the Mendocino submarine fracture zone, *Journal of Geophysical Research*, v. 64, pp. 49–59.

Talwani, M., X. Le Pichon, and M. Ewing, 1965, Crustal structure of the mid-ocean ridges: 2. Computed model from gravity and seismic refraction data, *Journal of Geophysical Research*, v. 70, pp. 341–352.

Telford, W. M., L. P. Geldart, R. E. Sheriff, and D. A. Keys, 1976, *Applied Geophysics*, Cambridge: Cambridge University Press, 860 pp.

Thomas, M. D., 1983, Tectonic significance of paired gravity anomalies in the southern and central Appalachians, in: *Contributions to the Tectonics and Geophysics of Mountain Chains*, edited by R. D. Hatcher, H. Williams and I. Zietz, Boulder, Colorado: Geological Society of America, Memoir 158, pp. 113–124.

Tomek, C., 1988, Geophysical investigation of the Alpine-Carpathian Arc, in: *Evolution of the Northern Margin of Tethys* (volume 1), edited by M. Rakus, J. Dercourt and A. E. M. Nairn, Paris: Mem. Soc. Geol., Nouvelle Serie No. 154, pp. 167–199.

Turcotte, D. L., and G. Schubert, 1982, *Geodynamics: Applications of Continuum Physics to Geological Problems*, New York: John Wiley and Sons, Inc., 450 pp.

Walcott, R. I., 1970, Flexure of the lithosphere at Hawaii, *Tectonophysics*, v. 9, pp. 435–446.

Watts, A. B., and J. R. Cochran, 1974, Gravity anomalies and flexure of the lithosphere along the Hawaiian-Emperor seamount chain, *Geophysical Journal of the Royal Astronomical Society*, v. 38, pp. 119–141.

Watts, A. B., and N. M. Ribe, 1984, On geoid heights and flexure of the lithosphere at seamounts, *Journal of Geophysical Research*, v. 89, pp. 11152–11170.

Woelk, T. S., and J. Hinze, 1991, Model of the midcontinent rift system in northeastern Kansas, *Geology*, v. 19, pp. 277–280.

Woodside, J. M., 1972, The Mid-Atlantic Ridge near 45° N.: The gravity field, *Canadian Journal of Earth Science*, v. 9, pp. 942–959, 1972.

Woollard, G. P., 1943, Transcontinental gravitational and magnetic profile of North America and its relation to geologic structure, *Geological Society of America Bulletin*, v. 44, pp. 747–789.

C H A P T E R 9

Magnetic Interpretation

***magnet** (mag'nit) **n.**,* [< *OFr.* < magnes < *Gr.* Magnētis, *stone of Magnesia*], *a piece of iron, steel, or lodestone that has the property of attracting iron, steel, etc.*

***force** (fôrs) **n.**,* [< *OFr.* < *LL.*, *L.* fortis, *strong*], *the cause, or agent, that puts an object at rest into motion or alters the motion of a moving object.*

***magnetic force** (mag net'ik fôrs) **n.**, the attracting or repelling force between a magnet and a ferromagnetic material.*

***field** (fēld) **n.**,* [*OE.* feld], *a space in which lines of force are active.*

***magnetic field** (mag net'ik fēld) **n.**, a region of space in which there is an appreciable magnetic force.*

***interpret** (in tur'prit) **vt.**,* [< *Mfr.* < *L.* interpretari < interpres, *negotiator*], *to explain the meaning of; to give one's own understanding of.*

***magnetic interpretation** (mag net'ik in tur'prə tā'shən) **n.**, an explanation of the distribution of magnetic materials within the Earth that causes observed changes in Earth's magnetic field.*

The magnetic field observed at Earth's surface varies considerably in both strength and direction. Unlike gravitational acceleration, which is directed nearly perpendicular to Earth's surface, magnetic field directions change from nearly horizontal at the equator, to nearly vertical at the poles (Fig. 9.1). The variation in strength of the gravity field is only about 0.5% ($\approx$ 978,000 *mGal* at the equator, 983,000 *mGal* at the poles), compared to doubling of the magnetic field ($\approx$ 30,000 nT at the equator, 60,000 nT at the poles).

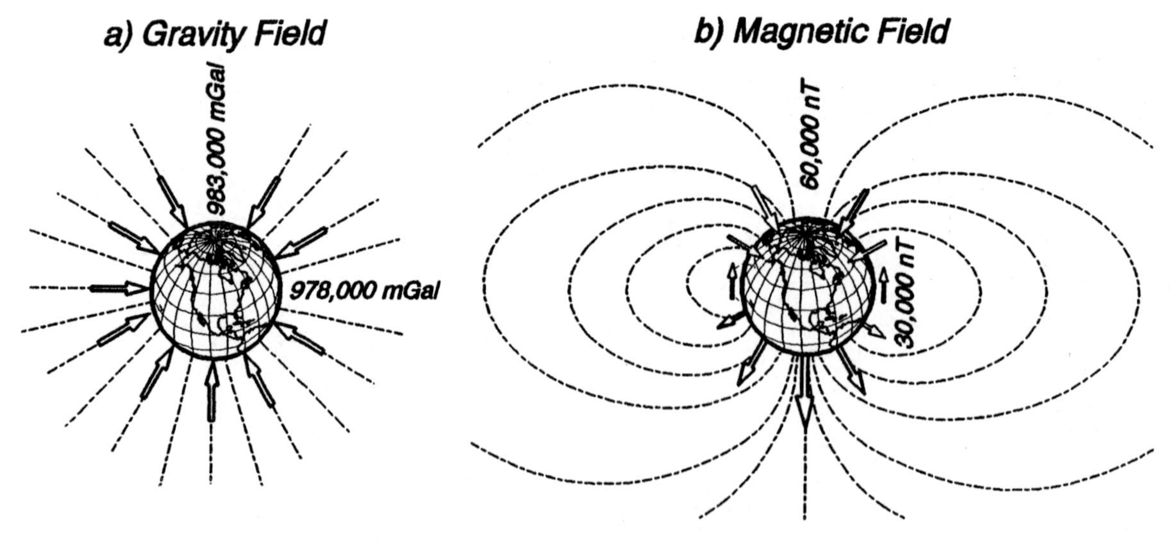

FIGURE 9.1 *Magnitudes and directions of Earth's gravity and magnetic fields.* a) The gravity field is approximately vertical, with only slight variation in magnitude from equator to pole. b) The magnetic field shows strong variations in both magnitude and direction.

The magnetic method has many important applications. Anomalies induced by Earth's natural field give clues to the geometry of magnetized bodies in the crust, and the depth to sources of the anomalies. The depth to the deepest sources of anomalies (Curie Depth) illustrates the depth below which rocks are too hot to retain strong magnetization (Curie Temperature). Studies of rocks that have been permanently magnetized (paleomagnetism) give clues to the ages of the rocks, the latitudes at which they formed, and to the relative positions of continents in the past.

EARTH'S MAGNETIC FIELD

About 98% of Earth's magnetic field is of internal origin, thought to be caused by motions of liquid metal in the core; the remaining 2% is external, of solar origin. Unlike the gravitational field, which is essentially fixed, the magnetic field has secular variations. Measurements in Europe since the 1600's show that the direction of the magnetic field has gradually drifted westward at rates up to 0.2° per year. The overall strength of the field has also decreased by about 8% in the last 150 years. In addition, several factors result in daily, monthly, seasonal, yearly, and longer period variations in the magnetic field. There are also sporadic variations ("magnetic storms") which momentarily disrupt the field.

Axial Dipolar Model

A *magnetic dipole*, inclined about 10.9 ° from Earth's rotational axis, can be used to describe about 90% of Earth's internal field. With such a model Earth's magnetic field is analogous to that of a bar magnet (Fig. 9.2). A homogeneous Earth with such a dipole would have north and south magnetic poles ("geomagnetic poles") exactly 180° apart (Fig. 9.3a).

The complex source of Earth's magnetism, however, produces lines of force that vary considerably from that of a simple dipole. The actual north and south magnetic poles (that is, positions where the field is vertical) are not 180° apart; they deviate considerably from both the geomagnetic and the geographic poles (Fig. 9.3b).

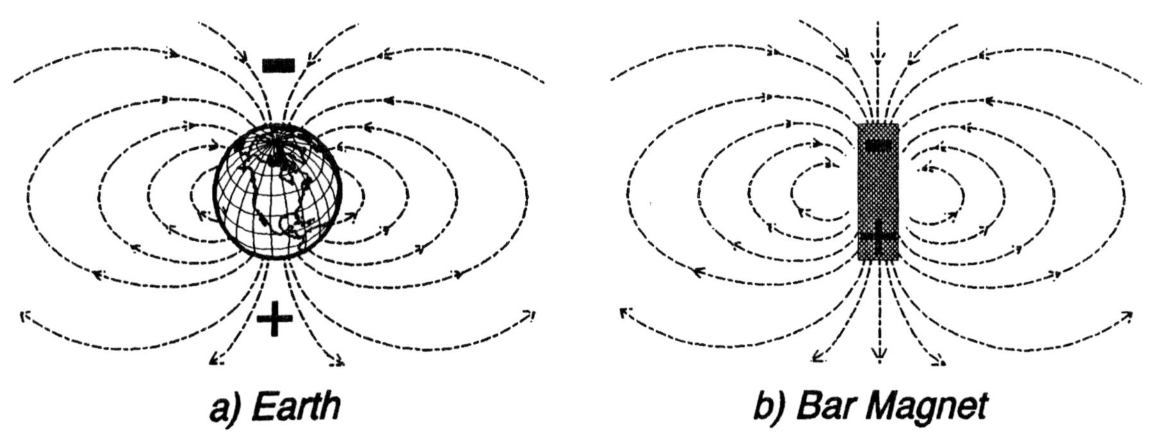

FIGURE 9.2 Earth's magnetic field is similar to that of a bar magnet, with the negative magnetic pole in the northern hemisphere and the positive magnetic pole in the southern hemisphere. The positive end of a compass needle (defined originally as "north-seeking") thus points roughly toward the geographic north pole.

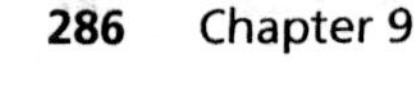

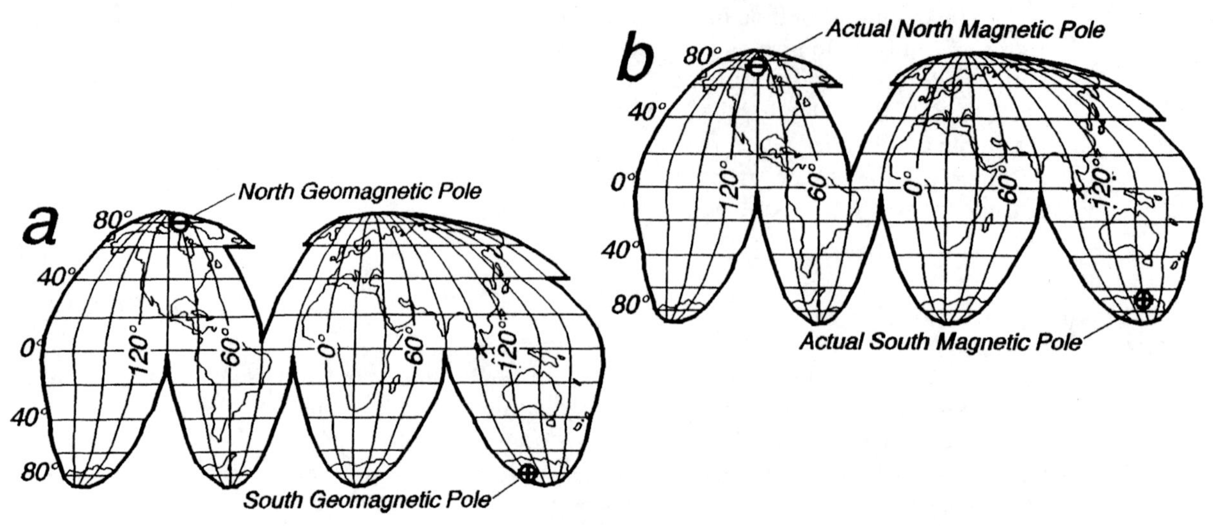

FIGURE 9.3 *Earth's magnetic poles.* a) *Geomagnetic poles*, according to a best-fit dipolar model. These poles are exactly 180° apart, deviating 10.9° from the geographic poles (North Geomagnetic Pole = 79.1° N, 71.1° W; South Geomagnetic Pole = 79.1° S, 108.9° E). b) *Actual magnetic poles*, where the magnetic field is vertical, are not 180° apart (North Magnetic Pole = 75° N, 101° W, on Bathhurst Island, Canada; South Magnetic Pole = 67° S, 143° E, in northeast Antarctica).

Strength and Direction of Magnetic Field

A magnetic field is composed of vectors, having both magnitude and direction. The orientation of a compass needle illustrates the direction of Earth's magnetic field (Fig. 9.4).

The *magnetic inclination* (i) is the angle that a compass needle makes with a horizontal ground surface (Fig. 9.4a). A compass needle points vertically toward the ground at the north magnetic pole (+90° inclination), straight upward at the south magnetic pole (−90° inclination). At the magnetic equator, a horizontal compass needle indicates an inclination of 0°.

The angle between a compass needle and true (geographic) north indicates the *magnetic declination* (δ). The north magnetic pole lies approximately along a line (roughly the arc of a great circle) running through the central United States to the geographic north pole; a compass needle would point close to true north along that line. In the northwest United States, a compass needle deviates about 20° eastward from geographic north (Fig. 9.4b).

Unlike gravity, where slight deviations from vertical are commonly not significant, knowledge of the direction of Earth's magnetic field is critical. A vector that describes the magnetic field strength and direction from a given position can be broken into its components as follows (Fig. 9.5):

$\vec{\mathbf{F}}$ = total magnetic field vector
$\vec{\mathbf{F}}_{\mathbf{H}}$ = horizontal component of total field vector
$\vec{\mathbf{F}}_{\mathbf{N}}$ = north component of horizontal vector
$\vec{\mathbf{F}}_{\mathbf{E}}$ = east component of horizontal vector
$\vec{\mathbf{F}}_{\mathbf{V}}$ = vertical component of total field vector
i = angle of magnetic inclination
δ = angle of magnetic declination.

The magnitude (F) of the total magnetic field vector (or total field *intensity*) is:

$$F = \sqrt{F_H^2 + F_V^2} = \sqrt{F_N^2 + F_E^2 + F_V^2}$$

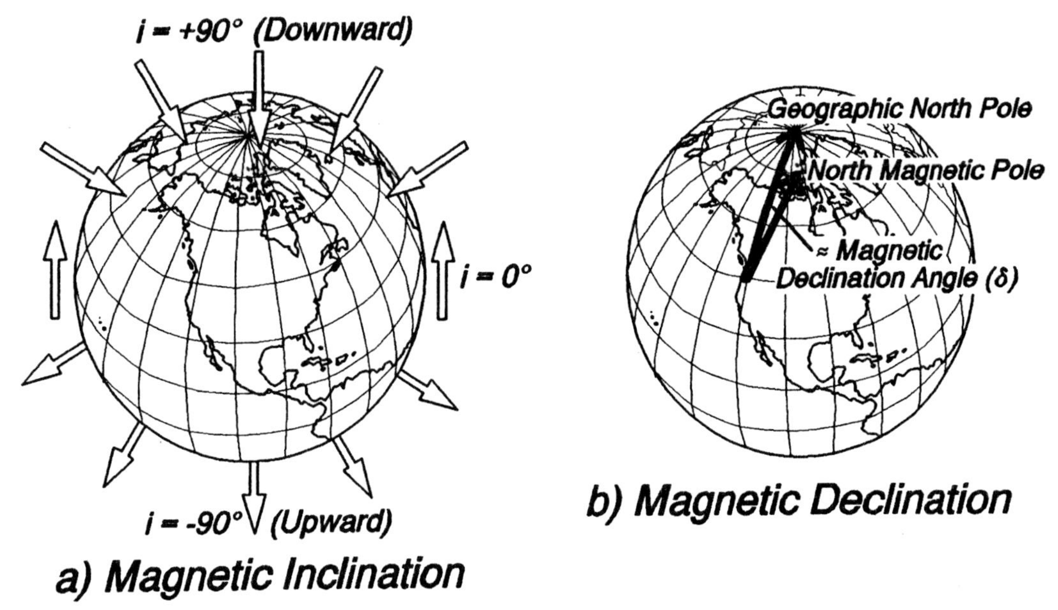

FIGURE 9.4 *Magnetic inclination and declination.* a) The angle between magnetic lines of force (Fig. 9.1b) and a horizontal ground surface is the magnetic inclination (i). b) The magnetic declination (δ) is the angle a compass needle deviates from geographic north.

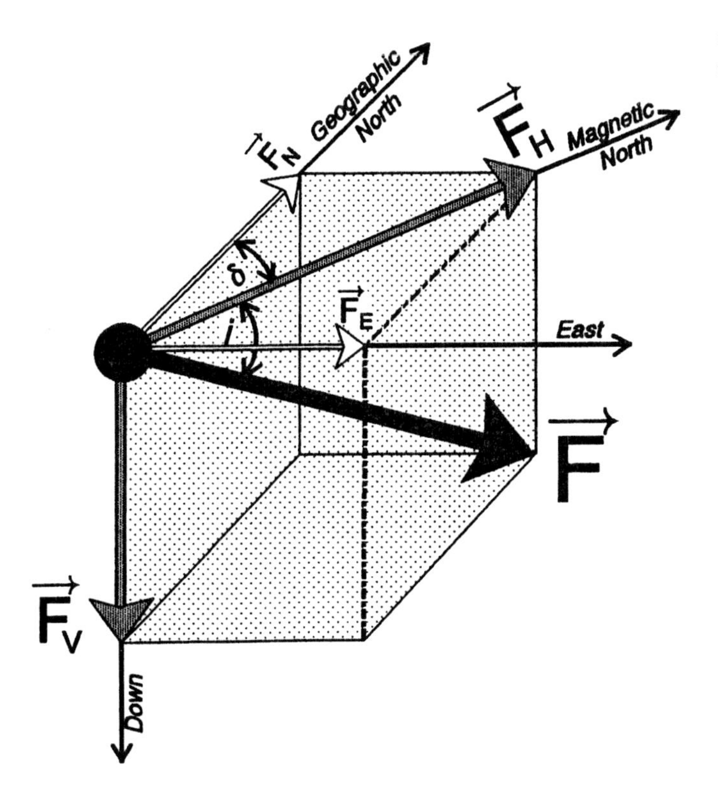

FIGURE 9.5 Components of total magnetic field vector defined in text.

where:

F_H = magnitude of horizontal component of total field vector
F_V = magnitude of vertical component of total field vector
F_N = magnitude of north component of total field vector
F_E = magnitude of east component of total field vector

The magnetic inclination and declination are:

$$i = \tan^{-1}(F_V/F_H)$$

$$\delta = \tan^{-1}(F_E/F_N)$$

The axial dipole model simplifies discussion of Earth's overall field because equations can be developed to describe the strength and direction of the field. With such a model the magnitudes of the horizontal, vertical, and total field vectors are (Butler, 1992):

$$F_H = \frac{M \cos\phi}{R^3}$$

$$F_V = \frac{2M \sin\phi}{R^3}$$

$$F = \frac{M\sqrt{1 + 3\sin^2\phi}}{R^3}$$

where:

R = radius of the Earth
M/R^3 = total field intensity at the magnetic equator
ϕ = magnetic latitude (for axis inclined 10.9° from the true rotational axis; Fig. 9.3a)

The magnetic inclination for an axial dipole also varies systematically with magnetic latitude as:

$$\tan i = 2\tan\phi$$

The equation expressing the total field intensity (F) illustrates that unlike the gravity field, which decreases by $1/R^2$, the magnetic intensity falls off by $1/R^3$. The same equation is analogous to the theoretical gravity formula discussed in Chapter 8. It thus illustrates that, for a value of $F = M/R^3 \approx 30{,}000$ nT at the magnetic equator ($\phi = 0°$), the total field intensity doubles to about 60,000 nT at the magnetic pole ($\phi = 90°$). Earth's field, though complex, approximates the axial dipolar model (Fig. 9.6a). Observed inclinations and declinations (Fig. 9.6b, c) also show general agreement with that idealized model.

MAGNETIZATION OF EARTH MATERIALS

Earth's magnetic field is perturbed locally by materials that are capable of being magnetized. Perturbations in the direction of the field can be illustrated by moving a magnet around a compass; the inclination and declination of the compass needle change in response to the position of the magnet (Fig. 9.7). Likewise, when magnetized rocks occur at or below Earth's surface, the direction and magnitude of Earth's overall magnetic field change slightly. It is thus important to understand the

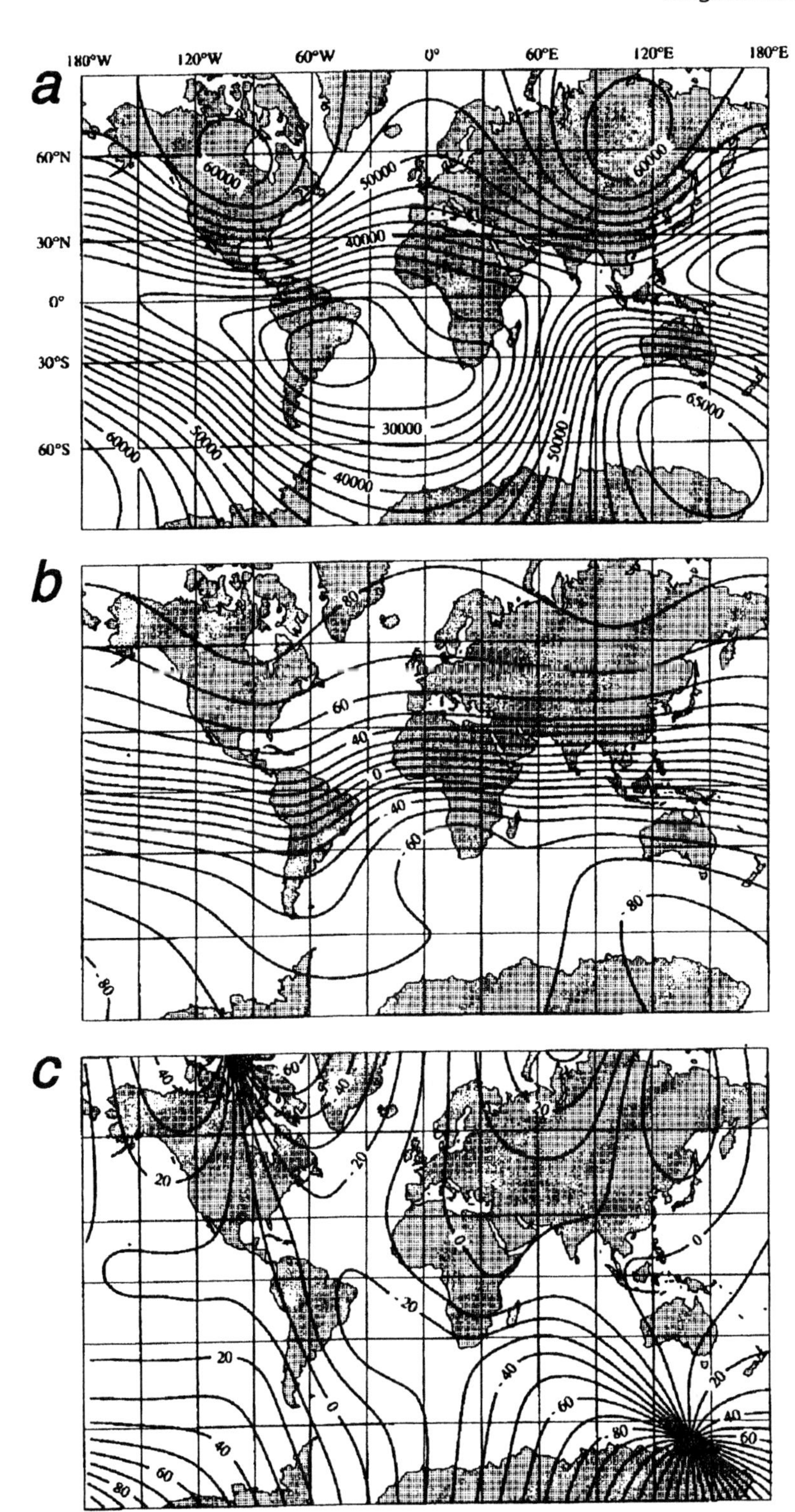

FIGURE 9.6 *Earth's magnetic field.* From *Potential Theory in Gravity and Magnetic Applications*, by R. Blakely, ©1995 Cambridge University Press. Reprinted with the permission of Cambridge University Press, New York. a) *Intensity* varies from about 30,000 nT near the magnetic equator, to about 60,000 nT at the poles. b) *Inclination* is roughly 0° near the magnetic equator, 90° near the poles. c) *Declination*, in degrees, is most pronounced near the magnetic poles.

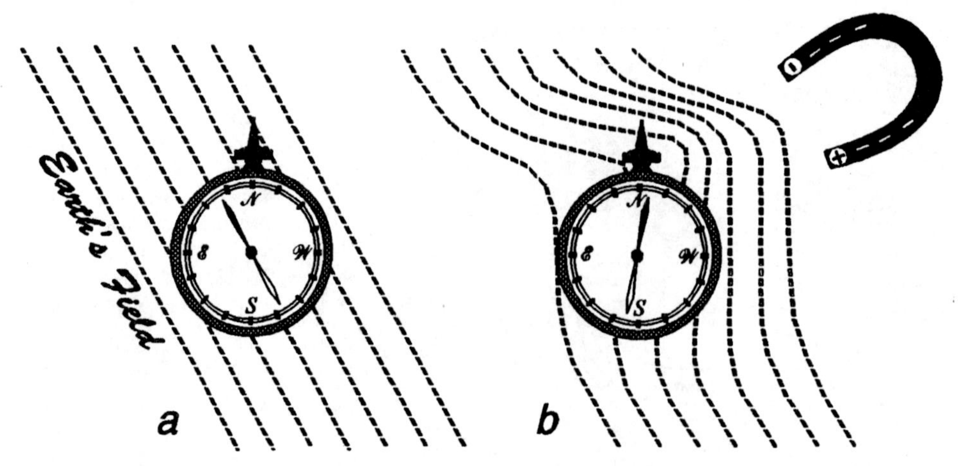

FIGURE 9.7 a) Compass responding to Earth's ambient magnetic field. b) Magnet causes a local deviation of Earth's field.

propensity of various types of materials toward magnetization, and how the magnetization locally affects Earth's field.

Units of measurement employed in magnetic field studies can be confusing to nonspecialists (see pp. 15–18 of Butler, 1992). For applications discussed here, it is sufficient to appreciate the relationship between the magnitude of Earth's total magnetic field ($\vec{\mathbf{F}}$) and the magnetization ($\vec{\mathbf{J}}$), induced within a body of magnetic susceptibility (χ). A typical magnetic survey employs a proton precession magnetometer, which measures the magnitude (intensity) of the total field vector, but not the direction. The International Standard (SI) unit for both magnetic field intensity and magnetization is the ampere/meter ($1\ \text{Am}^{-1} = 1\ \text{Cs}^{-1}\text{m}^{-1}$), while magnetic susceptibility is dimensionless. The unit of magnetic induction, the tesla (T) is also equivalent to $1\ \text{Am}^{-1}$. For convenience and to avoid confusion, the nanotesla ($1\ \text{nT} = 10^{-9}\text{T}$) is used in this text to express both magnetic field intensity and magnetization. In the older literature intensity is sometimes expressed in gammas ($1\ \gamma = 1\ \text{nT}$); intensity may appear in oersted ($1\ \text{Oe} = 10^{3}\text{T}/4\pi$), magnetization in gauss ($1\ \text{G} = 10^{3}\text{T}$).

The magnitude and direction of magnetization induced within a material depends on the magnitude and direction of the external (ambient) field, and the ability of the material to be magnetized:

$$\boxed{\vec{\mathbf{J}} = \chi\vec{\mathbf{F}}_{\mathbf{amb}}}$$

where:

$\vec{\mathbf{J}}$ = induced magnetization of the material
χ = magnetic susceptibility of the material
$\vec{\mathbf{F}}_{\mathbf{amb}}$ = magnitude and direction of the ambient field.

The *magnetic susceptibility* (χ, a dimensionless quantity) is a measure of the degree to which a substance may be magnetized. The overall susceptibility of a rock is roughly equivalent to the susceptibility of the magnetic mineral (or minerals) present, times the percentage of that mineral (or minerals), divided by 100. Table 9.1 illustrates that the amount of iron in a material, particularly in the form of the mineral magnetite (Fe_3O_4), strongly influences magnetic susceptibility. Ultramafic and mafic rocks (peridotite, basalt, gabbro), which are rich in magnetite, have high susceptibilities compared to felsic rocks (diorite, sandstone, granite).

TABLE 9.1 Typical magnetic susceptibilities of some common Earth materials.

Material	Magnetic Susceptibility
Magnetite	10000×10^{-5}
Peridotite	500×10^{-5}
Basalt/Gabbro	200×10^{-5}
Diorite	20×10^{-5}
Sandstone	10×10^{-5}
Granite	1×10^{-5}
Salt	-1×10^{-5}

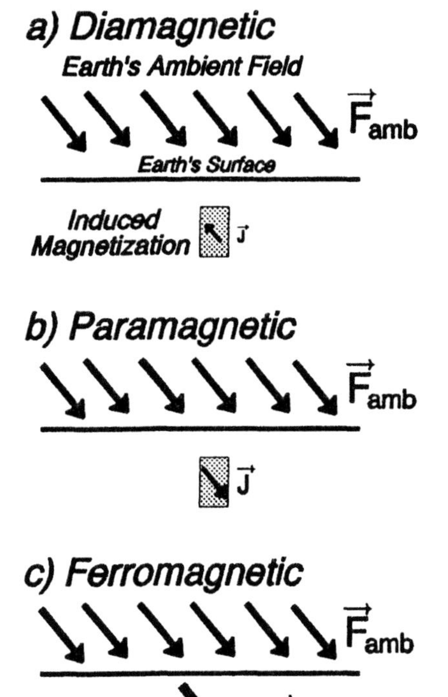

FIGURE 9.8 *Types of magnetic behavior.* a) *Diamagnetic* minerals acquire a weak magnetization ($\vec{\mathbf{J}}$) opposite to the external field ($\vec{\mathbf{F}}_{amb}$). b) The magnetization in *paramagnetic* minerals is weak but in the same direction as the external field. c) A strong magnetization, in the same direction as the external field, occurs in *ferromagnetic* minerals. [Note: In a strict sense, solids with coupling of atomic magnetic moments may be ferromagnetic (parallel magnetic moments), antiferromagnetic (antiparallel magnetic moments), or ferrimagnetic (antiparallel magnetic moments that do not cancel). As in Butler (1992) and Blakeley (1995) the term "ferromagnetic" is used here in a general sense for all three cases.]

Types of Magnetic Behavior

The type of magnetism exhibited by a mineral, in the presence of an external magnetic field, depends on the mineral's magnetic susceptibility. If a body containing the mineral is placed within an external (ambient) magnetic field ($\vec{\mathbf{F}}_{amb}$), the body acquires a magnetization ($\vec{\mathbf{J}}$) with intensity proportional to the overall magnetic susceptibility of the body.

Diamagnetism ($\chi \approx -10^{-5}$) A diamagnetic mineral, such as halite (rock salt), has negative magnetic susceptibility, acquiring an induced magnetization opposite in direction to an applied external field (Fig. 9.8a). The weak magnetization results from alteration of electron orbitals as force from the external field is applied to the material. Susceptibilities of only about -10^{-5} mean that magnetization is on the order of 1/100,000th the strength of the external field.

Paramagnetism ($\chi \approx +10^{-4}$) The magnetic susceptibility of paramagnetic minerals is positive; they acquire a magnetism parallel to an external field (Fig. 9.8b). The magnetism occurs as magnetic moments of atoms are partially aligned in the presence of the external field. Most magnetic minerals exhibit this type of weak magnetic behavior.

Ferromagnetism ($\chi \approx +10^{-1}$) In some metallic minerals rich in iron, cobalt, manganese, or nickel, atomic magnetic moments align strongly with an external field (Fig. 9.8c). Susceptibilities on the order of 10^{-1} indicate that the magnetization is in the same direction as, and about 1/10 the magnitude of, the external field. Under some circumstances induced magnetization may remain in ferromagnetic materials, even after the external field is removed (remanent magnetization).

Types of Magnetization

Magnetization of a rock occurs in two ways: it can be induced by Earth's present magnetic field, or it could have formed some time in the past, as the rock lithified.

Induced Magnetization Earth's overall (or ambient) magnetic field is an external field that can cause rocks to be temporarily magnetized (Fig. 9.9a). When taken out of the ambient field, this *induced magnetization* may be lost. The magnitude and direction of the induced field depends on the magnitude and direction of the ambient field and the magnetic susceptibility of the rock.

Remanent Magnetization When rocks form, the magnetic domains of some minerals (particularly magnetite) behave as compass needles, orienting themselves

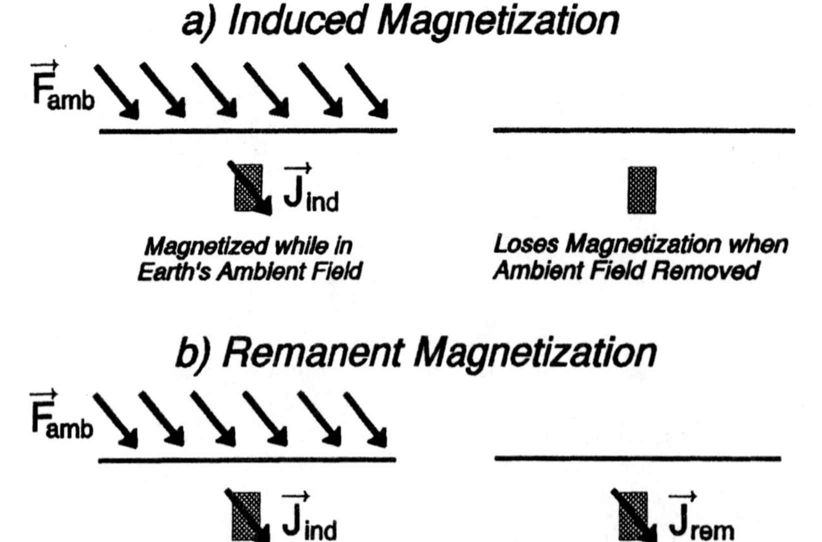

FIGURE 9.9 *Types of magnetization.* a) In the presence of an external magnetic field, magnetization may be *induced* in a material. Materials commonly lose their induced magnetization when the external field is removed. b) Some materials remain magnetized after the external field is removed, retaining a *remanent* magnetization.

in the direction of the ambient magnetic field. As the rock lithifies, the orientation of the magnetic domains may be frozen into the rock. This *remanent magnetization* remains even after the ambient field changes (Fig. 9.9b); it is commonly 5 times as great as the magnetization induced by the present field.

INTERPRETATION OF INDUCED MAGNETIC ANOMALIES

Induced magnetization results from application of an external magnetic field; the magnetization in turn leads to a local field about the magnetized body. In the absence of remanent magnetization, the total magnetic field observed in the vicinity of a magnetic body is the sum of Earth's ambient field at that location and the field induced within the magnetic body (Fig. 9.10):

$$\vec{\mathbf{F}} = \vec{\mathbf{F}}_{\mathbf{amb}} + \vec{\mathbf{F}}_{\mathbf{ind}}$$

where:

$\vec{\mathbf{F}}$ = total magnetic field
$\vec{\mathbf{F}}_{\mathbf{amb}}$ = Earth's ambient magnetic field in region
$\vec{\mathbf{F}}_{\mathbf{ind}}$ = induced magnetic field.

Local perturbations in Earth's field thus provide clues to the presence of magnetically susceptible materials in the subsurface. Such materials can be studied by subtracting the average value of the ambient field in the region from the total field, yielding the induced field surrounding the material:

$$\vec{\mathbf{F}}_{\mathbf{ind}} = \vec{\mathbf{F}} - \vec{\mathbf{F}}_{\mathbf{amb}}$$

Even though the total magnetic field is a vector ($\vec{\mathbf{F}}$), with both magnitude and direction, studies of subsurface susceptibility can be accomplished by taking simple measurements of only the magnitude of the total field (F). The *total field anomaly* (ΔF) is computed by subtracting the magnitude (intensity) of the ambient field (F_{amb}) from F (Fig. 9.10c):

$$\mathbf{\Delta F = F - F_{amb}}$$

ΔF is analogous to the gravity anomaly obtained by subtracting the observed gravity (g) from the theoretical gravity (g_t), according to the latitude of a station (see Chapter 8). A notable difference is that, while the overall gravity field ($\approx g_t$) has simple changes that correspond closely to latitude, the overall magnetic field ($\approx F_{amb}$) has changes that are complex; the values for F_{amb} must therefore be estimated empirically (for example, from a map, as in Fig. 9.6a).

Geometry of Magnetic Bodies in the Subsurface

Magnetization is induced in a material according to the direction and magnitude of Earth's ambient field, and the magnetic susceptibility (χ) of the material. Fig. 9.11 illustrates that, because the ambient field lies at varying angles (inclinations) with respect to Earth's surface, the forms of induced magnetic anomalies depend on latitude. The induced magnetization ($\vec{\mathbf{J}}$) within the body is parallel to the ambient field ($\vec{\mathbf{F}}_{\mathbf{amb}}$), according to $\vec{\mathbf{J}} = \chi\vec{\mathbf{F}}_{\mathbf{amb}}$. $\vec{\mathbf{J}}$ in turn leads to an induced field ($\vec{\mathbf{F}}_{\mathbf{ind}}$). Where ($\vec{\mathbf{F}}_{\mathbf{ind}}$) crosses the surface with a component in the same direction as $\vec{\mathbf{F}}_{\mathbf{amb}}$, the total field

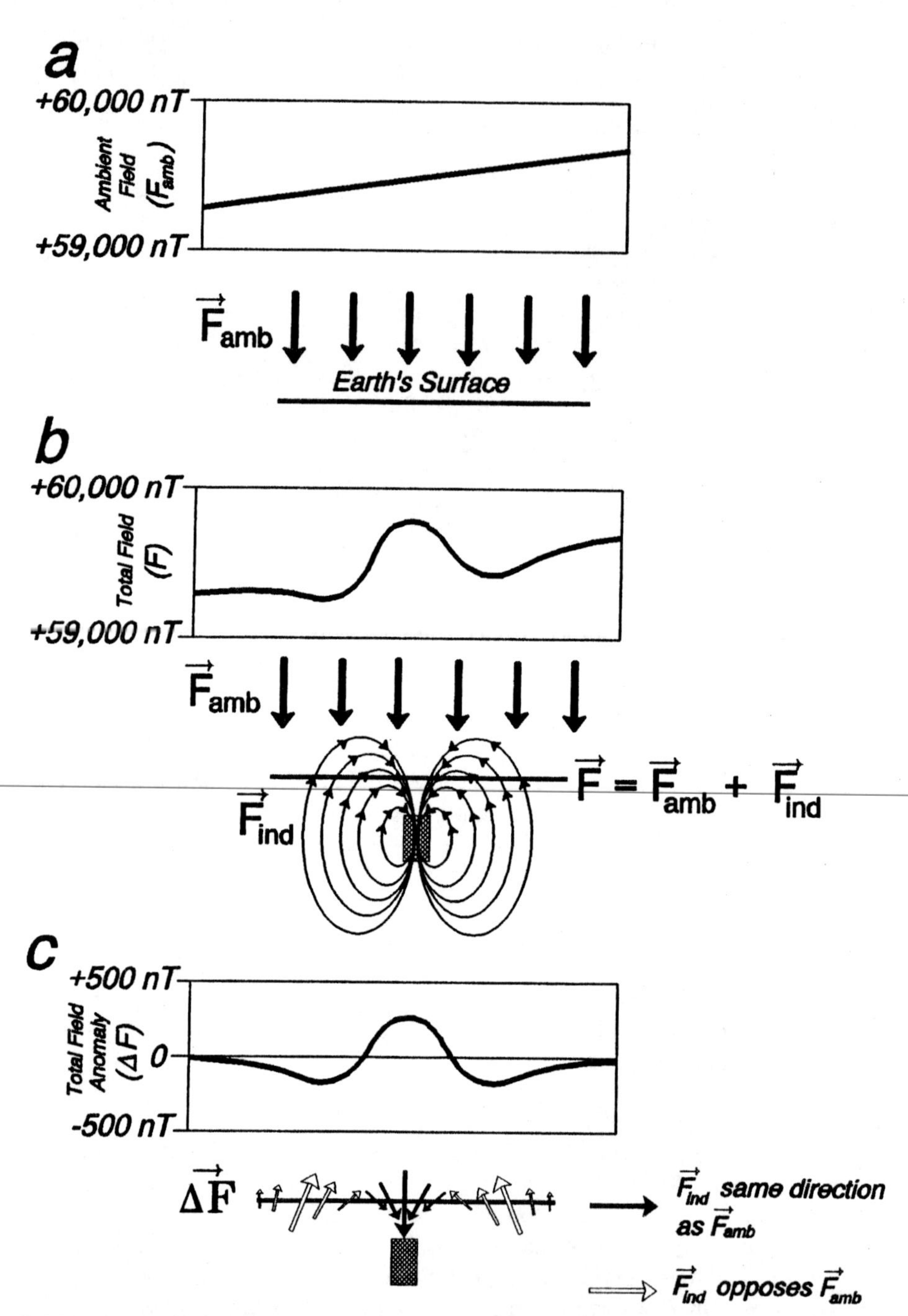

FIGURE 9.10 *Total field magnetic anomaly produced by local magnetic body.* a) Earth's ambient field ($\vec{\mathbf{F}}_{\mathbf{amb}}$) has magnitude of several thousand nT, with very long wavelength changes. b) A body of magnetization ($\vec{\mathbf{J}}$) is surrounded by an induced magnetic field ($\vec{\mathbf{F}}_{\mathbf{ind}}$), with amplitudes of perhaps a few hundred nT occurring over much shorter wavelengths. The total magnetic field ($\vec{\mathbf{F}}$) that results is the sum of the ambient and induced fields. c) Subtracting the magnitude of the ambient field (F_{amb}) from that of the total field (F) yields the total field anomaly (ΔF). A profile of ΔF thus reflects the effect of the induced field.

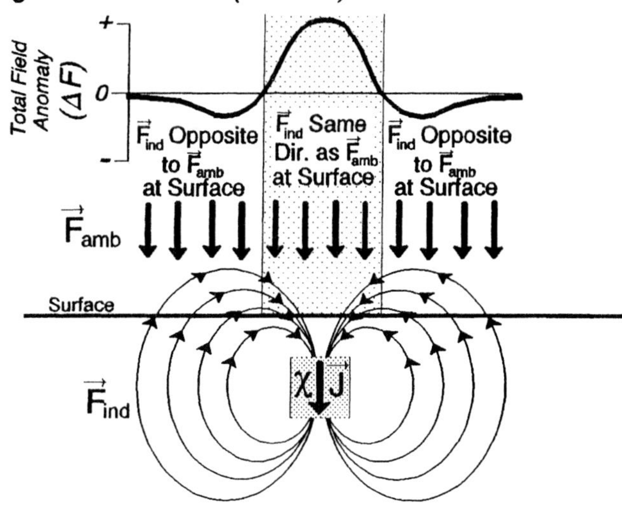

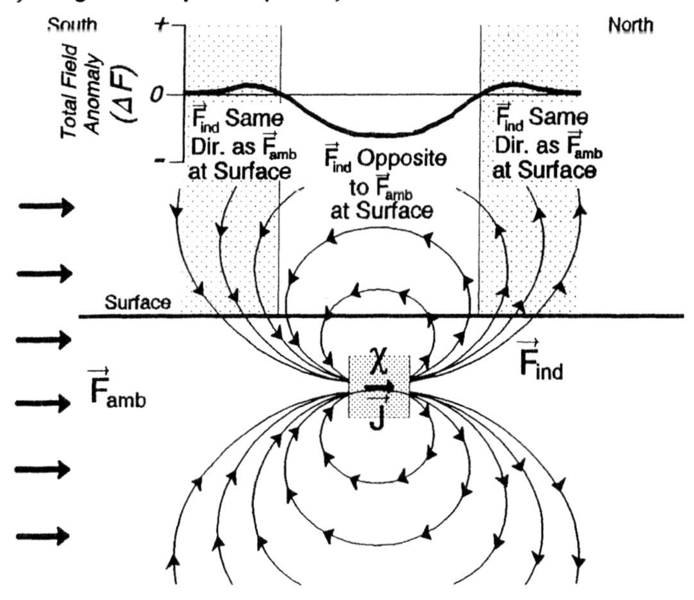

FIGURE 9.11 At different magnetic latitudes, magnetic anomalies from the same magnetic body (that is, a body of exactly the same shape, size, depth, and magnetic susceptibility, χ) are quite different. a) At the magnetic north pole, Earth's ambient field ($\mathbf{F}_{amb}$) is strong ($\approx$ 60,000 nT) and points downward (magnetic inclination i = 90°). Where the induced field ($\mathbf{F}_{ind}$) points in the same direction as $\vec{\mathbf{F}}_{amb}$ (gray shading), the total field anomaly (ΔF) is positive. Negative total field anomalies occur where the two fields oppose one another. b) At the magnetic equator, $\vec{\mathbf{F}}_{amb}$ is horizontal (i = 0°) and weaker ($\approx$ 30,000 nT). The magnetization ($\vec{\mathbf{J}}$) is smaller, leading to a weaker induced field. The total field anomaly is thus lower amplitude than in (a). $\vec{\mathbf{F}}_{ind}$ opposes $\vec{\mathbf{F}}_{amb}$ over the body, leading to negative ΔF values. (Modified from R. F. Butler, personal communication, 1996).

anomaly (ΔF) is positive. In areas where $\vec{\mathbf{F}}_{\mathbf{ind}}$ has a component opposite to $\vec{\mathbf{F}}_{\mathbf{amb}}$, ΔF is negative. For the same body, magnetic anomalies observed at the magnetic north pole (Fig. 9.11a) and the magnetic equator (Fig. 9.11b) are thus quite different. (The problem is even more complicated, because the *direction* of the *magnetic profile* is also important. Fig. 9.11b assumes a profile running in a south-north direction, cutting across induced field lines. East-west profiles would parallel induced field lines, resulting in a different total field anomaly profile).

Although some magnetic surveys are undertaken on the surface, an airplane is an effective means to cover large areas and to record large amounts of data. Unlike gravity surveys, which require a platform with no (or relatively small, predictable) motions, a standard proton precession magnetometer can be towed behind an airplane. Such *aeromagnetic surveys* measure the magnitude (F), but not the direction, of the total magnetic field. That measurement, along with knowledge of the local magnitude of Earth's ambient field (F_{amb}), is sufficient to determine the total field magnetic anomaly (ΔF). Thus, while magnetic anomalies are commonly more complex to interpret due to their latitude and direction dependence, adequate coverage of an area is generally easier to attain compared to gravity surveying.

Forward modeling of magnetic anomalies can be used to interpret the distribution of magnetic susceptibility contrasts ($\Delta\chi$) in the subsurface, analogous to estimating density contrasts ($\Delta\rho$) from gravity data. Changes in the magnitude and direction of Earth's ambient field, however, make the forms of magnetic anomalies far less intuitive than gravity anomalies. The same mass anomaly, buried at a given depth, will produce essentially the same gravity anomaly anywhere on the Earth; magnetic anomalies from the same body, however, vary in form and amplitude according to magnetic latitude (Fig. 9.12). It should be noted, too, that remanent magnetization is often much stronger than that induced by the ambient field. If the remanent and induced magnetizations are in the same direction, their net effect may be incorporated by choosing an equivalent susceptibility for modeling purposes.

Mapping of Magnetic Bodies Magnetics is a useful tool for mapping materials that have susceptibilities or remanent magnetizations that contrast with those of surrounding rocks, and for distinguishing between types of intrusive bodies. Igneous rocks often have large amounts of magnetite, inducing high magnetization (Table 9.1). Salt exhibits diamagnetic behavior, inducing a weak field opposite in direction to the ambient field (Fig. 9.8a). Like igneous intrusions, salt domes are recognized from seismic refraction experiments as high velocity material. Potential field studies can differentiate the two features (Fig. 9.13); the high density and high magnetic susceptibility of an igneous intrusion will produce pronounced gravity and magnetic anomalies, compared to a gravity minimum and subdued magnetic anomalies resulting from a low-density, low-susceptibility salt dome.

Depth to Magnetic Basement High ferromagnetic mineral content results in high levels of magnetization (Fig. 9.8c). Crystalline basement rocks, which are commonly more mafic than overlying sedimentary deposits, are thus the main source of magnetic anomalies in a region. Amplitudes and gradients for magnetic anomalies decrease as the sources get farther from the surface observation points, because magnetic force is a potential field (see examples for gravity, Figs. 8.28 and 8.31). The depth to the basement in a region can be estimated, therefore, by studying the pattern of magnetic anomalies. Regions with high magnetic anomaly amplitudes, short wavelengths, and steep gradients suggest shallow basement (Fig. 9.14a); in areas where the basement is deep, magnetic anomalies are subdued (Fig. 9.14b).

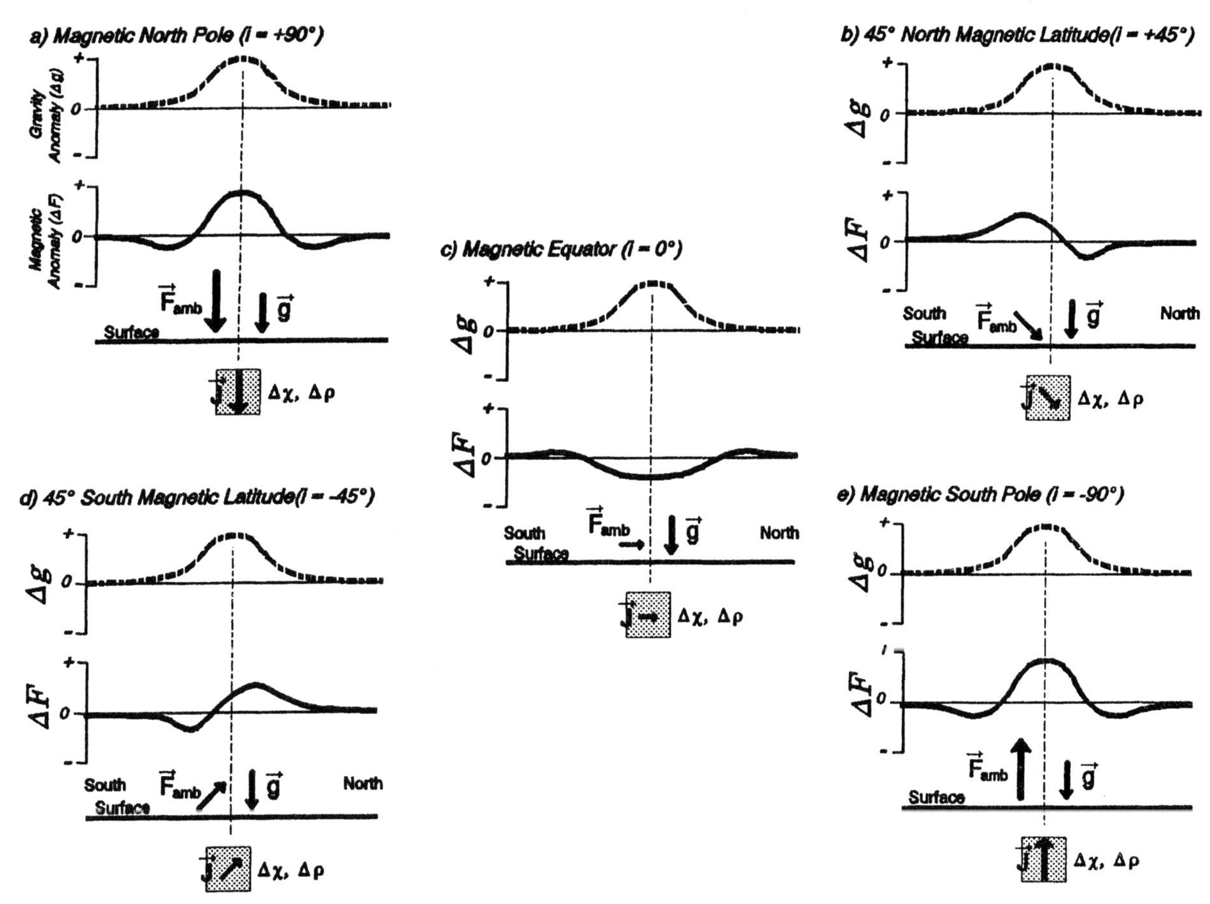

FIGURE 9.12 *Gravity and magnetic anomalies from the same body, at different magnetic latitudes.* At each latitude, the gravity anomaly (Δg) resulting from the body of density contrast (Δρ) is the same. For the same body, with magnetic susceptibility contrast (Δχ), the magnetic anomaly (ΔF) varies. At the magnetic north and south poles (a and e), the anomaly is a central high with flanking lows (Fig. 9.11a); a high magnitude ambient field ($F_{amb} \approx$ 60,000 nT) induces high magnitude magnetization (J), resulting in high amplitude magnetic anomalies (ΔF). At the magnetic equator (c), the induced field opposes the ambient field at the surface directly over the body, leading to negative magnetic anomalies (Fig. 9.11b); $F_{amb} \approx$ 30,000 nT leads to low amplitude ΔF. At mid-latitudes (b and d) the anomaly is asymmetric, with intermediate amplitude ΔF.

Curie Depth Minerals that exhibit strong (ferromagnetic) behavior at low temperatures have weaker (paramagnetic) properties when hotter than the *Curie temperature*. With increasing depth, temperature increases according to a geothermal gradient. Deeper than the *Curie depth*, rocks lose their strong magnetization as they heat up beyond the Curie temperature (≈ 600°C for most rocks). Areas with high geothermal gradient thus have a shallow **bottom** to magnetic basement, compared to colder areas (Fig. 9.15). The form of magnetic anomalies can thus be used to map the approximate depth where rocks reach about 600°C.

PALEOMAGNETIC STUDIES

The remanent magnetization in a rock records information about the direction to the north magnetic pole, at the time the rock lithified. Paleomagnetic studies may thus indicate the age of the rock, and perhaps suggest the latitude of the region at the time the rock formed.

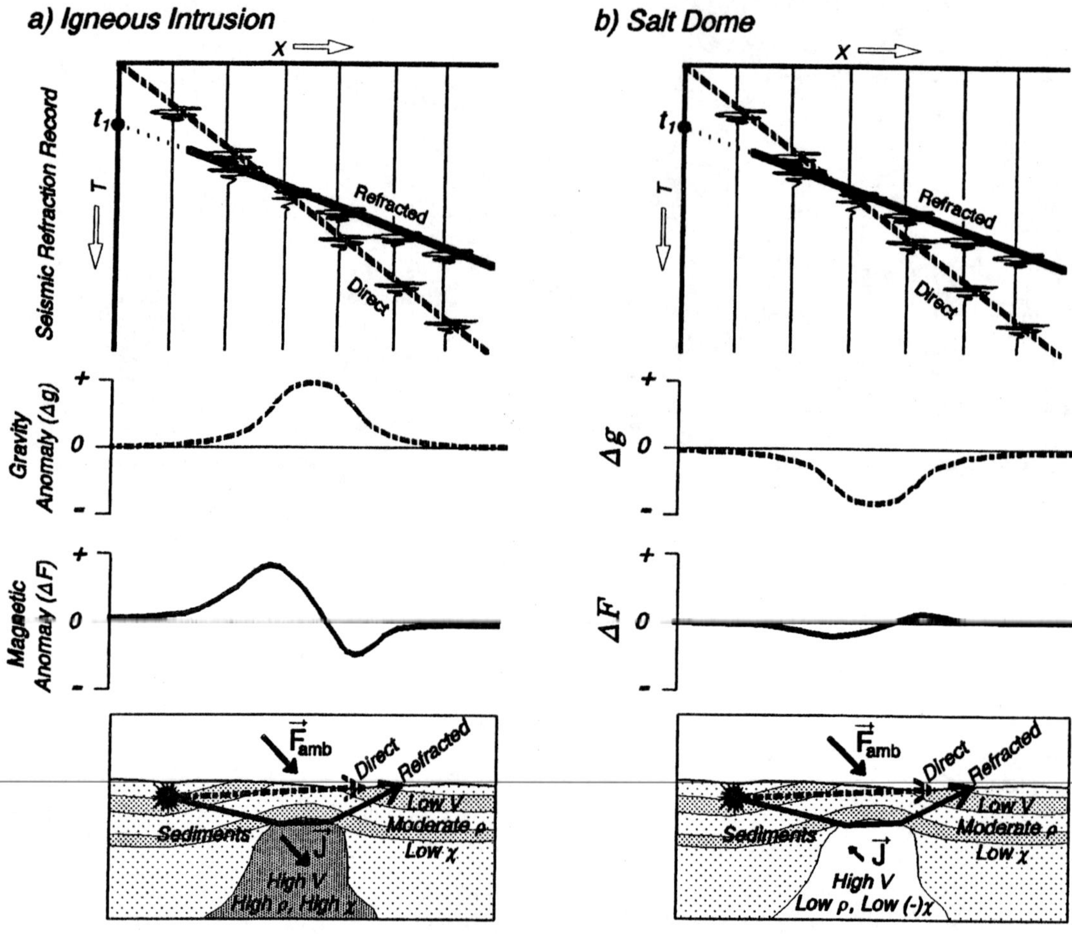

FIGURE 9.13 An igneous intrusion and salt dome might both be recognized as high-velocity material (V), leading to a critical refraction. a) An igneous intrusion generally has high density (ρ) and high magnetic susceptibility (χ), relative to surrounding sedimentary strata. A pronounced gravity maximum and high-amplitude magnetic anomaly result. b) A salt dome has low density, resulting in a gravity minimum. The small, negative susceptibility leads to a subdued magnetic anomaly.

The total magnetization of a material is the vector sum of the induced and remanent magnetizations:

$$\vec{\mathbf{J}} = \vec{\mathbf{J}}_{\mathbf{ind}} + \vec{\mathbf{J}}_{\mathbf{rem}}$$

where:

$\vec{\mathbf{J}}$ = total magnetization of the material
$\vec{\mathbf{J}}_{\mathbf{ind}}$ = induced magnetization of the material
$\vec{\mathbf{J}}_{\mathbf{rem}}$ = remanent magnetization of the material.

The total field at a given location is thus the local fields due to the induced and remanent magnetizations, added to the ambient field:

$$\vec{\mathbf{F}} = \vec{\mathbf{F}}_{\mathbf{amb}} + \vec{\mathbf{F}}_{\mathbf{ind}} + \vec{\mathbf{F}}_{\mathbf{rem}}$$

where:

$\vec{\mathbf{F}}$ = magnitude and direction of the total magnetic field
$\vec{\mathbf{F}}_{\mathbf{amb}}$ = Earth's ambient magnetic field in region

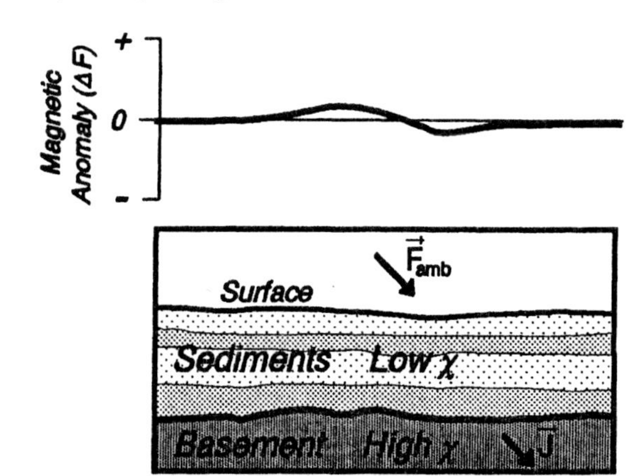

FIGURE 9.14 *Depth to magnetic basement.* Magnetic anomalies measured at Earth's surface depend on the depth of the magnetic sources, commonly located within crystalline basement. a) Where basement rocks are shallow, short-wavelength anomalies, with high amplitudes and steep gradients, occur. b) Basement buried deeply beneath sedimentary cover results in longer wavelength anomalies with smaller amplitudes and gradients.

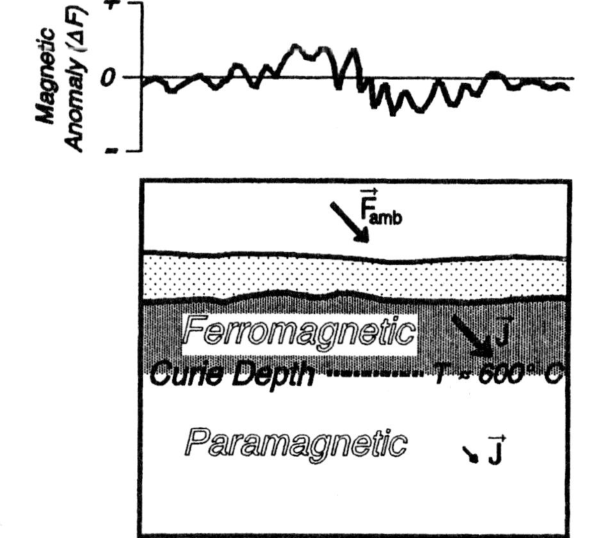

FIGURE 9.15 *Curie depth.* a) Regions with a normal geothermal gradient reach 600°C at depths of about 20 to 30 km. Their magnetic anomalies may exhibit high amplitudes because of the great thickness of basement rocks capable of ferromagnetic behavior. b) The Curie depth for hot regions (high geothermal gradient) may be considerably less than 20 km, resulting in a thinner layer with ferromagnetic behavior. Magnetic anomalies are potentially lower in amplitude compared to those in colder regions.

$\vec{\mathbf{F}}_{\mathbf{ind}}$ = magnetic field due to induced magnetization
$\vec{\mathbf{F}}_{\mathbf{rem}}$ = magnetic field due to remanent magnetization.

After the magnitude of the ambient field (F_{amb}) is subtracted from the magnitude of the total field (F), the total field anomaly (ΔF) is a function of the magnitudes of the induced (F_{ind}) and remanent (F_{rem}) fields. In many instances the remanent magnetization is several times stronger than the induced magnetization, so that the total field anomaly reflects the paleomagnetism component.

Types of Remanent Magnetization

Rocks are collections of diamagnetic, paramagnetic, and ferromagnetic minerals. Diamagnetic and paramagnetic materials acquire a magnetization when exposed to an external magnetic field (Fig. 9.8). At a given temperature, they have a constant susceptibility, acquiring a magnetization which is linearly proportional to the external magnetic field. The magnetization goes away when the external field is removed. Ferromagnetic materials, however, retain a permanent (or remanent) magnetization, even when the external field is removed. There are three ways by which materials can acquire remanent magnetization (see also chapter 3 of Butler, 1992).

1. Thermoremanent Magnetization ($\vec{\mathbf{J}}_{\mathbf{TRM}}$) At high temperature a ferromagnetic material exhibits paramagnetic behavior. As rocks cool below the Curie temperature, some minerals (particularly magnetite) change from the paramagnetic to the much stronger, ferromagnetic behavior. The rocks acquire a large, *thermoremanent magnetization* as magnetic domains orient themselves to Earth's ambient field (Fig. 9.16a).

A common misconception about thermoremanent magnetization is that mineral crystals orient themselves to Earth's field as a lava flow cools and hardens. Actually, the flow, though still hot, is quite hard (crystals randomly locked in place) before the remanent magnetization is acquired; the crystals cannot rotate to align with the ambient magnetic field. Instead, the magnetic moments of the collection of mineral grains acquires a bias in the direction of Earth's ambient field during cooling. Thermoremanent magnetization is thus a statistical, thermodynamic phenomenon; it results from the fact that a magnetite grain is in a slightly lower energy configuration when its magnetic moment is aligned with Earth's ambient field.

2. Detrital Remanent Magnetization ($\vec{\mathbf{J}}_{\mathbf{DRM}}$) When sediments settle in water, ferromagnetic mineral grains (particularly magnetite and hematite) tend to orient themselves along the ambient magnetic field of the Earth (Fig. 9.16b). The rock thus acquires a *detrital remanent magnetization*.

3. Chemical Remanent Magnetization ($\vec{\mathbf{J}}_{\mathbf{CRM}}$) As ions are precipitated from solution, forming ferromagnetic minerals, the magnetic domains in the mineral crystals preferentially orient themselves with Earth's ambient field (Fig. 9.16c). The rock thus acquires a *chemical remanent magnetization*. The paleomagnetism in iron-rich, continental sediments is generally formed in this manner. These "redbeds" account for the bulk of paleomagnetic observations on land, facilitating paleomagnetic stratigraphy and paleo-latitude studies (discussed below); the red color is derived from the ferromagnetic mineral hematite.

Paleomagnetic Interpretation

The paleomagnetism in rocks can lead to interpretations of the ages of the rocks (geochronology; paleomagnetic stratigraphy), and to the positions of crustal blocks when the rocks formed (paleolatitude).

Dating of rocks based on magnetic measurements revolves around the observation that Earth's magnetic field periodically reverses (Fig. 9.17). Magnetic domains in minerals, like compass needles, point toward the north magnetic pole in

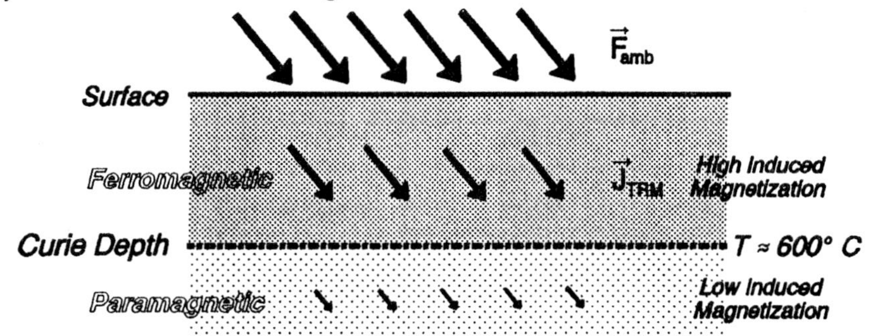

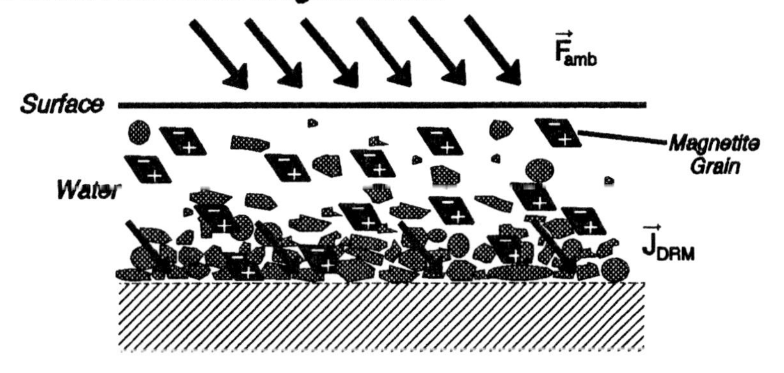

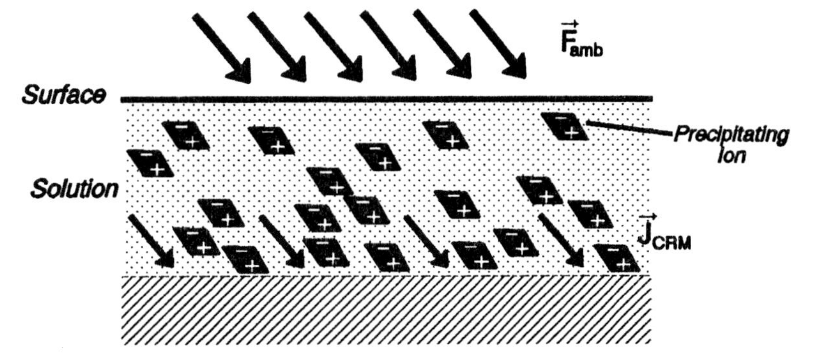

FIGURE 9.16 *Types of remanent magnetization.* For each type, the strength and direction of the magnetization remains, even after the ambient field is removed or changes orientation. a) *Thermoremanent magnetization* ($\vec{J}_{TRM}$). When the material (for example, a lava flow) cools below the Curie temperature, a strong ferromagnetization occurs, parallel to Earth's ambient field. b) *Detrital remanent magnetization* ($\vec{J}_{DRM}$). As sediments settle out in water, mineral grains rotate so that their magnetic domains preferentially orient with the ambient field. c) *Chemical remanent magnetization* ($\vec{J}_{CRM}$). As ions are precipitated from solution, their magnetic domains align with the ambient field.

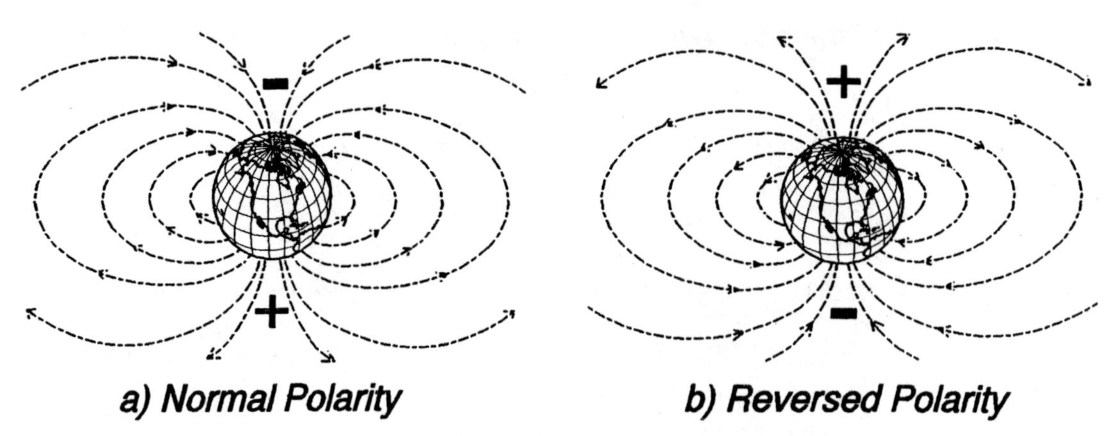

FIGURE 9.17 *Reversal of Earth's magnetic field.* a) "Normal" field, with lines of force pointing from a positive south pole to a negative north pole. b) "Reversed" field, with opposite polarities; a compass needle would point southerly.

the "normal" configuration that exists today. At times in the past, however, the magnetic field was reversed, so that mineral domains were magnetized in the opposite direction.

Utilizing observations of polarity reversals, a Geomagnetic Polarity Time Scale has been developed (Fig. 9.18). The scale is based on various methods, including: a) remanent magnetization studies of isotopically dated, young (< 5 million year old) igneous rocks; b) observations of marine magnetic anomalies, accompanied by paleontological dating of sediments and potassium/argon (K/Ar) dating of basalt recovered through the Deep Sea Drilling Project (DSDP); c) magnetic stratigraphy studies of sedimentary sections with unusually complete fossil records.

Geochronology Based on Paired Magnetic Anomalies across Mid-Ocean Ridges The ocean floors illustrate the utility of magnetic data to record the ages of certain Earth materials. As basaltic rocks at a mid-ocean ridge cool, they acquire a strong thermoremanent magnetization (Fig. 9.16a). A record of normal and reversed polarities is thus frozen into the basalt of oceanic layer 2 (Fig. 9.19b). The remanent magnetization of oceanic basalt results in additions and subtractions to Earth's ambient field. Profiles of total field anomalies recorded across mid-ocean ridges thus show alternating maxima and minima, representing times when oceanic layer 2 was normally or reversely polarized (Fig. 9.19c).

Magnetic polarity reversals observed across mid-ocean ridges agree with DSDP data showing that oceanic sediments and underlying basalts get progressively older away from mid-ocean ridge axes. The ages of oceanic rocks can thus be inferred from the pattern of observed magnetic anomalies, tied to the Geomagnetic Polarity Time Scale (Fig. 9.20).

The alternating bands of positive and negative magnetic anomalies show remarkable symmetry on opposite sides of mid-ocean ridges. This symmetry is testimony to the idea of creation of new lithosphere at divergent plate boundaries (Fig. 9.21). Anomaly widths indicate the rate at which plates diverge at mid-ocean ridges: narrow anomalies indicate slow spreading rates (Fig. 9.22a), wider anomalies faster rates (Fig. 9.22b).

Paleomagnetic Stratigraphy Sedimentary layers can acquire remanent magnetization through detrital and chemical mechanisms (Fig. 9.16). The ages of layers in a sequence of sedimentary strata may thus be determined by comparing the

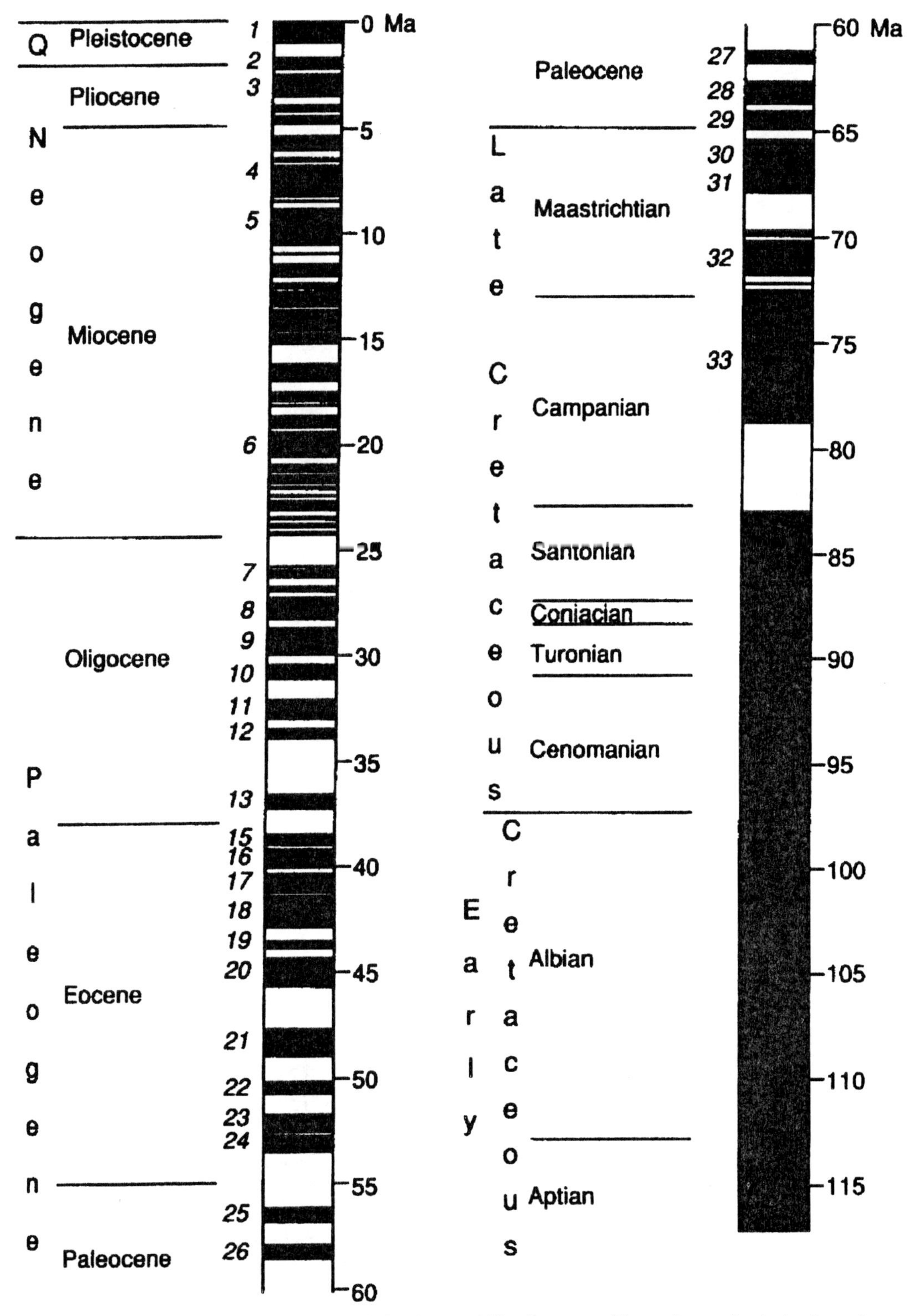

FIGURE 9.18 *Geomagnetic polarity time scale for the past 117 million years.* Times of normal polarity shown in black, reversed polarity in white. Magnetic anomaly numbers appear on the left sides of columns; corresponding ages in millions of years (Ma) are on the right. From "Magnetostratigraphic time scale," by A. Cox, in: *A Geologic Time Scale*, © 1982 Cambridge University Press. Reprinted with permission of Cambridge University Press, New York. Figure redrawn as in Butler (1992).

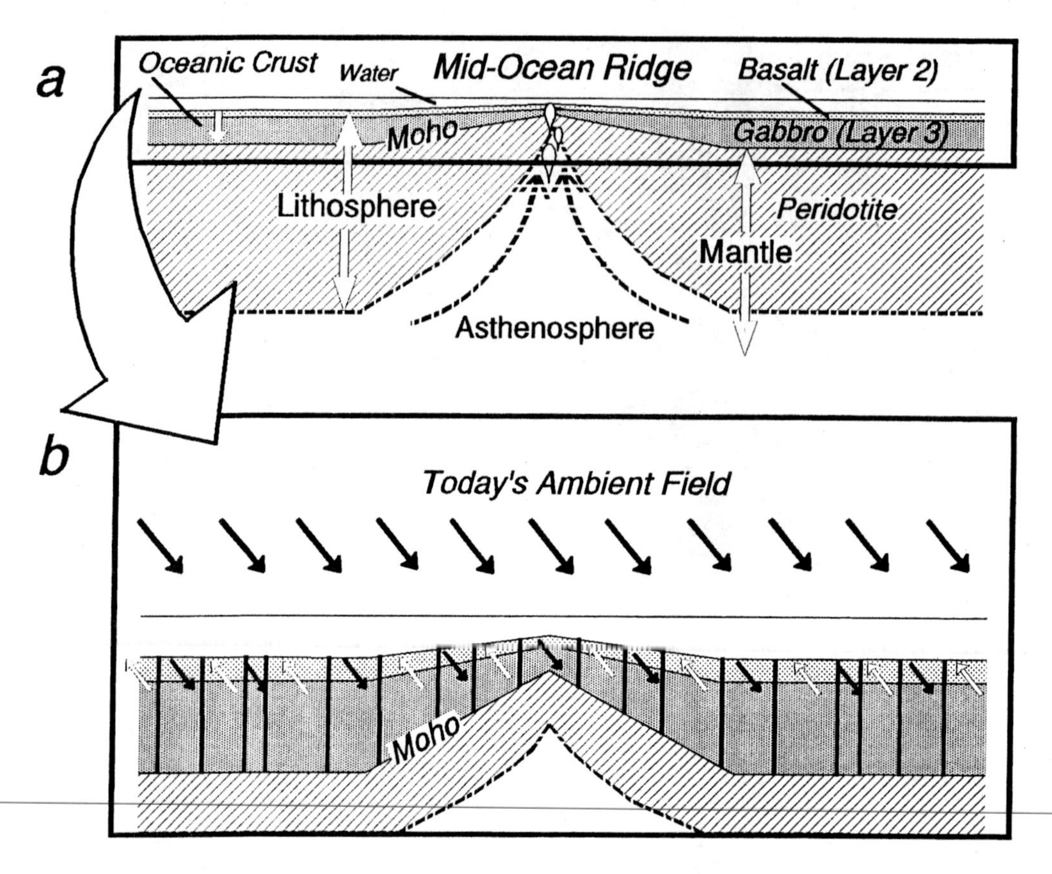

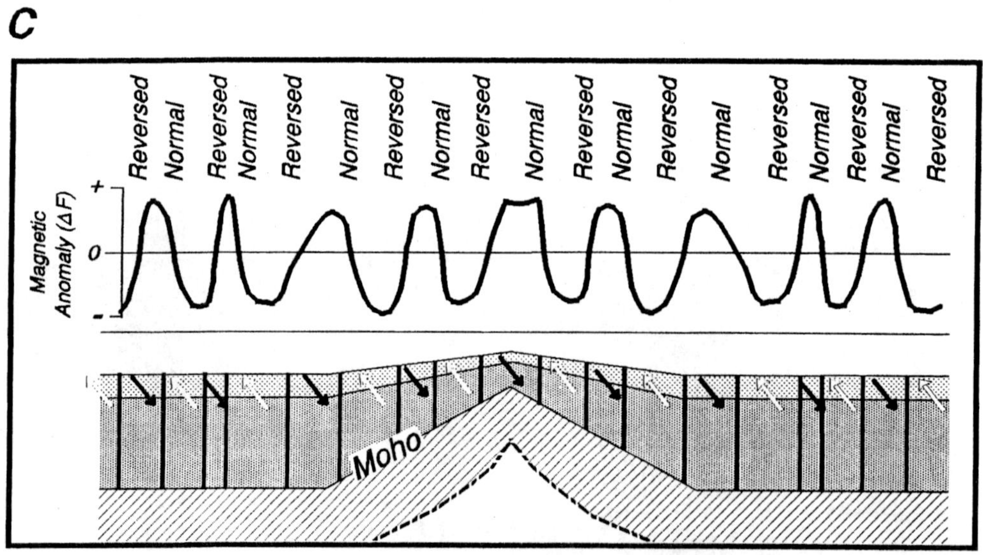

FIGURE 9.19 *Paleomagnetic alignment at a mid-ocean ridge.* a) Cross section of the lithosphere, showing formation of basalt (layer 2) and gabbro (layer 3). b) Closeup of the crust. Black arrow at the ridge axis illustrates magnetization aligned with a normal polarity, parallel to today's ambient field. Away from the axis, alternating bands of reversed (white arrows) and normal polarity reflect periodic reversals of the magnetic field. c) Magnetic anomaly pattern across a mid-ocean ridge. Alternating bands of oceanic crust with normal (black arrows) and reversed (white arrows) magnetization result in corresponding positive and negative total field magnetic anomalies (ΔF).

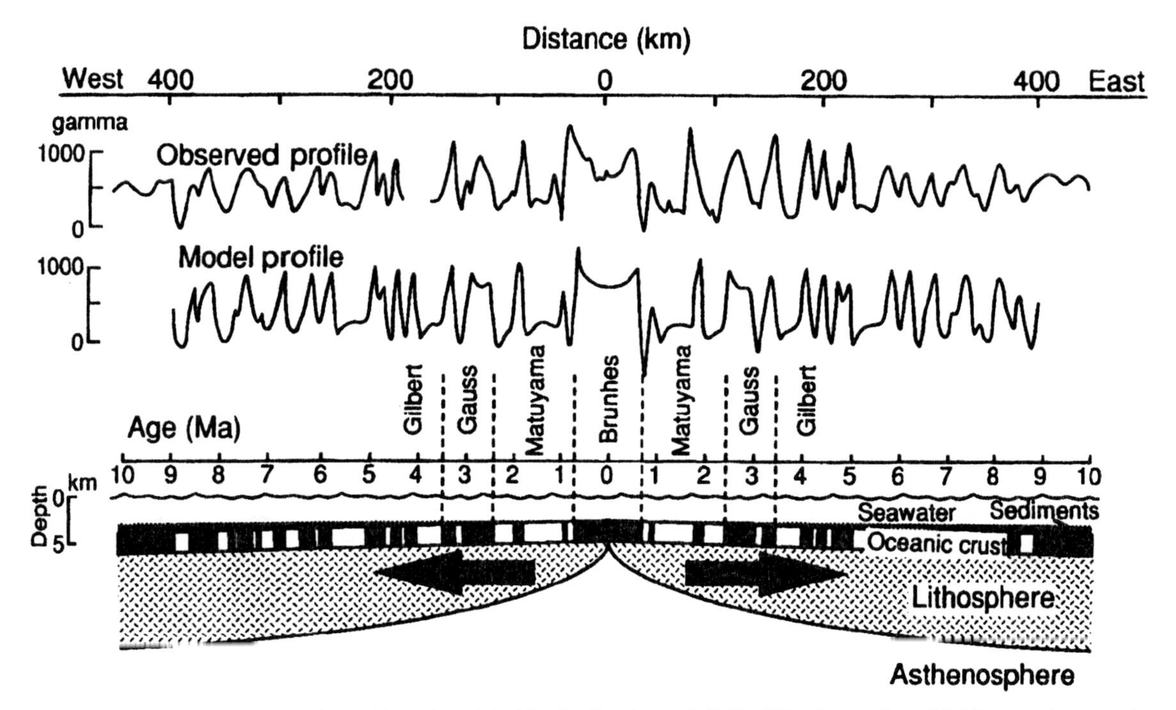

FIGURE 9.20 *Magnetic anomaly profile and model of the Pacific-Antarctic Ridge.* The observed total field magnetic anomaly profile (top) correlates well with a calculated profile (center) from the model of polarity reversals (bottom). The model incorporates times of normal (black) and reversed (white) paleomagnetism, according to the Geomagnetic Polarity Time Scale. Names refer to magnetic epochs in the time scale; numbers are ages in millions of years (Ma). Magnetic anomalies in gammas (1 gamma = 1 nT). From "Magnetic anomalies over the Pacific-Antarctic ridge," by W. Pitman, III, and J. Heirtzler, *Science*, vol. 154, pp. 1164–1171, © 1966. Figure redrawn as in Butler (1992).

observed polarity reversals with the Geomagnetic Polarity Time Scale. The technique is especially useful if parts of the section can be tied to isotopic dating of interbedded volcanic ash.

Fig. 9.23 shows an example from the Himalayan foreland basin deposits in Pakistan. Samples were taken from 10 stratigraphic sections, with results plotted in cross-sectional form on the diagram. According to *lithostratigraphy*, the sections are part of the Nagri Formation on the southwest, the Dhok Pathan Formation on the northeast. Samples analyzed for magnetic polarity show that there are correlations with positive and negative epochs of the Geomagnetic Polarity Time Scale. This *chronostratigraphy* thus reveals that parts of the Nagri and Dhok Pathan formations were deposited at the same time.

Paleolatitude Studies The remanent magnetization of a sedimentary or volcanic rock records the magnetic inclinations (i) at the time the rock formed. The inclination, in turn, can be related to the magnetic latitude (ϕ) of the rock at the time it formed, according to:

$$\tan \mathbf{i} = 2 \tan \phi$$

The equation assumes that the rock's magnetization records the original magnetic direction at the time of formation, that the field acts as a geocentric dipole, and that secular variation has been averaged out. Rocks formed near the magnetic equator

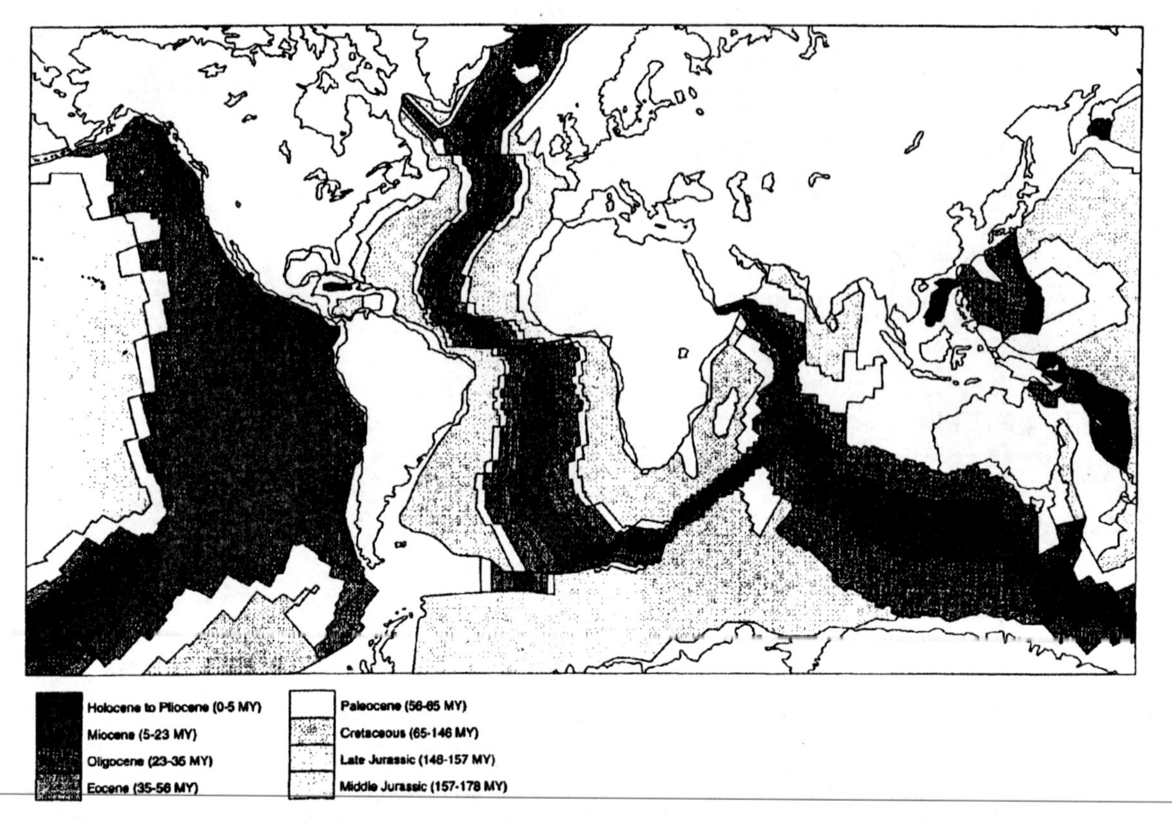

FIGURE 9.21 *World map showing the age of oceanic lithosphere determined from magnetic anomalies.* MY = Million years. From *Earth's Dynamic Systems,* 7th ed. by Hamblin/Christiansen, © 1995. Reprinted by permission of Prentice-Hall, Inc., Upper Saddle River, NJ.

would have remanent magnetizations oriented horizontally; magnetizations closer to the magnetic poles would be steeper (Fig. 9.4a). The magnetic inclination observed for a sedimentary layer thus indicates the magnetic latitude of the region at the time the layer was deposited (Fig. 9.24a). If the crustal block (terrane) on which the layer formed drifts to a different latitude, the inclination of remanent magnetization would differ from that of Earth's present magnetic field (Fig. 9.24b).

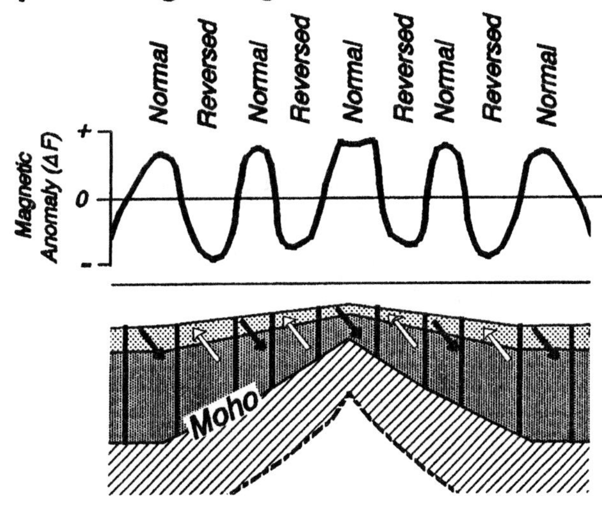

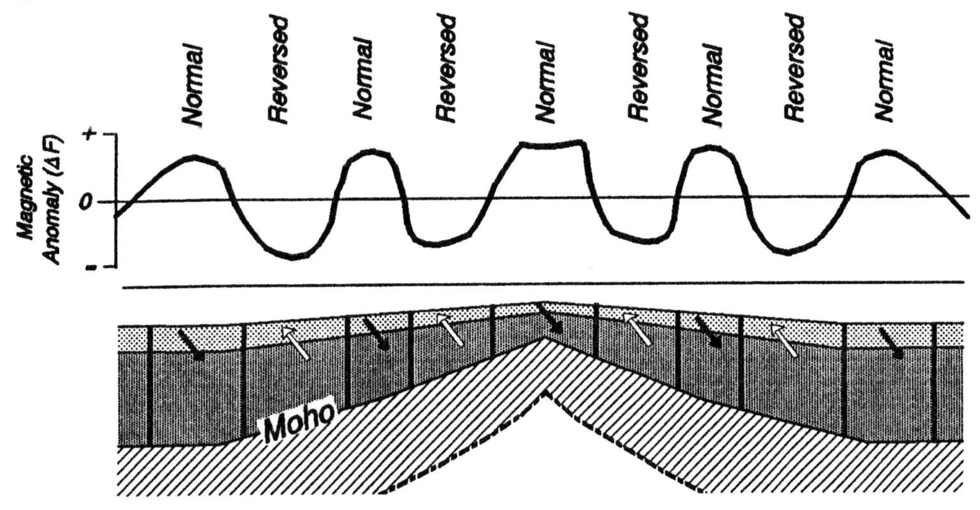

FIGURE 9.22 The widths of magnetic anomalies can be used to infer the spreading rate of mid-ocean ridges. a) *Mid-Atlantic Ridge*. Slow-spreading rate of 2–3 cm/a (20–30 km/Ma) results in narrow anomalies. b) *East Pacific Rise.* Broader anomalies result from spreading of 5–15 cm/a (50–150 km/Ma), creating a broader region of new lithosphere over the same time span as in (a).

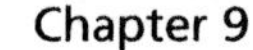

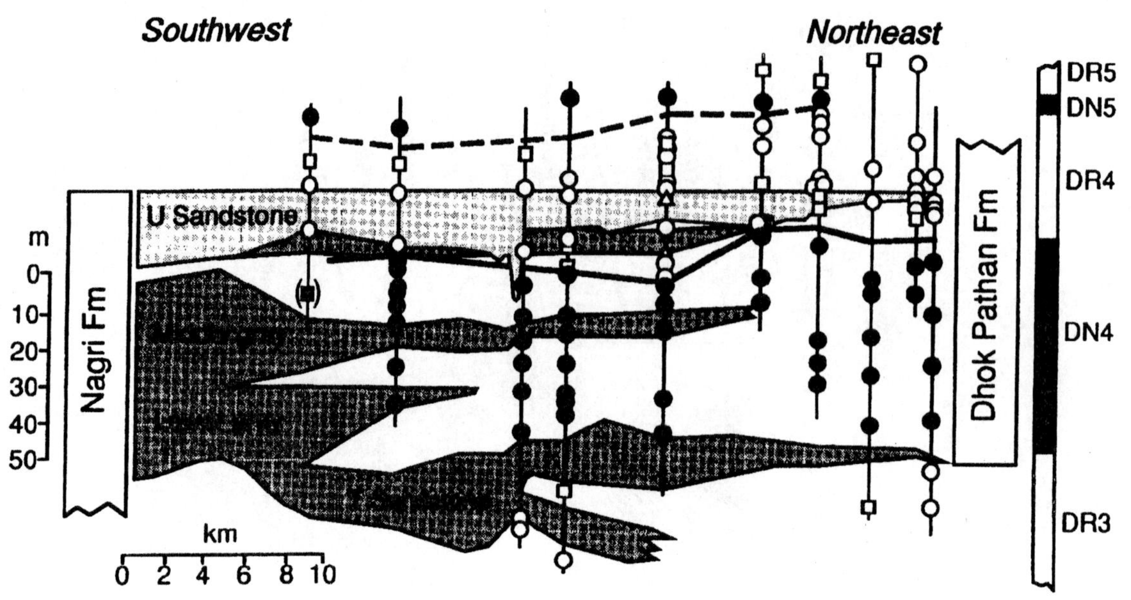

FIGURE 9.23 *Example of paleomagnetic stratigraphy*. Samples were taken from ten stratigraphic sections of Himalayan foreland basin deposits in Pakistan. Paleomagnetic normal polarities (filled circles and squares) and reversed polarities (open circles, squares, and triangles) were determined for each locality. The boundary between normal-polarity zone DN4 and reversed-polarity zone DR4 represents an 8.1 million-year-old time line, allowing age comparison of the various sedimentary deposits. From "Isochronous fluvial systems in Miocene deposits of northern Pakistan," by A. Behrensmeyer and L. Tauxe, *Sedimentology*, vol. 29, pp. 331–352, © 1982. Figure redrawn as in Butler (1992).

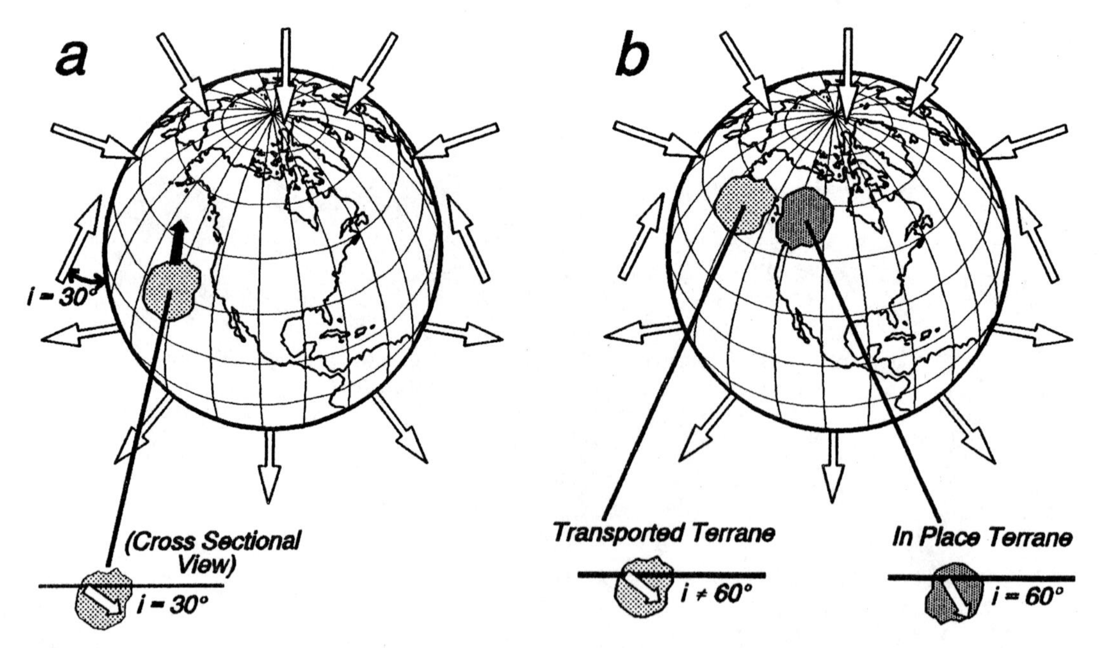

FIGURE 9.24 a) Direction of inclination of magnetization within a rock formed in the past at 30° magnetic latitude. b) The magnetic inclination (i) is retained when the rock drifts to a higher magnetic latitude, differing from the inclination of Earth's present field. The paleomagnetic inclination is thus a clue to the latitude at which the rock formed.

EXERCISES

9-1 The magnetic anomaly profile in Fig. 9.11b is drawn in a south-north direction, cutting across induced field lines. Sketch a cross section of the induced field and plot the total field anomaly for an east-west profile.

9-2 Redraw the gravity and magnetic profiles for the two models in Fig. 9.13, assuming the features occur at:

a) the magnetic equator;

b) the magnetic south pole.

9-3 For the model in Fig. 9.14, sketch the form of the magnetic anomaly profile, if the magnetic basement is at the surface.

9-4 **a)** For the stratigraphic section below, fill in the magnetic inclination for circles B-F, assuming the rocks were deposited at the following times and orientations:

A) 77 million years ago, 27° S magnetic latitude (answer provided; the black arrow represents normal polarity).

B) 53 million years ago, 12° S magnetic latitude.

C) 42 million years ago, magnetic equator.

D) 31 million years ago, 17° N magnetic latitude.

E) 16 million years ago, 25° N magnetic latitude.

F) Today, 15° N magnetic latitude.

b) Calculate the average (south to north) velocity of the continent as it moved from position A to position F.

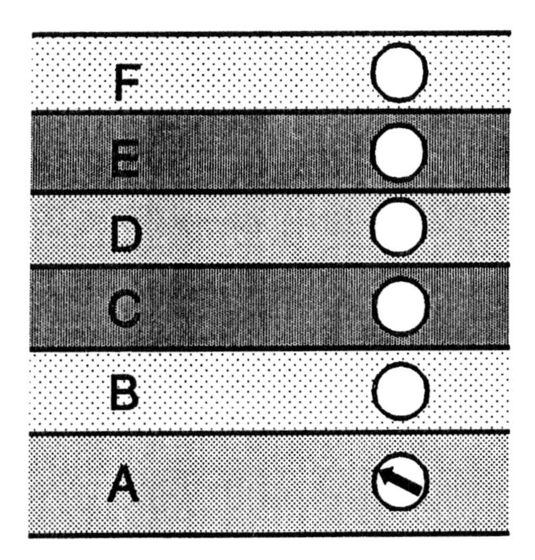

9-5 Across a mid-ocean ridge, the distance between total field anomaly pairs representing the Gilbert magnetic reversal epoch is 187 km. How fast has the ridge been spreading since that time?

9-6 A north-to-south magnetic profile is recorded across a magnetic body at 45° N magnetic latitude (model on next page). The body has a magnetic susceptibility of 0.1.

a) Assuming the body is shallow, sketch on the diagram: **i)** Earth's ambient magnetic field; **ii)** the direction and amplitude of magnetization within the body, relative to the magnitude and direction of the ambient field; **iii)** the induced magnetic field lines; **iv)** the total field magnetic anomaly profile.

b) Repeat (a) with the body very deep.

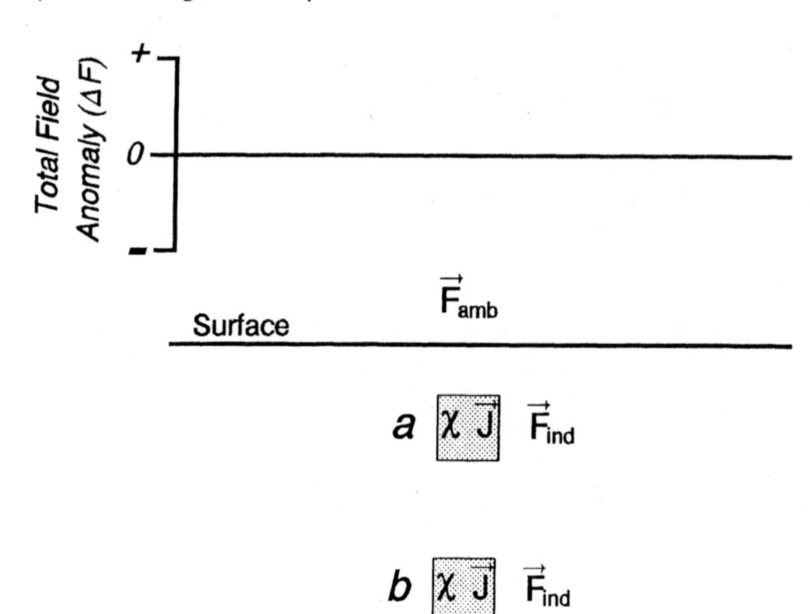

SELECTED BIBLIOGRAPHY

Behrensmeyer, A. K., and L. Tauxe, 1982, Isochronous fluvial systems in Miocene deposits of northern Pakistan, *Sedimentology*, v. 29, pp. 331–352.

Blakely, R., 1995, *Potential Theory in Gravity and Magnetic Applications*, Cambridge: Cambridge University Press, 441 pp.

Bott, M. H. P., 1982, *The Interior of the Earth: Its Structure, Constitution and Evolution* (2nd ed.), New York: Elsevier Science Pub. Co., 403 pp.

Burger, H. R., 1992, *Exploration geophysics of the shallow subsurface*, Englewood Cliffs, New Jersey: Prentice Hall, Inc., 489 pp.

Butler, R. F., 1992, *Paleomagnetism: Magnetic Domains to Geologic Terranes*, Boston: Blackwell Scientific Publications, Boston, 319 pp.

Cox, A., 1982, Magnetostratigraphic time scale, in: *A Geologic Time Scale*, edited by W. B. Harland and others, Cambridge: Cambridge University Press, pp. 63–84.

Fabiano, E. B., and N. W. Peddie, 1969, Total magnetic intensity map of the Earth in 1965, *National Oceanic and Atmospheric Administration Ocean Survey*, Technical Report C&GS 38.

Fowler, C. M. R., 1990, *The Solid Earth: An Introduction to Global Geophysics*, Cambridge: Cambridge University Press, 472 pp.

Garland, G. D., 1979, *Introduction to Geophysics* (2nd ed.), Toronto: W. B. Saunders Company, 494 pp.

Griffiths, D. H., and R. F. King, 1981, *Applied Geophysics for Geologists and Engineers: The Elements of Geophysical Prospecting* (2nd ed.), New York: Pergamon Press, 230 pp.

Hamblin, W. K., and E. H. Christiansen, 1995, *Earth's Dynamic Systems* (7th ed.), Englewood Cliffs, NJ: Prentice Hall, Inc., 710 pp.

Heirtzler, J. R., G. O. Dickson, E. M. Herron, W. E. Pitman, III, and X. Le Pichon, 1968, Marine magnetic anomalies, geomagnetic field reversals, and motions of the ocean floor and continents, *Journal of Geophysical Research*, v. 73, pp. 2119–2136.

Kearey, P., and M. Brooks, 1984, *An Introduction to Geophysical Exploration*, Boston: Blackwell Science Pub., 296 pp.

Merrill, R. T., and M. W. McElhinny, 1983, *The Earth's Magnetic Field*, London: Academic Press, 401 pp.

Nettleton, L. L., 1971, *Elementary Gravity and Magnetics for Geologists and Geophysicists*, Tulsa: Society of Exploration Geophysicists, Monograph Series 1, 121 pp.

Pitman, W. C., III, and J. R. Heirtzler, 1966, Magnetic anomalies over the Pacific-Antarctic ridge, *Science*, v. 154, pp. 1164–1171.

Smith, R. B., R. T. Shuey, Jr., R. Pelton, and J. P. Bailey, 1977, Yellowstone Hotspot: Contemporary tectonics and crustal properties from earthquake and aeromagnetic data, *Journal of Geophysical Research*, v. 82, pp. 3665–3676.

Telford, W. M., L. P. Geldart, R. E. Sheriff, and D. A. Keys, 1976, *Applied Geophysics*, Cambridge: Cambridge UniversityPress, 860 pp.

Vine, F. J., and D. H. Matthews, 1963, Magnetic anomalies over ocean ridges, *Nature*, v. 199, pp. 947–949.

C H A P T E R 1 0

Heat Flow

heat *(hēt)* **n.,** [*OE.* hætu], *the quality of being hot; energy manifest by the accelerated vibration of molecules within a material.*
flow *(flō)* **vi.,** [*OE.* flowan], *to move gently and smoothly; to pour out.*
heat flow *(hēt flō)* **n.,** *the rate of transfer of heat from the Earth across its surface.*
temperature *(tem' prə chər)* **n.,** [< *L.* < temperatus, *temperate*], *the degree of hotness; the amount of heat per unit mass of a material.*
geo- *(jē' ō),* [*Gr. geō-* < gaia, gē, *the Earth*], *pertaining to the Earth.*
thermal *(thər'm'l)* **adj.,** *having to do with heat.*
gradient *(grā'dē ənt)* **n.,** [< *L.* gradi, *to step*], *the amount of slope, as of a road; the rate of change of a parameter.*
geothermal gradient *(jē' ō thər'm'l grā' dē ənt)* **n.,** *the rate of change in temperature with increasing depth within the Earth.*

Heat flows outward from the interior through the surface of the Earth (Fig. 10.1). The rate of heat flow across a region of the surface is a function of the original temperature of the Earth, the production of heat within the Earth, the transfer of heat between different regions of the Earth, and the ability of different Earth materials to conduct heat (Fig. 10.2). Measurements of heat flow rates provide clues to the presence of hot material at shallow depths (young igneous intrusions; upwelling asthenosphere), cold material protruding into the Earth (subducting lithosphere), and the cooling of material through time (subsidence of the ocean floor away from a mid-ocean ridge).

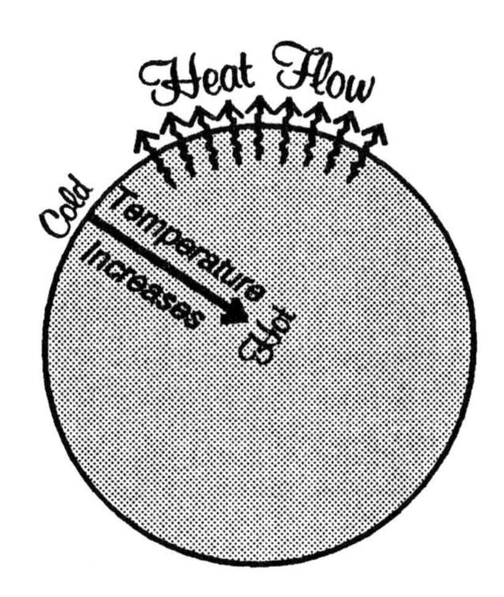

FIGURE 10.1 Heat flows outward, from Earth's hot interior across its cold surface. *Heat flow* refers to the amount of heat that crosses the surface in a given amount of time. The average rate is about 0.08 W/m^2, equivalent to 1.9×10^{-6} $cal/cm^2/s$, or 1.9 HFU.

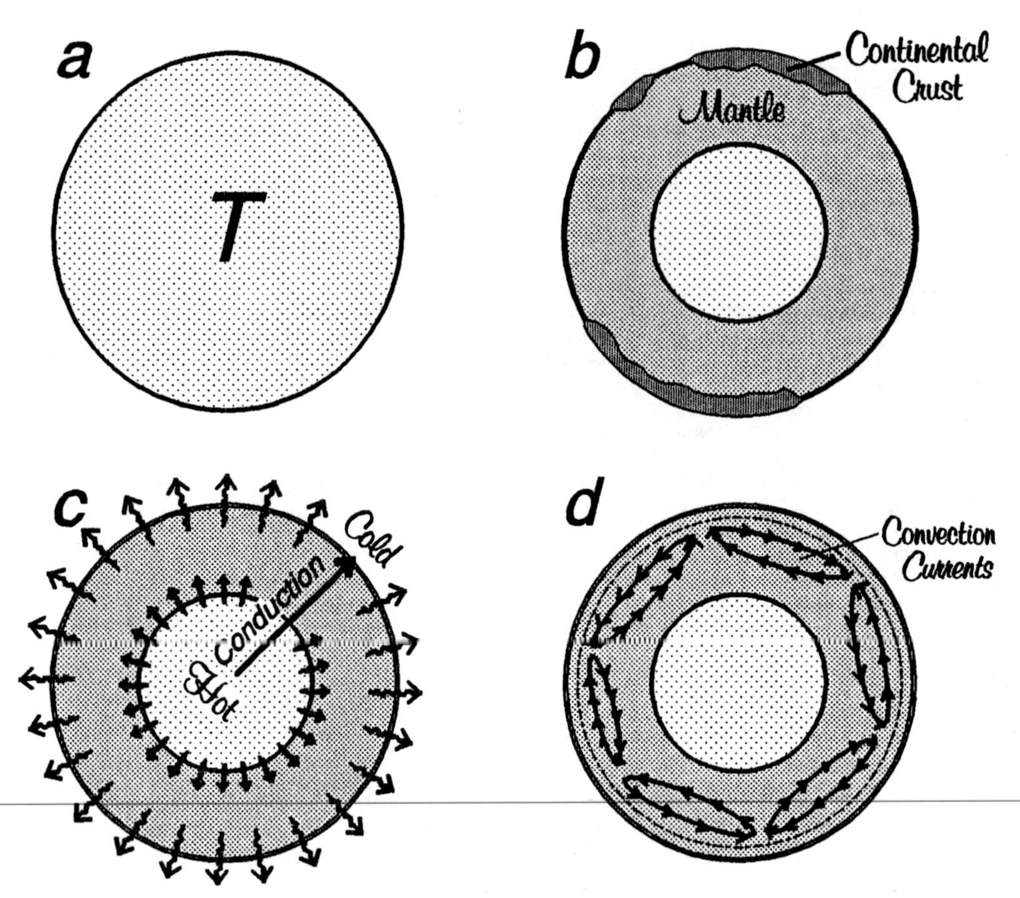

FIGURE 10.2 Several factors influence the rate that heat flows across different regions of Earth's surface. a) *Heat of formation.* A background level of heat for all regions is due, in part, to the original temperature distribution of the Earth (T). The overall temperature decreases as Earth cools through time. b) *Heat production.* New heat is a by-product of radioactive decay, primarily within Earth's mantle and continental crust. c) *Conduction.* Heat generally conducts outward, from Earth's hot interior to it's colder surface. Locally, some rocks conduct heat more readily than others. d) *Convection.* Large amounts of heat can be transferred from one region to another through convection. Convection can occur in the upper part of the asthenosphere and in the deeper mantle (see Elder, 1965).

HEAT WITHIN THE EARTH

Sources of Heat

Most of Earth's internal heat is due to the combined effects of the original temperature of the Earth (heat of formation) and new heat produced over time by the Earth itself (primarily through radioactive decay).

Heat of Formation In the 19th century, William Kelvin estimated the age of the Earth as between 20 and 40 million years, based on measurements of heat flowing across its surface (Fig. 10.3). Kelvin's assumptions were that all of Earth's heat originated when the Earth began to cool from a molten state, and that heat is transferred out of the Earth through conduction. This simple model can be visualized by taking a hot potato out of an oven and laying it on the kitchen counter (Fig. 10.4a). If someone were to enter the kitchen and feel that the potato is hot, they would reason that it was taken out of the oven very recently, within the last few minutes. A person touching a cold potato might conjecture it had been out of the oven for at

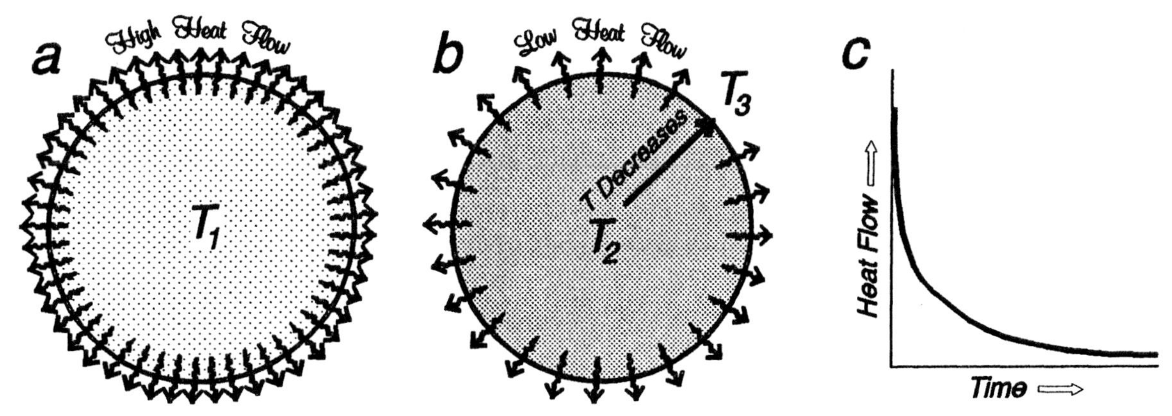

FIGURE 10.3 *Cooling of Earth through time.* a) Original temperature of the Earth (T_1), in a molten state at the time of its formation. b) As heat is lost through the surface, the temperature drops throughout; it is highest near the center (T_2) and decreases outward (to T_3). c) The rate of heat flowing across the surface also decreases through time; the average heat flow across the surface was thus used by Kelvin to estimate the age of the Earth.

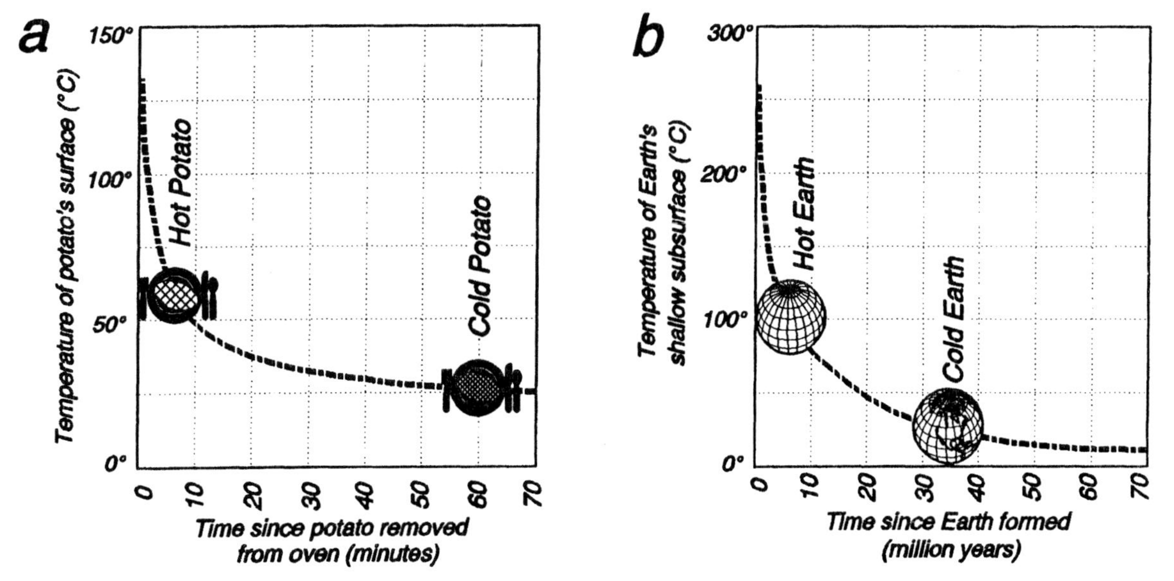

FIGURE 10.4 *Loss of heat through conduction.* a) The surface temperature of a potato suggests the time since the potato was removed from an oven. b) Assuming no new heat generation or convective heat loss, the temperature at about 1 km depth indicates Earth's age.

least an hour. Similarly, the average temperature near Earth's surface indicates Earth's age (Fig. 10.4b).

Modern dating techniques reveal that the Earth is about 4.6 billion years old, not 20 to 40 million. Kelvin's age estimate was low because both of his assumptions were inaccurate; the Earth has been generating new heat since its formation, and heat flows to the surface through convection as well as conduction.

Radioactive Decay Not all of Earth's heat originated at the time of its formation. New heat is produced when radioactive elements decay from one form to another, primarily in Earth's mantle. Continuing the above analogy, it would be like taking the hot potato from the normal oven and placing it in a microwave oven; the microwaves would excite new heat within the potato. On a low enough setting, the potato would cool, but not nearly as quickly as outside the microwave oven (Fig. 10.5a). Like the Earth (Fig. 10.5b), the potato would remain hot for a very long time, due to the continuous production of internal heat.

The production of heat depends on the decay of radioactive elements, like uranium, thorium, and potassium. Crustal rocks, especially continental, are high in those elements, and thus produce large quantities of heat per unit volume (Fig. 10.6a). The Earth's mantle is, however, far more voluminous than the crust (Fig. 2.3), so that most of the new heat is produced in the mantle (Fig. 10.6b).

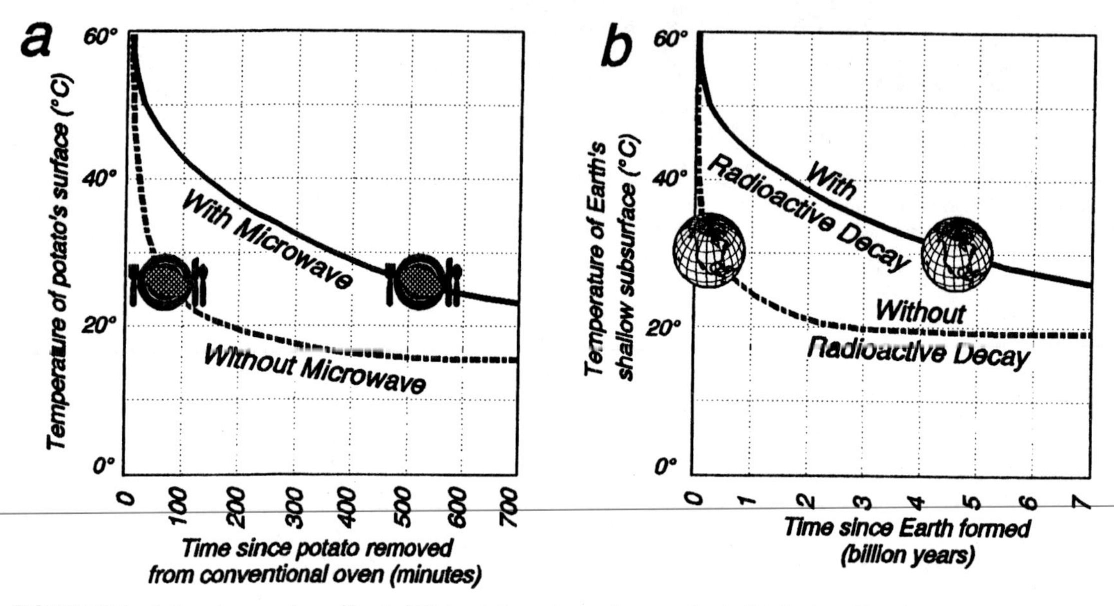

FIGURE 10.5 *Internal generation of heat.* a) Hot potato cools much more slowly if it is placed in microwave oven on low setting. b) The Earth has cooled much more slowly than might be expected, because new heat is generated during radioactive decay.

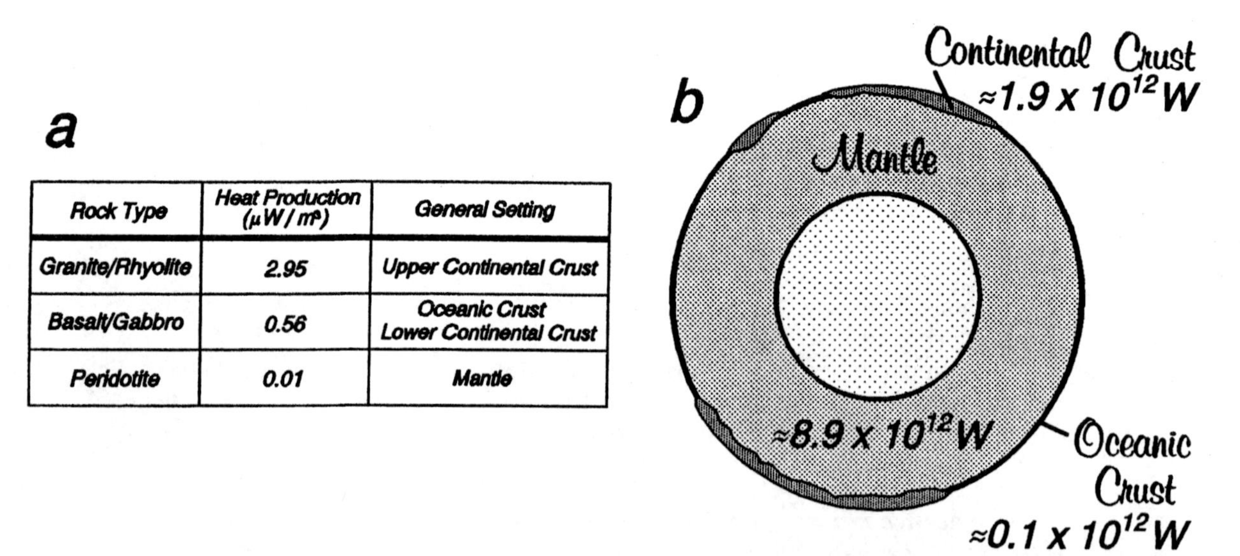
a

Rock Type	Heat Production ($\mu W/m^3$)	General Setting
Granite/Rhyolite	2.95	Upper Continental Crust
Basalt/Gabbro	0.56	Oceanic Crust Lower Continental Crust
Peridotite	0.01	Mantle

FIGURE 10.6 *Heat production in the Earth through radioactive decay.* a) Production of heat by igneous rock types (Philpotts, 1990). A Watt (W) is equivalent to one Joule of energy produced per second (J/s). A cubic meter of continental crust (granite/gabbro) produces heat much faster than a cubic meter of mantle (peridotite). b) The mantle is far more voluminous than the crust, however, so that it accounts for most of Earth's internal heat production.

Heat Transfer

The transfer of heat from one region to another can involve electromagnetic waves (*radiation*), atomic interaction between materials of different temperature (*conduction*), or the actual movement of hot materials (*convection; advection*).

Radiation A common and obvious source of heat at Earth's surface is *radiation* from the sun (Fig. 10.7). This radiation is on the order of 2×10^{17} W ($\approx 4 \times 10^2$ W m^{-2} average over the surface), resulting in an average surface temperature of about 20°C (Fowler, 1990). Solar heat drives atmospheric and oceanic circulation, which in turn drive surface geologic processes (weathering, erosion, stream flow, glaciation). Virtually all of the heat that shines on the Earth is radiated back and lost, however, so that solar radiation cannot explain the amount of heat that flows from Earth's deep interior ($\approx 8 \times 10^{-2}$ W m^{-2}). Other heat sources are therefore responsible for Earth's internal processes (earthquakes, volcanism, lithospheric plate motion).

Conduction Heat can be transferred by atomic vibrations, flowing from a region of higher to one of lower temperature. Such *conduction* is illustrated by a potato if it is dry; heat goes from the center of the hot potato to colder regions near its surface, at the contact with the cold air (Fig. 10.8a). A thermal gradient is thus established in the potato, with temperature increasing from the surface to the center. Likewise, conductive transfer of heat from Earth's interior to its surface sets up a *geothermal gradient*, with temperature increasing rapidly near the surface, more slowly with depth (Fig. 10.8b). Transfer of heat through conduction is a relatively slow process on a geologic time scale.

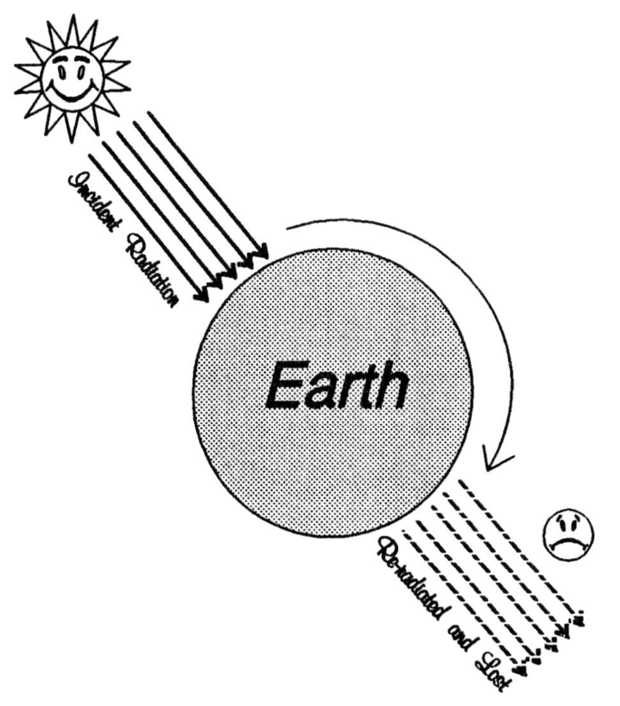

FIGURE 10.7 *Transfer of heat to Earth's surface through radiation.* The large amount of heat from the sun that is absorbed during the day is lost outward through radiation at night.

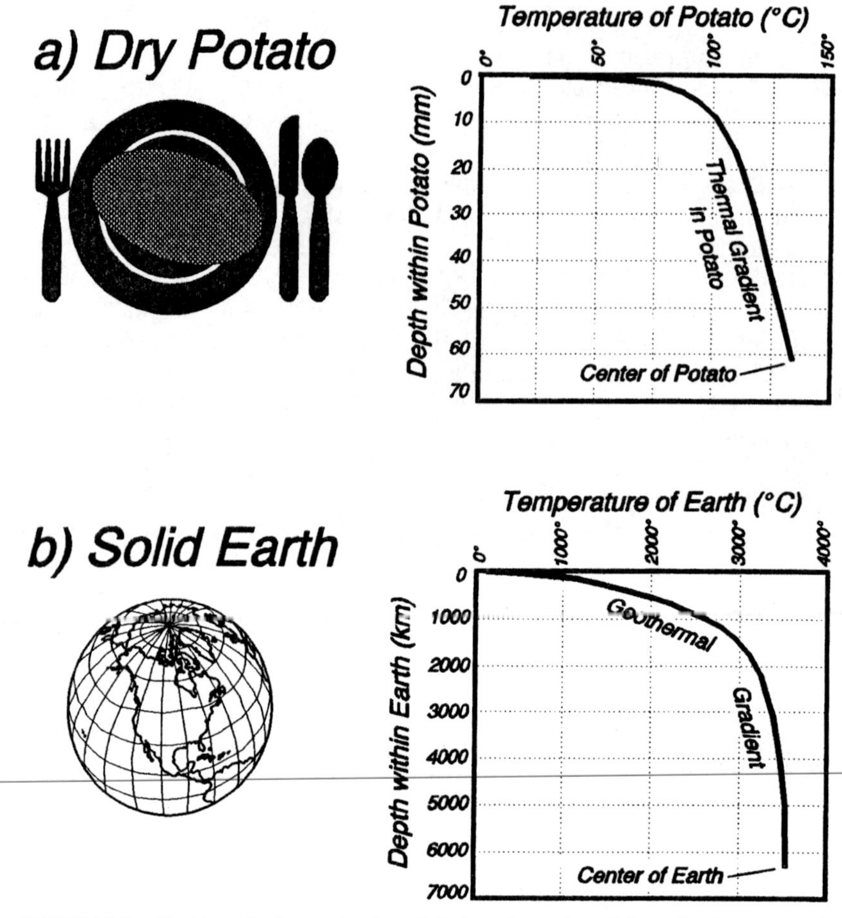

FIGURE 10.8 *Heat transfer by conduction.* a) As heat is conducted from the hot center of a potato to the cooler surface, a temperature gradient develops. The temperature increase is rapid near the surface, more gradual near the center. b) The Earth develops an analogous thermal gradient as heat conducts from its hot interior to its cool surface.

Convection The transfer of heat in conjunction with movement of material is called *convection*. In a pot of boiling water, convection currents move heat rapidly (Fig 10.9a). Hot water near the burner rises; colder, denser water at the surface descends to near the burner, where it is heated. A cherry pie baking might resemble the Earth's lithosphere/asthenosphere system (Fig. 10.9b); convection of the filling (asthenosphere) produces currents that rise and fall, in places deforming the pie crust (lithosphere).

Advection is a special type of convection, whereby heat is transferred through movement of material in a solid state. Advective heat flow generally proceeds slowly, so that conduction readjusts the temperature. In some instances, however, tectonic processes quickly move blocks of hot material large horizontal or vertical distances, so that advective heat transfer is significant. Examples include the horizontal and vertical movements along deep-seated thrust faults, as well as uplift of mountain ranges due to erosion and isostatic rebound (Fig. 10.10).

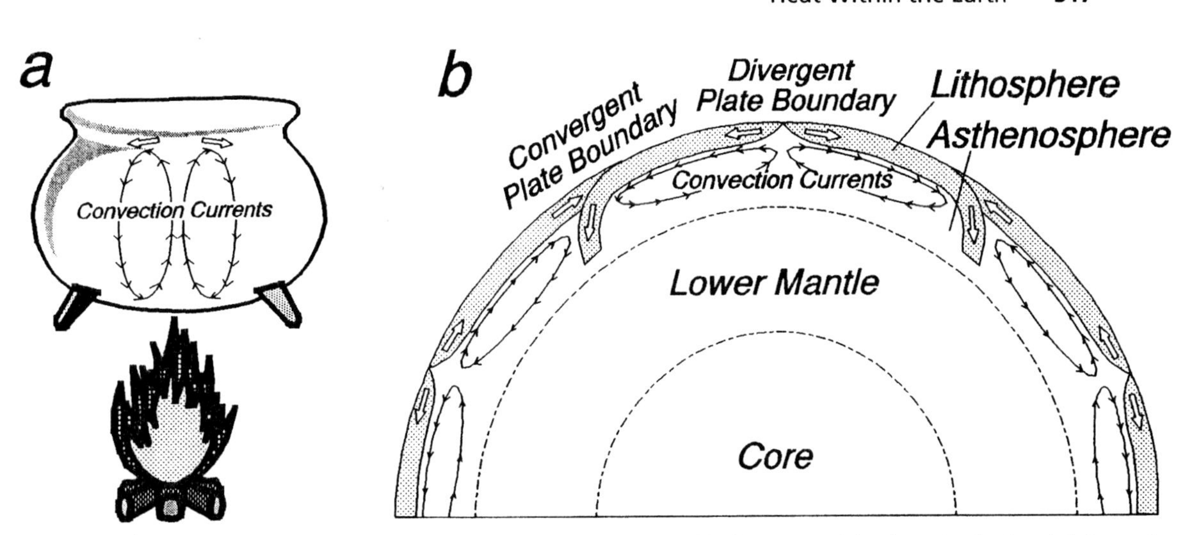

FIGURE 10.9 *Heat transfer by convection.* a) Heat moves rapidly in a pot of boiling water, rising from depth where it is hot and light, falling near the surface where it becomes cold and dense. b) At a very slow rate (a few cm/year) convection can occur in Earth's asthenosphere. In addition to redistributing heat, convection contributes to lithospheric plate motion: where convection currents rise, plates diverge; descending currents enhance plate convergence. Some convection is also thought to occur in the deeper mantle.

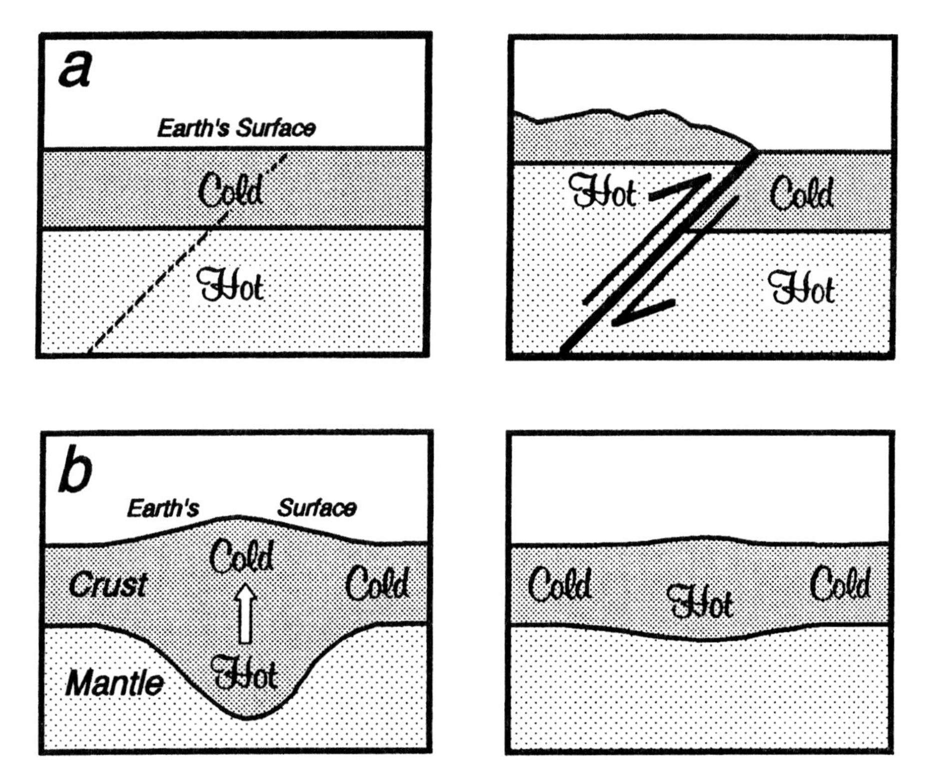

FIGURE 10.10 *Heat transfer by advection.* a) Thrust faulting. b) Erosion and isostatic rebound.

HEAT FLOW ACROSS EARTH'S SURFACE

Heat Flow Equation

The rate that heat flows by conduction from Earth's interior to its surface can be envisioned through a simple plate model (Fig. 10.11a). Heat energy is conducted from the level with highest temperature (T_2) to the lowest temperature surface (T_1), according to:

$$q = k\left(\frac{T_2 - T_1}{h}\right)$$

where:

q = rate of heat flow, per unit area, through the top of the plate (W m^{-2})
k = thermal conductivity of the plate (W m^{-1} °C^{-1})
T_1 = temperature at the surface of the plate (°C)
T_2 = temperature at the base of the plate (°C)
h = thickness of the plate (m).

If there is a linear increase in temperature from the surface downward (Fig. 10.11b), the heat flow can be expressed as a function of depth as:

$$q(z) = k\left(\frac{[T_1 + \Delta T] - T_1}{\Delta z}\right)$$

where:

$q(z)$ = heat flow at depth z (positive downward)
ΔT = change in temperature from the surface to depth z
Δz = distance from the surface to depth z.

Taking the limit, for very small Δz:

$$\lim_{\Delta z \to 0} q(z) = \lim_{\Delta z \to 0} k\left(\frac{\Delta T}{\Delta z}\right)$$

$$= k\left(\frac{\partial T}{\partial z}\right)$$

For constant increase in temperature with depth (Fig. 10.11c), the heat flow at Earth's surface is thus:

$$\boxed{\mathbf{q = k\,(\partial T/\partial z)}}$$

where:

q = rate that heat flows outward across Earth's surface (*heat flow*)
k = ability of rocks in the region to conduct heat (*thermal conductivity*)
$\partial T/\partial z$ = rate at which temperature increases from the surface downward in the region (*geothermal gradient*).

Table 10.1 shows that thermal conductivities for most rocks range from 2 to 3 W m^{-1} °C^{-1}. Values for rocks rich in quartz (sandstone, quartzite) can be significantly higher.

For a given thermal conductivity, higher heat flow results from higher geothermal gradient (Fig. 10.12a). If the geothermal gradient is constant, regions with rocks of higher thermal conductivity have higher heat flow (Fig. 10.12b).

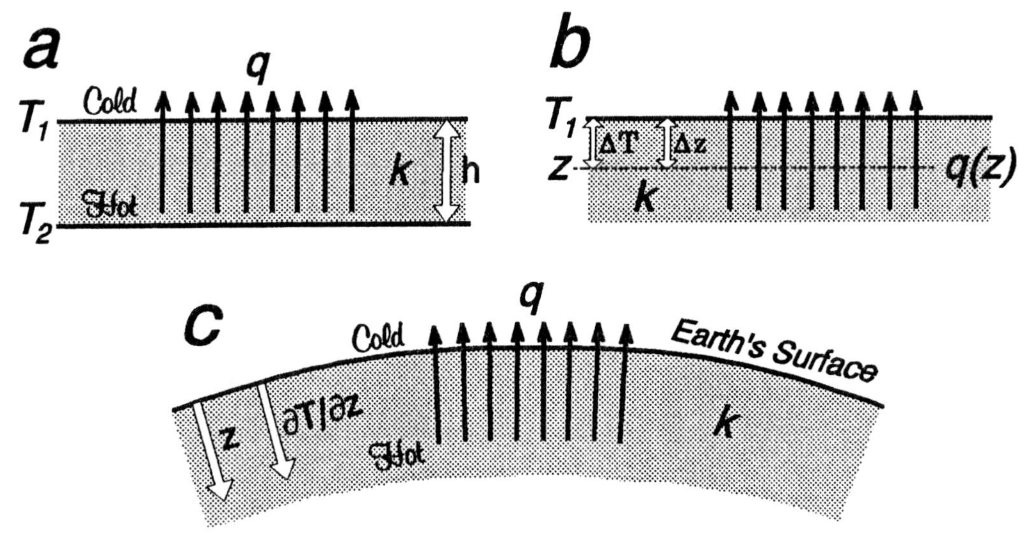

FIGURE 10.11 *Heat flow.* a) Transfer of heat across a thin plate of thermal conductivity (k). The temperature at the bottom of the plate (T_2) is hotter than at the top (T_1). b) For constant increase in temperature with depth, the heat flow [q(z)] can be determined from knowledge of k, T_1, and the change in temperature (ΔT) over a small change in depth (Δz). c) Heat flows through Earth's surface at a rate (q) according to the geothermal gradient ($\partial T/\partial z$) and the thermal conductivity (k) of rocks in the region.

TABLE 10.1 *Thermal conductivity of selected rocks* (averages from Jessop, 1990).

	Rock	Thermal Conductivity ($W\,m^{-1}\,°C^{-1}$)
Igneous	Rhyolite	2.6
	Granite	3.3
	Andesite	2.3
	Diorite	2.8
	Basalt	1.8
	Gabbro	2.8
Sedimentary	Shale	2.1
	Sandstone	3.7
	Limestone	3.4
Metamorphic	Amphibolite	3.0
	Serpentinite	3.5
	Quartzite	5.0

Measurement of Heat Flow

Fig. 10.13 illustrates that heat flow is measured by taking the temperature at different depths in a piston core or drillhole, yielding the geothermal gradient ($\partial T/\partial z$). If an average thermal conductivity (k) is known for rocks in the region, the heat flow (q) can be calculated.

International Standard (SI) units are stated above for parameters in the heat flow equation. Table 10.2 shows how other units commonly found in the literature

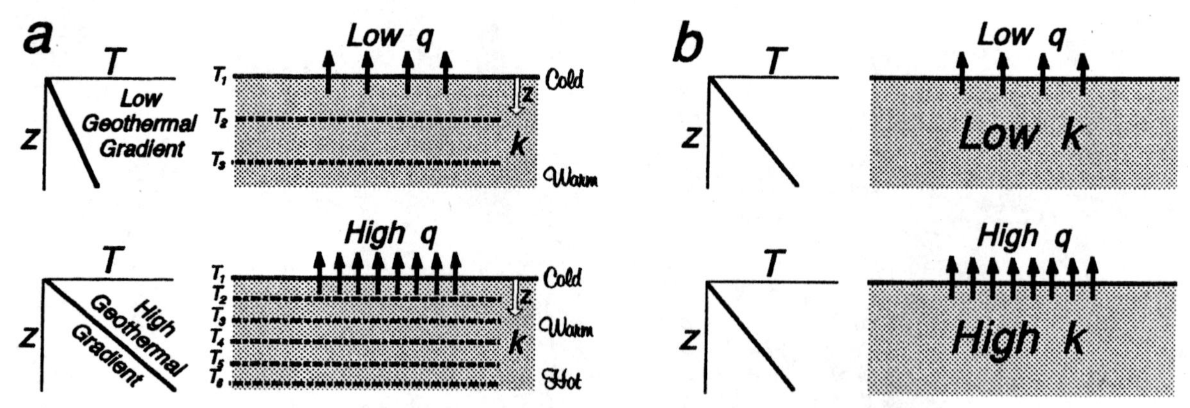

FIGURE 10.12 a) Change in heat flow (q) as a function of geothermal gradient; the thermal conductivity (k) is held constant. b) Change in heat flow as a function of thermal conductivity; the geothermal gradient remains constant.

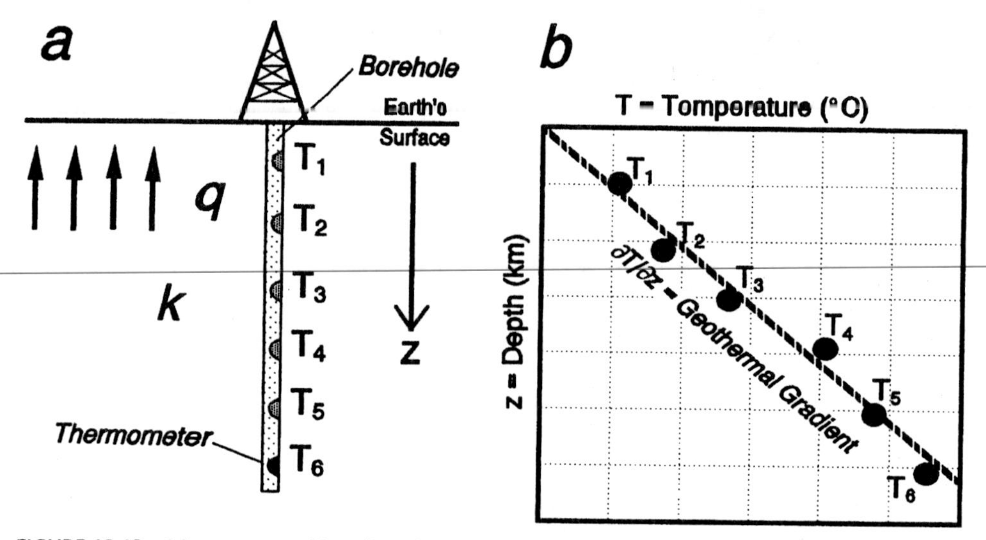

FIGURE 10.13 *Measurement of heat flow.* a) Temperatures ($T_1, T_2, \ldots, T_6$) measured at different depths (z) in a borehole. b) T plotted against z yields the geothermal gradient ($\partial T/\partial z$). Combining with the measured thermal conductivity (k) for the region, the heat flow is calculated as $q = k(\partial T/\partial z)$.

TABLE 10.2 *Conversion from International Standard (SI) to other units found in the literature.*

	SI	Equivalent Units
Thermal Conductivity (**k**)	**W m^{-1}°C^{-1}**	2.39 × 10^{-3} **cal cm^{-1}°C^{-1} s^{-1}**
Geothermal Gradient (**∂T/∂z**)	**°C m^{-1}**	10^3 **°C km^{-1}**
Heat Flow (**q**)	**W m^{-2}**	2.39 × 10^{-5} **cal cm^{-2} s^{-1}**
		2.39 × 10^1 **HFU**

relate to SI units. In particular, heat flow units (HFU) are often used on older maps ($1\ HFU = 10^{-6}\ cal\ cm^{-2}\ s^{-1} = 4.2 \times 10^{-2}\ W\ m^{-2}$). Some typical values measured for the Earth are:

$$\mathbf{k} \approx (2\ \text{to}\ 5)\ \mathbf{W\ m^{-1}\,^{\circ}C^{-1}}$$
$$\approx (0.005\ \text{to}\ 0.012)\ \mathbf{cal\ cm^{-1}\,^{\circ}C^{-1}\,s^{-1}}$$

and:

$$\partial \mathbf{T}/\partial \mathbf{z} \approx (0.01\ \text{to}\ 0.05)\mathbf{^{\circ}C\ m^{-1}}$$
$$\approx (10\ \text{to}\ 50)\mathbf{^{\circ}C\ km^{-1}}$$

so that:

$$\mathbf{q} \approx (0.03\ \text{to}\ 0.12)\ \mathbf{W\ m^{-2}}$$
$$\approx (1\ \text{to}\ 3)\ \mathbf{HFU}.$$

TECTONICS AND HEAT FLOW

The transfer of heat from one region of the Earth to another is enhanced by convection within the upper mantle; the convection facilitates movement of lithospheric plates (Fig. 2.10). Lateral transfer generally elevates the overall heat flow of oceanic regions relative to that of the continents. Vertical transfer makes shallow features hot, particularly where plates diverge (continental rifts, mid-ocean ridges). Where plates converge, the injection of cold lithosphere results in low heat flow at subduction zones and mountain ranges.

Continental Areas

Fig. 10.14 is a map of heat flow determined for the United States. Table 10.3 shows approximate geothermal gradients and heat flow values for three regions, one cold, one normal, and one hot. The geothermal gradients for the three regions are superimposed on phase diagrams for upper continental crustal rock (granite) and mantle (peridotite) in Fig. 10.15. The diagrams illustrate the consequences to rocks in those regions as a result of the different gradients.

Continental Craton On the stable interior of a continent (craton), there is normal increase in temperature with depth. Granitic rocks of the upper-to-middle crust are therefore too cold for partial melting to occur (Fig. 10.15a). The upper mantle is of normal temperature, resulting in typical refraction velocities of about 8.1 to 8.2 km/s (Fig. 4.15). Because the region is not tectonically active, it has had time to thermally equilibrate; a relatively flat lithosphere/asthenosphere boundary has thus been established at about 150 km depth (Fig. 10.16a).

TABLE 10.3 *Approximate geothermal gradients and heat flow values for three regions of North America.* ($1\ HFU = 4.2 \times 10^{-2}\ W/m^2$).

	Geothermal Gradient	Heat Flow	
Sierra Nevada Mountains	10°C/km	$0.03\ W/m^2$	0.75 HFU
Continental Craton	20°C/km	$0.06\ W/m^2$	1.5 HFU
Basin and Range Province	30°C/km	$0.1\ W/m^2$	2.5 HFU

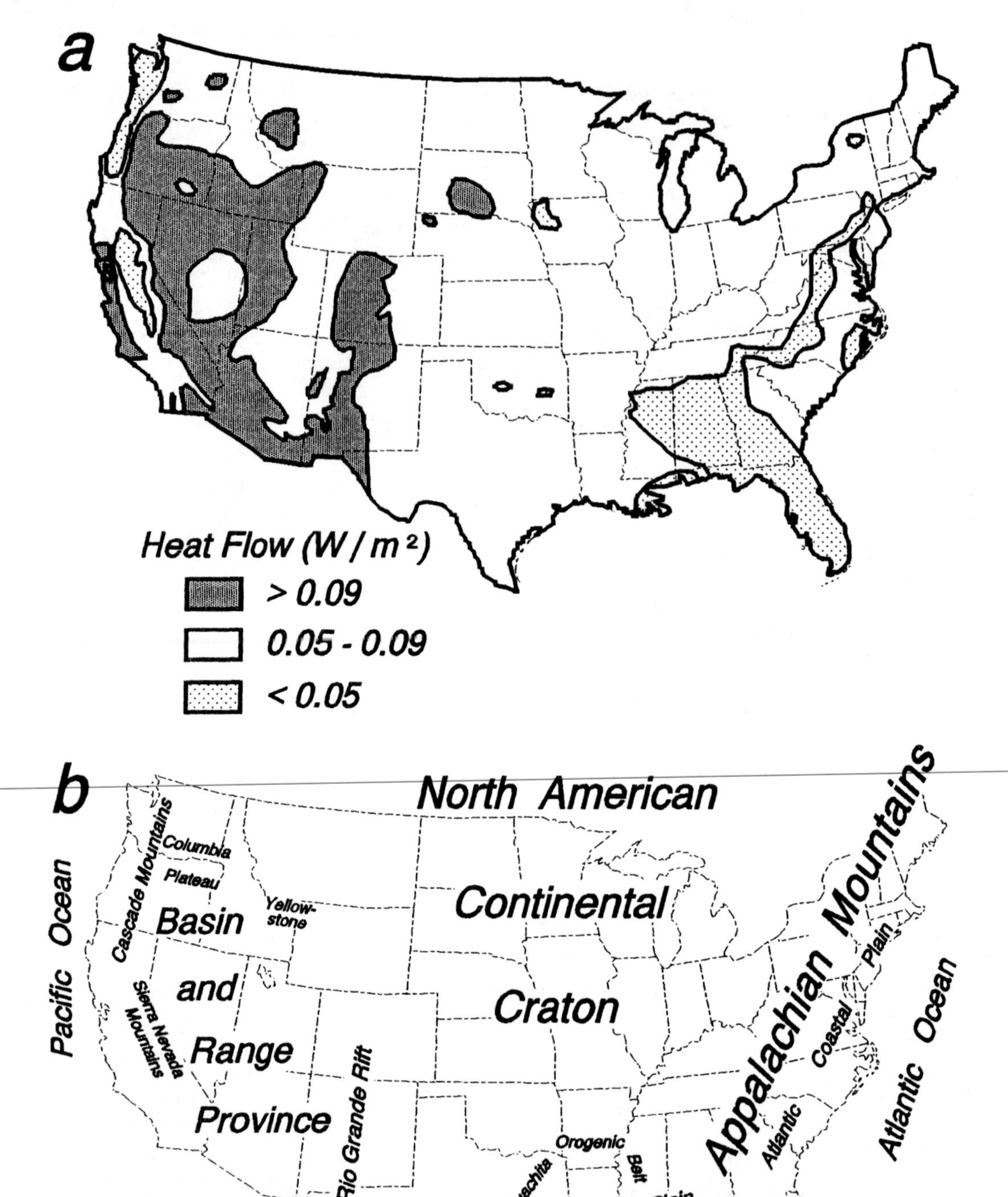

FIGURE 10.14 *Heat flow map of the United States* (simplified from U. S. Department of Energy Heat Flow Map, 1996; see also Blackwell and Steele, 1989, and Blackwell et al., 1991). Regions of high and low values, expressed in W/m^2, are highlighted. Older maps may show values in heat flow units (1 HFU = 4.2×10^{-2} W/m^2). High values (> 0.09 W/m^2) are observed in areas of continental rifting (Basin and Range Province; Rio Grande Rift) and at the Yellowstone hotspot. The continental craton is dominated by normal values (0.05 – 0.09 W/m^2). The region of ongoing subduction in the Pacific Northwest shows a low (< 0.05 W/m^2) in the forearc region (near the coast) and a high at the volcanic arc (Cascade Mountains). A prominent low marks the region of the Sierra Nevada Mountains, where plate convergence ceased about 20 million years ago. Low values are also observed in the Paleozoic continental collision zone in the East (Appalachian Mountains).

a) Dry Granite P/T Diagram

T = Temperature (°C)

0° 200° 400° 600° 800° 1000° 1200° 1400°

Z = Depth (km) 0 10 20 30 40 50 60 70

P = Pressure (MPa) 0 500 1,000 1,500 2,000

Basin & Range

Craton

Sierra Nevada

Solid

0%

Partial Melt

100%

Liquid

b) Wet Peridotite P/T Diagram

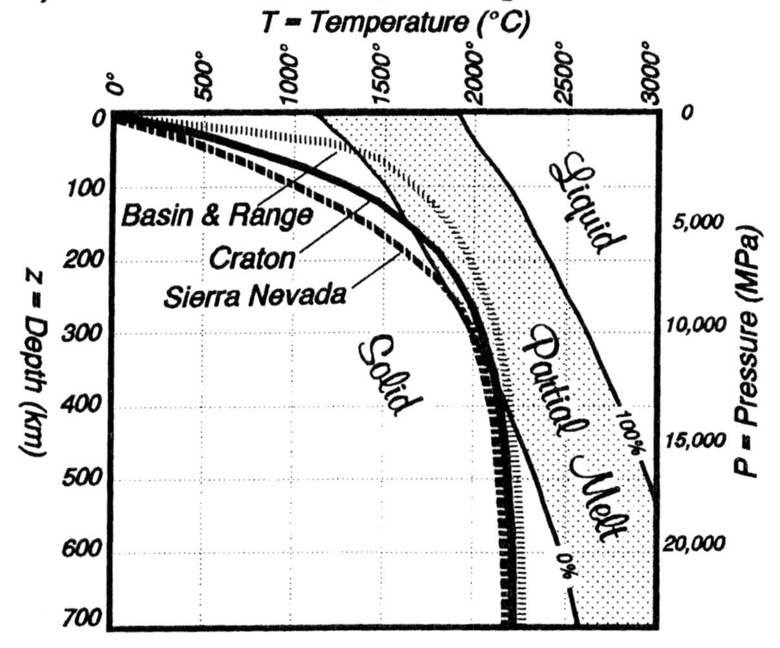

FIGURE 10.15 a) Geothermal gradients from Table 10.3, superimposed on phase diagram for dry granite. Note that temperatures are high enough at about 20 km depth in the Basin and Range Province to initiate partial melting; earthquakes occurring within the brittle upper crust thus terminate above that depth. b) Geothermal gradients superimposed on phase diagram for peridotite (Fig. 2.7). Notice the shallow depth (≈ 45 km) where partial melting may be encountered beneath the Basin and Range Province. Intersections of the geothermal gradients with the partial melt region suggest the base of the lithosphere at about 150 km depth for the craton, 250 km for the Sierra Nevada Mountains. Modified from *Physical Geology* by B. J. Skinner and S. C. Porter, © 1987. Used with permission of John Wiley and Sons, Inc., New York.

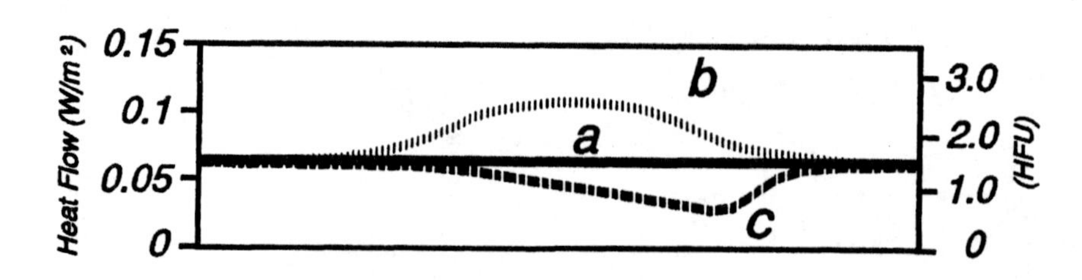

a) Normal Continental Lithosphere

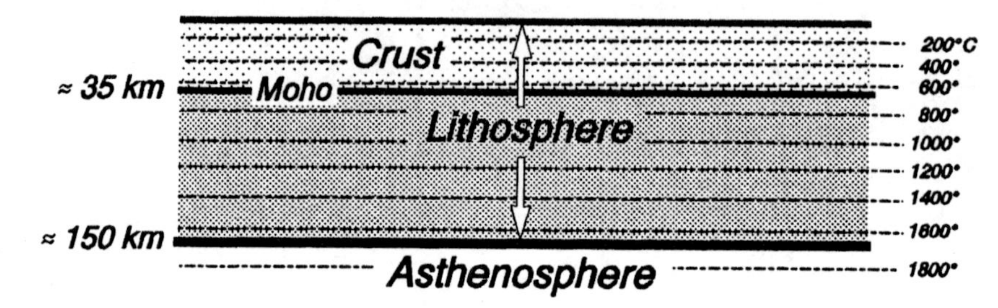

b) Continental Rift Zone

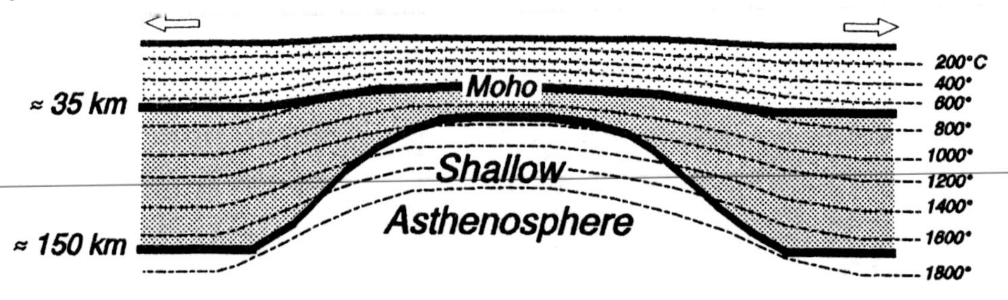

c) Remnant Lithosphere Slab

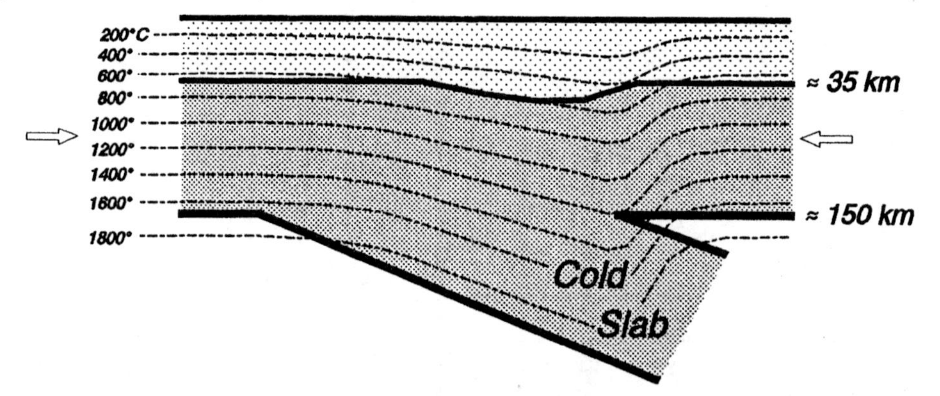

FIGURE 10.16 *Models explaining heat flow observed in three regions of North America* (Table 10.3; Fig. 10.14). Regions of high geothermal gradient have closely-spaced temperature contours; contours are far apart where the gradient is low. a) Typical values of heat flow (≈ 0.06 W m^{-2}, or 1.5 HFU) and geothermal gradient (≈ 20°C/km) on the craton result in temperature and pressure conditions that establish the base of the lithosphere at about 150 km depth (Fig. 10.15b). b) A high geothermal gradient elevates the lithosphere/asthenosphere boundary in the Basin and Range Province, enhancing partial melting of the asthenosphere (Fig. 10.15b). Temperatures may be hot enough so that a brittle-to-ductile transition occurs within the upper crust above 20 km depth (Fig. 10.15a). c) Subduction brought cold lithosphere to a deeper level beneath the Sierra Nevada Mountains, depressing the geothermal gradient and heat flow.

Continental Rift Hot asthenosphere is in a shallow position beneath a continental rift, heating overlying crustal rocks (Fig. 2.24b). At a depth of about 20 km beneath the Basin and Range Province, granitic rocks are hot enough to initiate slight partial melting (Fig. 10.15a). Earthquakes are thus limited to the upper 15 km or so because the high geothermal gradient results in a shallow brittle-to-ductile transition.

Low seismic velocities are observed for the upper mantle beneath the Basin and Range Province (Fig. 4.15), consistent with very shallow asthenosphere. The high geothermal gradient suggests that, at 45 km depth, the temperature is around 1300°C (Fig. 10.15b). The pressure on the hot asthenosphere is so low at that depth that significant partial melting occurs (Fig. 2.7); the lower crust may be underplated by new gabbroic material, as suggested by sub-horizontal events on seismic reflection profiles (Figs. 6.23, 6.24). The crust/mantle transition is new and flat (Fig. 10.16b), resulting from the magmatic differentiation of gabbro overlying peridotite; this new Moho is revealed by strong reflections from about 30 to 32 km depth throughout the Basin and Range Province.

Remnant Subduction Zone The Sierra Nevada Mountains in California and Nevada are the eroded roots of a volcanic arc that existed prior to the development of the San Andreas transform plate boundary (Fig. 2.17a). Low heat flow (Fig. 10.14) suggests the presence of thick, cold lithosphere, a remnant of the former subduction zone (Figs. 10.15b, 10.16c). Similar lithosphere slabs have been recognized through seismic delay-time studies of the eastern Alps and southern Carpathians in Europe (Fig. 7.34). Through time, these cold slabs will heat up to more normal temperatures for the upper mantle, and the lithosphere/asthenosphere boundary will return to a shallower, flatter configuration.

Oceanic Regions

Oceanic regions are generally hotter than the continents, because their lithosphere is younger. Through time, the upper mantle cools, lowering the lithosphere/asthenosphere boundary. Transects across mid-ocean ridges show correlations of crustal age, bathymetry, and heat flow consistent with the generation of new lithosphere. At subduction zones, heat flow profiles highlight regions of downgoing lithosphere and magma emplacement.

Mid-Ocean Ridge At mid-ocean ridges, both heat flow values and topography decrease exponentially with distance from the ridge axes (Fig. 10.17a,b). For a given spreading rate, the farther from the ridge, the older the lithosphere; decreasing heat flow values thus indicate cooling of the lithosphere as it ages (Fig. 10.17c).

At the same distance from ridge axes, fast-spreading ridges are more elevated than slow-spreading ridges (Fig. 10.17b). The elevation is a function of temperature; hotter regions have thermally expanded mantle and, hence, are more elevated. Fig. 10.17c thus shows that, for both slow- and fast-spreading ridges, the water deepens exponentially relative to the age of the lithosphere.

Collectively, the observations in Fig. 10.17 suggest that, as materials move away from a ridge axis, they cool and contract, causing the top of the crust to sink deeper below sea level. The cooling follows an exponential decrease with time, so that the top of a given age of lithosphere will be at a predictable water depth. Fast-spreading ridges remain hot and elevated some distance from the ridge axis, leading to a very broad profile (East Pacific Rise; Fig. 10.18a). Slow-spreading ridges are

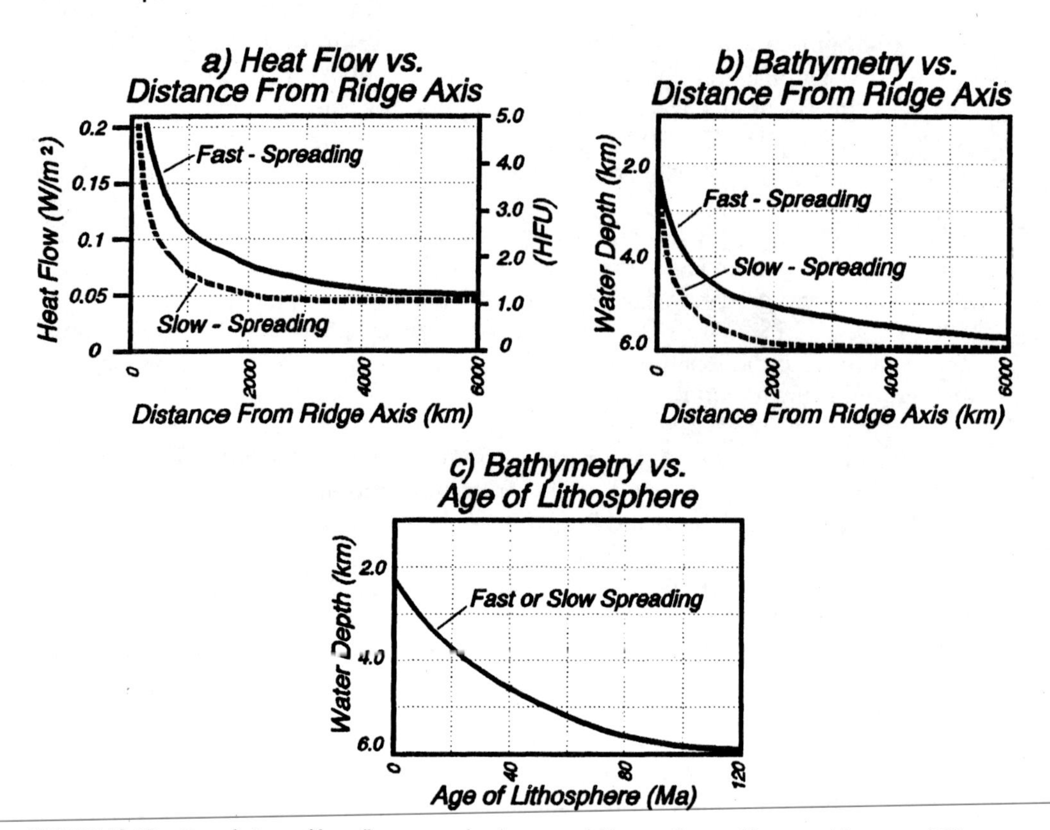

FIGURE 10.17 *Correlations of heat flow, water depth, age, and distance from mid-ocean ridge axes.* a) The faster the spreading rate, the broader the region of high heat flow. b) Fast-spreading ridges show a wider area of high elevation than slow-spreading ridges. c) The ocean floor subsides as mid-ocean ridge lithosphere cools through time.

older and cooler a similar distance from the ridge, producing steeper profiles (Mid-Atlantic Ridge; Fig. 10.18b).

Subduction Zone At a subduction zone, the thermal effect is much like putting an ice cube in a cup of hot coffee. It takes some time before enough heat is transferred from the coffee into the ice, causing melting. In the meantime, the temperature of the ice is less than that of the surrounding coffee.

When a slab of lithosphere subducts, it perturbs temperature contours, decreasing the geothermal gradient in the region of the subduction zone (Fig. 10.19). Heat flow measurements commonly show a broad low in the forearc region, with a local high at the volcanic arc. The low is a result of the cold slab, while the high reflects transfer of heat within the ascending magma. The map in Fig. 10.14a shows this low/high heat flow pattern at the Cascadia Subduction Zone in the Pacific Northwest (Fig. 2.17).

A consequence of rapid subduction is that cold material may descend to great depths, remaining rigid enough to fail brittely through earthquakes (Figs. 2.22, 7.23). Through time the slab heats up, but the cold may linger for tens of millions of years. Lithosphere roots have thus been recognized beneath areas of fairly recent plate convergence, like the Alps (Fig. 7.34) and Sierra Nevada Mountains (Figs. 10.14, 10.15b, 10.16c).

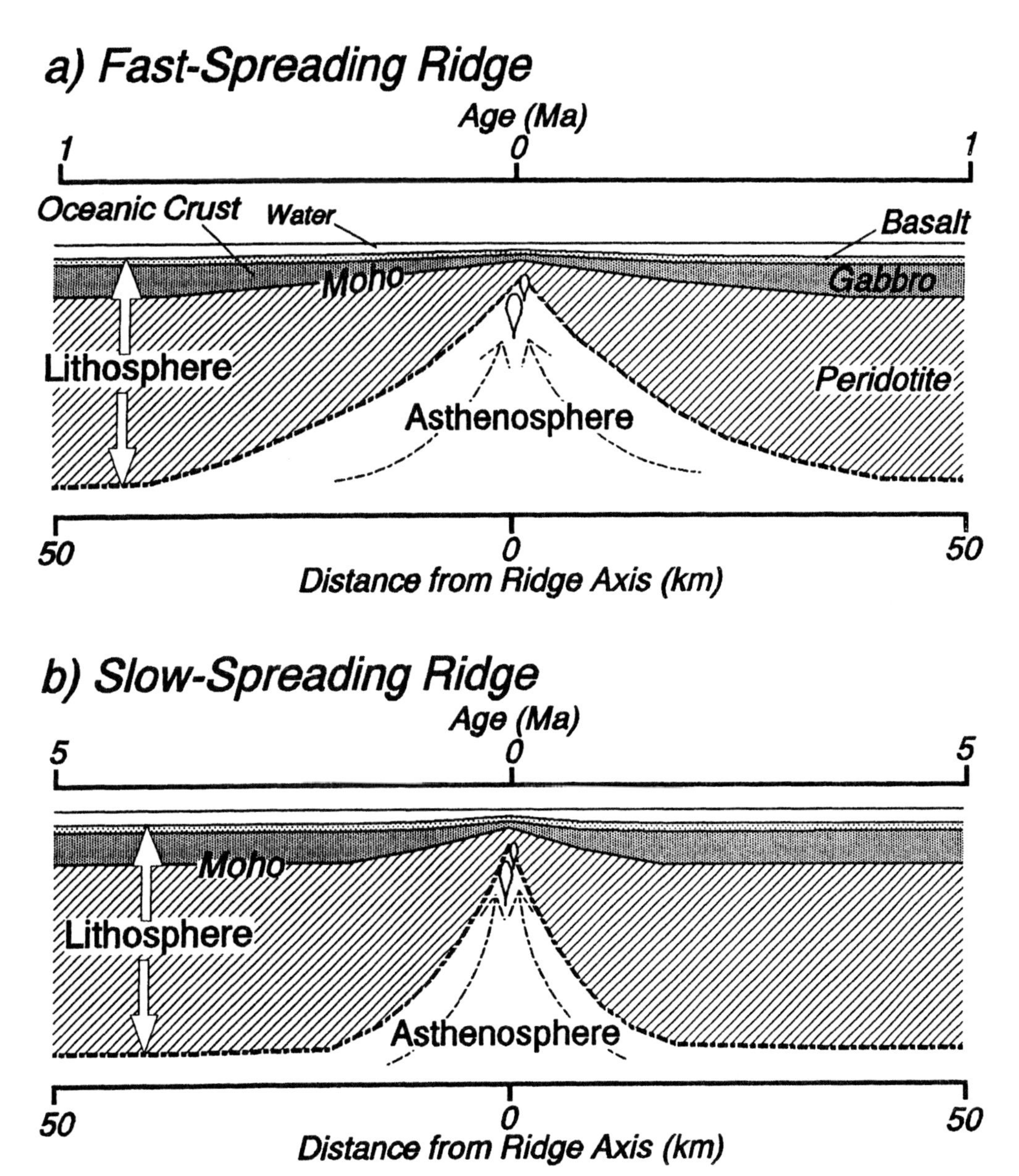

FIGURE 10.18 Fast-spreading leads to a broader mid-ocean ridge (a), with more gentle dips of the water bottom, Moho, and lithosphere/asthenosphere boundary, compared to slow spreading (b).

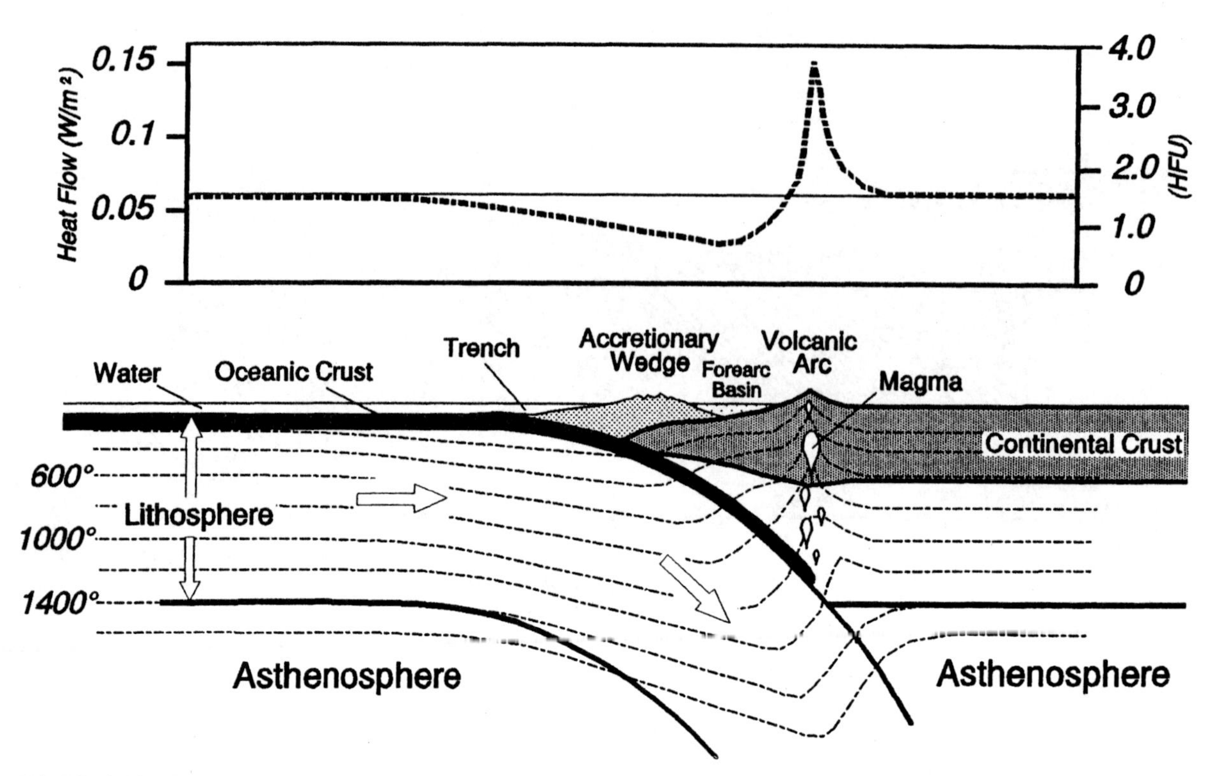

FIGURE 10.19 *Temperature contours and heat flow profile for a subduction zone.* Temperature contours are depressed as the cold slab subducts, leading to low heat flow in the forearc region. Hot fluids that migrate upward from the top of the subducting plate produce magma and high heat flow at the volcanic arc.

EXERCISES

10-1 Given the following temperatures encountered in a borehole:

582 m	17.1°C
1149 m	35.7°C
1951 m	57.0°C
2666 m	81.0°C
3262 m	98.9°C
4843 m	142.3°C
6397 m	190.9°C

a) Plot a temperature vs. depth graph and determine the geothermal gradient.

b) Assuming the well was drilled through a stratified sequence of sedimentary rock and basalt, calculate the heat flow.

c) Interpret the tectonic setting, giving geological as well as geophysical reasons for your interpretation.

10-2 **a)** From the values in Fig. 10.6, estimate the amount of heat (in Joules) produced per year by the total volumes of: **i)** continental crust; **ii)** oceanic crust; **iii)** mantle.

b) Explain why heat flow is so high at mid-ocean ridges.

10-3 **a)** For a geothermal gradient characteristic of a continental craton, estimate the depth where: **i)** granite will begin to melt; **ii)** granite will completely melt.

b) Repeat (a), but for granitic crust in a continental rift setting.

c) Based on results from (a) and (b), discuss the maximum depth of crustal earthquakes one might expect at each of the two settings.

10-4 Calculate the depth of, and discuss the reasons for, magma generation beneath collisional mountain ranges. Be sure to state your assumptions about rock type and geothermal gradient.

10-5 Using thermal considerations, explain the generation of magma and magmatic underplating at a continental rift zone. Include a discussion of the age and geometry of the Moho.

10-6 **a)** Using the graph in Fig. 2.7, explain the difference in the amount of magma that would be generated at a mid-ocean ridge if: **i)** the asthenosphere rises rapidly; **ii)** the asthenosphere rises more slowly.

b) To generate magma at a mid-ocean ridge, is it necessary that mantle material actually rise? Explain your answer.

10-7 At the ending phases of continental collision, the topography, Moho, and lithosphere/asthenosphere boundaries have the depth configurations illustrated in the cross section below. Draw sketches and explain how and why the depths to the boundaries will continue to change through time.

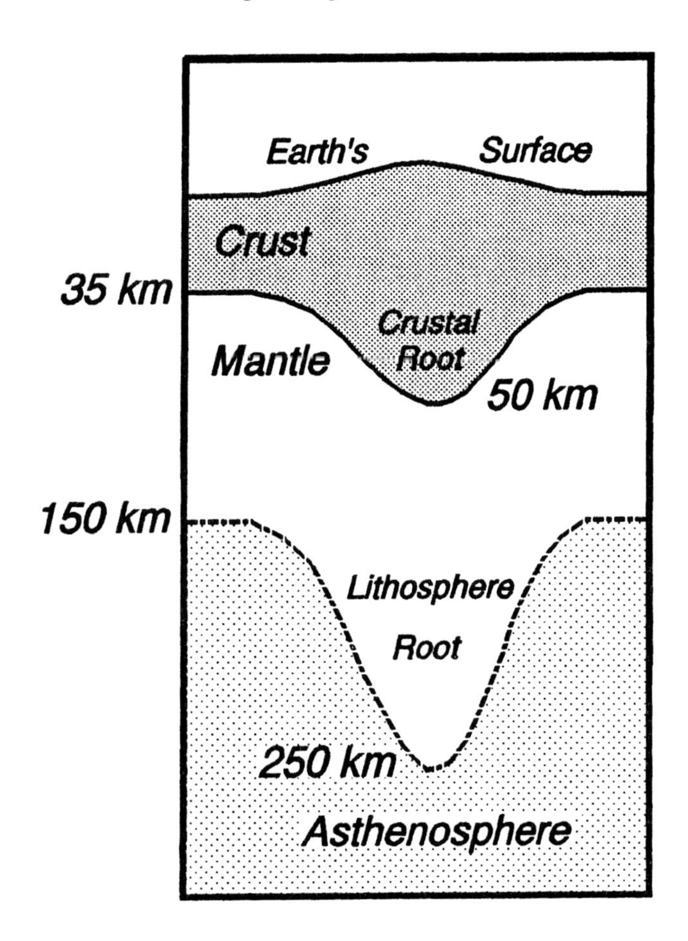

10-8 **a)** Draw a cross section and geophysical profiles of the United States along 40° N latitude, including the following information: **i)** depth to Moho from Fig. 4.16; **ii)** compressional wave velocities from Fig. 4.15; **iii)** gravity anomalies from the back, inside cover of this book; **iv)** positions of earthquakes from Fig. 7.20; **v)** heat flow from Fig. 10.14; **vi)** topography from the front, inside cover of this book.

b) Based on the information in (a), interpret the depth and configuration of the lithosphere/asthenosphere boundary.

c) Discuss how the heat flow profile relates to the configurations of the topography, Moho, and lithosphere/asthenosphere boundaries.

SELECTED BIBLIOGRAPHY

Blackwell, D. D., 1971, The thermal structure of the continental crust, in: *The Structure and Physical Properties of the Earth's Crust,* edited by J. G. Heacock, Washington: American Geophysical Union, Geophysical Monograph Series 14, pp. 169–184.

Blackwell, D. D., 1989, Regional implications of heat flow of the Snake River Plain, northwestern United States, *Tectonophysics,* v. 164, pp. 323–343.

Blackwell, D. D., and J. L. Steele (editors), 1989, *Geothermal Map of North America,* Boulder, Colorado: Geological Society of America, map CSM-007, 4 sheets, scale 1 : 5,000,000.

Blackwell, D. D., J. L. Steele, and L. S. Carter, 1991, Heat-flow patterns of the North American continent; A discussion of the geothermal map of North America, in *Neotectonics of North America,* edited by D. B. Slemmons, E. R. Engdahl, M. D. Zoback, and D. D. Blackwell, Boulder, Colorado: Geological Society of America, Decade Map Volume 1, pp. 423–436.

Blackwell, D. D., K. W. Wilson, and J. L. Steele, 1995, Geothermal regime in the central and eastern United States east of the Rocky Mountains, in: *Worldwide Utilization of Geothermal Energy: An Indigenous, Environmentally Benign Renewable Energy Resource,* Florence, Italy: Proceedings of the World Geothermal Congress, 1995, v. 1, pp. 649–653.

Bott, M. H. P., 1982, *The Interior of the Earth: Its Structure, Constitution and Evolution* (2nd ed.), New York: Elsevier Science Pub. Co., 403 pp.

Brune, J. N., T. L. Henyey, and R. F. Roy, 1969, Heat flow, stress, and rate of slip along the San Andreas Fault, California, *Journal of Geophysical Research,* v. 74, pp. 3821–3827.

Chapman, R. E., 1995, *Physics for Geologists,* London: UCL Press Limited, London, 143 pp.

De Bremaecher, J. C., 1985, *Geophysics: The Earth's Interior,* New York: John Wiley and Sons, Inc., 342 pp.

Elder, J. W., 1965, Physical processes in geothermal areas, in: *Terrestrial Heat Flow,* edited by W. H. K. Lee, Washington: American Geophysical Union, Geophysical Monograph Series 8, pp. 211–239.

Elder, J., 1981, *Geothermal Systems,* New York: Academic Press, 508 pp.

Fowler, C. M. R., 1990, *The Solid Earth: An Introduction to Global Geophysics,* Cambridge: Cambridge University Press, 472 pp.

Garland, G. D., 1979, *Introduction to Geophysics* (2nd ed.), Toronto: W. B. Saunders Company, 494 pp.

Jessop, A. M., 1990, *Thermal Geophysics,* Developments in Solid Earth Geophysics, No. 17, New York: Elsevier, 306 pp.

Lachenbruch, A. H., and J. H. Sass, 1977, Heat flow in the United States and the thermal regime of the crust, in: *The Earth's Crust, its Nature and Physical Properties,* edited by J. G. Heacock, Washington: American Geophysical Union, Geophysical Monograph Series 20, pp. 626–675.

Lachenbruch, A. H., J. H. Sass, and P. Morgan, 1994, Thermal regime of the southern Basin and Range Province: 2. Implications of heat flow for regional extension and metamorphic core complexes, *Journal of Geophysical Research,* v. 99, pp. 22121–22133.

Lee, W. H. K., 1965, (editor), *Terrestrial Heat Flow,* Washington: American Geophysical Union, Geophysical Monograph Series 8, 276 pp.

Morgan, P., and W. D. Gosnold, 1989, Heat flow and thermal regimes in the continental United States, in: *Geophysical Framework of the continental United States,* edited by L. C. Pakiser and W. D. Mooney, Boulder, Colorado: Geological Society of America, Memoir 172, pp. 493–522.

Philpotts, A. R., 1990, *Principles of Igneous and Metamorphic Petrology,* Englewood Cliffs, New Jersey: Prentice Hall, Inc., 498 pp.

Roy, R. F., D. D. Blackwell, and F. Birch, 1968, Heat generation of plutonic rocks and continental heat flow provinces, *Earth and Planetary Science Letters,* v. 5, pp. 1–12.

Roy, R. F., D. D. Blackwell, and E. R. Decker, 1970, Continental heat flow, in: *The Nature of the Solid Earth,* edited by E. C. Robertson, Harvard University symposium, pp. 506–543.

Saltus, R. W., and G. A. Thompson, 1995, Why is it downhill from Tonopah to Las Vegas?: A case for mantle plume support of the high northern Basin and Range, *Tectonics,* v. 14, pp. 1235–1244.

Sass, J. H., A. H. Lachenbruch, and A. H. Morgan, 1994, Thermal regime of the southern Basin and Range: 1. Heat flow data from Arizona and the Mojave Desert of California and Nevada, *Journal of Geophysical Research,* v. 99, pp. 22093–22119.

Sclater, J. G., C. E. Corry, and V. Vacquier, 1969, In situ measurements of the thermal conductivity of ocean-floor sediments, *Journal of Geophysical Research,* v. 74, pp. 1070–1081.

Sclater, J. G., and J. Francheteau, 1970, The implications of terrestrial heat flow observations on current tectonic and geochemical models of the crust and upper mantle of the Earth, *Geophysical Journal of the Royal Astronomical Society,* v. 20, pp. 509–542.

Sclater, J. G., E. J. W. Jones, and S. P. Miller, 1970, The relationship of heat flow, bottom topography and basement relief in Peake and Freen Deeps, north Atlantic, *Tectonophysics,* v. 10, pp. 283–300.

Schlater, J. G., R. N. Anderson, and M. L. Bell, 1971, Elevation of ridges and evolution of the central eastern Pacific, *Journal of Geophysical Research,* v. 76, pp. 7888–7915.

Skinner, B. J., and S. C. Porter, 1987, *Physical Geology,* New York: John Wiley and Sons, Inc., 750 pp.

Smith, R. B., and L. W. Braile, 1994, The Yellowstone Hotspot, *Journal of Volcanology and Geothermal Research,* v. 61, pp. 121–187.

Smith, R. B., R. T. Shuey, Jr., R. Pelton, and J. P. Bailey, 1977, Yellowstone Hotspot: Contemporary tectonics and crustal properties from earthquake and aeromagnetic data, *Journal of Geophysical Research,* v. 82, pp. 3665–3676.

Wheildon, J., P. Morgan, K. H. Williamson, T. R. Evans, and C. A. Swanberg, 1994, Heat Flow in the Kenya Rift Zone, *Tectonophysics,* v. 236, pp. 131–149.

APPENDIX A

Sequenced Writing Assignment For Whole Earth Geophysics Course

Geophysical observations comprise a wealth of information about Earth's crustal structure and tectonic evolution. A course in Whole Earth Geophysics can be used to train students to compile and interpret many aspects of a particular region, and to express their understanding of the region's structure and evolution through writing.

The course for which this book is designed differs from conventional geophysics courses, in that it focuses on writing as a method to learn course content. Many colleges and universities now require students to learn concepts in their own discipline through "writing intensive" (or "writing across the curriculum") courses. The assignment discussed below is thus designed for an upper division, writing intensive course in a geology or geophysics curriculum.

The assignment differs from a normal term paper in that it is *not* done all at once, near the end of the term (that traditional approach often results in disastrous quality and consequences!). Rather, the paper is developed gradually throughout the term, with critique from the instructor and revision at each step. Students thus learn that writing is an ongoing process, not something done at the last minute.

The writing sequence follows the content of this book. Students conduct literature research and write about geophysical observations from a selected region of the Earth at the same time each type of observation is discussed in lectures. Geophysical techniques thus have a context; they constrain specific aspects of geology in the region studied.

One of the problems with the traditional approach to term paper writing is that students are overwhelmed by large volumes of information. A student may write about the Mid-Atlantic Ridge, for example, with observations from seismic refraction, earthquake, gravity, magnetic, and heat flow data. The student may review so much literature all at once that the paper is a composite of facts and assorted opinions, with poor organization, insight, and comprehension. The instructor might offer comments and suggestions after the paper is turned in; that feedback is in vain, however, if the student is not required to revise. The incorporation of information a *little at a time*, accompanied by *critique* and *revision*, allows students to remain organized and focused on studying aspects of the Earth in their regions.

OUTLINE OF PAPER

During the course of the term each student is asked to write about the *crustal structure and tectonic evolution* of a region, as viewed through different types of geophysical data. Information is incorporated gradually, with feedback from the instructor at each step. The paper thus improves as it expands.

The assignment emphasizes *concise* writing. Students summarize the content of published articles, but in their own style and with their own thoughts included. Each step of the writing process has all the components outlined below; it is important that students visualize their final product, and that they revise each part as the paper evolves during the term.

- **TITLE**
- **ABSTRACT**
- **MAIN BODY (with appropriate subheadings):**
  - *INTRODUCTION*
  - *OBSERVATIONS AND INTERPRETATIONS* (added as paper evolves):
    - Seismic Refraction
    - Seismic Reflection
    - Earthquake
    - Gravity
    - Magnetics
    - Heat Flow
  - *DISCUSSION*
  - *CONCLUSIONS*
- **REFERENCES CITED**
- **FIGURES (with appropriate captions)**

POTENTIAL PAPER TOPICS

During the first week of the course, each student selects a feature on Earth's surface that fascinates him/her. The feature should be on a "tectonic scale," perhaps the Basin and Range Province or the Himalayan continental collision zone, or one of the other regions suggested in the list below. In consultation with the instructor, the student should insure that there are sufficient published articles presenting geophysical observations in the region (perhaps including some, or all, of *seismic refraction, seismic reflection, earthquake, gravity, magnetics,* and *heat flow*). The Selected Bibliography at the end of Chapter 2 may serve as a starting point.

1. **Extensional Tectonic Settings:**

   A) *Continental Rifts:*

   i) Basin and Range Province
   ii) Keweenawan Rift
   iii) East Africa
   iv) North Sea
   v) Rhine Graben
   vi) Reelfoot Rift

   B) *Passive Continental Margins:*

   i) Eastern North America
   ii) Eastern South America
   iii) West Africa

iv) Western Europe
v) Antarctica
vi) Margins of Gulf of Mexico
vii) Margins of Red Sea

C) *Mid-Ocean Ridges*:
i) Mid-Atlantic Ridge
ii) East Pacific Rise
iii) Central Red Sea

2. Convergent Tectonic Settings:

A) *Trench/Arc Systems*:
i) Japan
ii) Western South America
iii) U. S. Pacific Northwest
iv) Philippines
v) Aleutians

B) *Collisional Mountain Ranges*:
i) Himalayas
ii) Alps
iii) Carpathians
iv) Appalachians
v) Ouachitas
vi) Caledonides
vii) Canadian Rockies
viii) Australia/Banda Arc

3. Strike-Slip Settings:
i) San Andreas Fault
ii) New Zealand
iii) Southern Oklahoma Aulacogen

4. Hotspot Settings:
i) Hawaii/Emperor
ii) Columbia Plateau/Snake River Plain/Yellowstone

WRITING SEQUENCE

The sequence below is used for a one quarter course (10 weeks) that emphasizes refraction, earthquake, gravity, magnetic, and heat flow data; modifications would be necessary for classes of differing length and/or content. The instructor's time for reading and critique is about the same for each iteration—papers are generally easier to read as they expand in length during the term.

The percentage grades for each iteration reflect the overall value of the paper as *35% of the total course grade*. A value of 5% for the first iteration acknowledges that papers initially may be poorly researched, organized, and written. The instructor can send a message with grades of 1 to 3, out of the 5 percent, while still leaving the student opportunity to achieve a decent overall score for the assignment. By the last iteration many papers are of professional quality, earning scores of 11 or 12, out of the possible 12 percent.

First Iteration (Due 4th week of term; 5% of course grade).

Write an overview of one to three papers that discuss the *crustal structure* of the region based on *seismic refraction* observations. The paper should be typed, double spaced, including a *Title, Abstract, Main Body, Reference List,* and *Figures*. The *Main Body* of the paper should be about *2 pages* long at this stage.

Potential constraints now available:

1. *Seismic refraction:*
  a) Overall crustal thickness;
  b) Pattern of crustal thickness changes;
  c) Internal crustal velocities;
  d) Uppermost mantle velocities.

Potential new observations/interpretations:

- Thin crust at a continental rift or mid-ocean ridge?
- Thick crust at a mountain range?
- Transition from thick continental to thin oceanic crust at a continental margin?
- Low upper-mantle velocities at a continental rift or mid-ocean ridge?

Second Iteration (Due 6th week; 8% of course grade).

Rewrite the *entire* paper, considering the instructor's comments and adding information based on papers about *earthquake seismic* observations. (Main Body now *3 to 4 pages*).

Potential constraints now available:

1. Seismic refraction (see above).
2. *Earthquake:*
  a) Positions of lithospheric plate boundaries;
  b) Types of stresses in the region;
  c) Lithospheric thickness changes based on seismic delay times.

Potential new observations/interpretations:

- Shallow earthquakes at divergent and transform boundaries?
- Earthquakes to considerable depth at convergent boundaries?
- Normal fault focal plane solutions showing areas of extension?
- Reverse fault focal plane solutions showing areas of compression?
- Strike-slip fault focal plane solutions showing areas of shearing stress?
- Thin lithosphere at a continental rift or mid-ocean ridge?
- Thick lithosphere at a subduction zone or collisional mountain range?

Third Iteration (Due 8th week; 10% of course grade).

Revise entire paper, adding a part based on *gravity* interpretation. (Main Body now *5 to 7 pages*).

Potential constraints now available:

1. Seismic refraction (see above).

2. Earthquake (see above).
3. *Gravity:*
   a) Density distribution;
   b) Crustal thickness changes;
   c) Lithospheric thickness changes;
   d) State of isostasy of the region.

Potential new observations/interpretations:

- Thick crust beneath high topography of mountain ranges?
- Crustal thinning where water deepens at continental margins?
- Shallow asthenosphere supporting weight of elevated mantle and topography at continental rifts and mid-ocean ridges?
- Deviations from local isostasy, suggesting lithospheric strength?

Fourth Iteration (Due 10th week; 12% of course grade).

Revise and add material on *magnetics and/or heat flow*. At this stage, the DISCUSSION section should include many of the student's own interpretations and ideas on the crustal structure and tectonic evolution of the region, based on integration of many types of data. The Main Body is now *8 to 10 pages* long.

Potential constraints now available:

1. Seismic refraction (see above).
2. Earthquake (see above).
3. Gravity (see above).
4. *Magnetics*
   a) Susceptibility, size, and orientation of magnetic bodies;
   b) Depth to crystalline basement beneath sedimentary basins;
   c) Record of magnetic reversals.
5. *Heat Flow*
   a) Geothermal gradient;
   b) Depths to brittle/ductile transitions within the crust and upper mantle.

Potential new observations/interpretations:

Magnetics:

- Recognition of continental vs. oceanic basement?
- Age of seafloor created at mid-ocean ridges?
- Age of continental sedimentary deposits?
- Latitudes of rocks as they formed?

Heat Flow:

- Thickness of the lithosphere?
- Proximity to magma sources?

COMPONENTS OF PAPER

Each component of the paper accomplishes certain tasks. Although good writing is concise, some redundancy may be desirable. For example, you may wish to state your conclusions in the ABSTRACT and INTRODUCTION sections, as well as in the CONCLUSIONS; the reader may otherwise loose sight of the objective of your paper.

TITLE

- The title can be interesting, clever, and provocative, as long as it *tells the reader what your paper is about.*

ABSTRACT

- A *concise summary* of *your* paper.
- *Entices* the reader to want to read on; makes the reader *curious* about the details contained in the rest of your paper.

INTRODUCTION

- Introduces *geography* and *tectonic setting* of the region to the reader.
- Defines the *objective* of *your* paper ("to understand the *crustal structure and tectonic evolution* of . . .").
- Summarizes how you went about achieving your objective ("researched *constraints* on crustal structure and tectonic evolution offered by *geophysical observations*").
- States specific geophysical studies reported on in the literature.
- Offers *conclusions*, perhaps in the form of general models (or competing models), that can be developed from those studies.

OBSERVATIONS AND INTERPRETATIONS

- Gives specific *observations* for each type of geophysical data.
- Summarizes the *constraints* on crustal structure and/or tectonic evolution offered by each data type.
- Presents *interpretations* offered in the literature.

DISCUSSION

- Presents *your own* discussion on how the constraints contribute to understanding the *crustal structure and tectonic evolution* of the region.

CONCLUSIONS

- States what *you* conclude from *your* synthesis/analysis.

REFERENCES CITED

- Includes *all* papers cited in your paper.
- Includes *only* papers cited in your paper.

APPENDIX B

Units, Conversions, and Abbreviations

UNITS/CONVERSIONS

Fundamental Units

Quantity	Fundamental Unit and Abbreviation
distance	meter (m)
electric current	ampere (A)
mass	kilogram (kg)
temperature	degree Kelvin (°K) (increase 1°K = increase 1°C)
time	second (s)

Système Internationale (SI) Units

Quantity	SI Unit	Fundamental Units
electric charge	coulomb (C)	A s
energy	joule (J)	kg m^2 s^{-2}
force	newton (N)	kg m s^{-2}
frequency	hertz (Hz)	s^{-1}
magnetic field	tesla (T)	kg A^{-1} s^{-2}
pressure	pascal (Pa)	kg m^{-1} s^{-2}
power	watt (W)	kg m^2 s^{-3}

Geophysical Properties in SI Units and Conversions

Property and Abbreviation	SI Unit	Equivalent Units and Conversions
area (A)	m^2	10^4 cm^2 10^{-6} km^2
bulk modulus (k)	Pa	1 kg m^{-1} s^{-2} 1 N m^{-2}
density (ρ)	kg m^{-3}	10^{-3} g cm^{-3}
distance (x, y, z, h, l, L, r, R)	m	10^6 micron 10^3 mm 10^2 cm 10^{-3} km 39.37 inch (in) 3.281 feet (ft) 6.214×10^{-4} mile (mi) (1 in = 2.54 cm) (1 ft = 12 in = 30.48 cm = 0.3048 m) (1 mi = 5280 ft = 1609 m = 1.609 km) (1 cm = 0.394 in) (1 km = 0.6215 mi = 3282 ft)
energy (E)	J	1 kg m^2 s^{-2} N m 10^7 dyne cm 10^7 erg 2.39×10^{-1} cal 2.39×10^{-4} kcal
force (F)	N	1 kg m s^{-2} 10^5 dyne
geothermal gradient ($\partial T/\partial z$)	°C m^{-1}	°K m^{-1} 10^3 °C km^{-1} 1.8×10^3 °F km^{-1} 2.90×10^3 °F mi^{-1}
gravitational acceleration (g)	m s^{-2}	10^2 cm s^{-2} 10^2 Gal 10^5 mGal (1 Gal = 1 cm s^{-2}) (1 mGal = 10^{-3} cm s^{-2})
heat flow (q)	W m^{-2}	2.39×10^{-5} cal cm^{-2} s^{-1} 2.39×10^1 HFU (1 HFU = 10^{-6} cal cm^{-2} s^{-1} = 4.18×10^{-2} W m^{-2})

Property and Abbreviation	SI Unit	Equivalent Units and Conversions
heat production	$J\ kg^{-1}\ s^{-1}$	$W\ kg^{-1}$
		$10^{6}\ \mu W\ kg^{-1}$
		$10^{7}\ erg\ kg^{-1}\ s^{-1}$
		$10^{4}\ erg\ g^{-1}\ s^{-1}$
		$3.16 \times 10^{11}\ erg\ g^{-1}\ a^{-1}$
		$4.19 \times 10^{-3}\ cal\ g^{-1}\ s^{-1}$
		$1.32 \times 10^{5}\ cal\ g^{-1}\ a^{-1}$
magnetic susceptibility (χ)		Dimensionless
magnetic field intensity (F)	T	$kg\ A^{-1}\ s^{-2}$
		$kg\ C^{-1}\ s^{-1}$
		$C\ s^{-1}\ m^{-1}$
		Am^{-1}
		$10^{9}\ nT$
		10^{9} gamma (γ)
		$4\pi \times 10^{-3}$ Oersted (Oe)
magnetization (J)	T	$kg\ A^{-1}\ s^{-2}$
		$kg\ C^{-1}\ s^{-1}$
		$C\ s^{-1}\ m^{-1}$
		Am^{1}
		$10^{9}\ nT$
		10^{9} gamma (γ)
		$1/4\pi \times 10^{4}$ gauss (G)
mass (m)	kg	$10^{3}\ g$
power	W	$J\ s^{-1}$
		$10^{6}\ \mu W$
		$10^{-3}\ kW$
		$2.39 \times 10^{-1}\ cal\ s^{-1}$
		$2.39 \times 10^{-4}\ kcal\ s^{-1}$
pressure (P)	Pa	$kg\ m^{-1}\ s^{-2}$
		$N\ m^{-2}$
		$10^{-6}\ MPa$
		10^{-5} bar (b)
		10^{-8} kilobar (kb)
		9.87×10^{-6} atmosphere (atm)
shear modulus (μ)	Pa	$kg\ m^{-1}\ s^{-2}$
		$N\ m^{-2}$
temperature (T)	°C	°K − 273.16
		0.5556 (°F − 32)
thermal conductivity (k)	$W\ m^{-1}\ °C^{-1}$	$2.39 \times 10^{-3}\ cal\ cm^{-1}\ °C^{-1}\ s^{-1}$

Property and Abbreviation	SI Unit	Equivalent Units and Conversions
time (T)	s	10^3 ms 1.667×10^{-2} minute (min) 1.157×10^{-5} day 3.168×10^{-8} year (a) (1 min = 60 s) (1 day = 8.64×10^4 s) (1 a = 3.157×10^7 s) (1 Ma = 10^6 a)
velocity (V)	m s^{-1}	10^{-3} km s^{-1} (1 mm/a = 1 km/Ma) (1 cm/a = 10 km/Ma)
volume (V)	m^3	10^6 cm^3 10^{-9} km^3

ABBREVIATIONS USED IN TEXT

Units

a	annum (year)
C	coulomb
°C	degree centigrade
cm	centimeter
g	gram
Hz	hertz
in	inch
J	joule
km	kilometer
m	meter
Ma	million years
mGal	milligal
mm	millimeter
N	newton
Pa	pascal
s	second
T	tesla
W	watt
γ	gamma

Variables

a	acceleration
A	area
A	ground displacement
A_i	amplitude of incident wave

A_r	amplitude of reflected wave
A(t)	amplitude of seismic trace at time t
BC	Bouguer correction
BC_s	Bouguer correction at sea
CDP	common depth point
CMP	common midpoint
D	flexural rigidity of elastic plate
E	energy
E	Young's modulus
f	frequency
F	force
FAC	free air correction
g	gravitational acceleration
g_e	theoretical gravity at equator
g_t	theoretical gravity
h	elevation
h	layer thickness
h_a	thickness of air column
h_A	thickness of asthenosphere column
h_c	thickness of crust
$(h_c)_O$	thickness of oceanic crust
$(h_c)_C$	thickness of continental crust
$(h_c)_M$	thickness of crust at mountains
h_m	thickness of mantle column
h_w	thickness of water column; water depth
i	magnetic inclination angle
I	acoustic impedance
I(t)	amplitude of input signal at time t
k	bulk modulus
k	thermal conductivity
l	length, distance
L	length, distance
m	mass
m	body wave magnitude
M	mass of Earth
M_o	seismic moment
M_s	surface wave magnitude
M_w	moment magnitude
N(t)	amplitude of noise at time t
P	pressure
q	heat flow
q(x)	load applied to the top of elastic plate at x
q(z)	heat flow at depth z
Q	seismic attenuation (quality factor)
r	distance
R	distance from observation point to center of Earth
RC	reflection coefficient
RC(t)	amplitude of reflection coefficient at time t
t	time
t_0	T-axis intercept for reflected wave
t_1	T-axis intercept for critically refracted wave

t_{1d}	refraction T-axis intercept when shooting downdip
t_{1u}	refraction T-axis intercept when shooting updip
t_i	vertical, two-way travel time within the ith layer
t_{n-1}	T-axis intercept for refraction along layer n
T	travel time
T_{AB}	refraction travel time: shot at A, receiver at B
TC	terrain correction
T_d	travel time of direct wave
T_f	travel time of reflected wave
T_n	travel time for critical refraction along layer n
T_{NMO}	normal moveout time
T_p	arrival time of compressional wave
T_r	travel time of critically refracted wave
T_s	arrival time of shear wave
u	fault displacement
v_0	initial velocity
V	volume
V	velocity
V_{ap}	apparent velocity
V_{av}	average velocity
V_i	seismic velocity of ith layer
V_{int}	interval velocity
V_n	seismic velocity of layer n
V_P	compressional wave velocity
V_{rms}	root mean square velocity
V_S	shear wave velocity
V_t	true velocity
w	deflection of elastic plate
W	width
X	horizontal distance
X_c	critical distance
X_{cr}	crossover distance
z	vertical distance
α	dip angle of interface
α	correction factor for earthquake magnitude determination
δ	magnetic declination angle
Δ	earthquake epicentral distance (degrees)
Δg_B	Bouguer gravity anomaly
Δg_{fa}	free air gravity anomaly
Δg	magnitude of change in gravity
Δg_x	magnitude of horizontal change in gravity
Δg_z	magnitude of vertical change in gravity
$\partial T/\partial z$	geothermal gradient
θ	angle
θ_c	critical angle
θ_i	angle of incidence
θ_r	angle of refraction
λ	wavelength
λ	Lame's constant
μ	shear modulus

ρ	density
ρ_a	density of air
ρ_a	density of material above elastic plate
ρ_A	density of asthenosphere
ρ_b	density of material below elastic plate
ρ_c	density of crust
ρ_m	density of mantle
ρ_w	density of water
T	period
ϕ	latitude (degrees)
υ	Poisson's ratio
χ	magnetic susceptibility

Constants

G	Universal Gravitational Constant (6.67×10^{-11} Nm^2/kg^2)
π	ratio of circumference to diameter of circle (3.1416)

Vectors

$\vec{F}$	total magnetic field
$\vec{F}_E$	east component of horizontal magnetic field
$\vec{F}_H$	horizontal component of magnetic field
$\vec{F}_N$	north component of horizontal magnetic field
$\vec{F}_V$	vertical component of magnetic field
$\vec{F}_{amb}$	Earth's ambient field
$\vec{F}_{ind}$	induced magnetic field
$\vec{F}_{rem}$	remanent magnetic field
$\vec{J}$	total magnetization
$\vec{J}_{ind}$	induced magnetization
$\vec{J}_{rem}$	remanent magnetization
$\Delta\vec{g}$	change in gravity
$\Delta\vec{g}_x$	horizontal component of change in gravity
$\Delta\vec{g}_z$	vertical component of change in gravity

Mathematical Symbols

α	proportional to
Δ	change in
dT/dX	first derivative of T, with respect to x
$\partial T/\partial z$	partial derivative of T, with respect to z
$\int \Delta g_z\, dx$	integral of Δg_z, with respect to x
*	convolution operator

APPENDIX C

Igneous Rocks: Classification, Properties of Magmas, and Tectonic Occurrence

CLASSIFICATION CHART

The chart on the next page illustrates the classification of igneous rocks according to *silica content* (*chemical composition*) and *texture* (*grain size*). The seven rock names are a rough guide because igneous rocks commonly have compositions falling between the silica and heavy mineral percentages shown. Properties of the igneous rocks and the magmas that form them also exhibit gradations between the rock types shown.

Silica $(SiO_4)^{-4}$ Content

Silica is silicon (Si) and oxygen (O), the stuff that makes common *window glass*. An example of a mineral that is pure silica is *quartz* (chemical formula SiO_2). Common minerals found in igneous rocks that are *high in silica* include: quartz; orthoclase or potassium feldspar ($KAlSi_3O_8$); albite or sodium plagioclase feldspar ($NaAlSi_3O_8$); and micas [muscovite, $KAl_3Si_3O_{10}(OH)_2$; biotite, $K(Mg,Fe)_3Si_3O_{10}(OH)_2$]. Common minerals in igneous rocks *low in silica* are amphibole [hornblende, $NaCa(Mg,Fe)_5AlSi_7O_{22}(OH)_2$]; olivine [$(Mg,Fe)_2SiO_4$]; calcium plagioclase feldspar ($CaAl_2Si_2O_8$); and pyroxene (Ca,Fe,Mg-Silicate).

The percentage of silica is an important factor in the properties of magma. As magma cools, the silica begins to form molecules while the magma is still liquid. This early bonding of silica tends to make the magma more *viscous* (that is, more *sticky*), like grease or molasses. Magmas high in silica thus tend to flow sluggishly, while those with lower silica content flow more freely, like fountains or rivers of water. High silica minerals also tend to have lower melting temperatures than those with low silica, so that when rocks begin to melt, magmas with high silica content are generated before those low in silica.

When magma cools to rock, the amount of silica determines the minerals formed, hence affecting the rock's physical properties. High silica minerals tend to be light in color and weight; the rocks formed therefore have a pink to white appearance and low density (granite, rhyolite). When rocks are low in silica, they have a larger proportion of heavy, dark minerals that are high in iron and magnesium; those rocks have higher density and are dark brown, dark green, or black (basalt, gabbro, peridotite).

CLASSIFICATION and PROPERTIES of IGNEOUS ROCKS and MAGMAS

GRAIN SIZE		CHEMICAL COMPOSITION			
	Fine	Rhyolite	Andesite	Basalt	
	Coarse	Granite	Diorite	Gabbro	Peridotite

Cooling History		
Rate	Position	Rock Type
Fast	Above Surface	**Extrusive (Volcanic)**
Slow	Below Surface	**Intrusive (Plutonic)**

% Silica	**70%**	**60%**	**50%**	**30%**
% Dark (Heavy) Minerals	30%	40%	50%	70%
Names of Common Minerals	K-Feldspar Quartz	Plag.-Feldspar Amphibole	Plag.-Feldspar Pyroxene	Olivine Pyroxene
Color	Light Colors	Intermediate	Black	Dark Green
Density (gram/cubic centimeter)	2.7	2.8	2.9	3.3
Melting Temperature (degrees C)	800	950	1100	1400
Viscosity of Magma	High	Intermediate	Low	
Extent of Lava Flows	Very Small Area	Small Area	Large Area	
% Volotile Fluids	15%	8%	1%	
Type of Eruption	Very Explosive	Explosive	Quiet	
Type of Major Volcanoes	Composite	Composite	Shield	
Cinder Cones?	No	Sometimes	Often	
Tectonic Setting: 1. Continental Rift 2. Mid-Ocean Ridge 3. Convergent Plate Boundary 4. Hot Spot 5. Continental Collision Zone	Rhyolite 1a, 3a, 4a	Andesite 3a	Basalt 1a, 2a, 3a, 4a	
Where Solidified: a. Upper Crust b. Mid-to-Lower Crust c. Mantle	Granite 3b, 5b	Diorite 3b	Gabbro 1b, 2b, 4b	Peridotite 2c

Size of Mineral Grains (Texture)

As magma cools, mineral crystals begin to form. If the rock cools quickly, the crystals are so small that they cannot be seen with the naked eye. Slow cooling, however, gives crystals enough time to grow large. Igneous rocks are thus classified as *fine grained* (or ***aphanitic***, from Greek "invisible") and course grained (or ***phaneritic,*** from Greek "visible").

Texture can be used to understand the genesis of an igneous rock. Fine grained igneous rocks result from magma that cooled at or near Earth's surface, forming *extrusive* (volcanic) and *shallow intrusive* rocks. Magma that cooled slowly, deep within the Earth, formed coarse grained *intrusive* (plutonic) rocks.

DESCRIPTION OF IGNEOUS ROCKS

The descriptive classification scheme below relates to the chart on the previous page as follows:

- ***Capital letters*** **refer to the** ***texture*** **(grain size) of the rock.**
  - **A)** Coarse (phaneritic)
  - **B)** Fine (aphanitic)
- ***Numbers*** **correspond to the** ***chemical composition*****:**
  - **1)** High silica content (granite/rhyolite)
  - **2)** Intermediate silica content (diorite/andesite)
  - **3)** Low silica content (gabbro/basalt)
  - **4)** Extremely low silica content (peridotite)

A) Phaneritic (Coarse Grained) Textures:

1) Granite

a) Most common minerals:
- potassium (K) feldspar (orthoclase);
- quartz;
- some sodium (Na) feldspar (albite);
- minor amounts of mica (muscovite/biotite), which stand out as clear or black flakes.

b) Light colored (generally pink)

c) Relatively low density (~2.6 to 2.8 g/cm^3)

d) Common in upper and middle portions of continental crust.

2) Diorite

a) Most common minerals:
- sodium-calcium feldspar (plagioclase);
- only minor amounts of quartz;
- might have dark minerals amphibole and pyroxene.

b) Grayish color (looks like "salt and pepper")

c) Density between granite and gabbro (~2.8 g/cm^3)

d) Common in the mid-crust beneath volcanic arcs.

3) Gabbro

a) Most common minerals:
- calcium-rich feldspar (plagioclase);

- pyroxene;
- minor amounts of olivine.

b) Dark colored (dark green to black)

c) Relatively high density (~2.8 to 3.0 g/cm^3)

d) Common in lower part of both continental and oceanic crust.

4) Peridotite

a) Mostly composed of:
- olivine and pyroxene;
- perhaps some calcium-rich feldspar.

b) Dark green color

c) Very high density (3.3 to 3.4 g/cm^3)

d) Not common in the crust, but forms the mantle part of the lithosphere:
- generally formed from magmatic differentiation as hot asthenosphere rises at a mid-ocean ridge:
- low silica magma cools first, crystallizing as the heavy minerals pyroxene and olivine (the rock peridotite);
- the remaining silica-rich magma is lighter, so it moves upward to form the crust (gabbro and basalt).

B) Aphanitic (Fine Grained) Textures:

1) Rhyolite

(a), **(b)**, and **(c)** same as for granite

d) Occurrence:
- not found in ocean crust;
- generally occurs as volcanic eruptions on continental crust, where magma intrudes and melts silica-rich minerals.

2) Andesite

(a), **(b)**, and **(c)** same as for diorite

d) Common in volcanic rocks at convergent plate boundaries:
- the downgoing ocean crust and sediments are heated, releasing fluids;
- the fluids work their way to the surface, partially melting mantle and crust of the overriding plate;
- silica-rich minerals have lower melting temperatures, enriching silica content of the resulting magma.

3) Basalt

(a), **(b)**, and **(c)** same as for gabbro

d) Originates from cooling lava flows, commonly at divergent plate boundaries:
- upper part of ocean crust;
- continental rift zones.

4) Fine-grained equivalents to peridotite are very rare:
- too dense to extrude through the crust;
- normally, only the course-grained peridotite forms, because it cools slowly at depth, within the mantle;
- the rare examples, called *komatiites,* are very old rocks, formed when the Earth was much hotter.

PROPERTIES OF MAGMA

The physical properties of magma depicted in the chart depend, to a large degree, on the *silica content* of the magma.

Color

The less silica, the more dark colored minerals. Low-silica magmas therefore form igneous rocks that are dark green to black, while lighter colors (pink to white) occur as silica content increases.

Density

The less silica, the more heavy minerals (like those with iron and magnesium); the magma and resulting igneous rocks are thus more dense.

Melting (Solidification) Temperature

The more silica, the lower the melting temperature. As rock melts, high-silica magmas generally come out first. Conversely, as magma cools, lower silica minerals solidify first, followed progressively by those with higher and higher silica.

Viscosity (Resistance to flow; "stickiness")

Silica molecules (tetrahedra) begin to form while the magma is still liquid. High-silica magmas (rhyolite/andesite) are therefore more paste-like than low-silica magmas (basalt), which tend to erupt like fountains of water and flow like rivers.

Extent of Lava Flows

The viscosity determines how easily (and how far) lava is likely to flow. Low viscosity magmas (those with low silica) will therefore flow much farther (and are much thinner) than those with high viscosity (high silica).

Percent Volatiles

Volatiles (water vapor and carbon dioxide) escape easily from fluid magmas, but are trapped when magmas are sticky. High viscosity high-silica magmas therefore have a much higher percentage of trapped gasses than low-silica magmas.

Types of Volcanic Eruptions

As magma rises and cools, gasses escape from free-flowing, low-silica magma, but are trapped (under high pressure) within sticky, high-silica magma. Sudden release of pressure causes violent eruption of magmas with andesite to rhyolite composition.

Types of Major Volcanoes

Low-viscosity magmas (basalt) flow freely, forming volcanoes with gentle slopes of only 1° to 2° ("shield" volcanoes). High-viscosity magmas (andesite to rhyolite) stick to the sides of volcanoes, making the volcanoes much steeper (10° slopes; "composite" volcanoes).

Cinder Cones

Magma strewn high into the air generally is of the more fluid variety. The material that rains down, forming cinder cones, is generally of basalt to basaltic andesite composition.

TECTONIC SETTING

Numbers on the chart refer to the tectonic settings where specific igneous rocks commonly occur (Chapter 2). Lower case letters indicate where the magma solidified within the crust or mantle. The occurrence for each rock type is a rough guide, indicating settings where substantial amounts of that rock form.

APPENDIX D

Answers to Selected Exercises

Below are answers to some exercises that have specific solutions. The importance of each exercise is the steps taken to get to the solution. A solution alone is therefore not an appropriate response; the assumptions and steps that lead to the solution are also required. In all cases, sketches are essential; *illustrate how the problem relates to understanding the earth.*

2-7 a) 10–11 cm/year; N45° W–N50° W

b) 4.5–5.5 cm/year

3-3 a) The mantle is lower in silica (silcon and oxygen) and higher in heavy elements (iron and magnesium) than the crust. The mantle is denser (higher ρ) than the crust; the higher density actually tends to slow the waves down. Seismic waves travel *faster* in the mantle than in the crust because the mantle is *more incompressible* (higher k) and *more rigid* (higher μ) than the crust.

d) The outer core is heavy elements (iron, nickel) while the mesosphere is mantle material (iron-magnesium silicate). The outer core is denser than the mesosphere. Seismic waves travel *slower* in the outer core than in the mesosphere because the outer core is *denser* (higher ρ) and *has no rigidity* ($\mu = 0$).

3-5

	0 m	2,500 m	25,000 m
Direct P	2000 m/s	2000 m/s	2000 m/s
Refracted P	Doesn't Exist	Doesn't Exist	4000 m/s
Reflected P	∞	3774 m/s	2025 m/s

3-7 Compared to surrounding sedimentary strata, granite is higher velocity and higher density, while salt is higher velocity and lower density. Both materials result in critical refraction, but the granite intrusion produces a gravity high, while a gravity low occurs over a salt dome.

4-1 a) $h = 35$ km; $V_p = 6$ km/s

b) $V_p = 8$ km/s

h) $X_c = 79$ km

i) $t_1 = 7.7$ s

j) $X_{cr} = 185$ km

k) $L = 370$ km

4-3 b) Direct: $V_1 = 1.5$ km/s

1st Refraction: $V_2 = 3.3$ km/s; $t_1 = 4.1$ s

2nd Refraction: $V_3 = 5.3$ km/s; $t_2 = 5.8$ s

3rd Refraction: $V_4 = 8.3$ km/s; $t_3 = 8.0$ s

c) $h_1 = 3.4$ km

$h_2 = 3.1$ km

$h_3 = 6.3$ km

4-5 a) $V_1 = 3.9$ km/s; $V_2 = 6.8$ km/s

b) $\alpha = 14.5°$ (in B to A direction)

c) $z_u = 2.08$ km (below A); $z_d = 0.94$ km (below B)

5-1 a) Shale: $I = 5.50 \times 10^6$ kg m^{-2} s^{-1}

Tight Sandstone: $I = 9.12 \times 10^6$ kg m^{-2} s^{-1}

Porous Sandstone: $I = 6.30 \times 10^6$ kg m^{-2} s^{-1}

Shale: $I = 5.50 \times 10^6$ kg m^{-2} s^{-1}

c) 1st Interface: T = 0.800 s; RC = +0.248

2nd Interface: T = 0.826 s; RC = −0.183

3rd Interface: T = 0.859 s; RC = −0.068

5-2

	0 m	2,000 m	20,000 m
Direct P	2500 m/s	2500 m/s	2500 m/s
Rayleigh	1250 m/s	1250 m/s	1250 m/s
Refracted P	Doesn't Exist	3000 m/s	3000 m/s
P Primary	∞	2795 m/s	2503 m/s
P Multiple	∞	3536 m/s	2512 m/s

5-4 h = 2438 m

7-1 a) 10,000

b) $\approx$ 575,000

7-2 d) P-wave onset: $\approx$ 6:02:05; Z down; N–S to south; not on E–W

e) S-wave onset: $\approx$ 6:03:28

f) Love wave onset: $\approx$ 6:03:52; on E–W; not on Z or N–S

g) Rayleigh wave onset: $\approx$ 6:04:00; on Z and N–S; not on E–W

7-5 i) μ and k increase more than ρ

ii) μ and k decrease

iii) ρ increases and $\mu = 0$

iv) $\mu \neq 0$

8-2 a) i) $g_t = 980{,}900.91\ mGal$

ii) FAC = 150.27 $mGal$

iii) BC = 54.58 $mGal$

iv) $\Delta g_{FA} = -33.24\ mGal$

v) $\Delta g_B = -87.84\ mGal$

8-3 a) h_c(continent) = 32.07 km

b) i) $X = -\infty$: Δg_{FA}(water) = 0

$X = -2.5$ km: Δg_{FA}(water) = $-85.9\ mGal$

$X = 0$: Δg_{FA}(water) = $-171.8\ mGal$

$X = +2.5$ km: Δg_{FA}(water) = $-257.7\ mGal$

$X = +\infty$: Δg_{FA}(water) = $-343.6\ mGal$

ii) $X = -\infty$: Δg_{FA}(mantle) = 0

$X = -22.54$ km: Δg_{FA}(mantle) = $+85.9\ mGal$

$X = 0$: Δg_{FA}(mantle) = $+171.8\ mGal$

$X = +22.54$ km: Δg_{FA}(mantle) = $+257.7\ mGal$

$X = +\infty$: Δg_{FA}(mantle) = $+343.6\ mGal$

8-7 a) i) Maximum Δg(topo) = $+111.9\ mGal$

ii) Maximum $\Delta g(\text{Moho}) = +26.4\ mGal$

b) $h_A = 82.5$ km (lith/asth boundary 47.5 km below sea level)

c) **i)** Maximum $\Delta g(\text{lith/asth}) = -138.3\ mGal$

8-8 **a)** $\Delta z(\text{Moho}) = 11.2$ km (Moho 46.2 km below sea level)

b) **i)** Maximum $\Delta g(\text{Moho}) = -234.6\ mGal$; $z = 40.6$ km

ii) Maximum $\Delta g(\text{topo}) = +234.6\ mGal$; $z = 1$ km

9-4 **b)** ≈ 6 cm/year

9-5 ≈ 4.7 cm/year

10-1 **a)** $\partial T/\partial z \approx 29.7$ °C/km

b) $q \approx 0.08\ W/m^2$

10-3 **a)** **i)** ≈ 35 km

ii) ≈ 60 km

b) **i)** ≈ 20 km

ii) ≈ 35 km

Index

M

N

T